RISK, RELIABILITY AND SUSTAINABLE REMEDIATION IN THE FIELD OF CIVIL AND ENVIRONMENTAL ENGINEERING

RISK, RELIABILITY AND SUSTAINABLE REMEDIATION IN THE FIELD OF CIVIL AND ENVIRONMENTAL ENGINEERING

Edited by

THENDIYATH ROSHNI
National Institute of Technology Patna, Patna, Bihar, India

PIJUSH SAMUI
National Institute of Technology, Patna, Bihar, India

DIEU TIEN BUI
Geographic Information System group, University of South-Eastern Norway, Notodden, Norway

DOOKIE KIM
Department of Civil and Environmental Engineering, Kongju National University, Cheonan-daero, Republic of Korea

RAHMAN KHATIBI
GTEV-ReX Ltd, Swindon, United Kingdom

Elsevier
Radarweg 29, PO Box 211, 1000 AE Amsterdam, Netherlands
The Boulevard, Langford Lane, Kidlington, Oxford OX5 1GB, United Kingdom
50 Hampshire Street, 5th Floor, Cambridge, MA 02139, United States

Copyright © 2022 Elsevier Inc. All rights reserved.

No part of this publication may be reproduced or transmitted in any form or by any means, electronic or mechanical, including photocopying, recording, or any information storage and retrieval system, without permission in writing from the publisher. Details on how to seek permission, further information about the Publisher's permissions policies and our arrangements with organizations such as the Copyright Clearance Center and the Copyright Licensing Agency, can be found at our website: www.elsevier.com/permissions.

This book and the individual contributions contained in it are protected under copyright by the Publisher (other than as may be noted herein).

Notices

Knowledge and best practice in this field are constantly changing. As new research and experience broaden our understanding, changes in research methods, professional practices, or medical treatment may become necessary.

Practitioners and researchers must always rely on their own experience and knowledge in evaluating and using any information, methods, compounds, or experiments described herein. In using such information or methods they should be mindful of their own safety and the safety of others, including parties for whom they have a professional responsibility.

To the fullest extent of the law, neither the Publisher nor the authors, contributors, or editors, assume any liability for any injury and/or damage to persons or property as a matter of products liability, negligence or otherwise, or from any use or operation of any methods, products, instructions, or ideas contained in the material herein.

ISBN: 978-0-323-85698-0

For information on all Elsevier publications visit our website at
https://www.elsevier.com/books-and-journals

Publisher: Susan Dennis
Acquisitions Editor: Anita Koch
Editorial Project Manager: Howi M. De Ramos
Production Project Manager: Vijayaraj Purushothaman
Cover Designer: Matthew Limbert

Typeset by TNQ Technologies

Dedicated to my beloved grandfather Tarapada Samui.

Pijush Samui

Contents

List of contributors   xiii

1. A basic framework to integrate sustainability, reliability, and risk—a critical review

RAHMAN KHATIBI

1. Introduction   1
2. Critical insight into the structure of science   3
3. Governance   6
4. Goal orientation overarching organizations   12
5. Decision-making—3rd dimension of BSF   16
6. Discussions   23
7. Conclusion   24
8. Appendix—framing scientific concepts   25
References   27
Further reading   29

2. Principal component analysis of precipitation variability at Kallada River basin

BEERAM SATYA NARAYANA REDDY AND S.K. PRAMADA

1. Introduction   31
2. Study area and dataset   31
3. Methodology   32
4. Results and discussion   33
5. Conclusions   36
References   36

3. Healthcare waste management in Bangladesh: practices and future pathways

AFIA RAHMAN, BYOMKESH TALUKDER, AND
MOHAMMAD REZAUL KARIM

1. Introduction   37
2. Theoretical framework   40
3. Research design and methods of data collection   42
4. Results and discussion   44

5. Future ways to improve HCWM   49
6. Conclusion   51
References   51
Further reading   52

4. Seismic risk for vernacular building classes in the fertile Indus Ganga alluvial plains at the foothills of the Himalayas, India

M.C. RAGHUCHARAN AND SURENDRA NADH SOMALA

1. Introduction   53
2. Seismic demand   55
3. Population exposure and building inventory   56
4. Fragility functions   56
5. Results and discussion   64
6. Conclusion   69
References   70
Further reading   72

5. Comparative study between normal reinforced concrete and bamboo reinforced concrete

SUBHOJIT CHATTARAJ

1. Introduction   73
2. Methodology and materials   74
3. Experimentation   75
4. Results   76
5. Conclusion   78
Further reading   78

6. The power of the GP-ARX model in CO_2 emission forecasting

ELHAM SHABANI, MOHAMMAD ALI GHORBANI, AND
SAMED INYURT

1. Introduction   79

viii

2. Materials and method 81
3. Empirical results 84
4. Conclusion 89
References 89
Further reading 91

7. Integrated sustainability impact assessment of trickling filter

SELVARAJ AMBIKA

1. Introduction 93
2. Methodology 94
3. Results and discussion 96
4. Conclusion 106
5. Annexure A 107
Acknowledgments 108
References 109
Further reading 109

8. Critical soil erosion prone areas identification and effect of climate change in soil erosion prioritization of Kosi river basin

AADIL TOWHEED AND THENDIYATH ROSHNI

1. Introduction 111
2. Study area and data description 112
3. Methodology 114
4. Results and discussions 117
5. Conclusions 120
References 120
Further reading 121

9. Adaptive Kriging Monte Carlo Simulations for cost-effective flexible pavement designs

DEEPTHI MARY DILIP, ALEENA JOY, AND ANAMIKA VENU

1. Introduction 123
2. Numerical examples 127
3. Cost-effective flexible pavement structures 128
4. Conclusions 130
References 131

10. Aggregating risks from aquifer contamination and subsidence by inclusive multiple modeling practices

MARYAM GHAREKHANI, RAHMAN KHATIBI, ATA ALLAH NADIRI, AND SINA SADEGHFAM

1. Introduction 133
2. Study area 135

3. Methodology 137
4. Results 142
5. Discussion 145
6. Conclusion 147
Appendix I 147
References 152

11. Mapping and aggregating groundwater quality indices for aquifer management using Inclusive Multiple Modeling practices

ZAHRA SEDGHI, ALI ASGHAR ROSTAMI, RAHMAN KHATIBI, ATA ALLAH NADIRI, SINA SADEGHFAM, AND ALIREZA ABDOALLAHI

1. Introduction 155
2. Study area and data availability 159
3. Methodology 161
4. Results 171
5. Discussion 176
6. Conclusions 177
Appendix I 178
References 179
Further reading 182

12. Liquefaction hazard mitigation using computational model considering sustainable development

SUFYAN GHANI AND SUNITA KUMARI

1. Introduction 183
2. Study area and data collection 184
3. Theoretical details of empirical and computational model 185
4. Advanced first order second moment reliability method 188
5. Data processing and analysis 189
6. Results and discussion 190
7. Conclusion and summary 194
References 195

13. Probabilistic risk factor—based approach for sustainable design of retaining structures

ANASUA GUHARAY

1. Introduction 197
2. Articulation of probabilistic risk factor 199
3. Cantilever retaining wall 199
4. Gravity retaining wall 203
5. Conclusion 206
References 206

Contents

14. Blast-induced flyrock: risk evaluation and management

AVTAR K. RAINA AND RAMESH MURLIDHAR BHATAWDEKAR

1. Introduction 209
2. Flyrock definition and causes 211
3. Brief analysis of data in literature 215
4. Impact of geology on flyrock and associated risk 221
5. Models for flyrock distance prediction 223
6. Use of intelligent techniques in flyrock prediction 226
7. Flyrock risk and management measures 226
8. Maturity model for flyrock risk assessment 235
9. Need for future research 238
10. Conclusions 241
References 241

15. The importance of environmental sustainability in construction

BESTE CUBUKCUOGLU

1. Introduction 249
2. Environmental issues, their causes, and sustainability 250
3. The role of engineers in sustainable development 253
4. Conclusions 254
References 254

16. Rock mass classification for the assessment of blastability in tropically weathered igneous rocks

RAMESH MURLIDHAR BHATAWDEKAR,
EDY TONNIZAM MOHAMAD, MOHD FIRDAUS MD DAN,
TRILOK NATH SINGH, PRANJAL PATHAK, AND
DANIAL JAHED ARMAGAHNI

1. Introduction 255
2. Literature review 256
3. Blastability index 256
4. Comparative function based RMC for blastability 259
5. Assessment of slope stability with rock mass classification 263
6. Development of weathering classification systems for tropically weathered igneous and andesite rocks 267
7. Site study of tropically weathered igneous rocks 268
8. Comparison of tropically weathered igneous rocks in Indonesia, Thailand, Cambodia, and Malaysia 276
9. Conclusion 279
References 279
Further reading 283

17. Best river sand mining practices vis-a-vis alternative sand making methods for sustainability

RAMESH MURLIDHAR BHATAWDEKAR, TRILOK NATH SINGH,
EDY TONNIZAM MOHAMAD, RAJESH JHA,
DANIAL JAHED ARMAGAHNI, AND
DAYANG ZULAIKA ABANG HASBOLLAH

1. Introduction 285
2. Global sand scenario and environment accountability 287
3. Environmental impacts of sand mining 287
4. Best river sand mining practices 290
5. Sustainability 301
6. The alternatives 301
7. Comparison of river sand and manufactured sand 305
8. A case study on sand from waste rocks 306
9. Need of future research 309
10. Conclusions 310
Acknowledgments 311
References 311
Further reading 312

18. Learning lessons from river sand mining practices in India and Malaysia for sustainability

B.R.V. SUSHEEL KUMAR, MUHAMMAD FAIZ BIN ZAINUDDIN,
DATO CHENGONG HOCK SOON, AND
RAMESH MURLIDHAR BHATAWDEKAR

1. Introduction 315
2. Objectives 315
3. Gap evaluation depending on need-supply evaluation based on district survey report 316
4. Availability of sand and regulatory mechanism to meet local requirements 320
5. Replenishment of sand 322
6. Curbing illegal sand mining 322
7. Manufactured sand (crushed stone sand) 324
8. Sand mining in Malaysia 328
9. Conclusion 330
References 330

19. Probabilistic response of strip footing on reinforced soil slope

KOUSHIK HALDER AND DEBARGHYA CHAKRABORTY

1. Introduction 333
2. Methodology 334

3. Problem statement: Probabilistic bearing capacity of strip footing on reinforced soil slope 342
4. Problem statement: Probabilistic load carrying capacity of strip footing on geocell reinforced soil slope 349
5. Problem statement: Probabilistic stability analysis of reinforced soil slope subjected to strip loading 354
6. Conclusions 356
References 357

20. Multivariate methods to monitor the risk of critical episodes of environmental contamination using an asymmetric distribution with data of Santiago, Chile

CAROLINA MARCHANT, VÍCTOR LEIVA, HELTON SAULO, AND ROBERTO VILA

1. Symbology, introduction, and bibliographical review 359
2. Uni and multivariate fatigue-life distributions 362
3. Fatigue-life statistical process control 365
4. Illustrations 371
5. Conclusions and future investigation 376
Acknowledgments 377
References 377

21. A combined sustainability-reliability approach in geotechnical engineering

DIPANJAN BASU AND MINA LEE

1. Introduction 379
2. Sustainable practices in geotechnical engineering 384
3. Life cycle assessment 390
4. Reliability and resilience 395
5. An integrated sustainability framework 405
6. Concluding remarks 409
References 410

22. Safety risks in underground operations: management and assessment techniques

PARTHIBAN KATHIRVEL

1. Introduction 415
2. Potential risks in underground operation 416
3. Identifying the risks in underground construction 418
4. Development and progression of safety management system 419
5. Approaches in assessing the safety risks 420
6. Conclusions 432
References 432

23. Sustainability: a comprehensive approach to developing environmental technologies and conserving natural resources

HOSAM M. SALEH AND AMAL I. HASSAN

1. Introduction 437
2. Sustainable development goals 439
3. Recent environmental technologies to reach sustainability 441
4. Mechanisms for activating solar energy applications 443
5. Conclusion 445
References 446

24. Effectiveness and efficiency of nano kaolin clay as bitumen modifier: part A

RAMADHANSYAH PUTRA JAYA, HARYATI YAACOB, NORHIDAYAH ABDUL HASSAN, AND ZAID HAZIM AL-SAFFAR

1. Introduction 449
2. Preparation of NKC 450
3. Chemical analysis 451
4. Determination size of NKC 453
5. Conclusions 459
Acknowledgments 460
References 460

25. Nano kaolin clay as bitumen modifier for sustainable development: part B

RAMADHANSYAH PUTRA JAYA, KHAIRIL AZMAN MASRI, SRI WIWOHO MUDJANARKO, AND MUHAMMAD IKHSAN SETIAWAN

1. Introduction 461
2. Rheological properties of the asphalt binder incorporating nanoclay 462
3. Asphalt binder characterization 463
4. Penetration 463
5. Softening point 463
6. Storage stability 464
7. Rutting resistance 464
8. Failure temperature 466
9. Phase angle 468
10. Fatigue resistance 469
11. AFM analysis 470
12. XRD analysis 472
13. Summary 474
Acknowledgments 474
References 474

26. Prediction of rutting resistance of porous asphalt mixture incorporating nanosilica

KHAIRIL AZMAN MASRI, RAMADHANSYAH PUTRA JAYA, CHIN SIEW CHOO, AND MOHD HAZIMAN WAN IBRAHIM

1. Introduction   477
2. Nanosilica and mixing process   478
3. Porous asphalt mix design   478
4. Rutting resistance   478
5. Results and discussions   479
6. Summary   486
Acknowledgments   486
References   486

27. Policy options for sustainable urban transportation: a quadrant analysis approach

ANISH KUMAR AND SANJEEV SINHA

1. Introduction   487
2. Data and methods   488
3. Data analysis   492
4. Results and discussion   492
5. Conclusions   496
Funding information   496
References   496

28. Pavement structure: optimal and reliability-based design

PRIMOŽ JELUŠIČ AND BOJAN ŽLENDER

1. Introduction   499
2. Optimizing algorithm   500
3. Optimization model for pavement structure   501
4. Application of PAVEOPT model—case study   503
5. Failure probability of an optimally designed pavement structure—case study   505
6. Conclusion   507
Acknowledgment   508
References   508

29. Assessment of factors affecting time and cost overruns in construction projects

J. VIJAYALAXMI AND UMAIR KHAN

1. Introduction   511
2. Literature review   512
3. Methods and materials   513
4. Factors causing cost and time overruns   515
5. Method of analysis   515
6. Results and discussion   516
7. Conclusion   517
References   521

Index   523

List of contributors

Alireza Abdoallahi Department of Water Engineering, Bu-Ali Sina University, Hamedan, Iran

Zaid Hazim Al-Saffar Building and Construction Eng. Technical College of Mosul, Northern Technical University, Iraq

Selvaraj Ambika Department of Civil Engineering, Indian Institute of Technology Hyderabad, Telangana, India; Department of Climate Change, Indian Institute of Technology Hyderabad, Telangana, India

Danial Jahed Armagahni Department of Civil Engineering, Faculty of Engineering, University of Malaya, Kuala Lumpur, Selangor, Malaysia; Department of Urban Planning, Engineering Networks and Systems, Institute of Archi-tecture and Construction, South Ural State University, Chelyabinsk, Russia

Dipanjan Basu Department of Civil and Environmental Engineering, University of Waterloo, Waterloo, ON, Canada

Ramesh Murlidhar Bhatawdekar Geotropik-Centre of Tropical Geoengineering, Department of Civil Engineering, Universiti Teknologi Malaysia, Johor Bahru, Johor, Malaysia; Department of Mining Engineeing, Indian Institute of Technology, Kharagpur, West Bengal, India

Debarghya Chakraborty Department of Civil Engineering, Indian Institute of Technology Kharagpur, Kharagpur, West Bengal, India

Subhojit Chattaraj Department of Civil Engineering, MCET, Berhampore, West Bengal, India

Chin Siew Choo Department of Civil Engineering, College of Engineering, University of Malaysia Pahang, Kuantan, Pahang, Malaysia

Beste Cubukcuoglu Institute of Building Materials Research, RWTH Aachen University, Aachen, Germany

Deepthi Mary Dilip BITS Pilani Dubai Campus, United Arab Emirates

Sufyan Ghani Department of Civil Engineering, National Institute of Technology Patna, Patna, Bihar, India

Maryam Gharekhani Department of Earth Sciences, Faculty of Natural Sciences, University of Tabriz, Tabriz, East Azerbaijan, Iran

Mohammad Ali Ghorbani Water Engineering Department, Faculty of Agriculture, University of Tabriz, Tabriz, Iran

Anasua GuhaRay Department of Civil Engineering, BITS-Pilani Hyderabad Campus, Telangana, India

Koushik Halder Department of Civil Engineering, University of Nottingham, Nottingham, United Kingdom

Dayang Zulaika Abang Hasbollah Geotropik-Centre of Tropical Geoengineering, Department of Civil Engineering, Universiti Teknologi Malaysia, Johor Bahru, Johor, Malaysia

Amal I. Hassan Radioisotopes Department, Nuclear Research Center, Egyptian Atomic Energy Authority, Giza, Egypt

Norhidayah Abdul Hassan Faculty of Engineering, School of Civil Engineering, University of Technology Malaysia, Skudai, Malaysia

Dato Chengong Hock Soon MQA and Training Development Committee, Sectorial Training Committee, Malaysia Quarry Association, Kuala Lumpur, Selangor, Malaysia

Samed Inyurt Department of Geomatics Engineering, Faculty of Engineering and Architecture, Tokat Gaziosmanpasa University, Tokat, Turkey

Ramadhansyah Putra Jaya Department of Civil Engineering, College of Engineering, University of Malaysia Pahang, Kuantan, Pahang, Malaysia

Primož Jelušič University of Maribor, Faculty of Civil Engineering, Transportation Engineering and Architecture, Slovenia

Rajesh Jha Aggregate Innovations Pvt Ltd, New Delhi, India

Aleena Joy BITS Pilani Dubai Campus, United Arab Emirates

Mohammad Rezaul Karim Department of International Programme, Bangladesh Public Administration Training Centre, Dhaka, Bangladesh

Parthiban Kathirvel School of Civil Engineering, SASTRA Deemed University, Thanjavur, Tamil Nadu, India

Umair Khan School of Planning and Architecture, Vijayawada, Andhra Pradesh, India

Rahman Khatibi GTEV-ReX Limited, Swindon, United Kingdom

Anish Kumar Department of Civil Engineering, Rajkiya Engineering College, Azamgarh, Uttar Pradesh, India; Department of Civil Engineering, National Institute of Technology, Patna, Bihar, India

B.R.V. Susheel Kumar Mining Engineers Association of India, Hyderabad, Telangana, India

Sunita Kumari Department of Civil Engineering, National Institute of Technology Patna, Patna, Bihar, India

Mina Lee Department of Civil and Environmental Engineering, University of Waterloo, Waterloo, ON, Canada

Víctor Leiva School of Industrial Engineering, Pontificia Universidad Católica de Valparaíso, Valparaíso, Chile

Carolina Marchant Faculty of Basic Sciences, Universidad Católica del Maule, Talca, Chile; ANID-Millennium Science Initiative Program-Millennium Nucleus Center for the Discovery of Structures in Complex Data, Santiago, Chile

Khairil Azman Masri Department of Civil Engineering, College of Engineering, University of Malaysia Pahang, Kuantan, Pahang, Malaysia

Mohd Firdaus Md Dan Department of Infrastructure and Geomatic, Faculty of Civil and Environmental Engineering, Universiti Tun Hussein Onn Malaysia (UTHM), Johor, Darul, Takzim, Malaysia

Edy Tonnizam Mohamad Geotropik-Centre of Tropical Geoengineering, Department of Civil Engineering, Universiti Teknologi Malaysia, Johor Bahru, Johor, Malaysia

Sri Wiwoho Mudjanarko Narotama University, Sukolilo, Surabaya, Indonesia

Ata Allah Nadiri Department of Earth Sciences, Faculty of Natural Sciences, Institute of Environment, University of Tabriz, Tabriz, East Azerbaijan, Iran; Traditional Medicine and Hydrotherapy Research Center, Ardabil University of Medical Sciences, Ardabil, Iran; Medical Geology and Environmental Research Center, University of Tabriz, Tabriz, Iran

Beeram Satya Narayana Reddy Department of Civil Engineering, National Institute of Technology Calicut, Kozhikode, Kerala, India

Pranjal Pathak Department of Mining Engineeing, Indian Institute of Technology, Kharagpur, West Bengal, India

S.K. Pramada Department of Civil Engineering, National Institute of Technology Calicut, Kozhikode, Kerala, India

M.C. Raghucharan Indian Institute of Technology Hyderabad, Department of Civil Engineering, Hyderabad, Telangana, India

Afia Rahman Department of Research and Development, Bangladesh Public Administration Training Centre, Dhaka, Bangladesh

Avtar K. Raina CSIR-Central Institute of Mining and Fuel Research & AcSIR, Nagpur, Maharashtra, India

Thendiyath Roshni Department of Civil Engineering, National Institute of Technology Patna, Patna, Bihar, India

Ali Asghar Rostami Department of Water Engineering, University of Tabriz, Tabriz, East Azerbaijan, Iran

Sina Sadeghfam Department of Civil Engineering, Faculty of Engineering, University of Maragheh, Maragheh, East Azerbaijan, Iran

Hosam M. Saleh Radioisotopes Department, Nuclear Research Center, Egyptian Atomic Energy Authority, Giza, Egypt

Helton Saulo Department of Statistics, Universidade de Brasília, Brasília, Brazil

Zahra Sedghi Department of Earth Sciences, Faculty of Natural Sciences, University of Tabriz, Tabriz, East Azerbaijan, Iran

Muhammad Ikhsan Setiawan Narotama University, Sukolilo, Surabaya, Indonesia

Elham Shabani Department of Agriculture Economics, Faculty of Agriculture, University of Tabriz, Tabriz, Iran

Trilok Nath Singh Earth Science Department, Indian Institute of Technology Bombay, Mumbai, Maharashtra, India

Sanjeev Sinha Department of Civil Engineering, National Institute of Technology, Patna, Bihar, India

Surendra Nadh Somala Indian Institute of Technology Hyderabad, Department of Civil Engineering, Hyderabad, Telangana, India

Byomkesh Talukder Dahdaleh Institute for Global Health Research, York University, Toronto, ON, Canada

Aadil Towheed Department of Civil Engineering, National Institute of Technology Patna, Patna, Bihar, India

Anamika Venu BITS Pilani Dubai Campus, United Arab Emirates

J. Vijayalaxmi School of Planning and Architecture, Vijayawada, Andhra Pradesh, India

Roberto Vila Department of Statistics, Universidade de Brasília, Brasília, Brazil

Mohd Haziman Wan Ibrahim Faculty of Civil Engineering and Built Environment, Tun Hussein Onn University of Malaysia, Batu Pahat, Johor Bahru, Malaysia

Haryati Yaacob Faculty of Engineering, School of Civil Engineering, University of Technology Malaysia, Skudai, Malaysia

Muhammad Faiz Bin Zainuddin KSSB Consult Sdn Bhd, Tingkat LPH, Shah Alam, Selangor Darul Ehsan, Malaysia

Bojan Žlender University of Maribor, Faculty of Civil Engineering, Transportation Engineering and Architecture, Slovenia

CHAPTER 1

A basic framework to integrate sustainability, reliability, and risk—a critical review

Rahman Khatibi
GTEV-ReX Limited, Swindon, United Kingdom

1. Introduction

Research activities and practices on sustainability, reliability, and risk are largely driven by policymaking and goal orientation in learning organizations. This chapter contextualizes these activities by introducing a Basic Sustainability Framework (BSF) in three dimensions: (i) governance (policymaking and planning); (ii) goal-oriented learning organizations (sustainability); and (iii) decision-making (including reliability and risk). Notably, the dimensions of a framework are selected consensually, which together should give rise to a new capability, where each dimension, in turn, frames together a set of working ideas or models. The least benefit of BSF would be to integrate scientific activities through a bottom-up learning process nurtured among scientific communities and to make visible the ongoing fragmentation in the landscape of science, which is often sensed as barriers against knowledge transfer.

This chapter promotes critical thinking toward identifying more dimensions for a high-level theorization of science.

The critical review presented in this chapter is not an exhaustive consideration of research or practice works and as such paper-by-paper reviews are not intended. Its thrust is on using the above three dimensions in terms of outlining their emergence, contexts, growth, and methodologies. Each of the three dimensions follows some sort of a best practice procedure to allow free flow of information. The three dimensions of BSF are introduced below.

Governance emerged in the 1990s with the key shift on designating the dynamics of *inclusion of the citizens and the society* within the political processes of decision-making, see Lavelle (2013). Stoker (1998) presents five propositions for governance and argues that the term applies to a wide range of activities, institutions, and organizations, which are explored in due course. The term governance is not treated as an alternative to the term government referring to the coercive formal and institutional processes functioning at the nation state level, which maintains public order and facilitates collective actions. Arguably, governments, institutions, or organizations share a common feature that they all have a legal mandate for their existence; and since the emergence of the concept of governance in the 1990s, there is an implied adherence to transparency, accountability, and ethical commitments to environmental statements or vision statement.

Goal orientation or goal-seeking refers to systems capable of responding differently to events to reach desired states (OSG, 1981). Mullins (1993) describes the essentials of such organizations as open systems comprising inputs, processes, outputs, and feedbacks from goals to objectives, which can be applied to all systems. Kaplan and Maehr (2007) describe the importance of the goal orientation theory in achieving motivation, particularly in academia since 1980. Since the emergence of the concept of sustainable developments in the 1980s, the United Nations (UN) adopted the Agenda for Sustainable Development in 2015, which is a program until 2030. It comprises 17 Sustainable Development Goals (SDGs) to balance economic and social development with environmental sustainability (The 2030 Agenda, 2015). Biermann et al. (2017) discuss their implications toward the emergence of "governance through goals" as a novel mechanism for world politics and are explicit that such global governance is new. SDGs are now taken as global drivers for goal orientation and enshrine human aspirations for good governance.

This chapter puts emphasis on the role of systems science and regards goal orientation as the dimension capable of overarching the other two dimensions. Its concept emerged in the 1950s and galvanized various movements in science, as to be reviewed in due course. In essence, systems science integrated a set of concepts/ principles such as positive/negative feedback, feedforward (goal orientation), performance, entropy, hierarchy, and so on. Systems make up the context of goal-oriented organizations necessary for implementing SDGs. These are the key to the necessity of and urgency for the SDGs, which arise out of entropic impacts of the Industrial Revolution (1750–1950) and fragmented scientific methodologies. As such, the full lifecycle management policies are not available for many systems, but this is an indication of ignoring their entropic impacts. The delivery of any goal-orientated system depends on the proper implementation of the appropriate principles of systems science, and this chapter discusses their subsequent haphazardly potentials.

Decision-making is contextualized over systems to create a purposeful space, where reliability, risk, and uncertainty tools are used for decision-making. This chapter integrates decision-making with the other dimensions of BSF, where sustainability is the full function of a system but decision-making activities maintain its homeostasis state by formulating appropriate strategies using reliability, risk, uncertainty, and similar studies. These decision-making studies become more meaningful within the context of systems, in which inputs, processes, and outputs are clearly defined. Systems serve as platforms, where devising best practice procedures become feasible to study proactively the reliability to ensure that operational systems at their failure remain safe, as well as devise strategies to cater for risks with an integrated insight of performance—risk—uncertainty. These attributes are spinoffs of systems science and topical research drivers, which have already given rise to best practice procedures in many disciplines. Other attributes include vulnerability, resilience, recoverability, accessibility, availability, maintainability, transferability, serviceability, durability, extensibility, scalability, and tolerability. Potentially, these attributes can contribute to sustainable systems toward their robust performances.

This chapter is minded on the connectivity between good research outputs and their uptakes. If a research output does not specify its legal basis or if it is not helping decision-making to steer toward SDG, its return is at risk of being wasted unless it serves increased knowledge. It is often up to the researcher to ensure the uptake of their outputs but without inclusivity. Uptakes of research outputs take place in competitive environments and this gives rise to the removal of many good ideas from reaching their targets. Also, uptakes of research works are not readily measurable and those in terms of citation are academic without any practical significance. However, BSF can make a difference by its inclusivity and creating a hierarchy of techniques, each with its appropriate reliability and caveats.

2. Critical insight into the structure of science

Scientific methodologies are driven by data, evidence, and continual refinements (negative feedback) and these give rise to an evolving world of science often driving almost continual changes. Arguably, science acts as an agent of change, but there are hardly any scientific concepts to explain the structure of science. The subsequent gap is filled by the philosophy of science through "exclusionary" doctrines taking retrospective and incoherent views. This section aims to stimulate critical thinking by a critical review of the subject by simplifying the time dimension by dividing it into (i) the modern era and (ii) the "dim past," see Fig. 1.1. The modern era literally reconstructed the

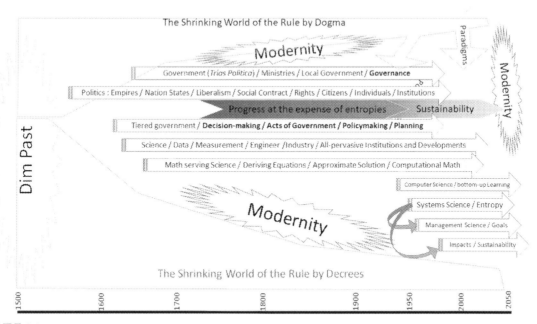

FIGURE 1.1   Illustrating modern concepts crack opening the dim past driven by dogma/decree. Note: The time base and the movements are highly place dependent and variable.

knowledgebase of the dim past from the first principles. Notably, the chapter avoids various niceties, as historic elaboration is out of scope of the chapter. Some of the issues are expanded in the next sections, where necessary.

2.1 Pivotal scientific concepts

The term "dim past" is introduced to refer to medieval times to emphasize that the thinking in science bears almost nothing from the pre-modern era and this is illustrated in Fig. 1.1. The dim past is associated with philosophical discourses on the supremacy of mind over matter or vice versa and determines the thinking of the time, ostensibly driven by justifications for religions and/or empires, though, on occasions; it also challenged the status quo. The dim past is also strongly associated with dogma and metaphysics. Arguably, an uncontroversial overview of the dim past includes the following: (i) the intellectual world was thriving on metaphysical philosophies to justify the deep grip of imperial decrees or religious dogmas; (ii) philosophy failed to uncover the role of data, evidence, and negative feedback; (iii) philosophy regenerated itself within its own limits and constraints of decrees and/or dogmas, although dissidents appeared from time to time; and (iv) philosophy bears historic significance but surprisingly still reemerges from time to time to interfere with science.

Modernism was triggered as past corruptions made a case for seeking alternatives, which emerged out of the thinking of humanists and secularists who wanted to take charge of their lives and gave rise to a new mindset through waves of movements, see Russell (1971) and Zahoor (2019). Scientific thinking emerged exclusively in the modern era, with main characteristics that scientific concepts of: (i) the past were deconstructed by constructing new theories from first principles; (ii) substances of new findings are overwhelmingly evidence-based; (iii) scientific theories are mutually inclusive and build on existing concepts; and (iv) scientific knowledge evolves, although the rules driving evolutionary processes are not yet evident.

Once the movement for evidence-based science formed a *critical mass* in the 18th century, it created a *snowball effect* through exploring different areas and forming different disciplines. The emerging science *crack opened* a gap between dogma and decree, as depicted in Fig. 1.1 and widened it in time to the extent that past beliefs were shaken off and social orders were reshaped within the gap. The emerging new cultures and mindsets are testing grounds for new ideas or shaking off governments or orders sometimes through violent means.

2.2 The framing scientific concepts

The gist of this section is that today there is a wide range of terms to denote the particular scientific activities, which include the use of "analysis" for any scientific framing activities prior to the 20th century (see Stewart, 1989). Traditional terms of laws, theory, hypothesis/proposition, postulate, and conjecture were coined or redefined in due course and more have emerged in recent years, including cluster sciences, frameworks, paradigms, principles, and heuristics. These terms have emerged over time to capture the particular aspects of science and to represent movements at the time. Without the sense of these changes, scientific activities would seem ontological, i.e., timeless or *as if they have always been like this*. Arguably, an ontological view of science is counterproductive. These terms are explained in the appendix.

2.3 Diminishing role of philosophy in science

Losee (1987, p. 1) introduces the philosophy of science as a second-order interpretation of a first-order subject matter. If this reflects the mindset of philosophical communities, it can hardly be shared by scientists. It is well known that while philosophers strive for cross-fertilization with science, scientists are rarely interested in philosophy. This gives rise to the following problems: (i) philosophy remains stimulating by asking critical questions but of why-questions that scientists are hardly interested; (ii) philosophical premises are invariably loaded with some sort of dualism that contravenes science founded on sanitized and measurable terms. For instance, such terms are not used in science: universal truth, the supremacy of something over another or ontology, and (iii) the horizon of science is unlimited as it keeps expanding, and as such, scientists do not engage with explaining the expansion of science, whereas philosophers thrive on such explanations with counterproductive outputs.

Up to the end of the enlightenment times, early modern philosophy was the only dissident voice, which lubricated the environment for the emergence of science. Since the explicit eminence of science from the 19th century, science overshadowed philosophy, but philosophical doctrines to explain the expansions of science received attention in terms of exclusionary doctrines including confirmation, positivism, falsification, paradigm shift, Scientific Research Program, "Against Methods," and phenomenalism. These landmark doctrines in philosophy are not further elaborated here, although paradigm shifts were outlined in the appendix. They all share one premise that they are all loaded with hidden dualism but without full regard to their hidden message, they can be misleading.

The philosophy behind postmodernism is touched on here due to the clout that it keeps creating to overshadow science. A postmodernist premise is not of that of a scientist to report inconsistencies in science for the promotion of refinement or to restart a theory (paradigm shifts) but a trojan horse with a philosophical motivation to revive the mindset of the Middle Ages, the Age of Deference. It targets modernism in terms of associating modernism with grand narratives and ideologies, criticizing rationality promoted by the age of enlightenment, and questioning ideology underpinning political or economic power. These are all legitimate issues, and while raising issues is a positive contribution, their solutions in terms of reviving the medieval mindset or the claim by Forman (2007) on the supremacy of technology to science are quite disconcerting. He claims that modernism gave supremacy to science prior to circa 1980 and since then postmodernism has given that to technology. Arguably, any claim on any supremacy ought to be regarded as pseudoscience. Arguably, failing to question the universal truth is a fundamental shortfall in any intellectual movement.

Scientific terms are often sanitized before their adoption as opposed to philosophical terms. The most obvious example is the terms with the prefix of "meta," e.g., metaphysics and as such one prefers to use data warehousing to metadata. This is not a call for "political correctness" but a safeguard against the dualism of matter and mind, whereas science normally regards matter-related notions to form the context to the phase of a system and by the same token, mind-related notions to form its state. As such, it studies both the phase and state of a system within one scientific platform under different configurations with no supremacy attached to them. It may be that a scientific method is purely focused on phase or state, but such methods are branded from the beginning as approximate methods to ensure plurality, to cut off any need for the supremacy of one over others, so no need for political correctness. Supremacy is simply a presumption, an opinion, or a dogma with no place in science.

It is not enough for scientists just to maintain the status quo of broadening the horizon of science by innovations, but there is a desperate need for the scientists to learn the way the enterprise of science is evolving and offer its route maps and foresight.

2.4 Evolution of science

Systems science emerged in the 1960s and galvanized various movements in science. Its rise is discussed by Lilienfeld (1978) to encompass (i) Ludwig von Bertalanffy presenting the General Systems Theory; (ii) Norbert Wiener's cybernetics, W. Ross Ashby's related works leading to feedback and automation; (iii) Shannon, Weaver, and Cherry developing the information and communication theory; and (iv) Neumann and Morgenstern's games theory. It was also promoted through the Society for General Systems Research in 1957 (https://www.bcsss.org) and the Club of Rome (https://www.clubofrome.org/about-us/history/). Its uptake is now all-pervasive and for their outline, reference may be made to appropriate sources (e.g., https://www.ecology.gen.tr/general-systems-theory/34-what-is-systems-theory.html).

Permeating all disciplines of science and system science is now the mainstream science and serves as its multiperspectival, transdisciplinary, and interdisciplinary tools or concepts.

Complexity science is the outgrowth of systems science and builds on it (e.g., hierarchy, interconnectivity, feedback, or emergence), and in simple terms, it refers to the interconnection of many systems. It also gained an impetus from chaos theory to focus on small changes triggering large changes. While systems science is within the remit of causality (cause and effect), complexity science goes beyond by focusing on large changes in response to small causational changes and vice versa. In this way, complexity science triggered expectations for opportunities. For an overview of complexity science, see Weisbuch and Solomon (2007), Mitchell (2009), among others.

Science now pools together a host of disciplines and practices and seeks their inclusivity, toward "sustainability science," as the next stage of its evolution. This is unchartered territory, and it seems that the development of new indicators is emerging as its driver. If so, there is an urgent demand for the theorization of the structure of science, its evolutionary transitions, and vision.

3. Governance

The three dimensions presented in this chapter can serve as a model for a more holistic view of individual research works to make them transparent to one another through the integration of governance, goal orientation (organizations), and decision-making. This is similar to the separation of powers (*trias politica*) model of governments presented by Montesquieu (1689–1755) in 1748, which he argues for the necessity of breaking down state governments into branches of power comprising a legislature, an executive, and a judiciary, although this is now questioned for its inability to pay due debts of a human generation to the future generations (see Tremmel, 2014) and other species. There is a difference in the approach of BSF with *trias politica*, as BSF integrates at least three dimensions, but the separation of powers reduces power to three dimensions. Each dimension is presented in one section. The particular focus of this section is on critical issues of the new thinking driven by governance and its roots, in which inclusion is the keyword to bring a new meaning to democracy as carrying all opposed to its polarized meaning of the dictatorship of the majority.

3.1 Past political orders toward a new governance

When mainstream intellectual discourses during the dim past were driven by decree, religion, and philosophy by indulging with the supremacy of mind over matter or vice versa, the main preoccupation was to justify the prevailing decrees of the empires of the time and to preserve religious dogmas. The ineffectiveness of religion and secular authorities against waves of disasters in the late middle age in Europe was likely to have paved the ground for the emergence of modernity, see Benedictow (2006). Modernity was established through a process, as depicted in Fig. 1.1, and the *crack opened* intellectual freedom led to the gradual widening of its subsequent space. The agents of change were individualism (which was impossible under empires and religions favoring obedience), the proliferation of institutions, and possibly the uptake of the vernacular languages spoken by the commons in contrast to religious languages. The space evolved and now it is the age of sustainable development, giving rise to governance. The background is that the crack-opened space is not universal and not uniform but prevailed among each culture, country, or nation states in different ways and formats. Can the governance for sustainable development find a universal architecture? This fundamental question forms the root of this chapter, else diversity of techniques on sustainability, reliability, risk, and uncertainty may not have a direct bearing on the needs of the time.

The ages of humanism and enlightenment together with the subsequent evidence-based science triggered a process to moderate autocratic or dogmatic orders to what is now called decision-making through the individuals taking responsibilities and through proliferated institutions, often referred to as liberalism in Britain and the United States of America. Decisions were gradually taken by parliaments in many countries, which acquired a wider basis throughout the industrial world as the instrument of policymaking and replaced command-driven mindsets of rulers and religious authorities. Policymaking and decision-making by individuals in social scale are not existential but emerged and have been evolving within the living memory of modernism, where (i) the top-down agents of the change were the three constituents of the separation of power; (ii) the concept of checks-and-balance between them

removed the need for any supreme being; and (iii) the bottom-up agents of change increasingly became evidence-based science. The author argues that while one is often inclined to admire the age of enlightenment, the thinking at the age was often a mixed bag of ideas, containing unsavory ideas by today's standards.

Liberalism is a complex reading in history and often opposes radicalism such as those by socialism. It remains as a prevailing ideology, but its historiography is outside the scope of this chapter (for more, refer to https://plato.stanford.edu/entries/utilitarianism-history/ and Russell, 1971). Generally, Loptson (1995) states that central enlightenment positions in Britain and France favored freedom from state control and came to be viewed as liberal. It is relevant to know that it has gone through many shifts outlined as follows: (i) Loptson (1995) states that the 18th century currents prefigured advocacy of freedom from control as liberalism; (ii) he also states that the 19th century philosophers are identified with the advocacy of minimal government restrictions on trade, movement, and ideas without opposing traditions; (iii) in the 20th century, neoliberalism revived the 19th century ideals on *laissez-faire* economics giving more power to the private sector through privatization, deregulation, and reduced government spending in the private sector [neoliberalism | Definition, Ideology, & Examples | Britannica]; and (iv) since the 1950s, liberal democracy prevailed as a political ideology for the advocacy of elections between multiple political parties, the separation of powers, the rule of law as prerequisite for an open society, a market economy with private property, and the equal protection of human rights, civil rights, civil liberties, and political freedoms for all people (Harpin, 1999) by drawing upon a constitution (Lührmann et al., 2020).

It may be pointed out that governance is cognate to government, but their meanings are different. Anglo-American political theories use the term "government" to refer to the formal institutions of the state and their monopoly of legitimate coercive powers (Lavelle, 2013). A government is characterized by its ability to make decisions and its capacity to enforce them and to refer to the formal and institutional processes at the level of the nation state to maintain public order and facilitate collective action (Lavelle, 2013). Theoretical work on governance reflects a shifting pattern in styles of governing through the inclusivity of the citizens and the society in decision-making acts. One of the massive shortfalls of liberalism is the dismantlement of the sense of community as the price for individualism, a form of grouping together that acts as a mechanism to cope with disasters. Arguably, the emerging governance is sufficiently sensitive to past shortfalls.

3.2 The new paradigm of governance

The emergence of the term governance in politics in the 1990s and the creation of the narrative of collective identities in political discourses of institutions are underlined by researchers including Skogstad and Schmidt (2011). They state that the process of governance designates the dynamics of *inclusion of the citizens and the society within the political processes of decision-framing and decision-making*. Blyth (2001) remarks that governance acts as "cognitive locks" that create "an intellectual path dependency in policymaking." Prior to modernity, there was just one significant organization, the autocratic government in each empire, but their absence was filled by clanship and/or patriarchy/matriarchy, where the latter forms were existential. Modernism gave rise to nation states and the gradual proliferation of organizations and institutions with ad hoc systems of administrations at their best or some form of management system, perhaps not much different than clanship and patriarchy. Over the years since the last

decade of the 20th century, the concept of governance became all-pervasive.

Lavelle (2013) argues that governance is an outcome of opposition between the democratic and the nondemocratic entities in terms of technocratic (skilled-based power), the ethocratic (virtue-based power), and the epistocratic (wisdom-based power) mindsets (poles). Outcomes of each interaction and counteraction between these mindsets or other mindsets create or assume the condition for social rules or decisions to be valid; they reflect, discuss, and make by an elite of experts, virtuous or wise individuals, or groups, practitioners, or experts. Inevitably, there is also the issue of trust and distrust in relation to the ability of the people to take charge of public affairs, to cope with appropriate standards and norms, and to comply with rules, regulations, and conducts (Lavelle, 2013).

Stoker (1998) presents governance through five propositions, each illustrating an aspect of governance to avoid any claim on their generality and they are as follows: (i) governance refers to a set of institutions and actors drawn from but also beyond government; (ii) it identifies the blurring of boundaries and responsibilities for tackling social and economic issues; (iii) it identifies the power dependence involved between institutions in collective action; (iv) it is about autonomous self-governing networks of actors; and (v) it recognizes the capacity to get things done without the power of government to command or use its authority, where governments steer and guide toward new tools and techniques.

3.3 Governance in framing policymaking and planning

Up to the 19th century, political orders were top-down and it would have not been surprising if the decisions were arrived at off-the-cuff, as there was hardly any social learning. Moving on from the dim past in the world of decree, dogma, and the age of deference, emerging politicians were persuaded by moral philosophers to legislate and one of the most attractive moral doctrines in the 18th and 19th centuries was utilitarianism (https://plato.stanford.edu/entries/utilitarianism-history/). While liberalism became the political paradigm in industrial countries, the emergence of utilitarianism in the 19th century turned into a movement for policymaking. Utilitarianism, as moral philosophy, influenced early legislations by seeking to maximize happiness by returning the greatest utility, but it ignored impacts and accelerated anthropogenic changes at an unprecedented rate; for more background, see Jennings (2009). Notably, other political paradigms are not discussed for brevity.

Greenhalgh and Russell (2009) divide different approaches in policymaking into three broad schools: (i) positivist, a philosophical doctrine that places a high value on experiment and observation and on drawing inferences about a phenomenon from a sample to a stated population by formal hypotheses, which put more emphasis on methods (the controlled experiment, the randomized trial, and the "standardized" and "validated" questionnaire) than the theory; (ii) interpretivist (hermeneutic), with an emphasis on producing and reproducing social reality through the actions and interactions of people, as social reality can never be known objectively or studied unproblematically; and (iii) critical research, which seeks to reveal the inherent contradictions and conflicts.

The emergence of moral philosophies influencing early policymaking is arguably an indication of a reaction to excessive insensitivities of early liberalism to social justice. The emergence of philosophical doctrines and finding their ways to science may be indicative of the preoccupation of science with fundamental deeper issues than wider issues. However, each of these doctrines has strong philosophical predicaments and arguably incapable of coping with the impending environmental crises, indicative of the need for paradigm shifts in the near future.

As impacts of the Industrial Revolution (1750—1950) have become increasingly measurable since 1950 through the proliferation of systems science, alternative policymaking rationales were formulated based on reality but not on the predicaments of philosophical doctrines. These are already in the mainstream and known as evidence-based policymaking, which is a process of "social-learning." Hall (1993, pp. 275—276) argues that change in policies occurs through a process of social learning.

Overall, two positive aspects of social democracy and liberal democracy in modern times are (i) the erosion of the age of deference and (ii) the aptitude to change the tune of its politics, where developed countries under liberal democracy are rather progressive within their own countries and with their allies but often Machiavellian with the developing and stateless countries. Liberal democracy also gave rise to a number of crises including (i) the global financial crisis of 2007—09 but this was just one of its recent instances, as there have been numerous similar crises in the past; (ii) liberalism is known to have been insensitive to the extreme poverty and issues such as slavery; (iii) there are major ongoing environmental issues with impending crises but it is unlikely that liberal ideologies can resolve any of these serious issues without a new dimension of rethinking; and (iv) females are increasingly contributing to politics but normally through the masculine mindset of politics, characterized by punishing and banishing, and yet there is an increasing discovery of feminine mindset of politics, characterized by inspiration and forgiveness, aspects of which are elaborated by Antonescu (2015). These impending crises are of a different kind and the emphasis is on rethinking toward urgent actions needed now but not on another iteration.

3.4 Policymaking and planning in actions driven by governance

Governance has methodologies, processes, and procedures, but these are technical skills and outside the intended scope of this chapter. However, the general overview on governance is that it includes the central governments, local governments, agencies, institutions, and stakeholders, which work hand-in-hand to carry out their functions. Central governments in industrialized countries took shape with the separation of powers giving rise to ministries, but the emergence of local government goes back only to the middle of the 19th century and the other groupings are often of more recent decades.

Hallsworth et al. (2011) argue that the strength of policymaking is integral to that of government and that of the country, so that when policies fail, the costs can be significant. The procedure for policymaking is outside the scope of this chapter, but the governmental function of policymaking is often tiered as (i) policymaking, at the level of central governments; (ii) planning, at the level of local governments; and (iii) various governmental agencies or intelligent authorities are empowered with specific tasks.

Central governments often go through a procedure to issue their policies, which express drivers, scopes, limitations, the interconnectivity of professional partners, policy owners, responsibilities, processes, and procedures as well as policy documents to be developed by various authorities to be entitled as intelligent authorities. Local governments are not so much involved with policymaking but with carrying out policies using a planning system, often through a special department allocated as planning departments. Intelligent authorities are various agencies in the central government or local governments. In the United Kingdom, these authorities can be planning authorities, flood authorities, land drainage authorities, coastal protection authorities, land use authorities, highway agencies, and/or many similar agencies. They are empowered with various scopes to fulfill predefined responsibilities.

Intelligent authorities look after their policy designations as well as they sponsor appropriate studies to produce reports to establish a proactive understanding of the domain of their responsibilities. Any individual or organization making any change has to comply with policy requirements by liaising with intelligent authorities through following a format that identifies changes and proposes mitigation solutions. Intelligent authorities use their local knowledge and either consent to the proposals or challenge them, or they go through iterations until full compliance. A glimpse of such organizations is shown in Fig. 1.2.

Both policy and planning strategies undergo periodic reviews, which are designed to allow social learning and make appropriate procedural changes. The procedure includes consultations to take on board the views of the participants. Over the years since 2000, policies have normally been overhauled to embed the requirements for sustainable development, climate change, risk-based decision-making, and other environmental issues ad best practice procedures. In contrast, in many developing and underdeveloped countries the social learning is often nonexistent or very poor.

3.5 Toward future of governance, policymaking, and planning

In the absence of risk-based impact analysis, the arguments against bad laws and practices were that they lacked utility with a tendency to misery and unhappiness and did not result in any happiness. The rationale was then that if a law or an action does not *do any good, then it is not any good* (https://plato.stanford.edu/entries/utilitarianism-history/). Then, up to the

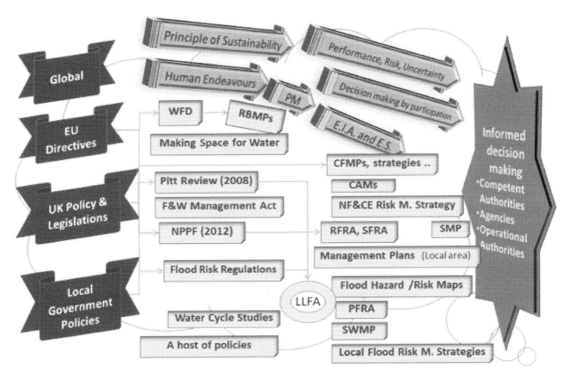

FIGURE 1.2 A broad overview of complex flood risk management in England in 2010. Note: EU directives are no longer applicable in the United Kingdom and some of the policy documents are now superseded, which illustrate the nature of variations in policymaking. Abbreviations: *CAMS*: Catchment Abstraction Management Strategy; *CFMP*: Catchment Management Plans; *EU*: European Union; *E.I.A*: Environmental Impact Statement; *E.S*: Environmental Statement; *F&W*: Flood Water; *LLFA*: Lead Local Flood Authority; *PFRA*: Preliminary Flood Risk Assessment; *RBMP*: River Basin Management Plans; *RFRA*: Regional Flood Risk Assessment: *SFRA*: Strategic Flood Risk Assessment; *SMP*: Shoreline Management Plans; *SWMP*: Surface Water Management Plans; *WFD*: Water Framework Directive.

1960s, the procedures and policies ignored impacts and accelerated anthropogenic changes at an unprecedented rate. Philosophy-driven past policymaking acted as a normative instrument and affected policies but only reinforced intended positive outcomes with no views on impacts of the Industrial Revolution (1750–1950) or bad social policies reinforcing poverty and social inequality.

Impacts of policymaking have become increasingly measurable since 1950 through the alternative views created by the proliferation of systems science. If policies do not deliver their required outcomes, they will be measured, reviewed, and refined. This is the scientific approach and now it is evidently the only credible way. Philosophical doctrines are resurrected from time to time, e.g., Gustafson (2013) argues for the usefulness of utilitarianism. However, this chapter does not justify philosophy-driven doctrines to interfere with science and argues that the evidence-based policymaking together with the evidence-driven planning systems befits goal orientation in a more appropriate way, as discussed in the next section.

4. Goal orientation overarching organizations

Sustainable development can only be delivered through organizations with flexible operating systems through systems science, as presented in the section.

4.1 Emergence of systems science and its uptake

Science from its emergence in the early 17th century till now thrives in terms of theories and empirical techniques to understand nature and life through data-driven explanations, which sometimes discovers laws and sometimes remains in the grip of lower-grade conjectures, hypotheses, or heuristics. These were the main building blocks of reductive science (circa from 1700 to 1950); but systems science synthesized a raft of new building blocks to explain life, scientific systems, social organizations, the environment, and ecology.

Since its emergence, science has gone through evolutionary transitions of reductive science (early 17th century to 1950) and systems science (since 1950). Without dwelling on historiography, the pivotal issues of reductive science include the following: (i) it did not build on the past knowledge but created new knowledge by undoing the past knowledge and by creating evidence-based new findings through a data-driven and bottom-up learning process; (ii) it gave mankind evidence-based facts owing to its scientific methodology and led to great discoveries, such as gravity, atomic structures, evolution by natural selection; (iii) it was an all-pervasive agent of change but more like a sculptor and sculpture, most of which were positive changes but also inflicted a host of entropic changes only to emerge after the transition to systems science, e.g., climate change; and (iii) the computational requirements of reductive science were largely met by manual processes, which also acted as the main entropy and triggered the need for a change.

Reductive science reduced a problem to parts in terms of their definable properties and analyzed them using the "scientific method," but systems science emerged in reaction to this by focusing on the structure of systems, which share several basic organizing principles. The organizing principles of systems science are now all-pervasive by selectively permeating through diverse disciplines since the 1960s, which include positive feedback, negative feedback, and feedforward loops; performance and failure; information and entropy; hierarchy, emergent property, and purpose; and dynamic equilibrium and homeostasis equilibrium.

4.2 Goal orientation

Each system throughout the global, social, technical, environmental, and cultural spheres is a selection from the above principles/concepts. If any of these entities qualify for being a goal-oriented system, it needs to be explicit in their inputs, processes, and outputs. Without the connection from outputs to inputs, the system follows positive feedback loops, see Fig. 1.3A. Characteristically, these systems are inflationary with no facilities for regulation. Many entities may pretend to be a system but are likely to be only a positive feedback system and tend toward an eventual death due to their increasing disorder (entropy). Current social and environmental problems can all be traced to this type of system. They are still in abundance and unlikely to contribute to the delivery of SDGs.

The author argues that only the systems with negative feedback capabilities become flexible enough to adapt and this is a pivotal requirement for becoming a goal-oriented system. Negative feedback is one of the most fundamental concepts developed in systems science in explicit terms. However, it has been invented

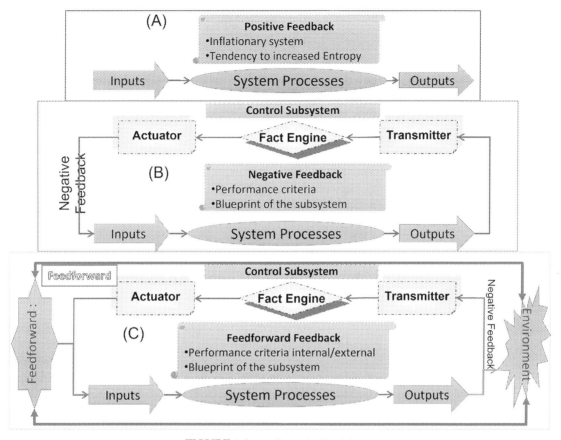

FIGURE 1.3 Different feedback loops.

by natural selection in diverse contexts and configurations. It was also embedded tacitly in the scientific methodology of reductive science, the first-ever instantiation in human endeavors. Its prerequisite is a module, depicted in Fig. 1.3B, with at least having detector, transmitter, fact engine, and actuator units and a flexible system capable of accommodating the actuated changes. Negative feedback loops are at risk of stagnation, as the fact engines are not capable of "self-modification" to modify their settings as the environmental conditions are dynamic and capable of radical changes. Thus, environmental conditions would render systems ineffective and irrelevant or stagnant forces if tuned to past conditions.

Stagnation in negative feedback loops can be prevented by feedforward loops. The definition of feedforward is not often clear-cut, but it is taken as a way of anticipating future conditions and responding to them to ensure the consistency of the operational systems with the external environment. As the environment is not static in the long run, and patterns of change in the environment are significant, feedforward loops can be a way of learning goals and resetting negative feedback loops. For instance, the transport industry is expected to undergo revolutionary changes in the near future to reduce impacts on the environment. Fig. 1.3C illustrates an interpretation of a feedforward loop.

Feedforward loops are essential for ensuring goal-oriented systems and to safeguard such systems against "goal displacement" and "goal fixation." Thus, goal displacement is defined by Merton (1968) as fixations and lack of adaptability, where the means become ends and more important than the actual goals. Klein (2009) remarks: "In complex settings, most of the goals are vague and ambiguous rather than clear-cut. Goal fixation—that is, continuing to pursue goals that have become irrelevant—affects too many managers." "Yet many people take the opposite route. When confronting complex situations, they try to increase their control over events. They try to increase the details in their instructions instead of accepting the unpredictability of events and getting to adapt. They stumble into goal fixation," Klein (2009).

The author argues that feedback loops are basic all-pervasive concepts and need to be understood by researchers, professionals, and practitioners of any discipline. Without a good understanding of these concepts, scientific premises would transform into a heap and many islands of disciplines. In particular, feedforward loops are the key for goal-oriented organizations, essential for the delivery of sustainable developments, as discussed next. If a system does not have seamless flows of information in its appropriate units of positive/negative feedback and feedforward loops, its goal orientation is doubtful.

4.3 Sustainable development

Sustainable development is a response to the outgrowth of entropic and often discarded impacts of technology-pushed and science-pulled changes. When the use of the concentrated energy contained in coal began driving industrial machines and carbon-based liquid fuel began to drive newly invented means of transport, no one was mindful of their impacts on air pollution and the onsetting risk of climate change and global warming. What is called *progress* has been at the expense of giving rise to a set of antitheses, now known as impacts or system entropy. Technological progress characterizes the mindset of the era of reductive science. Only after 1950, the accumulation of technology-pushed and science-pulled entropic impacts surfaced out. The author argues that reductive science is prized with progress, but the price paid for its achievements were entropic impacts. Thus, 1950 may be set as the transition toward the new era of sustainability, as critically reviewed below.

The period from the 1950s up to 1987 may be regarded as the interim period of the Green Movement for building up scientific evidence. One such study by the Club of Rome reported on "The Limits of Growth" (Meadows et al., 1972) by identifying five major trends of global concern: accelerating industrialization, rapid population growth, widespread malnutrition, depletion of nonrenewable resources, and a deteriorating environment. The increasing knowledge of this nature was brought to its conclusive delivery by the 1987 report "Our common future" or Brundtland Report, as the pinnacle of the World Commission on Environment and Development (WCED), set up in 1983. It laid down guiding principles for sustainable development as generally understood today (https://www.sustainable-environment.org.uk/Action/Brundtland_Report.php).

The term sustainable development was used by the Brundtland Report and defined it in the unequivocal term as "Development that meets the needs of the present without compromising the ability of future generations to meet their own needs." This is unmistakably a feedforward loop, the implementation of which requires goal orientation.

The report called for a strategy to bring development and the environment onto one platform, but the formulation of the strategy went through a number of landmark activities until the current status. These include (i) Earth Summit in 1992, during which 172 nations at the UN Conference on Environment and Development (UNCED) sought solutions to poverty, the growing gap between industrialized and developing countries, and growing environmental, economic, and social problems toward the objective of sustainable development around the world. (ii) The three pillars of sustainable development became topical (economic viability under the focus of economists; environmental protection under the focus of environmentalist; and social equity under the scrutiny of ecologists to integrate humanity to our common natural world) to deliver three agreements: (1) the legally nonbinding Agenda 21 (Rio Declaration) and Statement of Forest Principles; (2) the legally binding of Framework Convention on Climate Change; and (3) legally binding Convention on Biological Diversity. (iii) In between the first Earth Summit (1992) and up to the MDG 2000, the focus shifted from *needs* to *rights*, as the principal line of inquiry (Redclift, 2005), which was linked to the neoliberal economic agendas of the 1990s, and the growth of interest in congruent areas, including human security and the environment, social capital, critical natural capital, and intellectual property rights, which strengthened the linkages between "natural" and "human" systems, including attention to questions of environmental justice, where global environmental justice was gaining importance.

Millennium Development Goals (MDGs) were adopted by the United Nations Millennium Declaration in 2000 during the Millennium Summit in September 2000. This was a program for the period of 2000—15 and comprised eight goals for "Shaping the 21st Century Strategy" and showed a historic shift toward a sustainable future to integrate the three pillars of sustainability to exemplify a novel global governance through goal-setting features or "governance through goals." It brought to the research agenda critical global environmental problems and primarily related them to the result of poverty and unsustainable patterns of consumption and production.

Agenda 2030 brought all the ongoing initiatives and activities under one framework and replaced Agenda 21 and was adopted unanimously in 2015 to enshrine human aspirations for good governance. Its 17 SDGs make up a program that is an urgent call for action. There is an overwhelming realization that the world needs to pace up for greater efforts if the solutions for meeting SDGs are to be delivered. The environment makes up the direct core of most of these SDGs (11 out of 17) and the UN Environment Program helps countries achieve the SDGs for sustainability and resilience through science-based policymaking, global advocacy, and partnership building.

The above account represents the author's particular viewpoint on the transition from the age of "progress" overlooking entropy of industrial progress to the age of sustainable development. Khatibi and Haywood (2002) use Eq. (1.1) below to capture the transition in the following terms:

$$Development = function\_of \\ (cost - benefit\_with\_respect \\ to\ component)$$

(1.1)

$$Sustainable\_Development = fuction\_of\_\left(past\_ \\ experience\_and\_all\_future\_system\_wide\_impacts\right)$$

(1.2)

Consideration of impacts and the future were shaped through the following characterizing features: (i) the age of progress was not sustainable, as the generations after 1950 started paying the price of the complacency of the anthropocentric mindset of the age of progress; (ii) the new age is itself going through the identification of past entropies with the pinnacle of integrating environmental entropies (industrial complacency) with social entropies (poverty in the south; gender issues and inequality within the north); (iii) the transformation of the expression of needs into rights; (iv) inclusivity gained a recognition against past assimilation or exclusionary practices; and (v) the framing of SDGs in the service of bottom-up goal orientation organizations. But more entropic features are emerging that have not yet found their framing dimensions, e.g., the menace of the onsetting *posttruth*.

5. Decision-making—3rd dimension of BSF

The new thinking in terms of the emerging governance and goal-oriented systems is implemented by decision-making, as outlined in this section. The focus is on monitoring the implementation of SDGs through assessment/appraisal and a derivation of appropriate indices also supported by decision-making processes. The enterprise of SDGs is complex and often seems overwhelming but the following building block suggest a helpful architecture to understand the complexity. An SDG enterprise may comprise (i) the *phase*, in terms of physical layout, feedback loops, and hierarchies; (ii) the *state*, in terms of response to actions set at the boundaries of the system and the undergoing actions and processes within the system; (iii) configurations in terms of the physical arrangements and the settings of feedback loops; (iv) a tiered approach to managing the enterprise often described in terms of transnational, regional, national, local, and site scales; and (v) problem-solving approaches in terms of producing high-level assessments/appraisals, analysis techniques, and theories.

5.1 The enterprise of the SDGs

The lifecycle of any system (inception, design, operation, management, direction, and decommissioning) is through processes, but are also punctuated with making choices and hence decision-making, as acts capable of making a difference in the future course of the system by setting appropriate courses to achieve the goals.

There are multiple perspectives on each of these phases, including production systems, operational systems, maintenance systems, services systems, resources systems (human, water, energy …), and emergency response systems. The state of the system can be considered under a range of time resolutions, including dynamic time scale of seconds, real-time scales of hours or days, operational time scale of weeks, months years, planning/resource horizon of 20–30 years, and long-term time-scales of a century or more. Khatibi (2013) argues that there are also time resolutions for evolutionary and adaptive processes.

The landscape outlined above is just a glimpse of the world of decision-making, where there is no known single generic architecture for the methodologies to carry out analysis of these systems. Therefore, diverse techniques are inevitable and a few examples include G.O.F.E.R, given by Janis and Mann (1977), and D.E.C.I.D.E given by Guo (2008) and Turpin and Marais (2004), who review various decision theories. These reflect on complex decision landscapes and tools.

One perspective on decision-making is presented by Pekka et al. (2020), who divide the landscape into vision-based and rule-based decision-making. Decision-making based on vision or intuition is a perceptive model, which taps on rich contributions by individuals, but this is inducive to autocratic relationships and not amenable to programming and challenges. This chapter is concerned with informed decision-making using rule-based approaches, which are programmable, reproducible, and challengeable. The theoretical basis for problem-solving in decision-making is outlined below.

5.2 Sustainability appraisals and indices to measure decision-making

Delivering SDGs is now a common purpose in the global agenda, which steers toward an inclusive model of the world. As argued by Fukuda-Parr et al. (2014: 105), it has become feasible to use "global goals and target-setting as a central instrument defining the international development agenda." Following the lessons learned from MDGs, the 17 SDGs are broken down to some 169 targets (each of them are described in UN sites, see: https://unstats.un.org/sdgs/metadata/or https://www.concernusa.org/story/sustainable-development-goals-explained/). The goals and targets are transformed into several hundred indicators which are being used to measure progress toward achieving the goals and targets. The indices are used to monitor progress toward the SDGs at the local, national, regional, and global levels.

Sustainability is becoming the requirement to set the processes to scrutinize systems, actions, and changes toward identifying decisions to ensure that economic, social, and environmental considerations are in balance and have positive benefits to prove that the Earth and future generations are treated with continuity. The 17 SDGs are now seen as the backbone or roadmaps and the targets and indicators are regarded as tools. Each country would allocate resources to develop implementation strategies with a report card to measure progress toward sustainable development to ensure accountabilities. These reports are known as Voluntary National Review, often reported every 2 years (see https://sustainabledevelopment.un.org/vnrs/).

The SDGs and targets form a new complexity enterprise supported actively by Research and Development (R&D) to ensure their delivery in several ways including (i) R&D organizations, experts, professionals, or practitioners advise governments to embed SDGs in policymaking and planning systems; (ii) they appraise Voluntary National Reviews (VNR); and (iii) they develop various indicators in relation to SDGs to understand interdependence among the SDGs and targets, to delineate bottlenecks and to encourage synergies. For instance, in the UK and EU countries, the requirements for the delivery of SDGs are continually embedded in policies and directives. There is a policy document known as Strategic Environmental Assessment

and local government advises development proposals on the submission of SEAs to ensure the sustainability of the project.

Each member state undersigning the SDGs is required to submit VNR database by conducting regular and inclusive reviews of progress at the national and subnational levels to serve as a basis for the regular reviews by the high-level political forum. These provide platforms to share experiences, including successes, challenges, and lessons learned, to accelerating the delivery of the 2030 agenda.

Already, there are several hundred sustainability indices to heed for the various aspects of 17 SDGs and 169 targets. One example is outlined here, given by Sandoval-Solis et al. (2011) for water resources, who refer to it as Sustainability Index (SI). This is devised to evaluate and compare different water management policies with respect to their sustainability, assuming that all the components in the system are in balance, where these systems are seen as those designed and managed to contribute fully to the objectives of society, now and in the future, while maintaining their ecological, environmental, and hydrological integrity. They present a model to express SI in terms of three dimensions of reliability, resilience, and vulnerability. Simplified SI is expressed as follows:

$$Sustainability\ Index\ for\ Water\ Resources\ during$$
$$the\ period\ of\ simulation = (Reliability: probability$$
$$that\ the\ available\ water\ supply\ meets\ the\ water$$
$$demand) \times (Resilience: probability\ of\ successful$$
$$periods\ following\ failure\ periods\ for\ all\ failure$$
$$periods) \times (Vulnerability: 1 - average$$
$$failure\ or\ deficit\ period\ in\ the\ simulation\ period)$$
$$(1.3)$$

The above definition of reliability is by no means universal, but this is covered in the next subsection. Resilience was originally defined by Holling (1973) to describe the ability of a dynamic multispecies ecological system to persist with the same basic structure when subjected to stress. Also, vulnerability is one definition as it has various versions.

Sandoval-Solis et al. (2011) show results and conclude that SI differentiates between scenarios that reflect operational characteristics of a basin and may be used to evaluate, compare, and identify operational options. The SI indicator above exemplifies several hundred indicators that are being devised to understand the balance between social, economic, and environmental contributions to the sustainability of any systems.

5.3 Decision-making tools

The author argues that there is an implicit hierarchical architecture in sustainability problems, which may be articulated through systems thinking. The system requires to have an internal coherence (analogous to biology) and needs to react to, or interact with, the external environment (analogous to psychology). The internal coherence of the systems is catered for by a series of problems under the discipline of reliability problems. Interactions of the systems with the environment may be catered for through a set of problems, which may be organized hierarchically as (i) probability-based decision-making, which studies the system as a black box in terms of events and makes probabilistic/stochastic inferences on system behaviors; (ii) risk-based decision-making in terms of taking on board adverse events and their outcomes to gain considerable insight into the system; (iii) decision-making problems under uncertainty to study the system when the data are insufficient to understand outcomes; (iv) decisions problems driven by the principle of precaution, when there is not enough data or knowledge to study adverse events and their outcomes. All these decision problems may be considered tiered as (i) high-level or broad-scale studies for the assessments of decision-making problems; (ii) medium-level analysis

for developing strategies; and (iii) high-resolution investigations supported by theories and models.

5.3.1 Within system reliability—decisions tools within systems

There is a wealth of skill, experience, and knowledge to make sustainable operational systems coherent and reliable. These approaches tap on existing practices, expertise, and knowledgebase, which have grown in scale and scope since the 1970s. These are tools to discover the system, prior to which designs were based on deterministic approaches and improvements were slow and driven by trial-and-error. Reliability problems together with a set of other variables (e.g., availability, maintainability, validity …) have transformed past fragile systems into safe-fail systems. Reliability is defined in various ways including the ability to operate under specified operating conditions for a specified period in a cycle.

Best practice procedures for studying the reliability of sustainable systems are carried out at the following resolutions. At a professional or practice level, *reliability assessments* are often carried out by reliability engineers. The assessments systematize the procedure to seek robust evidence that a system is not associated with unacceptable failures by studying various operational activities including worst-case scenarios and most catastrophic failures. Thus, undesirable modes of operation are identified, including catastrophic modes of failure. These lead to understanding failure costs and their trade-off with reasonably balanced cost of improved design. Assessments are carried out through steps, often including (i) identify unreliable or hazardous events including failure modes and root causes and errors; (ii) tap on the available knowledgebase to identify their possible solutions; (iii) work out mitigation options; and (iv) select the most appropriate option. The process is studied qualitatively or quantitatively.

Reliability analysis considers two dimensions of failure in terms of (i) probability of occurrence of failure and (ii) magnitude of the consequence of failure. The analysis normally does not consider detailed processes, but the data from a particular system or its components are analyzed by a raft of techniques, outside the scope of this chapter, as detailed in textbooks, e.g., Modarres (1993).

The difference between the terms reliability analysis and reliability theory seems a matter of semantics and interchangeable. However, some seem to use analysis for black box treatments based on probabilistic approaches but theory for methods giving a sense of processes. Therefore, reliability theory here is used in the sense of system treating failure through the following steps: (i) use the definition given above with respect to load-resistance model and (ii) use system-specific data to study fail-safe solutions. A reliable system is defined in terms of the probability of nonfailure and failure, both assessed by considering the interaction of Load (L) and Resistance (R). The system is reliable when the load does not exceed the resistance but fails if exceeds and renders the system unreliable. The reliability of a system is defined mathematically as follows (Tung et al., 2005):

$$P_s = 1 - P_f = P(L \leq R) \qquad (1.4)$$

where $P(\cdot)$ is probability function, P_f is the probability of failure, and P_s is the probability of non-failure (reliability). Eq. (1.4) is reformulated by using the concept of performance function (Z) as follows:

$$P_s = 1 - P_f = P(Z \geq 0) \qquad (1.5)$$

The performance function $P(Z \geq 0)$ is defined in terms of load and resistance. Treating these equations uses various indices various probability density functions, and incorporate Monte Carlo Simulation techniques but their mathematical treatments are outside the scope of this chapter.

There are also techniques, which use the Taylor expansion theorem to treat the performance function. These give rise to various techniques including (i) the First Order Second Moment (FOSM); (ii) the Second Order Reliability Method (SORM); and (iii) there are other variations, as well but their descriptions are not intended in the chapter.

5.3.2 Decision-making tools based on probability

Probability-based decision-making tools are applicable to systems, in which some recorded or real-time data are available, but the state of the system cannot be described by cause-and-effect processes as either the behavior is complex, or the variables are random. Beyond a statistical description of the data in terms of estimating their expected values for random variables, probabilistic models take a black box view of the system and provide additional tools to make inferences with respect to the population, as outlined below.

Best practice procedures for probability-based decision-making of sustainable systems are carried out at the following resolutions. Decision-making tools by using probabilistic techniques include the following: (i) at a broader scale, statistical descriptions of the data are used to gain an insight into the problem; (ii) at an intermediate level, there are various probability distributions techniques to make inferences on the data; and (iii) at a detailed level, there are ensemble representations of the system using such techniques as Monte Carlo Simulation. A detailed account of these techniques is outside the scope of this chapter.

5.3.3 Risk-based decisions

Risk-based decision-making is used when adverse events and their outcomes can be identified to gain considerable insight into the system. They reflect the current scientific thinking. It is a simple scientific admission that there is no certainty or determinism in philosophical terms! Risk-based decision-making is conjugated with the future and Beck (1992) argues that "As soon as we speak in terms of 'risk', we are talking about calculating the incalculable, colonizing the future." According to him, the modern concept of risk inherently contains the concept of control and is driven by decision-making, whereas at premodern times, dangers were attributed to nature, gods, or demons. Giddens (2006) argues that fate and destiny have no formal place in systems operated via open human controls of natural and social worlds.

As reviewed by Khatibi (2011), the basic definition of risk is in three dimensions and it is taken as products of (i) hazards, which express local adverse effects; (ii) likelihood, which express the probability of adverse impacts; and (iii) responsibility, as reviewed by Rehmann-Sutter (1998). Some practices on risk use only one dimension and these are often emergency-based practices, e.g., in flood risk management, there are flood maps, which indicate risk in terms of frequency. Most risk-based decision-making activities are based on the first two dimensions and Rehmann-Sutter (1998) refers to this as the economic concept of risk. Yet, there are judicial practices, which consider all these three dimensions. Risk-based decisions cope with the analysis of actions and systems, which acts as the basis for performances. Thus, a system (as described in Sections 2.4) is envisaged to be performing, but risk-based decisions address the particular system configurations to be triggered should failures occur at any time in the future. Hence, performance and risk are conjugated and these are conjugated with uncertainty, as discussed later in this section.

The review by the Defra and Environment Agency to overhaul their functions from "flood defenses" to risk-based flood management took place later in the first decade of 2000 after considerable reviews and R&D works, see Defra (2003). The report presented a rethinking for (i) considering the management of floods as systems in the framework of performance, risk, and uncertainty; (ii) identifying various functions of flood risk management with respect to the tiering of the functions at national, regional, and local levels; and (iii) clarifying the roles at various levels of high-level and bottom-up responsibilities for serviceable, tolerable, reliable flood risk management systems. R&D works then closed the cycle of responsibility by the local authorities required to lead on the recoverability and resilience by working with the at-risk population when exposed. Various aspects of risk-based decisions are illustrated in Fig. 1.4 on systems' context.

While risk perception goes back to prehistory, risk-based decision-making is recent and goes back to circa the 1990s, which built on existing culture of risk assessment. As stated by Aven (2016), risk assessment and management emerged in the 1970s through the development of principles and methods to conceptualize, assess, and manage risk. Since then, research and modeling serve the consolidation and refinement of the principles and methods and their applications have been all-pervasive. Aven (2016) illustrates this by citing the range of specialty groups of the Society for Risk Analysis (www.sra.org) covering interalia: dose-response, ecological risk assessment, emerging nanoscale materials, engineering and infrastructure, exposure assessment, microbial

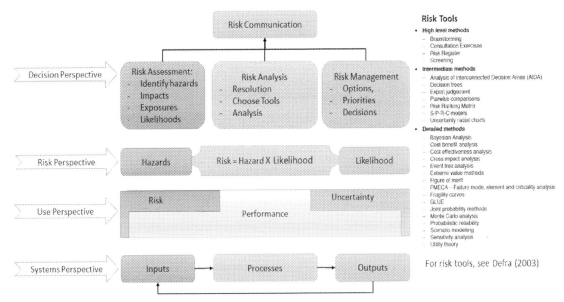

FIGURE 1.4 Basic features of risk-based decision-making.

risk analysis, occupational health and safety, risk policy and law, and security and defense.

Best practice procedures for risk-based decision-making of sustainable systems are carried out at the following resolutions. Professional risk practices often use *risk assessments* (see Fig. 1.4) to study and treat the risk of specific activities (for example, the operation of an offshore installation or an investment). However, the tools are continually honed by R&D works on theories, frameworks, approaches, principles, methods to understand, assess, characterize, communicate, and (in a wide sense) manage/govern risk (Aven and Zio, 2014; www.sra.org).

The full range of techniques at high, intermediate, and detailed levels are displayed in Fig. 1.4 but their details are outside the scope of this chapter. Not all of the risk-based decisions use tools based on the above definition of risk, which require a considerable amount of past data to quantify absolute values of risk. There are other techniques, which estimate relative values of risk in terms of vulnerability and hazards, see Sadeghfam et al. (2020). However, the terminology becomes rather slack and therefore the author argues that relative risk may be estimated as follows:

Copied the first part of the equation to its remaining part.

$$\begin{aligned}
Risk\ Index(RI) &= Vulnerability \times Hazard \\
&= \Big(PVI_{Passive\ Vulnerability\ Index} \\
&\quad to\ express\ local\ properties \\
&\quad encouraging\ adverse\ effects\Big) \\
&\quad \times \Big(AVI_{Active\ Vulnerability\ Index} \\
&\quad to\ express\ system\ wide\ impacts\Big)
\end{aligned}$$
(1.6)

Here, PVI maps to hazard and AVI to likelihood and avoids semantic problems.

5.3.4 Decision-making under uncertainty

Decision-making under uncertainty is used when adverse events can be identified but the probability of their outcomes cannot be quantified. Uncertainty limits the scope to implement the intended changes on sustainable systems and Young (2001) argues that uncertainty is pivotal to the discourse of sustainable development. He outlines a range of elements of the environment that warrants decision-making under uncertainty, including (i) the complexity and interconnectedness of the environment leading to difficult issues of threshold and resilience; (ii) the environment is a public good; (iii) damages are often irreversible but decisions can be reversed. The first quadrennial UN Global Sustainable Development Report (https://sdgs.un.org/sites/default/files/2020-07/24797GSDR_report_2019.pdf) categorizes challenges in socio-political systems concerning SDGs into (i) simple challenges, in terms of existence of scientific evidence; (ii) complex challenges, in terms of needing evidence where there are gaps; (iii) complicated challenges, in terms of the need for societal consensus; (iv) wicked challenges when solutions seem insurmountable and difficult; and (v) chaos, in terms of the full harm may arise by crossing tipping points.

Sustainable systems, by-and-large, are now complex systems and often uncertain. The strength of uncertainty thinking is in being prepared to set the limits of expectations for decision-making. Even if it is possible to develop performance models or risk models, uncertainty studies can supplement them by providing additional insights. There are various definitions of uncertainty, including a lack of sureness but this is tautological. There are various references in the literature to "perfect" solutions or "complete" solutions that conform with certainty. However, the author argues that these are philosophical notions related to so-called the "principle of unknowable," which are not scientific notions. Experience shows that there is no perfect solution and if it exists, it is not that interesting in a world where there are multiple agents and perspectives for change. A more satisfactory definition is in terms of the inherent variability of each and all the components and of the system, as a whole, in terms of their performances, models, and knowledge-base, including aspects not yet anticipated. This is particularly useful for ex ante models, where perturbations of model variables and parameters expose possible models of behavior.

The techniques to study uncertainty are wide but there are two broad approaches, those which treat the systems as reducible and those which treat each of them as unique. The reducible approaches are wide and their best practice procedures for sustainable systems may include the following: (i) at a high level: brainstorming, consultation, and register adverse effects; (ii) at an intermediate level: expert judgment, knowledge generation, stakeholder involvement, adaptive management, sensitivity tests if feasible, and statistical summary if data are available; and (iii) at a detailed level: probabilistic modeling including Bayesian Analysis and Monte Carlo simulations. For cases with performance or risk models model, still uncertainty analysis is applicable by studying their inherent variabilities. Finally, Young (2001) argues the sustainable system is unique and therefore need to be treated by belief theories as Shackle's theory. The author argues claims on uniqueness are often driven by philosophical doctrines, which are likely to be resolved the scientific thinking.

5.3.5 Decision-making based on the principle of precaution

The precautionary principle is an important decision-making criterion when activities are highly uncertain, which render none of the above decision tools useable for decision-making in sustainable systems, and yet the protection of the environment and humans are key issues. These may arise when (i) scientific understanding is yet incomplete; (ii) there is a threat to human life or health; (iii) possible damages to the environment are irreversible; (iv) changes are inequitable to the present or future generations;

or (v) the human rights of those impacted are not considered adequately. One strategy to cope with possible adverse outcomes is to resort to the precautionary principle, which is defined in a UNESCO document as follows (https://unesdoc.unesco.org/ark:/48223/pf0000139578): "When human activities may lead to morally unacceptable harm that is scientifically plausible but uncertain, actions shall be taken to avoid or diminish that harm."

Bodde et al. (2018) discuss the cases where the principle of precaution is implemented and these include the following: (i) it is difficult to balance competing goals, where there is a perceived need to protect humans and the environment against the decision; (ii) the threat of serious damage is irreversible; (iii) the lack of full scientific knowledge shall not be used as a reason for postponing cost-effective measures to prevent environmental degradation; or (iv) human activities lead to morally unacceptable harm. The precautionary principle is expected to strengthen the robustness of the decision-making process.

Best practice procedures are being developed for the implementation of the precautionary principle in various practices of sustainable practice (e.g., Bodde et al., 2018) and some of the measures include (i) carry out worst-case scenarios; (ii) use moratorium for certain requirements; (iii) adopt proportionate response or incremental options with a portfolio of possible options; (iv) invest on increasing knowledge, limiting the probability of harm, and damage or increasing resilience; and (v) explore incremental changes until science-based evidence is improved.

6. Discussions

It is argued that even if a "perfect" system is in place in terms of its positive/negative feedback and feedforward loops, as well as other premises promoted by systems science, its performances will effectively depend on the prevailing governance within the society, law reinforcement, and the general culture of respect to the law. Good and democratic governances are often a continuous media with multiple directions of information flows and the rule of law, which encourage good performances driven by accountability, transparency, the culture of lifelong learning, empowerment, gender equality, pluralism, equity among competing interests, respect to others, and inclusivity, whereas draconic and autocratic governances are by-and-large rigid entities shrouded by hypocrite, which would undermine good governance and the delivery of SDGs. Goal orientation and good governance through evidence-based policymaking and localism of planning systems are the right steps toward sustainability, but this is not everything. Still, decision-making processes need to be defensible.

Reductive science was an important agent of change during its era of 1700–1950 and deconstructed the mindset inherited from the dim past (the past up to 1500 that does not have any significant role in the thinking of modern men). The gains by reductive science are often labeled as "progress" and this can be justifiable when compared with the mindset in the dim past. However, the reconstruction of the modern era by reductive science and its heritage of progress are associated with entropy, some of which include (i) an anthropocentric complacency, which puts man at the center of every change and at the expense of the environment, the Earth, and species; (ii) environmental entropies accumulated including acid rain, ozone hole, climate change, global warming, and mountains of wastes; (iii) social complacency grew, e.g., poverty and inequality, as well as policy changes, which aimed to meet the needs but not the rights; (iv) the main drivers were exclusive freedom for elites; (v) and now the emergence of posttruth. Evidently, entropy is an omnipresent force and an agent of change, which cannot be overlooked. These entropic relics triggered a movement toward new changes in the form of the 2030 Agenda for Sustainable Development or 17 SDGs.

This chapter argues for the need to form a framework to organize better the emerging concepts, initiatives, and theory. It presents a framework with a minimum of three dimensions: governance, goal orientation, and decision-making. The concept of governance deconstructs the past exclusions at their core by a new culture of inclusion. The concept of goal orientation can be very haphazard if all-pervasive systems in the landscape of every aspect of human endeavor are cherrypicked without putting in place appropriate levels of feedback loops and creating flexible systems, which can adapt to changes by bridging those put in place by policymaking and planning systems and those put in place during the lifecycle of the systems through decision-making. In effect, the author is arguing for the development of Total Science composed of a series of modular frameworks, each catering for aspects of 'evolving entities' from their prehistory to the aftermath of their entropic death with a particular focus on the architects of evolution and various time resolutions as discussed by Khatibi (2013).

There is always something new on the horizon of science and culture, where the positive aspects are known as innovations and SDGs are classic examples of innovation, but what about negative aspects? Coronavirus is something new, which brought the highly dynamic modern world to the point of standstill, during which only the machinery for the basic needs was allowed to function without interruption. However, not everything was reduced to the point of a standstill, as the wheels of debts kept quietly turning with no critical questions asked on the accruing interests. Similarly, the scale of "false information" in social media is reaching astounding proportions. Are these invisible entropic aspects of the modern way of life not yet to reach the point of crisis? These remain to be seen.

Attention is drawn to the fact that this chapter opts to a generic presentation without referencing to a particular country. The concepts act as an envelope for all countries and cultures, but specific situations can be mapped out for each culture or country.

7. Conclusion

The emerging sustainable systems are goal oriented owing to 17 SDGs promoted by the United Nations (UN). However, the horizon within which these systems have currently come to being is uneven and haphazard for lacking a best practice of minimum standards. It is not surprising that these systems may be seen as unique by some researchers. In the wake of impending environmental crises, some solutions have emerged, and this chapter outlined a framework for better integration. It comprises the following: (i) governance with an emphasis on the inclusion of citizens and the society within the political processes of decision-making; (ii) goal orientation with an emphasis on the importance of goal orientation to be underpinned by appropriate concepts of systems science (positive/negative feedback and feedforward loops; performance and failure or information and entropy; hierarchy, emergent property, and purpose …); and (iii) decision-making in terms of tools necessary for sustainability and risk-based decision-making and a raft of other system attributes (reliability, validity, vulnerability, resilience, recoverability, availability, durability, maintainability, serviceability, tolerability).

When these emerging tools are integrated together as a framework, it then becomes feasible to identify loose ends, bottlenecks, instabilities, barriers, and efficacies of the solutions to safeguard the sustainable future. Their integration as a framework can also serve as the basis to transfer knowledge from one discipline to another by having a grip over the differences on the complexity of the systems. Taking a critical historic overview of the systems, it becomes feasible to identify additional barriers stemming from the inflexibility of philosophical doctrines impacting policymaking. The aim is to deliver a sustainable future by an effective and reliable delivery of the SDGs.

8. Appendix—framing scientific concepts

A wide range of terms are mentioned in Section 2.2 to denote the particular scientific activities, but these are elaborated in this appendix.

A theory is the main currency of science, but a theory is not one if it is not using data, is not evidence-based, and/or does not have the capability to learn from its application for eventual refinements. Theories, by definition, are not permanent features, but scientists are always ready to refine or abandon any theory in favor of a more reliable alternative. In this culture, the propensity to criticize a theory and to replace it with an alternative is overwhelming. This suggests that theories are never perfect, but as long as they are workable, they are useable and this point is emphasized by Walsham (1997, p. 478) in a different way: "There is not, and never will be, a best theory. A theory is our chronologically inadequate attempt to come to terms with the infinite complexity of the real world. Our quest should be for improved theory, not best theory, and for a theory that is relevant to the issues of our time."

The issue of a permanent change in science reflects a special culture, in which there is a presumption of adherence to a range of principles, e.g., the principle of the integrity of the scientist and the transparency among scientists. However, if the scientist is not an individual due to being obedient to decrees or dogmas, scientific integrity and transparency will be undermined. This explains why science could have not existed in the dim past, when the persons were subject to no legal basis for being full individuals. It is in this context that the statement by Ziman (1991, p. 42) makes sense, who remarks that "The fundamental principle of scientific observation is that all humans are interchangeable as observers."

Other currencies of science (hypothesis, proposition, postulate, and conjecture) refer to lower grade scientific activities but are not discussed further, as there is little controversy on their common usage. Nonetheless, consider a hypothesis, which refers to cases involving no data, no evidence, or no opportunity yet for testing. Its science content is, therefore, poorer than the theories already tested. More framing concepts have emerged since the 1950s and these include cluster sciences, frameworks, paradigms, principles, and heuristics and some (cluster sciences, frameworks, paradigms, principles, and heuristics) are elaborated below.

The bottom-up learning from the inclusive accumulation of science revealed its hierarchical structure through *Comte's Theory of Science*, following Auguste Comte (1798–1857) and now comprise quantum physics, physics, chemistry, biology, psychology, sociology, and anthropology, see Khatibi (2013). Scientometrics, emerged in the 1970s, studies the development of science by research as an informational process in the form of a quantitative research method based on citations, see Šubelj et al. (2016), Mingers (2015). Often millions of citations are analyzed through bottom-up data-driven cluster modeling techniques using artificial intelligence. These studies justify and confirm such emerging terms as environmental sciences, life sciences, management sciences, and computer sciences. These are often an indication of a mosaic appearance of the body of science and confirm an inherent industrial scale in scientific outputs.

Since the 1990s, there has also been a top-down process of forming frameworks for pooling together a wide range of activities under one umbrella. For instance, the DPSIR framework (Drivers, Pressures, State, Impact, and

Response) refers to describing the interactions between society and the environment. This was originally presented as a "stress-response" model (Rapport and Friend, 1979) and was adopted by the Organization for Economic Cooperation and Development (OECD, 1991, 1993), and European Environment Agency transformed it into a DPSIR framework, which considers individual behavior change, societal responses to these changes overarched by policies (e.g., environmental, economic) or instruments (e.g., regulatory) to prevent, reduce, or mitigate pressures or environmental damage, Reis et al. (2015). Notably, each activity is referred to as a dimension and the dimensions are selected by consensus and therefore there may be wide variations.

There are literally 100s of frameworks in science, education, and management science, but their critical review is outside the scope of this chapter. These are largely introduced by public organizations after a comprehensive review and strategic research, an example of which is DRASTIC presented by Aller et al. (1987). Sometimes, a framework may be introduced by individual researchers, e.g., ALPRIFT, see Nadiri et al. (2018), Total Information Management, see (Nadiri et al., 2021), and IMM-RHEO, see Khatibi et al. (2020) and Khatibi and Nadiri (2020). The latter one addresses the problems in modeling practices, as follows: (i) Exclusionary Multiple Models (EMMs), which by-and-large seek to select a superior model from the models under study, but in reality, the results are only fit-for-purpose; and (ii) IMM practices, which put an emphasis on learning. The learning is carried out by modeling practices, which puts together the following four dimensions: (i) Model Reuse; (ii) Hierarchy and/or Recursion; (iii) Elastic Model Learning; and (iv) Goal Orientation, which together form the acronym of RHEO. They have produced a significant number of proof-of-concept papers to demonstrate that the results through IMM become defensible.

The concept behind the term paradigm by Thomas Kuhn (1922—96) is rather philosophical, which defines a paradigm as recognized scientific model problems and solutions for a community of practitioners for a time (Kuhn, 1962). For instance, Weiß (2002) outlines three paradigms for implementing agent systems: *agent-oriented* programming, *market-oriented* programming, and *interaction-oriented* programming. Its rationale strongly derives from the Marxian-Hegelian dialectics but replaces violent overthrows of thesis and antithesis by Kantian constructivism. Thus, according to Kuhn, scientists rally around an emerging paradigm to be engaged in puzzle-solving practices (thesis) but keep reporting inconsistencies (antithesis) until new paradigms emerge to resolve inconsistencies. This definition seems quite benign and without the usual trojan horse, but when philosophical doctrines are imported to science, the terms have their dark sides as well, as discussed in due course.

Heuristics in problem-solving for decision-making activities is the strategy of using and reusing experience in various forms including trial-and-error, rule-of-thumb, best practice procedure, a wide range of strategies for extracting correlation in data, or a clearly defined scheme of doing things. The case for heuristics is exploited at any opportunity that the development of a theory, a hypothesis, or a paradigm is not conceived, but there is a conceived approach, which is seemingly reasonably sound and not arbitrary. The outcome is a procedure for informed decisions but liable to bias and in order to make the results fit-for-purpose, various mathematical procedures may be needed. The procedure is old as discussed by Pólya (1945) but has been one of the mainstream activities since the 1960s.

Khatibi (2013) argues that changes take place at three types of time resolutions: (i) slow evolutionary changes driven by mutations causing small internal structure (phase changes); (ii) medium-term adaptive changes without any structural changes but through seeking adaptation to the environment; and (iii) dynamic changes through state changes. A study procedure for one resolution is not normally applicable to the other resolutions. The emergence of an idea and its development to its mature state may be treated as analogous to these three types of the above changes as studied by Khatibi (2013).

References

Aller, L., Bennett, T., Lehr, J.H., Petty, R.J., Hackett, G., 1987. DRASTIC: A Standardized System for Evaluating Ground Water Pollution Potential Using Hydrogeologic Settings. US Environmental Protection Agency, Washington, DC, p. 455.

Antonescu, M.V., 2015. The concepts of "great feminine governance" and "age of holy spirit". In: The Axiological Re-ordering of the XXIst Century Global Society. Publ. by Logos Universality Mentality Education Novelty, Section: Philosophy and Humanistic Sciences ISSN: 2284–5976 (print), ISSN: 2284–5976 (electronic) Covered in: CEEOL, Index Copernicus, Ideas RePEc, EconPapers, Socionet. http://lumenjournals.com/philosophy-and-humanistic-sciences/.

Aven, T., 2016. Risk assessment and risk management: review of recent advances on their foundation. Eur. J. Oper. Res. 253, 1–13.

Aven, T., Zio, E., 2014. Foundational issues in risk assessment and risk management. Risk Analysis 34 (7), 1164–1172. https://onlinelibrary.wiley.com/doi/abs/10.1111/risa.12132.

Beck, U., 1992. From industrial society to the risk society. Questions of survival, social structure and ecological enlightenment. Theor. Cult. Soc. 9, 97–123.

Benedictow, O.J., 2006. The Black Death 1346–1353: The Complete History Paperback – Illustrated, 2004. Cambridge University Press, p. 134 XX. Cited on April 2021 in. https://dailyhistory.org/How_did_the_Bubonic_Plague_make_the_Italian_Renaissance_possible%3F.

Biermanna, F., Kanieband, N., Kim, R.E., 2017. Global governance by goal-setting: the novel approach of the UN Sustainable Development Goals. Environ. Sustain. 26–27, 26–31.

Blyth, M., 2001. The transformation of the Swedish model: economic ideas, distributional conflict, and institutional change. World Polit. 54, 1–26.

Bodde, M., van der Wel, K., Driessen, P., Wardekker, A., Runhaar, H., 2018. Strategies for dealing with uncertainties in strategic environmental assessment: an analytical framework illustrated with case studies from The Netherlands. Sustainability 10 (7), 2463. https://www.mdpi.com/2071-1050/10/7/2463.

Defra, Sayers, P.B., Gouldby, B.P., Simm, J.D., Meadowcroft, I., Hall, J., 2003. R&D Technical Report FD2302/TR1, Risk, Performance and Uncertainty in Flood and Coastal Defence – A Review. Defra, 2005.

Forman, P., 2007. The primacy of science in modernity, of technology in postmodernity, and of ideology in the history of technology, history and technology. Int. J. 23 (1–2), 1–152.

Fukuda-Parr, S., 2014. Global goals as a policy tool: intended and unintended consequences. J. Hum. Dev. Capab. 15 (2–3), 118–131. https://doi.org/10.1080/19452829.2014.910180.

Giddens, A., 2006. Fate, risk and security pp. 29–60 and 29–66. In: Cosgrave, J.F. (Ed.), The Sociology of Risk and Gambling Reader. Taylor and Francis Group, LLC.

Greenhalgh, T., Russell, J., 2009. Evidence-based policymaking – a critique. Perspect. Biol. Med. 52 (2), 304–318. The Johns Hopkins University Press.

Guo, K.L., June 2008. DECIDE: a decision-making model for more effective decision making by health care managers. Health Care Manag. 27 (2), 118–127. https://doi.org/10.1097/01.HCM.0000285046.27290.90.

Gustafson, A., 2013. In defense of a utilitarian business ethic. Bus. Soc. Rev. 118 (3), 325–360.

Hallsworth, M., Parker, S., Rutter, J., 2011. Policy Making in the Real World - Evidence and Analysis. A Report Published by Institute for Government.

Harpin, R., 1999. Liberalism, Constitutionalism, and Democracy. Oxford.

Holling, C.S., 1973. Resilience and stability of ecologic system. Annu. Rev. Ecol. Systemat. 4, 1–23.

Janis, I.L., Mann, L., 1977. Decision Making: A Psychological Analysis of Conflict, Choice, and Commitment. Free Press, New York, ISBN 978-0029161609.

Jennings, B., 2009. Public health and liberty: beyond the millian paradigm, center for humans and nature. Publ. Health Ethics 2 (2), 123–134.

Kaplan, A., Maehr, M.L., 2007. The contributions and prospects of goal orientation theory. Water Resour. Manag. 21, 699–715.

Khatibi, R., 2011. Evolutionary systemic modelling for flood risk management practices. J. Hydrol. 401 (1–2), 36–52. https://doi.org/10.1016/j.jhydrol.2011.02.006.

Khatibi, R., 2013. Learning from natural selection in biology: reinventing existing science to generalise theory of evolution – evolutionary systemics. In: Lynch, J.R.,

Williamson, D.T. (Eds.), Chapter 1: Natural Selection: Biological Processes, Theory and Role in Evolution. Published by Nova Publishers. Google availability at. https://www.novapublishers.com/catalog/product_info.php?products_id=41521. https://www.researchgate.net/publication/285771385.

Khatibi, R., Haywood, J., 2002. The role of flood forecasting and warning on sustainability of flood defence. Proc. Inst. Civ. Eng. 151 (4), 313—320. http://www.icevirtuallibrary.com/content/article/10.1680/muen.2002.151.4.313.

Khatibi, R., Nadiri, A.A., 2020. Inclusive Multiple Models (IMM) for predicting groundwater levels and treating heterogeneity. J. Geosci. Front.

Khatibi, R., Ghorbani, M.A., Naghshara, S., Harun, A., Karimi, V., 2020. A Framework for 'Inclusive Multiple Modelling' with Critical Views on Modelling Practices—Applications to Modelling Water Levels of Caspian Sea and Lakes Urmia and Van.

Klein, G.A., 2009. Streetlights and Shadows: Searching for the Keys to Adaptive Decision Making. Published by Massachusett Institute of Technology.

Kuhn, T.S., 1962. The Structure of Scientific Revolution, third ed. The University of Chicago Press.

Lavelle, 2013. Paradigms of Governance: From Technocracy to Democracy: An IGI Book Chapter: Social Sciences & Humanities IGI Global. https://doi.org/10.4018/978-1-4666-3670-5.ch009. igi-global.com. https://pdfs.semanticscholar.org/f454/cb32c03308c4e336ecee5bddf39f6ec8a5f6.pdf.

Lilienfeld, 1978. The Rise of Systems Theory: An Ideological Analysis. Wiley.

Loptson, P., 1995. Theories of Human Nature.

Losee, J., 1987. Philosophy of Science and Historical Enquiry. Clarendon Press, Oxford.

Lührmann, A., Maerz, S.F., Grahn, S., Alizada, N., Gastaldi, L., Hellmeier, S., Hindle, G., Lindberg, S.I., 2020. Autocratization Surges – Resistance Grows. Democracy Report 2020. Varieties of Democracy Institute (V-Dem).

Meadows, D.H., Meadows, D.L., Randers, J., Behrens III, William, W., 1972. The Limits to Growth: A Report for the Club of Rome's Project on the Predicament of Mankind Paperback. Universe, New York.

Merton, R.K., 1968. Social Theory and Social Structure, Revised edition. Collier Macmillan.

Mingers, J., 2015. A review of theory and practice in scientometrics. Eur. J. Oper. Res.

Mitchell, M., 2009. In: Complexity: A Guided Tour Oxford University Press Understanding the Nexus, Background Paper for the Bonn2011 Nexus Conference. http://www.water-energy-food.org/en/news/view__255/understanding_the_nexus.html.

Modarres, M., 1993. Reliability and Risk Analysis. Marcel Dekker, Inc.

Mullins, L.J., 1993. Management and Organisational Behaviour, third ed. Pitman Publishing, pp. 78—79.

Nadiri, A.A., Taheri, Z., Khatibi, R., et al., 2018. Introducing a new framework for mapping subsidence vulnerability indices (SVIs): ALPRIFT. J. Sci. Total Environ. 628—629, 1043—1057. https://www.sciencedirect.com/science/article/pii/S0048969718304194.

Nadiri, A.A., Sedghi, Z., Khatibi, R., 2021. Qualitative risk aggregation problems for the safety of multiple aquifers exposed to nitrate, fluoride and arsenic contaminants by a 'Total Information Management' framework. J. Hydrol.

OECD, 1991. Environmental Indicators. A Preliminary Set. Paris, France.

OECD, 1993. OECD Core Set of Indicators for Environmental Performance Reviews. OECD Environment Monographs No. 83. OECD.

OSG, 1981. Systems Behaviour, Open Systems Group. Harper and Row Publishers, London, p. 18.

Pekka, J., Korhonen, P.J., Wallenius, J., 2020. Making Better Decisions—Balancing Conflicting Criteria. Springer Nature Switzerland AG.

Pólya, G., 1945. How to Solve it: A New Aspect of Mathematical Method. Princeton University Press, Princeton, NJ, ISBN 0-691-02356-5. ISBN 0-691-08097-6.

Rapport, D.J., Friend, A.M., 1979. Towards a Comprehensive Framework for Environmental Statistics: A Stress Response Approach. Minister of Supply and Services, Canada, Ottawa. Statistics Canada Catalogue 11—510.

Redclift, M., 2005. Sustainable development (1987—2005): an oxymoron comes of age; sustainable development. J. Sustain. Dev. 13, 212—227.

Rehmann-Sutter, C., 1998. Involving others: towards an ethical concept of risk. Risk Health Saf. Environ. 9, 119.

Reis, S., Steinle, S., Morris, G., Fleming, L.E., Cowie, H., Hurley, F., Dick, J., Smith, R., Austen, M., White, M., 2015. Developing an integrated conceptual model for health and environmental impact assessment. J. Public Health 129 (10), 1383—1389.

Russell, B., 1971. A History of Western Philosophy. George Allen and Unwin Limited, London. Ruskin House, Museum Street.

Sadeghfam, S., Khatibi, R., Dadashi, S., Nadiri, A.A., 2020. Transforming subsidence vulnerability indexing based on ALPRIFT into risk indexing using a new fuzzy-catastrophe scheme. J. Environ. Impact Assess. Rev 82.

Sandoval-Solis, S., D. C. McKinney, M., Loucks, D.P., 2011. Sustainability Index for water resources planning and management. J. Water Resour. Plann. Manag. 137, 381—390.

Skogstad, G., Schmidt, V., 2011. Chapter 1—Introduction: policy paradigms. Transnationalism, and Domestic Politics in the Book Policy Paradigms. https://doi.org/10.3138/9781442696716-003 (edited by).

Stewart, I., 1989. Does God Play Dice? The Mathematics of Chaos - Basil Blackwell Inc.

Stoker, G., 1998. Governance as Theory: Five Propositions, ISSJ 155/1998, UNESCO 1998. Blackwell Publishers, 108 Cowley Road, Oxford OX4 1JF, UK and 350 Main Street, Malden, MA 02148, USA. Published by.

Šubelj, L., van Eck, N.J., Waltman, L., 2016. Clustering scientific publications based on citation relations: a systematic comparison of different methods. PLoS One 11 (4), e0154404. https://doi.org/10.1371/journal.pone.0154404. https://doi.org/10.1371/journal.pone.0154404.

The 2030 Agenda, 2015. United Nations General Assembly: Transforming Our World: The 2030 Agenda for Sustainable Development. Draft Resolution Referred to the United Nations Summit for the Adoption of the Post-2015 Development Agenda by the General Assembly at its Sixty-Ninth Session. UN Doc. A/70/L. 1 of 18 September 2015.

Tremmel, J.C., 2014. An extended separation of powers model as the theoretical basis for the representation of future generations, presented to forthcoming Birnbacher, Dieter/Thorseth, May. In: Roads to Sustainability. Earthscan, London, 2014.

Tung, Y.K., Yen, B.C., Melching, C.S., 2005. Hydrosystems Engineering Reliability Assessment and Risk Analysis. McGraw−Hill Professional, New York.

Turpin, S.M., Marais, M.A., 2004. Decision-making: theory and practice. Orion 20 (2), 143−160. http://www.orssa.org.za.

Walsham, G., 1997. Actor-network theory and IS research: current status and future prospects. In: Lee, A.S., Liebenau, J., DeGross, J.L. (Eds.), Information Systems and Qualitative Research. Chapman and Hall, London, UK, pp. 466−480.

Weisbuch, G., Solomon, S., 2007. Tackling Complexity in Science, the Think Tank Report of the FP6; General Integration of the Application of Complexity in Science (GIACS).

Weiß, G., 2002. Agent orientation in software engineering. Knowl. Eng. Rev. January 2002.

Young, R.A., 2001. Uncertainty and the Environment. Edward Elgar publ. by.

Zahoor, M.A., 2019. The emergence of individual rights in Europe: a historical recapture. J. Eur. Stud. 35/1.

Ziman, J.M., 1991. Reliable Knowledge. Cambridge University Press, p. 42.

Further reading

Aven, T., Renn, O., 2010. Risk Management and Governance: Concepts, Guidelines and Applications, vol. 16. Springer Science & Business Media, Heidelberg.

Pagel, H.R., 1991. Uncertainty and complementarity. In: Tomothy, F. (Ed.), The World Treasury of Physics, Astronomy and Mathematics. Little, Brown and Company, pp. 97−110. Published by.

Saylor Academy, 2012. Principles of Sociological Inquiry: Qualitative and Quantitative Methods. Principles of Sociological Inquiry.pdf (saylor.org).

SM ICG, 2013. Risk Based Decision Making Principles by the Safety Management International Collaboration Group (SM ICG Risk Based Decision Making Principles v2.doc - Google Drive).

CHAPTER 2

Principal component analysis of precipitation variability at Kallada River basin

Beeram Satya Narayana Reddy, S.K. Pramada
Department of Civil Engineering, National Institute of Technology Calicut, Kozhikode, Kerala, India

1. Introduction

Precipitation is a significant source for the availability of water for various purposes. The fluctuation quantity of precipitation is varied in temporal scale. The identification of precipitation variation is done by applying the principal component analysis (PCA) and finding out the variance explained based on the Eigen values. The PCA is a technique that transforms the correlated data into new orthogonal axis and explains the maximum possible variance in each components.

The temporal variation of rainy season is studied by the application PCA for Karnataka (Iyengar, 1991) and west Bengal (Basak, 2014). Many similar studies have been conducted using PCA in finding the coherent zones of precipitation, clustering rainfall station data, spatial variability of seasonal precipitation (Goossens, 1985; Othman et al., 2015; Singh, 2006; Stathis and Myronidis, 2009), precipitation regionalization (Mills, 1995), and characteristics of rainy season precipitation (Camberlin and Diop, 2003). The errors of spatial rainfall data derived from INSAT are assessed using PCA (Roy Bhowmik and Sen Roy, 2006). The outgoing long wave radiation and summer precipitation variability is studied using PCA (Haroon and Rasul, 2007). In the current study, the precipitation variability of single station data is assessed using PCA for extraction features contributing the major variation over the period.

2. Study area and dataset

Kallada is one of the rivers which is flowing toward the west and passes through the Kollam district, Kerala. It travels 121 km with a watershed area of 1699 sq.km and drains into the Ashtamudi Lake and later joins at the Arabian Sea. The basin consists of a reservoir so called as Thenmala reservoir at the confluence of three tributaries namely Shendurney, Kulathupuzha, and Kalthuruthi and a hydroelectric power generation unit with 15 MW installed capacity. Monthly observed precipitation data are collected from Indian Meteorological Department for Punalur station which is located at Kallada River basin. The details of the study area are shown in Fig. 2.1. The collected data are available for the period of 33 years, i.e., 1981−2013. The basic statistical details of dataset are shown in Table 2.1.

3. Methodology

The deletion of features, the extraction of features, and the reduction of dimensions of the original variables were performed using the PCA methodology. Firstly the Punalur monthly precipitation dataset is arranged and the PCA technique is performed to find the variability of precipitation for the period of 33 years. The analysis is performed on 17*33 matrix. The original variables are transformed into uncorrelated dimensionality space, called principal components, based on the correlation matrix, covariance matrix, and Eigen values of the corresponding Eigen vectors without losing much of the original details (Othman et al., 2015).

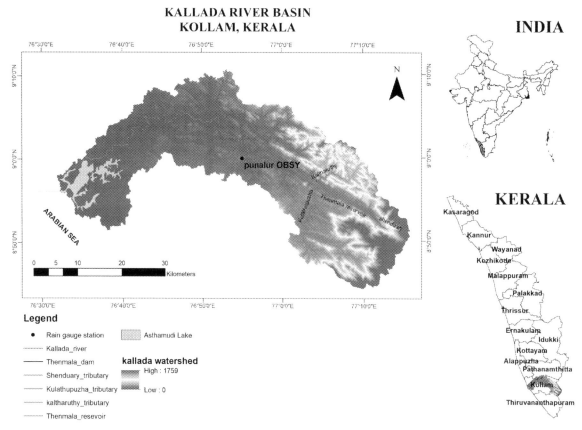

FIGURE 2.1 Details of study area.

TABLE 2.1 Descriptive statistics of monthly, annual, and seasonal dataset.

Variable	Mean (mm)	Std Dev (mm)	Std Err	N
Jan	15.939	27.505	4.788	33
Feb	47.873	52.889	9.207	33
Mar	85.842	61.385	10.686	33
Apr	216.715	81.446	14.178	33
May	231.100	144.249	25.111	33
Jun	444.700	203.695	35.459	33
Jul	375.233	131.044	22.812	33
Aug	281.136	109.033	18.980	33
Sep	235.858	131.951	22.970	33
Oct	413.867	205.369	35.750	33
Nov	222.427	122.147	21.263	33
Dec	48.594	66.504	11.577	33
Annual	2619.285	516.072	89.837	33
Winter	115.264	103.744	18.060	33
Summer	533.658	180.318	31.389	33
SW monsoon	1336.927	353.559	61.547	33
NE monsoon	636.294	264.804	46.096	33

TABLE 2.2 Details of Eigen values and percentage of variance.

Components	Eigen value	% of Var.	Cum. %
PC 1	4.192	24.661	24.661
PC 2	2.834	16.669	41.330
PC 3	2.742	16.132	57.462
PC 4	1.487	8.746	66.208
PC 5	1.221	7.185	73.393
PC 6	0.975	5.736	79.129
PC 7	0.972	5.719	84.848
PC 8	0.690	4.058	88.906
PC 9	0.675	3.973	92.879
PC 10	0.418	2.457	95.336
PC 11	0.394	2.320	97.656
PC 12	0.232	1.365	99.021
PC 13	0.167	0.979	100.000
PC 14	0.000	0.000	100.000
PC 15	0.000	0.000	100.000
PC 16	0.000	0.000	100.000
PC 17	0.000	0.000	100.000
Total	16.999		

4. Results and discussion

The precipitation variability of Punalur station is studied using PCA. From the analysis, a total of 17 principal components are created and the total variance according to Eigen values is 16.999 out of which the last four principal components reflect nil variance. The Eigen values and percentage of variance of each principal component are tabulated in Table 2.2. The Eigen values of principal component are presented graphically in descending order using scree plot shown in Fig. 2.2. The individual and cumulative percentages of variance explained by the principal components are plotted in Fig. 2.3. First five PCs are considered with Eigen value greater than 1. The cumulative percentage of variance explained by first five PCs is 73.393%.

The first principal component is reflecting the largest variance of precipitation with 24.66% and the fifth PC is reflecting 7.185% since further components account less than 27% of variance and can be ignored. The scores of selected principal components are estimated and tabulated in Table 2.3. From Table 2.3, it is clear that the six variables are significantly correlated with PC 1, six with PC 2, three with

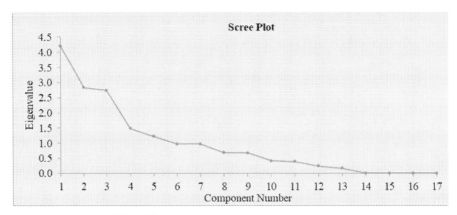

FIGURE 2.2 Scree plot of principal components.

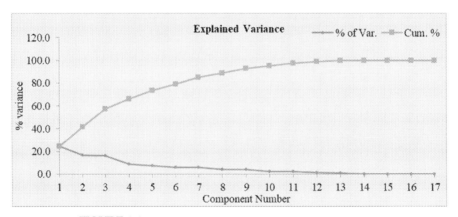

FIGURE 2.3 Explained variance plot of principal components.

TABLE 2.3 Correlation of variables with the principal components.

Variable	PC 1	PC 2	PC 3	PC 4	PC 5
Jan	−0.147	0.026	−0.339	0.022	−0.058
Feb	0.078	−0.356	−0.365	−0.090	−0.470
Mar	0.040	−0.065	−0.424	−0.303	0.244
Apr	−0.075	0.155	−0.492	−0.135	−0.457
May	0.348	**0.583**	**−0.534**	0.090	0.235
Jun	**0.622**	−0.347	−0.089	**0.677**	0.087
Jul	**0.621**	−0.020	0.023	−0.228	0.345
Aug	0.302	−0.449	0.163	−0.460	−0.272
Sep	0.373	**−0.580**	0.112	−0.452	−0.146

TABLE 2.3 Correlation of variables with the principal components.—cont'd

Variable	PC 1	PC 2	PC 3	PC 4	PC 5
Oct	**0.658**	**0.570**	0.265	0.218	−0.270
Nov	0.411	0.347	0.371	−0.362	0.481
Dec	−0.158	−0.048	0.184	−0.075	−0.228
Annual	**0.991**	0.062	−0.082	−0.052	−0.024
Winter	0.051	−0.202	**−0.528**	−0.077	**−0.552**
Summer	0.258	**0.514**	**−0.794**	−0.092	0.065
SW monsoon	**0.821**	**−0.562**	0.049	−0.005	0.040
NE monsoon	**0.700**	**0.602**	0.377	0.002	0.012

Bold specifies the significant correlation.

4. Results and discussion

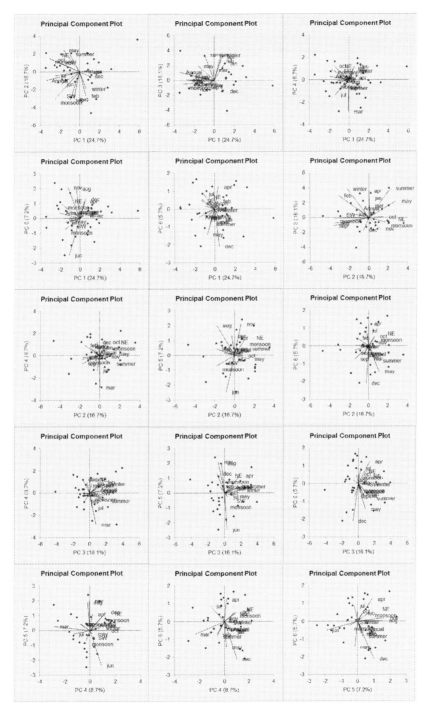

FIGURE 2.4 Principal components pairwise pattern comparison plots.

PC 3, one with PC 4, and one with PC 5. The cut-off value of 0.5 is considered as significant. PC 1 surges with increase in the precipitation of June, July, October, annual, SW, and NE monsoon periods. In a similar way, PC 2 surges with increase in the precipitation of May, October, summer, and NE monsoon and plunges if the precipitation of September and SW monsoon increases or vice versa. If the precipitation of May, winter, and summer decreases, then PC 3 surges or vice versa due to negative correlation. The PC 4 surges if the precipitation of June increases (+ve correlation). The PC 5 and winter precipitation are correlated negatively; if one increases the other decreases or vice versa. The pairwise comparison plots of components with original variables are shown in Fig. 2.4.

5. Conclusions

The variation of observed precipitation data of Punalur is assessed with the application of PCA for a period of 33 years, i.e., 1981−2013. The real variables are transformed into orthogonal axis which are uncorrelated to each other, so-called principal components. The first component described the highest variation, the second component explained the next highest variance, and so on to the lowest possible variance. The conclusions drawn from the study are as follows:

1. The assessment using PCA extracted 17 principal components out of which 5 principal components are considered.
2. The first five PCs are finalized based on Eigen value and percentage of variance explained.

3. The total cumulative percentage of variance explained by the first five PCs is 73.393%.
4. The majority of rainfall variation is observed during the period of SW and NE monsoon.

References

Basak, P., October 2014. Variability of south west monsoon rainfall in West Bengal: an application of principal component analysis. Mausam 4, 559−568.

Camberlin, P., Diop, M., January 2003. Application of daily rainfall principal component analysis to the assessment of the rainy season characteristics in Senegal. Clim. Res. 23, 159−169. https://doi.org/10.3354/cr023159.

Goossens, C., 1985. Principal component analysis of mediterranean rainfall. J. Climatol. 5, 379−388.

Haroon, M.A., Rasul, G., 2007. Principal component analysis of summer rainfall and outgoing long-wave radiation over Pakistan. Pak. J. Meteorol. 5 (10), 109−114.

Iyengar, R.N., 1991. Application of principal component analysis to understand variability of rainfall. Earth Planet. Sci. 100 (2), 105−126.

Mills, G.F., 1995. Principal component analysis of precipitation and rainfall regionalization in Spain. Theor. Appl. Climatol. 183, 169−183.

Othman, M., Hanan, Z., Diyana, N., 2015. Long-term daily rainfall pattern recognition: application of principal component analysis. Proc. Environ. Sci. 30, 127−132. https://doi.org/10.1016/j.proenv.2015.10.022.

Roy Bhowmik, S.K., Sen Roy, S., 2006. Principal component analysis to study spatial variability of errors in the INSAT derived quantitative precipitation estimates over Indian monsoon region. Atmósfera 19 (4), 255−265.

Singh, C.V., 2006. Pattern characteristics of Indian monsoon rainfall using principal component analysis (PCA). Atmos. Res. 79, 317−326. https://doi.org/10.1016/j.atmosres.2005.05.006.

Stathis, D., Myronidis, D., 2009. Principal component analysis of precipitation in Thessaly region (Central GREECE). Glob. Nest J. 11 (4), 467−476.

CHAPTER

3

Healthcare waste management in Bangladesh: practices and future pathways

Afia Rahman[1], Byomkesh Talukder[2], Mohammad Rezaul Karim[3]

[1]Department of Research and Development, Bangladesh Public Administration Training Centre, Dhaka, Bangladesh; [2]Dahdaleh Institute for Global Health Research, York University, Toronto, ON, Canada; [3]Department of International Programme, Bangladesh Public Administration Training Centre, Dhaka, Bangladesh

1. Introduction

In the Anthropocene, huge amounts of waste are generated across the world due to consumption by the ever-increasing population. As human needs and activities overload the carrying capacity of the biosphere, the question of waste management has become prominent (Marchettini et al., 2007). Various environmental threats and human health risks occur due to the mishandling and ignorance of waste management (Khan et al., 2019). The World Health Organization (WHO) reported that almost 80% of the waste produced by healthcare establishments (HCEs) is comparable to domestic waste, but 10%−15% is infectious, toxic or both and needs special treatment (WHO, 2013; GoB, 2016). In terms of worldwide public health, healthcare waste (HCW) is a major environmental concern due to its infectious and hazardous nature (Wen-Chuan et al., 2017). HCW is the second-most hazardous waste after radiation waste (Wafula et al., 2019). The hazardous

portion of HCW can cause occupational health risks, while improper disposal creates overall environmental hazards. In 1988, WHO argued that proper management of HCW is a major obstacle in most developing countries, especially in those countries where regular municipal solid waste is poorly managed (WHO, 2005; WHO, 2007). The appropriate management of HCW is the responsibility of health centers but also involves the participation of other actors (Mugambe et al., 2012). Therefore, the risk associated with HCW and its management has gained attention in national and international forums.

The need for proper healthcare waste management (HCWM) has been a particularly crucial issue in many developing countries in Asia and Africa (Chiplak and Kaskun, 2015). The healthcare sector is growing at a very rapid pace in these countries, which in turn has led to a tremendous increase in the quantity of HCW generated by hospitals, clinics, and other HCEs (Zafar, 2019). Evidence from different reports indicates that approximately 1−5 kg/bed/day

Risk, Reliability and Sustainable Remediation in the Field of Civil and Environmental Engineering
https://doi.org/10.1016/B978-0-323-85698-0.00008-3

© 2022 Elsevier Inc. All rights reserved.

waste is generated in developed countries (Tabish et al., 2018), whereas it is 0.5—2.5 kg/bed/day in developing countries (Zafar, 2019). For instance, India generates 2.0 kg/bed/day (Tabish et al., 2018). In Bangladesh, the rate of waste generation is 2.6 kg/bed/day (Wen-Chuan et al., 2017). National 3R Strategy, the Department of Environment, Bangladesh (2010) stated that, approximately 12,271 metric tons of HCW is generated in the capital city, Dhaka; wherein about 20% of which is infectious or hazardous. WHO (2018) also stated that high-income countries generate on average up to 0.5 kg of hazardous waste per bed/day from each hospital, while on an average 0.2 kg/bed/day hazardous wastes from a hospital are generated by low-income countries. However, no matter how much waste is produced, there is little information about what is done with it. Thus, the availability of data on HCW generation and disposal is also a crucial concern in developing countries (Zafar, 2019).

The WHO describes all waste generated by HCEs as HCW. This HCW is classified, as *nonrisk or general healthcare waste* (like domestic waste) and *hazardous waste*. Hazardous waste has the potential to pose a variety of health risks and threats to the environment. Hazardous waste includes pathological and chemical waste (Chartier et al., 2014). HCW is generated by healthcare activities that make use of a broad range of materials, from used needles and syringes to soiled dressings, body parts, diagnostic samples, blood, chemicals, pharmaceuticals, medical devices, and radioactive materials and therefore contains potentially harmful microorganisms that can infect hospital patients, healthcare workers, and the community (WHO, 2015). In particular, concerns about infection with HIV and hepatitis B and C arise because of strong evidence of disease transmission via HCWM (Chaerul et al., 2008). For instance, unsafe injection residuals were responsible for as many as 33,800 new HIV infections, 1.7 million hepatitis B infections, and 315,000 hepatitis C infections in developing countries in 2010 (WHO, 2018).

As in other developing countries around the world, proper management of HCW has been a matter of critical concern in Bangladesh (Wafula et al., 2019). A study revealed that most of the HCEs in Dhaka city do not separate their generated waste and do not collect the waste properly. Further, the waste is disposed of with municipal waste in municipal dustbins (Hassan et al., 2008). Another study found that all the HCEs of Dhaka city discharge their liquid waste into the general sewers or drains in the absence of a proper liquid waste management system (Nuralam et al., 2017). The same study noted that none of the HCEs in Dhaka city has any waste management treatment plant yet. An NGO named PRISM Bangladesh collects HCW from a few HCEs in Dhaka city, but the coverage is very limited. The government of Japan donated an incinerator, but it has limited use. Nuralam and his colleagues acknowledged that the government has taken a few initiatives to develop a modern and sustainable approach to deal with HCWs properly, such as a law about proper HCWM established in 2008. However, there is a lack of national policy on HCW management in Bangladesh, and initiatives that have been taken are not sufficient (Nuralam et al., 2017).

Hasan and Rahman (2018) observed that several stakeholders also have the role for initiating and applying service delivery mechanisms for safe HCWM. The argument endorsed by Patwary et al. (2009) is that a layered arrangement of networks is required, which informs the action of actors in managing HCW in Dhaka city. Patwary et al. further stated that different networks can work at each level from national governance to waste operatives, but the interaction between the networks is often not constructive; for instance, corruption is evident in most of the cases. Moreover, the management strategy of HCW relies heavily on the government for setting and implementing policies and allocating financial resources (Hassan et al., 2008). Thus, it is essential to identify how the networks are structured and work to manage HCW properly.

1.1 Problem statement

WHO (2003) reported that HCWM is a *management* issue before being a technical one. Safe management of HCW involves three key principles: reduction of unnecessary waste, separation of general waste from hazardous waste, and waste treatment that reduces risks to healthcare workers and the community. Therefore, this management issue completely depends on the commitment of the entire staff within the HCEs. Further, other stakeholders from government and nongovernment organizations play a vital role in managing HCW. This dedication will only be possible if people are first made aware of the risks that this particular type of waste poses (WHO, 2018).

In developed countries, specific rules and regulations have already been formulated and implemented for the handling of HCW. For example, several countries in the Western Pacific Region, such as Australia, Fiji, China, Japan, and Malaysia, have national policies, guidelines, action plans, and best practices for HCWM. In developing countries, however, handling HCW is still a growing concern. The government of Bangladesh has undertaken some initiatives and formulated rules and regulations such as the Medical Waste (Management and Processing) Rule 2008. Nevertheless, the rule is still not being implemented in many HCEs in the country.

The management of HCWs is different in large cities and medium-sized or small cities. The small and medium-sized cities of Bangladesh tend to be less concerned about managing HCW. In 2004, a local NGO named "Prism Bangladesh Foundation" introduced an HCW management system including waste separation at the source to dispose of the waste in Dhaka city. They also established a waste treatment plant near Dhaka. However, this organization provides services to only a few hospitals and diagnostic centers. A large number of hospitals in Dhaka do not have appropriate HCW management systems (Hassan et al., 2008).

Mymensingh Medical College Hospital (MMCH), a government hospital, and 116 clinics and private hospitals provide healthcare services in the Mymensingh municipal area. The Annual Report of Mymensingh Municipality (2016) revealed that around 142 tons of waste are generated in the municipal area each day, mainly from households and hospitals. Among the total solid waste of the municipality, HCW comprises about 4.54 tons/day, which includes 2.68 tons of kitchen/organic waste, 1.14 tons of inorganic waste, 673 kg of infectious waste, and 54 kg of hazardous or injurious waste. A total of 133 municipal dustbins store around 168.5 tons of waste per day. However, the report stressed that the municipality can collect only 126 tons/day. The remaining waste is kept on the roadside or in dustbins. This certainly creates an unpleasant situation for the local citizens by spreading bad odors and transmitting diseases. All the solid waste from the municipality is discharged into the Shambhuganj dumpsite near the bank of the river Brahmaputra. Almost 95% of the dumpsite is filled, with a severe environmental effect on the locality (Islam et al., 2015).

WHO (2018) reported that improper planning, lack of communication and coordination, and gaps in information sharing among stakeholders are responsible for inadequate HCWM systems. Sometimes the process of waste collection is irregular due to the lack of cleaners and transports. In addition, inadequate training in proper waste management, insufficient financial and human resources, absence of waste management and disposal systems, and lack of awareness about the health hazards associated with HCW are common issues (WHO, 2018). The WHO also noted that many countries either do not have appropriate regulations or do not enforce them. In Bangladesh, the Department of Environment (DoE) developed a pocketbook in 2004 for safe management of HCW which was revised in June 2010. For proper HCWM the government formulated a law in 2008. However, the existing law is outdated and imposes low penalties for the offenders; sometimes penalties are overlooked (Nuralam et al., 2017). This is one of the reasons for the lack of appropriate, safe, and cost-effective HCWM strategy.

Furthermore, the management practice in the medium-sized cities of Bangladesh is more vulnerable. Evidence shows that appropriate HCWM practice has not been addressed in the capital city, and the situation in other cities is even more miserable with regard to managing HCW. A few studies have focused on HCWM in medium-sized and small cities in Bangladesh (Wen-Chuan et al., 2017), but there has been little effort to identify the relationship among the actors in dealing with HCW and management strategies for handling HCW in developing countries, especially in Bangladesh. This study focuses on this knowledge gap. Thus, to explore how the top-row public hospitals in a middle-sized city deal with and manage the large amount of HCW, this study investigated the Mymensingh Medical College Hospital (MMCH) in Mymensingh city of Bangladesh.

1.2 Objectives of the study

The objectives of the study are to

(i) assess the current practices of managing the HCW generated by MMCH, and
(ii) identify gaps and pathways for better HCWM.

2. Theoretical framework

The 21st century is facing serious environmental issues across the world. Considering those issues, the theory of Ecological Modernization (EMT) describes recent changes in environmental policy making and assumes a positive correlation between the economy and the environment of the nations. Therefore, EMT has emerged as one of the leading theories in environmental sociology. The theory has evolved through various stages of development with the aim of analyzing how contemporary industrialized societies deal with environmental crisis (York and Rosa, 2003).

During the early 20th century, concern about the environment increased among environmental social scientists in Western societies. In the 1960s, environmental concern was centered on the demand for the protection of natural resources which were deteriorated because of industrialization. In the 1960s and early 1970s, environmental sociologists began to analyze environmental concerns, social changes, and social struggles for the environment (Mol, 1995). The central concern during this time was the fundamental reorganization of the social order that academics considered to be inevitable for an ecologically sound society. Consequently, scholarly efforts resulted in the establishment of government departments for the protection of the environment, the formulation of environmental legislation, and the emergence of nongovernmental environmental organizations (ENGO) in most industrialized societies. This concept was recognized as fulfilling the demands for "social change" and "ecological reform." However, ecological reform at the institutional level was not popular enough in 1960—1970s (Jänicke, 2020; Mol, 1995).

A major turn occurred in the 1980s, when institutional transformation became a topic of concern (Mol and Sonnenfeld, 2000). This concern led to fundamental environmental reform of modern institutions to protect nature. The environment as an issue began to move from the periphery to the center of the social development of industrialized societies during the 1980s and onward. This movement emphasized the notion of the "ecological restructuring of industrial society" in the constant reformation of modern societies. During this time, EMT emerged as a concept in environmental sociology focusing on the relations between institutional development and the environment (Mol, 1995).

To restore the balance between nature and modern society based on a capitalist economy, the concept of ecological modernization emerged in the field of environmental sociology. EMT was first developed in the early 1980s, primarily in a small group of western European countries, remarkably Germany, the Netherlands and the United Kingdom (UK). The German sociologist, Joseph Huber, is recognized as the founder of this theory, which was revised by Arthur P.J. Mol and Gert Spaargaren later. Empirical studies of this theory have also been carried out more recently by Jokinen and Koskinen (1998), Rinkevicius (1999), Gille, Frijns, and Sonnenfeld (Mol and Sonnenfeld, 2000). In its first phase, the theory emphasized the role of technological innovations in environmental reform. Later, theorists looked for the social roots of environmental problems in the institutional design of production and consumption in modern societies. Proponents of the theory argued that economic practices need to be reorganized into more ecologically sound practices within the institutional context rather limiting technological innovation to minimize environmental degradation. Gradually, the theory of ecological modernization focused on the ecological restructuring of production and consumption. Based on this view, the scholarship of ecological modernization puts more focus on the roles of the state and the market in ecological transformation rather than technological innovation and more attention was given to the institutional and cultural dynamics of ecological modernization (Jänicke, 2020).

The core theme of the theory valued science and technology not only for their role in the emergence of environmental problems but also their potential role in preventing those issues (Jänicke, 2020). Traditional repair options are replaced by socio-technological approaches from the design stage of technological and organizational innovations. The theory also considered the importance of market dynamics and economic agents such as producers, consumers, and credit institutions as carriers of ecological reform. In addition, transformation in the governance style of the state was emphasized. For instance, more decentralized and flexible styles of governance emerged as political modernization. Modifying the role of social movements regarding environmental reform

became another focus of theoretical development. Finally, intergenerational solidarity in dealing with the environment (i.e., sustainable development) has emerged as a core principle (Mol and Spaargaren, 2000).

In order to assess the value of ecological modernization, scholars argued for an analytical model that can provide the tools for a systematic evaluation of the hypothesis derived from EMT. In the literature on environmental sociology, scholars have used various models to analyze the process of environment-induced change and its effect on the society. Arthur Mol, one of the theorists of EMT, has used the triad network approach to identify institutional changes (in economic sector) in the social systems with regard to ecological performance (Mol, 1995). This approach provides three networks (economic, political, and social) that provide the benefits of uniting both the structural properties of institutions and the interactions between the actors constructing a network. In accordance with the approach, networks are characterized as a social system in which actors remain engaged in more or less permanent, institutionalized interactions. These networks have a distinctive institutional arrangement and a restricting number of interacting actors that are considered important with regard to the particular issue. In each network, several actors interact with each other interdependently. Economic networks underlie the economic interaction among the actors, social networks emphasize the relationships between civil society, and the economic sector and policy networks highlight the cooperation between an organization and government bodies (Mol, 1995).

The structural properties of an institution include its locational profile (e.g., geographical location of the organization), socio-economic profile (e.g., organizational history and culture), and environmental profile (e.g., effects of the activities of that organization on the environment). The profile definition helps to provide fundamental information for the analytical perspective of the triad network model. The model is used to understand the environment-induced transformation that makes a difference in the society and underpins the point of intervention for policy initiatives.

This study followed the model developed by Arthur Mol (1995) to systematically examine the process of environment-induced social change in an industrial society and improve the theory further.

3. Research design and methods of data collection

This section describes the study design and research methods.

3.1 Research design

The study was conducted in Mymensingh district under Mymensingh division. The city is located on the banks of the Brahmaputra River, about 120 km north from Dhaka. Mymensingh city is the fourth-largest city in the country. The city is a major financial center and educational hub of north central Bangladesh. The district covers an area of 4363.48 km^2, bordered on the north by Garo Hills.

In order to explore the present HCWM system in Bangladesh, a case study approach was chosen for the study. Mymensingh Medical College Hospital (MMCH), situated in Mymensingh district, was selected as the case. MMCH was established in 1924 and is currently the third-largest public hospital in the country and the highest-ranked government-owned hospital in Mymensingh city. A significant number of private hospitals are also located in the city area (see Fig. 3.1, Mymensingh Sadar). However, the study focused on the public hospital MMCH because of the number of patients, cost of treatment facilities, and the service delivery system of the public institution. The study

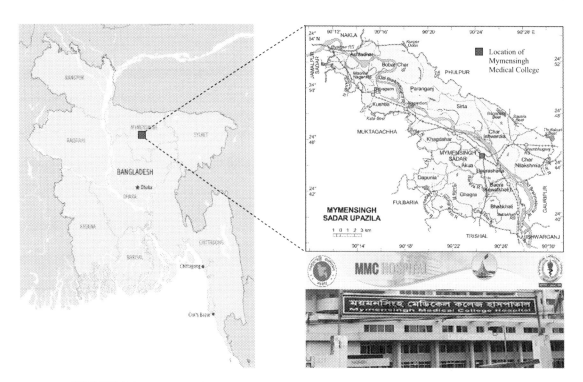

FIGURE 3.1 Mymensingh Medical College Hospital (MMCH) in Mymensingh, Bangladesh.

followed a mixed-method approach emphasizing the qualitative aspects (e.g., network analysis) to meet the research objectives.

3.2 Methods of data collection

Primary data were collected by administering structured questionnaires among the nurses and doctors of MMCH. Further, semistructured interviews were conducted with key informants. Two sets of questionnaires were administered: one for cleaners and waste pickers and another for doctors and nurses. The respondents were asked about their perception of HCWM, the amount of waste produced, source of HCW from different wards (e.g., Neurology, Cardiology, Gynecology), methods of waste segregation, waste collection process in the ward, frequency of waste collection, storage of waste in the hospital, transportation process of waste inside the hospital, transportation of waste outside the hospital, health risks of the waste, and reasons for improper HCWM. A total of 100 questionnaires were distributed, and 87 were completed and used for analysis.

A Likert scale (1 = Strongly disagree to 4 = Strongly agree) was used to identify the agreement level of the respondents regarding waste collection practices of the hospital: (1) separate wastes at the source, (2) do not separate wastes at the source, (3) collect wastes from the patient's bedside, (4) separate wastes from a common bin of the ward, (5) collect wastes from the hospital's dustbin for disposal, and (6) collect wastes from the municipal dustbin for disposal. A Weighted Average Score[1] (WS) was calculated for each response to prioritize the features of respondents' views.

Apart from the questionnaire, 12 interviews (KII, key informant interview) were carried out with the city mayor, the Civil Surgeon, three ministry officials, the director of MMCH, two mid-level doctors, two senior nurses, the ward master, and the sanitary inspector. A checklist was prepared for each KII with a focus on waste generation rate, waste collection process, storage and transportation of waste, waste disposal process of the hospital, role of different actors for handling HCW, and the way of safe and proper management of HCW in the hospital.

Data were analyzed using simple quantitative methods and presented using graphs and frequency tables. Qualitative data gathered mainly from interview were analyzed through triad network model analysis. The following section presents the network analysis method that enabled the exploration of in-depth information.

3.2.1 Triad network model analysis

Data gathered from the interviews were analyzed by the triad network model. This model is used to explore the interactions among actors located in different networks in dealing with HCW. This model was also applied to figure out the way forward for improving the HCW management strategy. Following the model developed by Arthur Mol (1995), this study identified several actors who are closely associated with the HCWM at MMCH. The study also explored the connections among the actors to handle HCW and presented their interaction through social, economic, and policy networks.

For the analysis of triad network model, the study identified that the economic network consists of the actors who provided the healthcare services and who managed the HCW. Actors from the hospital (e.g., doctors, patients, nurses, cleaners) and from the municipality (e.g., sanitary inspector, municipality cleaners, waste

---

[1] WS = A weighted average is the average of a data set that recognizes certain numbers as more important than others. **Weighted scoring** is a prioritization framework designed to help decide how to prioritize features and other initiatives on a product roadmap.

pickers) are part of the economic network. The policy network highlights the relationship of the government organizations (e.g., Ministry of Health and Family Planning, Directorate General of Health Services, Ministry of Local Government and Rural Development) and the actors from the institutions (e.g., Director of MMCH, City Mayor, Civil Surgeon) who look after the administrative issues. The regulatory aspects, rules, and systems are related to the policy network of the HCWM that is also consulted for the model analysis. These documents are also part of the structural profile of the model. In the social network of HCWM, relations between the actors from the hospital and the civil society (e.g., citizens, NGOs, concerned private sector associations) are investigated.

4. Results and discussion

4.1 Practices in healthcare waste management

4.1.1 Waste generation

The hospital logbook indicates that approximately 350–400 kg of waste is generated per day at MMCH. The ward master also validated these figures. During the interview, he explained that HCW recorded 250 kg of waste per truck in 1996 with the support of an NGO. The interviewee also noted that the waste is not measured in a scientific way at present. The rate is recorded haphazardly. Nonavailability of a system as well as the lack of equipment are the reasons for not measuring the HCW in a scientific manner. As a result, the actual rate of waste generation can only be approximated.

4.1.2 Waste collection

Waste is collected from each ward by the cleaners and stored at a specific point on each floor of each building. Nurses and cleaners mentioned that waste is separated at the source only in some wards such as surgery and the neonatal unit, but this is very uncommon in other wards of the hospital such as the medicine ward and the gynecology/maternity ward. The cleaners and nurses of those wards commented that they never segregate the wastes at their source. After being stored in a common bin on each floor, the HCW is transported to the internal dustbin of the hospital. The respondents noted that only the paper, empty bottles, boxes, and empty pharmaceutical packets are separated from other wastes for recycling.

The study collected data on the waste collection process by applying a 4-point scale to identify respondents' agreement with several statements. Most of the respondents reported that waste is collected from the bedsides of the patients. Responses from the nurses and the cleaners differed regarding the waste collection. There was a strong disagreement concerning the statement regarding waste separation from a common bin of the ward. Most of the nurses (WS 59) agreed that waste is separated from a common bin, but the cleaners (WS 29) and doctors (WS 48) disagreed with the view of nurses in this regard. This indicates that a knowledge gap prevails among the respondents on the issue of waste separation. Furthermore, a large number of respondents (doctors WS 92.59, nurses WS 75.63, cleaners WS 90) reported that waste pickers (from the municipality) collected waste from the municipal dustbin located at the entrance of the MMCH (Fig. 3.2). This indicates that the respondents are aware of the collection of HCW from the municipal dustbin for final disposal. It is observed that waste pickers from the municipality have collected HCW along with the municipal waste from the municipal dustbin.

4.1.3 Transportation of HCW

Transportation of waste from MMCH is a vital issue of HCWM. In general, cleaners who are employed by the hospital carry the waste to the municipal dustbin. The sanitary inspector from the municipality stated that cleaners from

4. Results and discussion 45

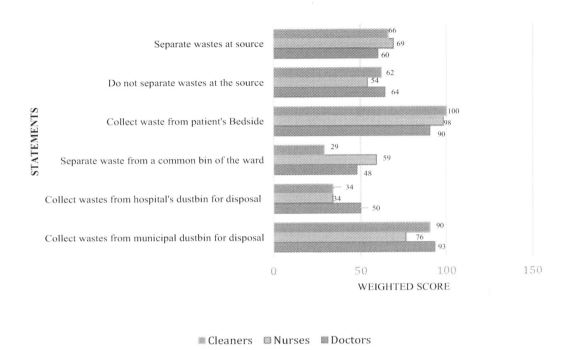

FIGURE 3.2 Current practice of waste collection reported by cleaners, nurses, and doctors.

the municipality are responsible for collecting waste from the municipal dustbins for final disposal. During collection and transportation of the HCW, cleaners and waste pickers are frequently injured by sharp materials (e.g., needles, blades) and infected by contagious equipment. In the interviews, 56% of the doctors, 96% of the nurses, and 100% of the cleaners reported being injured by sharp materials. Furthermore, animals are also exposed to toxic contamination from the open disposal of wastes (Fig. 3.3A and B). Most of the respondents (doctors 56%, nurses 88%, cleaners 90%) stated that HCW was transported to the municipal dustbin and directly dumped with the municipal waste (Fig. 3.3B). The final disposal of the wastes was at the Shambhuganj site close to the Puratan Brahmaputra River.

4.2 Findings of triad network model analysis

Network analysis emphasizes the interactions among the actors in an institutional setting. The actors are grouped into three networks: (1) policy, (2) economic, and (3) social. The following sections present the connections between the actors among the three networks.

4.2.1 Policy networks

In the policy network of proper HCWM in Mymensingh, all actors are from government bodies and hospital authorities, including the Ministry of Health and Family Welfare (MOHFW), Ministry of Local Government, Rural Development and Cooperatives (MLGRDC), Ministry of Environment, Forest and Climate

FIGURE 3.3 (A) Animals are exposed to infectious healthcare waste. (B) Healthcare wastes are disposed of by the riverside.

Change (MoEF), Department of Environment (DoE), the City Mayor, the Civil Surgeon, the Director of MMCH, and doctors at the MMCH.

The MOHFW, MLGRDC, and MoEF did not meet regularly with each other to know the status of HCWM all around the country, formulate policy or law for managing these wastes, monitor the role of the DoE to handle HCW, or allocate financial resources for HCWM (from interviews). The interviewees (Civil Surgeon, Director MMCH, and the City Mayor) specified that regular meeting in a certain interval can create a scope to identify the gap of the law at the implementation level, to reallocate the financial resources and employing the human resources for proper HCWM in MMCH.

The policy document "Bangladesh Medical Waste Management and Processing Rules 2008" was formulated to provide guidance for different stakeholders to handle HCW. This document was drafted in consultation with several ministries, doctors, and hospital owners and adapted the policy documents of some neighboring countries. In accordance with the rule, an HCW treatment plant has to be established in each district with the support of the municipality. However, respondents in this study (e.g., City Mayor, Civil Surgeon, and Director of the MMCH) stated that they are not well informed about the rule, so it has not been implemented by the MMCH authorities or by the municipality yet. This result highlights the knowledge gap about the policy document among the actors.

The interviewees noted that there is no HCW treatment plant in Mymensingh and pointed out that the establishment of such a facility is yet to be included in the waste management plan of the municipality as well as the MMCH. Furthermore, the allocation of financial resources is insufficient for technology-based HCW management. Lack of funding and the gap in planning mentioned by both the hospital authorities and the municipality are hindrances to appropriate HCWM. In their opinion, coordination among stakeholders must be emphasized for proper handling of HCW.

The civil surgeon as a stakeholder of waste management does not have the authority to put pressure on the municipality to handle HCW, and the MMCH authorities have not met with municipal personnel to work out the appropriate way to deal with HCW. In addition, the Department of the Environment (DoE) has been less than active in monitoring HCWM though it has the legal role of supervision (GoB, 2008). In interview, the DoE officials reported that lack of manpower is one of the reasons for the shortage of supervision of HCWM in Mymensingh.

4.2.2 Economic networks

Actors from hospitals and municipalities who are directly involved in healthcare services (e.g., doctors, nurses, patients, cleaners) and HCWM are part of the economic networks. Apart from general duties such as providing medication and necessary treatment to a large number of patients, doctors are also involved in the decision-making process for the betterment of the health services. They also play managerial roles in the hospital as they supervise the nurses. Nurses are responsible to take care of the patients and monitor the activities of the cleaners. The role of cleaners is to clean the patients' beds and rooms and to collect waste. Therefore, doctors, nurses, patients, and cleaners are interconnected in performing their roles.

Doctors and nurses are very focused on the treatment of patients, using their knowledge and skills and sharing their experience to perform their duties. The doctors who participated in this study are more concerned about the care given by the nurses than about ensuring the roles of nurses in HCWM. Nevertheless, nurses and cleaners are very close to the HCWM process. The nurses and cleaners indicated that all the empty pharmaceutical packets, used syringes, cotton balls, and bandages are kept in bins under the patients' beds. Then the cleaners collect all the waste from the room and deposit it in the general dustbin for the floor. Another group of cleaners then collect the waste from each floor and transport it to a central dustbin outside the hospital building, and yet another group of cleaners transport the waste to the municipal dustbin. Waste operators from the municipality then carry the HCW along with the municipal waste to Shambhuganj for final disposal. Fig. 3.4 shows the current workflow of HCWM at MMCH.

The nurses and cleaners stated that waste is not separated at the source because of the pressure of the huge number of patients and visitors and the small number of cleaners. Above all, segregation of waste at the source does not occur because there are no strict roles and practices of waste management. The nurses commented that wastes were separated at the source a few years ago and kept in the different colored bins according to their nature. At that time, an NGO supplied four types of bins for each ward of the hospital and provided training for the nurses

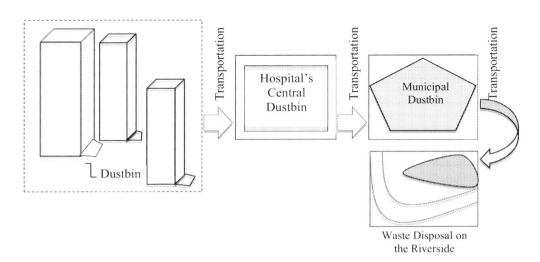

FIGURE 3.4 Current HCWM practice at MMCH.

and cleaners (though not for doctors) about separating wastes at the source. However, the practice has not continued. Furthermore, the MMCH authorities have not integrated waste segregation as part of HCWM, so the system has failed. Currently, nurses monitor the activities of the cleaners inside the wards or cabins/rooms, and doctors are more concerned about the treatment of the patients. Doctors are also engaged in other private business. Nurses play a vital role in administering medication with the advice of the doctors. Sometimes doctors take advice from the nurses regarding medication as some of them were highly experienced through practice. In addition, nurses take the responsibility for monitoring the work of cleaners to see whether they are disposing of the waste in a timely fashion. Doctors are dependent on nurses to care for the patients and oversee the work of cleaners. Thus, the doctors—nurses—patients—cleaners relation is found to be very supportive.

In line with the theory of EMT by Mol (1995), it is apparent that one way of interaction among the actors is resource allocation to provide services to the customer. In considering this view, financial allocation for managing HCW becomes very important. The Director of MMCH opined those financial allocations are mainly for the treatment of the patients and these resources are insufficient even for that. Therefore, financial resource allocation for HCW management has not been a matter of concern to that extent.

The city mayor also noted that the municipality lacks adequate financing for HCWM and added that the allocation of budget is very low for infrastructure (e.g., construction of road, bridge, culvert, drain, etc.) for city development. Furthermore, there is no waste management treatment plant inside the hospital (MMCH) or in the city itself. All the waste from MMCH and other HCEs in the city is dumped near the Brahmaputra along with the municipal wastes. Overall, respondents noted that a system of HCW collection and transportation was absent

at MMCH, and various stakeholders such as the Director, Civil Surgeon, City Mayor, and Sanitary Inspector mentioned that a scientific method of HCW treatment and HCW dumping was unavailable in MMCH as well as in the city of Mymensingh.

4.2.3 Social networks

The social networks consisted of NGOs and civil society organizations dealing with the environmental and human health effects of HCW. Media and the local citizens are identified as the actors in the social networks of HCWM. Respondents commented that a local NGO named "Gramhouse" is focused only on the solid waste management (e.g., collection and disposal) of the city. The study found that there is no involvement of any NGO in handling HCW in the locality. Most of the interviewees believe that NGOs can play a strong role in the process of HCWM together with the municipality and the MMCH. In particular, they felt that NGOs can arrange training for nurses and cleaners and provide logistics, such as different colored bins for waste segregation at the source, in collaboration with the government. Respondents stated that an international NGO organized training for nurses and doctors and provided different colored bins a couple of years ago, but those are unavailable now.

Furthermore, the director of MMCH and the city mayor emphasized that the media can also play role to raise awareness among citizens about the risks of inappropriate HCWM, including the health hazards and consequences of HCW mismanagement to the environment. For example, a report was published in a national daily newspaper in 2013 (Bangladesh) entitled "an arrangement by municipal authorities related to garbage disposal near the bank of the Brahmaputra River" highlighted that regular dumping of garbage near the river badly affects the environment of the river, which is almost dying. This report grabbed the attention of the negative impacts of the miss management of the hospital waste among the city dwellers and authorities. This type of report will create a pressure among the authority to manage hospital waste properly.

Respondents commented that although this news created severe protests by the local citizens, it is not reinforced, and the authorities have shown less sensitivity about this issue. Local citizens perceived that their voice cannot reach the authorities and they have felt disappointed. Similarly, other media (e.g., television, movies) do not emphasize the issue of HCW mismanagement and the loss of the environment. From the view of the interviewees, media can make the people aware of HCWM, and NGOs can contribute in managing HCW in Mymensingh city.

Analyzing all the networks for appropriate HCWM (Fig. 3.5), the findings highlighted that coordination among ministries is infrequent and interaction among the MMCH authority, Civil Surgeon, and municipal authorities is irregular. These actors need to meet regularly to enhance the management system for handling HCW and monitor the status of HCWM. The ministries have to be more concerned about the implementation of HCWM rules. The connection between doctors, nurses, patients, and cleaners is very consistent. However, local citizens and the media had infrequent interaction with other actors in the other networks.

5. Future ways to improve HCWM

The respondents highlight a number of strategies to improve the waste management process from waste collection, storage, and transportation to disposal. In accordance with WHO, the management of healthcare waste includes the process of waste segregation at the source, collection of waste from different points, storage of waste at a particular place, waste transportation, and final disposal of waste. The study identified that reducing the number of visitors in the hospital (doctors 94%, nurses 98%) and building awareness among the health workers, patients, and citizens (doctors 62%, nurses 92%)

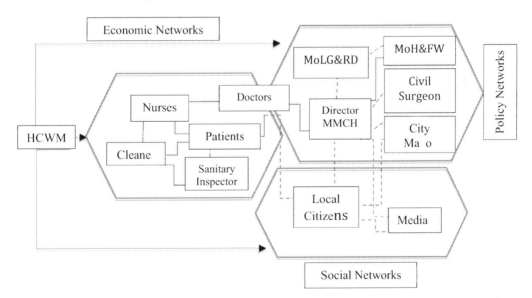

FIGURE 3.5 Network analysis for appropriate HCWM. Note: Here, the solid lines indicate frequent interaction between the actors and the broken lines denote infrequent interactions. *Modified from Mol, A.P.J., 1995. The Refinement of Production: Ecological Modernization Theory and the Chemical Industry, International Books. The University of Michigan.*

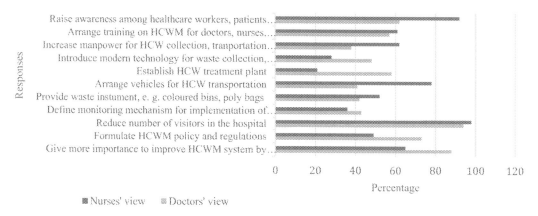

FIGURE 3.6 Feasible measures for improving HCWM at MMCH.

are the topmost priorities to improve the current practice of HCW management (Fig. 3.6). Other interviewees (e.g., Director, MMCH, Civil Surgeon) agreed with this argument made by the doctors and nurses. Doctors also emphasized the role of policy makers to look at the issue deeply and place more importance on introducing an appropriate system of HCWM across the country. In that case, MMCH can be a model hospital for managing HCW. Doctors also highlighted the formulation of HCWM policy and regulations (73%), establishment of a waste treatment plant (58%) and introducing modern technology for waste collection, transportation, and disposal (48%). The view of doctors reflects that they prefer government initiatives and technological establishment for HCWM. On the other hand, nurses emphasized arranging vehicles for waste transportation (78%), increasing manpower for waste collection, transportation, and waste disposal (62%) and providing waste segregation and storage materials (52%). Both doctors (57%) and nurses (61%) highlight that training for doctors, nurses, and cleaners on HCWM is a matter of concern. Further, nurses suggested that NGOs can provide waste segregation and storage material for better management (52%).

In addition, the director of MMCH, the city mayor, and the Civil Surgeon revealed that nurses have direct supervision of the collection and storage process; cleaners perform their job following the instructions of nurses and the ward master of the hospital. Thus, nurses have hands-on knowledge in managing HCW. It is also evident from the interviews that nurses have a critical role in HCWM and should be part of policy formulation. Furthermore, doctors have a close linkage with nurses as well as technical knowledge. The respondents thought that doctors' technical knowledge can be utilized in identifying an appropriate HCWM system for MMCH.

In particular, a systematic approach is absent in managing HCW at MMCH. The study also identified that more financial and technological resource allocation is required for establishing a treatment plant and framing the overall management process. Concerning the significance of the issue, respondents stated that separate section is required within the hospital to monitor HCWM activities at MMCH and in Mymensingh. Thus, institutional transformation is urgent. This transformation will require direct communication and coordination among the actors to reduce the adverse effect of

HCW mismanagement. Moreover, most of the respondents emphasized the need for more importance to be placed at the policy level to deal with HCW which poses a serious threat to human health as well as the environment.

6. Conclusion

Safe management of HCW is fundamental for providing safe and people-centered care and safeguarding the environment. The findings of this study demonstrate that MMCH has managed the HCW inside the hospital in a very traditional way, generally disposing of HCW along with municipal waste. MMCH's HCW was not separated at the source but kept in a single bin. Furthermore, patients' bedsides were the main source of waste, and the large number of visitors contributed to generating the large volume of waste.

Identifying the different roles in managing HCW was another objective of this study. Coordination among the various ministries was found to be infrequent in the policy network analysis. The government formulated a rule for proper management of HCW in 2008, but this rule has not yet been implemented at MMCH. In addition, an HCW treatment plant has not been established in the city with the support of the municipality. The underlying factor of the nonestablishment of a system of proper HCWM was that the Civil Surgeon and MMCH authorities were not strong enough to put pressure on the municipality to handle the HCW of the city.

The triad network analysis shows that doctors have a dual role in decision-making and treatment of the patients in both the policy networks and economic networks. Even though doctors were less aware of HCWM at MMCH, they have a significant role in two networks. This importance needs to be considered during policy formulation. However, nurses and cleaners were

close to the HCWM process. Thus, doctors and nurses have to work together in the HCWM process. Moreover, the respondents wanted NGOs to be involved in the process of HCWM. Lastly, actors in different networks need to cooperate frequently to ensure the process of HCWM in the MMCH as well as in the municipality.

References

Chaerul, M., Tanaka, M., Shekdar, A.V., 2008. A system dynamics approach for hospital waste management. Waste Manag. 28 (2), 442–449.

Chartier, Y., Emmanuel, J., Pieper, U., Pruss, A., Rushbrook, P., Stringer, R., Zghondi, R., 2014. Safe Management of Wastes from Health-Care Activities, Second ed. WHO Library.

Ciplak, N., Kaskun, S., 2015. Healthcare waste management practice in the West Black Sea Region, Turkey: a comparative analysis with the developed and developing countries. J. Air Waste Manag. Assoc. 65 (12), 1387–1394.

GoB, 2016. Medical Waste Management Policy. Ministry of Health and Family Welfare, Dhaka.

Government of Bangladesh (GoB), 2008. The Medical Waste (Management and Processing) Rule, 2008. Ministry of Health and Family Welfare, Dhaka.

Hasan, M.M., Rahman, M.H., 2018. Assessment of healthcare waste management paradigms and its suitable treatment alternative: a case study. J. Environ. Res. Public Health.

Hassan, M.M., Ahmed, S.A., Rahman, K.A., Biswas, T.K., 2008. Pattern of medical waste management: existing scenario in Dhaka City, Bangladesh. BMC Public Health 8 (1), 36.

Islam, M.A., Mobin, M.N., Baten, M.A., Hossen, M.A.M., Islam, M.J., 2015. Hospital waste generation and management in mymensingh municipality. Int. J. Environ. Sci. Nat. Resour. 8 (1), 135–138.

Jänicke, M., 2020. Ecological modernization—a paradise of feasibility but no general solution. In: The Ecological Modernization Capacity of Japan and Germany. Springer VS, Wiesbaden, pp. 13–23.

Khan, B.A., Cheng, L., Khan, A.A., Ahmed, H., 2019. Healthcare waste management in Asian developing countries: a mini review. Waste Manag. Res. 37 (9), 863–875.

Marchettini, N., Ridolfi, R., Rustici, M., 2007. An environmental analysis for comparing waste management options and strategies. Waste Manag. 27 (4), 562–571.

Mol, A.P.J., 1995. The Refinement of Production: Ecological Modernization Theory and the Chemical Industry, International Books. The University of Michigan.

Mol, A.P., Sonnenfeld, D.A., 2000. Ecological modernisation around the world: an introduction. Environ. Policy 9.

Mol, A.P., Spaargaren, G., 2000. Ecological modernisation theory in debate: a review. Environ. Polit. 9 (1), 17–49.

Mugambe, R.K., Ssempebwa, J.C., Tumwesigye, N.M., Van Vliet, B., Adedimeji, A., 2012. Healthcare waste management in Uganda: management and generation rates in public and private hospitals in Kampala. J. Publ. Health 20 (3), 245–251.

National 3R Strategy, 2010. National 3R Strategy - the Department of Environment. http://old.doe.gov.bd/publication_images. (Accessed 25 February 2021).

Nuralam, H.M., Xiao-lan, Z., Dubey, B.K., Wen-Chuan, D., 2017. Healthcare waste management practices in Bangladesh: A Case Study in Dhaka City, Bangladesh. Int. J. Environ. Ecol. Eng. 11, 534–539.

Patwary, M.A., O'Hare, W.T., Street, G., Elahi, K.M., Hossain, S.S., Sarker, M.H., 2009. Quantitative assessment of medical waste generation in the capital city of Bangladesh. Waste Manag. 29 (8), 2392–2397.

Tabish, S.A., Amir, S.K., Bhat, S., 2018. Knowledge, attitude and practice regarding biomedical waste management among the health care workers in hospitals of Kashmir. Intl. J. Sci. Res. 7 (5), 926–930.

Wafula, S.T., Musiime, J., Oporia, F., 2019. Health care waste management among health workers and associated factors in primary health care facilities in Kampala City, Uganda: a cross-sectional study. BMC Public Health 19 (1), 203.

Wen-Chuan, D., Nuralam, H.M., Xiao-lan, Z., Dubey, B.K., 2017. Healthcare waste management practices in Bangladesh: a case study in Dhaka City, Bangladesh. Int. J. Environ. Eng. 11.

World Health Organization (WHO). (2003). Fundamentals of health-care waste management. United Nations Environment Programme/SBC National Health-Care Waste Management Plan, Guidance Manual, 7–23.

World Health Organization. (2005). Management of solid health-care waste at primary health-care centres: A decision-making guide., 2005. WHO.

World Health Organisation (WHO), 2013. Safe Management Waste from Health-Care Activities, second ed. Department of Public Health, Envronmental and Social Determinants of Health, World Health Organization, Geneva.

World Health Organization. (2015). Status of health-care waste management in selected countries of the Western Pacific Region, 2015. http://iris.wpro.who.int. (Accessed 1 August 2021).

WHO, 2015. Retrieve Status of Health-care Waste Management—World Health apps.who.int)iris)rest)bit streams).

WHO, 2018. Fundamentals of Health-Care Waste Management — World. www.who.int)medicalwaste)guidance manual1.

York, R., Rosa, E.A., 2003. Key challenges to ecological modernization theory: Institutional efficacy, case study evidence, units of analysis, and the pace of eco-efficiency. Organ. Environ. 16 (3), 273–288.

Zafar, S., 2019. Medical Waste Management in Developing Countries. Bioenergy Consult (Accessed 12 December 2017).

Further reading

Islam, M.A., Mobin, M.N., Baten, M.A., Hossen, M.A.M., Islam, M.J., 2015. Hospital waste generation and management in Mymensingh municipality. J. Environ. Sci. Nat. Res. 8 (1), 135–138.

Massive Garbage Dumping Pollutes Old Brahmaputra. https//www.thedailystar.net/news/massive-garbage-com (Accessed 20 February 2021).

Mol, A.P., 2006. Ecological modernization. In: Encyclopedia of Globalization (No. A to E. A Taylor and Francis Company, Routledge, pp. 354–357.

Rahman, A., 2017. Healthcare Waste Management in Bangladesh: A Case of Mymensingh Medical College Hospital (MMCH) (Unpublished Master's Dissertation). Wageningen University, Wageningen.

Salmar, Z., 2015. Medical Waste Management in Developing Countries. Solid Waste Management in Kuwait. Municipal Solid Wastes Management in Oman. India.

World Health Organization. (2007). *Safe health-care waste management* (No. WHO/SDE/WSH/07.10). World Health Organization. Safe health-care waste management, 2007.

York, R., Rosa, E.A., Dietz, T., 2010. Ecological modernization theory: theoretical and empirical challenges. Int. Handbook Environ. Sociol. 77–90.

WHO, 2018. Health-care waste - WHO | World Health Organization. https://www.who.int) Newsroom) Fact sheets) Detail (Accessed 18 June 2021).

CHAPTER

4

Seismic risk for vernacular building classes in the fertile Indus Ganga alluvial plains at the foothills of the Himalayas, India

M.C. Raghucharan, Surendra Nadh Somala

Indian Institute of Technology Hyderabad, Department of Civil Engineering, Hyderabad, Telangana, India

1. Introduction

The Indus Ganga plains (IGPs) are one of the most populated basins in the world, covering 25 urban clusters with a population of more than a million (Bagchi and Raghukanth, 2017). The major cities like Karachi, Delhi, Kolkata, and Dhaka, which lie in this region, have more than 15 million populations. The Indian part of IGP from Punjab in the east to Assam in the west has an urban population of more than 125 million, which is close to one-third of the total population of the nation (www.censusindia. gov.in), and is under severe threat in the event of a great earthquake. In addition to the large population, the region is covered with sediments with thickness varying from 0.5 to 3.9 km, which not only amplify the seismic wave energy but also pose a threat to liquefaction (Chadha et al., 2016). Bilham (2009) has estimated casualties of around 0.3 million if an earthquake of Mw ~ 8 occurs near an urban agglomeration of IGP.

Violation of seismic design rules and poor quality of construction in most parts of India have led to calamitous after-effects due to an earthquake. The events 1991 Uttarkashi, 1999 Chamoli, 2001 Bhuj, and 2005 Kashmir are only few examples to unveil. With an annual population growth rate of 1.2% (Yeole and Curran, 2016), a moderate magnitude of Mw > 7.0 in a developed area can destroy the economy of India to a great degree. Thus, in the Central Indo-Gangetic Plain (CIGP) region of India, an estimation of the potential damage to life and assets is prerequisite, by virtue of high population density, poor soil characteristics, and the possibility of occurrence of a major earthquake.

Seismic hazard assessment in the Indian subcontinent started in the 1970s by Basu and Nigam (1977), using a probabilistic concept to develop seismic hazard maps of PGA for a recurrence interval of 100 years. Khattri et al. (1984) prepared a PGA-based hazard map for a 10% probability of exceedance in 50 years. The Global Seismic Hazard Assessment Program developed probabilistic seismic hazard maps between 1992 and 99 showing contours of maximum PGA for 10% probability of exceedance in 50 years' time

Risk, Reliability and Sustainable Remediation in the Field of Civil and Environmental Engineering
https://doi.org/10.1016/B978-0-323-85698-0.00025-3

© 2022 Elsevier Inc. All rights reserved.

window. Bhatia et al. (1999) conducted a probabilistic seismic hazard estimation for India under this framework. Parvez et al. (2003) computed hazard for India using deterministic approach.

Hazard estimations were also carried out for smaller regions of India. Petersen et al. (2004) investigated seismicity models for Gujarat province considering three fault sources in the north-western region. Sharma and Malik (2006) performed PSHA for north-east India. Jaiswal and Sinha (2007) reported a seismic hazard in Peninsular India, employing the probabilistic method. Mahajan et al. (2010) conducted PSHA for the north-western part of the Himalayan region. Kumar et al. (2013) conducted the seismic hazard analysis for the Lucknow region, considering local and active faults.

National Disaster Management Authority report headed by Prof. R. N. Iyengar along with several eminent professors from IITs, IMD, NGRI, SERC, and GSI of India derived probabilistic hazard maps for 32 seismogenic zones covering entire India (NDMA 2011). With the success of NDMA's effort for hazard estimation all over India, more studies have been taken up by other researchers along similar lines (Nath and Thingbaijam, 2012; Seetharam and Kolathayar, 2013).

The seismic zoning map given of India given in IS 1893:2016 was prepared based on the earthquake catalog up to 1993 and has not been updated till 2021. As per this code, the design acceleration for the highest area, Zone V, for the Design Basis Earthquake (DBE) is 0.18 g, and the Maximum Considered Earthquake (MCE) is 0.36 g for the service life of the structure.

Remembering the several instances in the last century in the Himalayas, the Indian code design acceleration for DBE is too optimistic and may underestimate the seismic loading in such a high seismicity province (Ghosh et al., 2012). Hence, an updated seismic hazard map for the country is imperative considering new data, novel findings, and methodological advancements (Nath and Thingbaijam, 2012). Das et al.

(2006) identified that one zone factor indicated for the entire north-eastern India is unreliable, and Jaiswal and Sinha (2007) demonstrated that hazard in few regions of the Indian shield region is greater than BIS (2016); Mahajan et al. (2010) also came up with similar findings for the north-western Himalayan region.

The Himalayas and the Indo-Gangetic plains of India are principally susceptible to higher levels of seismic hazard due to the occurrence of four great earthquakes of magnitude >8 (i.e., Assam, 1950; Bihar—Nepal, 1934; Kangra, 1905; and Shillong, 1897). The danger of Indian cities to seismic hazards has also raised considerably, demanding an appropriate hazard evaluation, especially for localities with high population density in higher seismic zones (Verma and Bansal, 2013). Hence, adopting site-specific seismic hazard for a region is appropriate than a generalized response spectrum given by IS 1893: 2016, at least for areas under high seismic threat.

Seismic risk can be defined as the risk of damage and loss from an earthquake to a structure, system, and entity, which is often computed in terms of probability of damage, economic loss, and casualties. Seismic risk is often used synonymously with a more specific term called earthquake loss estimation (Lang, 2013).

Early works by McGuire (2004) conducted earthquake loss estimation studies using empirical data based on the macroseismic intensity scale due to the scarce instrumental ground motion records. Most of the seismic loss estimation studies based on intensity used data from the observed damage with expert opinion after an earthquake (Porter and Scawthron, 2007). With the development of the nonlinear pushover analysis by Krawinkler and Seneviratna (1998), as well as the derivation of the Capacity Spectrum Method (Freeman et al., 1975; Freeman, 1978; ATC-40, 1996), and the Displacement Coefficient Method (FEMA 273, 1997; FEMA 356, 2000; FEMA 440, 2005), analytical approaches came in to existence in the field of seismic damage and loss estimation.

Seismic vulnerability assessment of vernacular residential buildings was effectively taken up by Duzgun et al. (2011) and Bahadori et al. (2017) both at urban and city level, respectively. Lang et al. (2012) compared the earthquake loss estimates by empirical and analytical methods for Dehradun city. The economic losses and casualties varied between 12 and 36 billion INR and 915—2138 across two methods. Nanda et al. (2015) compared several loss estimation tools and estimated loss of life and property for institutional buildings of NIT Durgapur campus. Building damage, economic losses, and casualty were estimated for Kolkata city by Ghatak et al. (2017). PGA and Pseudo Spectral Acceleration (PSA) values with a recurrence interval of 475 years were used for computation of losses. The expected economic losses and casualties were 231 billion INR and 3300, respectively. Wyss et al. (2017) estimated casualties for two Himalayan earthquakes, the M 7.9 Subansiri 1947 and M 7.8 Kangra 1905 earthquakes. The number of fatalities reported was 1.0 and 2.0 lakh for Subansiri and Kangra events, respectively. Wyss et al. (2018) computed the number of human losses and affected people in the case if Himalayan earthquakes of 1555 and 1505 should reoccur with the same magnitude and location. The estimates for 1555 and 1505 were 2.2 and 6.0 lakhs, respectively.

In this work, a site-specific seismic hazard for 2475 and 475 years return period, at 54 districts of Uttar Pradesh state (CIGP), is computed employing the ground motion equations of Raghucharan et al. (2019). Utilizing the computed hazard levels at each district, seismic risk in terms of damage probability of buildings, economic losses, and casualties is reported. Further, the sensitivity analysis is conducted to assess the relative influence of both parametric and model uncertainties.

2. Seismic demand

The seismic hazard for a region involves the identification of earthquake source models, which characterize the magnitude—frequency distribution, scaling relationships, and potential finite rupture geometries (Rong et al., 2017). The potential seismic sources around the study region, which can cause damage to the building and casualties, are identified and demarcated into six area source zones, as shown in Fig. 4.1. The areas are discretized into grids of 50 km, and the source parameters are selected from literature (NDMA, 2011; Nath et al., 2019) to model distributed seismicity. The intensity measure predicted was PGA and Sa at 25 periods [0.1, 0.2, 0.3, 0.4, 0.5, 0.55, 0.6, 0.65, 0.67, 0.7, 0.8, 0.9, 1.0, 1.1, 1.2, 1.3, 1.4, 1.5, 1.6, 1.7, 1.8, 1.9, 2.0, 3.0, and 4.0 s]. Table 4.1 shows the seismicity parameters of each source zone.

It also requires the selection of a reliable and accurate attenuation relationship for the region. Hence, region-specific ANN ground-motion model derived by Raghucharan et al. (2019) is considered for computing seismic hazard for return periods of 475 and 2475 years for CIGP, India. The GMPEs are scripted in Python language within the OpenQuake engine platform for calculating the hazard. Later the results are compared with the existing hazard values calculated by other researchers in CIGP to ascertain the predictions made in this study.

Seismic demand for the study region is obtained from the site-specific response spectrum for each geographical unit (district) from the seismic hazard. The spectral acceleration (Sa in g) values at 0.0, 0.1, 0.2, 0.3, 0.5, 0.75, 1.0, 1.5, and 2.0 s at bedrock are taken from the site-specific response spectrum for 2475 years return period, corresponding to MCE level. Soil amplification is addressed using the Vs30 parameter,

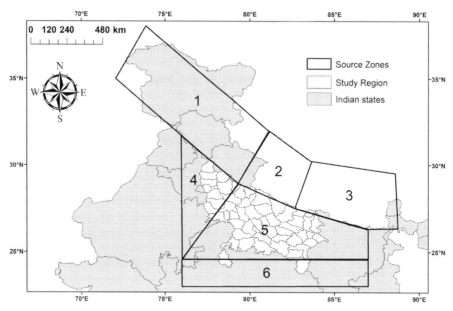

FIGURE 4.1 Map showing the study region (yellow [Light gray in print version] color) and source zones considered for computing hazard in CIGP.

TABLE 4.1 Seismicity parameters selected for each zone.

Zone	a value	b value	Mmax	References
1	5.37	0.86	8.8	NDMA (2011)
2	3.15	0.69	7.8	
3	2.30	0.74	8.8	
4	2.79	0.69	7.2	Nath et al. (2019)
5	2.81	0.69	7.2	
6	2.86	0.73	7.0	

which was obtained from Wald and Allen (2007). Fig. 4.2 shows the map view of Vs30 values across all districts of the study region. Topographic amplification is neglected, as the CIGP is mostly flat territory with a slope <10 degrees.

3. Population exposure and building inventory

The building inventory and population information of CIGP (54 geo-units of Uttar Pradesh state of India) are acquired from the Census of India (2011). The population density of all districts of the study region is represented in Fig. 4.3. Building inventory details by predominant material used for roof and walls from Census (2011) are presented in Tables 4.2 and 4.3. Considering the construction practices prevailing in India, Prasad et al. (2009) designated 40 building classes based on wall, roof, framing types, and story numbers. These building classes are accredited to 10 types and given nomenclature along with replacement cost of each building in rupees in Table 4.4 (Raghucharan and Somala, 2018b). Also, Tables 4.2 and 4.3 are united to give the percentage of houses at all geo-units to these model building types (MBTs), as shown in Table 4.5. The MBTs are categorized into 10 types for assigning vulnerability functions based on the material and type of construction followed.

4. Fragility functions

HAZUS-MH database provides fragility functions for different model building classes that are

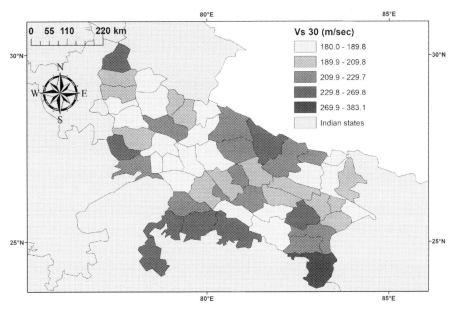

FIGURE 4.2 Shear wave velocity (Vs30) across 54 geo-units of the study area (Wald and Allen, 2007). Light shade indicates loose soils and thick shade represents rock strata.

useful for computing damage probability and socio-economic losses. As the construction methodologies for building inventory in India vary noticeably from buildings of the United States (US), fragility functions recommended in HAZUS-MH (FEMA, 2003) cannot be applied directly. Analytical capacity and fragility functions shall be meticulously chosen since they will bear a substantial impact on the risk and loss results (Lang et al., 2012). The capacity and fragility functions (vulnerability functions) of the 10 MBTs are obtained from Cattari et al. (2004), Kappos and Panagopoulos (2008, 2010), and Prasad (2010).

For buildings typologies made of adobe materials and rubble stone (MBT 1-2 in Table 4.4), vulnerability functions are unavailable for India. Hence, these curves are adopted from Italian conditions (Cattari et al., 2004) due to the resemblance they exhibit in terms of ingredients used in construction. For building types made of mud mortar and clay in brick masonry, MM, ML, and MC, curves derived by Prasad (2010) by suitably consideration of construction practices in North India are adopted for MBTs 3-8. For MBTs 9-10, the curves provided by Kappos and Panagopoulos (2008, 2010) with fully infilled reinforced concrete (RC) frame buildings are chosen for RC low (1–3 stories) and Mid-rise (4–7 stories). Since a large number of the buildings in the CIGP region are constructed before code or by breaching the IS code guidelines, the low-code/precode conditions can be applied. Table 4.6 presents the capacity and fragility function values for four damage states: slight, moderate, extensive, and complete, for all 10 building types, along with their reference.

NORSAR and the University of Alicante have developed SELENA, a MATLAB (2017)-based tool with a fundamental methodology adopted from HAZUS. SELENA computes damage probability of buildings, economic outlay, and human casualties using an analytical method, as explained in Molina et al. (2010).

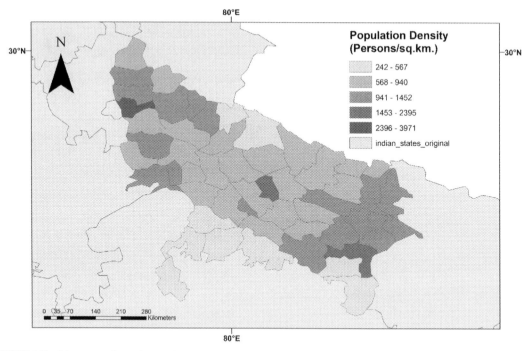

FIGURE 4.3 Map view of population density across 54 geo-units of the study area. The density of the region increases with an increase in color shade.

TABLE 4.2 Distribution of households by predominant material of roof as per the Census of India (2011).

India/ State/ Union territory	Total no. of households (excluding institutional households)	Grass, thatch, bamboo, wood, mud, etc.	Plastic, polythene	Tiles Handmade tiles	Machine-made tiles	Total	Brick	Stone/ slate	G.I. Metal, asbestos sheets	Concrete	Any other material
Uttar Pradesh	32,924,266	23.3	0.5	8.7	8.1	0.6	32.8	13.9	2.9	17.6	0.4

4.1 Seismic risk

Seismic risk in terms of seismic damage and loss estimation has been ascertained for the first time in CIGP by combining seismic demand, population exposure, and vulnerability (fragility) of buildings exposed.

The CIGPs lying between the Himalayas and India shield region are contemplated as a zone of high risk due to thick soil deposits and adjacency to the highest seismic activity in India, the Himalayas. The sediments not only amplify the ground motions but also make foundations vulnerable to liquefaction.

TABLE 4.3 Distribution of households by predominant material of wall as per the Census of India (2011).

		Distribution of households by predominant material of wall										
						Stone			G.I. Metal, asbestos sheets	Burnt brick	Concrete	Any other material
India/ State/ Union territory	Total no. of households (excluding institutional households)	Grass, thatch, bamboo, etc.	Plastic, polythene	Mud, unburnt brick	Wood	Total	Not packed with mortar	Packed with mortar				
Uttar Pradesh	32,924,266	5	0.3	20	0.1	3.2	1.2	2	0.1	68.4	0.6	2.4

TABLE 4.4 Classification of each Model Building type along with their replacement cost selected in the study (10 in total).

Model building class	Nomenclature	Replacement cost in rupees, per m²
AMM	Adobe Mud Mortar walls with Temporary roof	4500.0
ALC	Adobe lime and cement Mortar walls with Temporary roof	4500.0
MMB	Mud Mortar Bricks with Temporary roof	8937.5
BTR	Bricks and tiles roof	8937.5
BSR	Bricks and stone roof	8937.5
BCM	Bricks in cement mortar for walls and roof	8937.5
BMS	Bricks wall with metal sheet roof	8937.5
BCS	Bricks wall with concrete slab	10,350.0
RCL	Reinforced wall and slab-low rise	10,350.0
RCM	Reinforced wall and slab-medium rise	10,350.0

TABLE 4.5 Distribution of Houses in all districts of Uttar Pradesh (Census, 2011) to Standard Model Building types adopted from Prasad et al. (2009).

Total houses/ households	Distribution of households by predominant material of the wall									
32,924,266	Grass, Thatch, Bamboo, Plastic, Polythene, Wood, Mud, Unburnt bricks, etc.	Stone Packed with mortar	Burnt Bricks							Concrete
% of Households	29	2	68.4							0.6
Distribution based on the material of Roof	NIL	NIL	Grass, Thatch, Bamboo, Wood, Mud, Plastic, Polythene, others, etc.	Tiles: Handmade and Machine made	Stone/ Slate	Bricks	G.I, Metal, Asbestos Sheets	Concrete	NIL	NIL
% of Households	29	2	24.1	8.7	13.9	32.8	2.9	17.6	25	75
Model Building Type (MBT) classification	AM11, AM21	AL11, AL21, AL31, AC11, AC21, AC31	MM11, MM21, MM31	ML11, ML21, ML31	ML12, ML22, ML32	MC11, MC21, MC3L1	MC12, MC22, MC3L2	MC3M	RC1L	RC1M
% of Total Houses	29.00	2.00	16.48	5.95	9.51	22.44	1.98	12.04	0.15	0.45
MBT code	AMM	ALC	MMB	BTR	BSR	BCM	BMS	BCS	RCL	RCM

TABLE 4.6 Vulnerability functions adopted for different types of building in the CIGP region. Capacity values are specified at yield and ultimate, while the fragility values are reported as mean spectral displacement and standard deviation for four damage states (slight, moderate, extensive, and complete). The reference is provided in the last column.

Model building class	Building type	Stories	Capacity curve parameters (yield and ultimate points)				Fragility function parameters for damage states (slight, moderate, extensive, and complete)								References
			Dy (mm)	Ay (m/s2)	Du (mm)	Au (m/s2)	$Sd_{,slight}$ (mm)	β_{slight}	$Sd_{,mod}$ (mm)	β_{mod}	$Sd_{,ext}$ (mm)	β_{ext}	$Sd_{,comp}$ (mm)	β_{comp}	
AMM	AM11, AM21, AM12, AM22	1	1.01	1.408	10.70	1.408	0.64	1.11	1.36	1.11	5.85	1.11	10.80	1.11	Cattari et al. (2004)
ALC	AL11, AL21, AL31, AC11, AC21, AC31	1	1.01	2.288	11.10	2.288	0.64	1.13	1.37	1.13	6.06	1.13	11.20	1.13	
MMB	MM11, MM21, MM31	1	1.50	1.079	8.40	1.472	1.30	0.80	2.30	0.90	4.10	0.90	8.00	1.05	Prasad (2010)
BTR	ML11, ML21, ML31	1	1.00	1.570	8.30	2.158	0.90	0.80	1.80	0.90	4.00	0.90	8.00	1.05	
BSR	ML12, ML22, ML32	2	2.60	1.275	14.60	1.766	2.20	0.80	4.00	0.90	7.80	0.90	14.40	1.05	
BCM	MC11, MC21, MC3L1	1	1.30	1.962	8.00	2.453	1.10	0.80	2.00	0.90	4.10	0.90	8.00	1.05	
BMS	MC12, MC22, MC3L2	2	2.60	1.570	14.60	2.158	2.20	0.80	4.50	0.90	7.70	0.90	14.00	1.05	
BCS	MC3M	3+	2.60	1.570	14.60	2.158	2.20	0.80	4.50	0.90	7.70	0.90	14.00	1.05	
RCL	RC1L	1–3	4.90	3.237	51.10	3.924	3.40	0.41	7.30	0.62	28.00	1.04	51.10	1.32	Kappos and Panagopoulos (2008, 2010)
RCM	RC1M	4–7	10.30	2.256	68.00	2.943	7.20	0.38	15.40	0.54	39.10	0.86	68.00	1.09	

4.1.1 Damage probability

In order to evaluate the probability of damage of a building, the spectral displacement value, which is the point of intersection (performance point) of capacity and demand curve, shall be identified (Molina et al., 2015). At present, SELENA (Molina et al., 2010) has three methods for computing the performance point, the capacity spectrum method (CSM) (ATC, 1996; FEMA, 1997), the modified capacity spectrum method (MADRS), and the improved displacement coefficient method (I-DCM) (FEMA, 2005). Since the I-DCM methodology is an improved version of the three, this methodology is selected for computing the performance point of each model building class in the CIGP region.

Seismic performance point based on I-DCM, for a given capacity and demand curve, is determined according to the following methodology. The capacity curve for a given MBCs is generated, and the effective time period of the system is computed as follows:

$$T_{eff} = 2\pi\sqrt{\frac{D_y}{A_y}} \qquad (4.1)$$

The spectral acceleration for the corresponding linear SDOF system, S_a^{es}, can be read from the demand curve for the obtained effective time period (T_{eff}). From the obtained S_a^{es}, the peak elastic spectral displacement demand S_d^{es} is computed from the following:

$$S_d^{es} = \frac{T_e^2}{4\pi^2} \cdot S_a^{es} \qquad (4.2)$$

The performance point is then computed as follows:

$$\Delta_p = C_1 \cdot C_2 \cdot S_d^{es} \qquad (4.3)$$

where C_1 and C_2 are modification factors given in FEMA-440 (FEMA, 2005).

After computing the maximum displacement demand (performance point) for each model building class, the fragility functions, which represent spectral displacement (median value)

and standard deviation, for each structural damage case (slight, moderate, extensive, and complete) of each model building class are selected from literature (Table 4.6). By overlaying the maximum displacement demand (performance point value) on fragility curves, damage probability for five different damage states (including None) can be obtained, as illustrated in Fig. 4.4. From the damage probability of each damage state, the damage results are obtained with respect to the number of damaged buildings, or building floor area of each damaged building class.

4.1.2 Economic losses

The economic losses induced by structural damage at a given geo-unit, in user-defined input currency, are computed when the building inventory data are provided in terms of building floor area. The economic losses covering slight and moderate damage states (for the repair of the building) and extensive and complete (for replacement) are computed using Eq. (4.4) (Molina et al., 2010).

$$E_{loss} = R_{cm} \cdot \sum_{k=1}^{N_{ot}} \sum_{l=1}^{N_{bt}} \sum_{m=1}^{N_{ds}} BA_{k,l} \cdot DP_{l,m} \cdot CR_{k,l,m}$$

$$(4.4)$$

Not indicates the types of occupancy; N_{bt} refers to building typologies, and N_{ds} refers to damage states ds, in numbers. R_{cm} indicates the regional cost multiplier, which is introduced to account for cost variation across geographical units; the term $BA_{k,l}$ denotes the built area of the MBT, l in the occupancy type k; $DP_{l,m}$ indicates the probability of damage for each structural damage m for MBT, l; and $CR_{k,l,m}$ indicates repair or replacement cost in the given input currency for structural damage m, for the type of occupancy k and MBT, l.

The cost of replacement of building model classes was adopted from Lang et al. (2012), which was derived based on the field data collected in the different socio-economic clusters

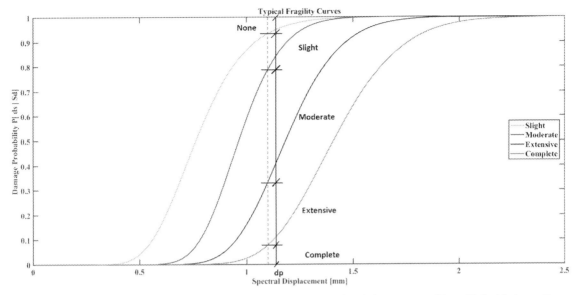

FIGURE 4.4 Typical fragility curves showing the damage probability of each damage cases (None, Slight, Moderate, Extensive, and Complete) for a given maximum displacement demand, Dp. *After Raghucharan, M.C., Somala, S.N. 2018a. Generating site-specific ground motions for Delhi region for seismic vulnerability assessment of buildings—promoting disaster resilient communities. In: Urbanization Challenges in Emerging Economies: Resilience and Sustainability of Infrastructure. American Society of Civil Engineers, Reston, VA, pp. 290–299.*

in Dehradun. Since Dehradun, lying in Uttarakhand state, was part of Uttar Pradesh state of India till its bifurcation in the year 2000, the same replacement cost is considered appropriate for this study.

Since most of the economic clusters that prevail in India are occupied by low-income and middle-income groups, the total replacement cost (sum of structural, nonstructural, and contents cost) is taken as the average of those income groups and is represented in Table 4.4 along with nomenclature of buildings. As the cost variation across geo-units was not considered in the present study, R_{cm} is set to 1.0 in the calculations. Due to lack of data, the cost for other damage states, slight, moderate, and extensive is adopted from HAZUS-MH (FEMA, 2003) as 2%, 10%, and 50% of complete damage, respectively.

Another useful parameter for comparing the risk across geo-units is the Mean Damage Ratio (MDR) (FEMA 2003). It is the ratio of cost corresponding to each damage state, to the cost of new construction. The damage ratio is computed in two forms: (i) for each geo-unit for all model building classes known as MDR and (ii) for each model building class and all geo-units known as MDR total.

4.1.3 Human casualties

The human casualties are computed based on the simplified formula given below and accordingly modified to incorporate the injury level as follows:

$$C_i = C_l^{sd} + C_l^{nd} + C_l^{in} \tag{4.5}$$

The terms C_l^{Sd} denotes number of injuries/deaths occurring due to structural damage for each injury level, l; C_l^{nd} denotes number of injuries/deaths occurring due to nonstructural damage for each injury level, l; and C_l^{in} indicates number of injuries/deaths due to earthquake induced threats, such as fires, floods, landslides, etc., for each injury level, l where l represents the four levels of severity, light ($l = 1$), moderate ($l = 2$), heavy ($l = 3$), and death ($l = 4$).

SELENA v6.5 (Molina et al., 2010) estimates human losses generated from structural damage only. Human losses due to nonstructural damage imply death due to falling off shelves, fans, woodwork, electronic appliances, etc., are assumed to be rare and hence neglected. In view of considering the occupancy cases which depend strongly on the portion of the day, the number of injuries/deaths is calculated at early hours (2:00 a.m.), day time (10:00 a.m.), and rush hours (5:00 p.m.) to generate highest number of casualties for the people at home (early hours), the people at work (day time), and the people during commuting time (rush hour).

5. Results and discussion

The seismic hazard is computed for 2% and 10% probability in 50 years, covering 54 districts of the CIGP region by employing the two GMPE models derived by authors. Fig. 4.5 shows the seismic hazard across the study region for 10% and 2% probability of exceedance in 50 years in terms of PGA and PSA at 0.2 and 1.0 s, respectively. These results are validated using the previous works on hazard available in the literature at certain cities (Meerut, Agra, Kanpur, Lucknow, Allahabad, and Varanasi) in the CIGP (Table 4.7). The prediction of hazard from the ANN-based GMPEs derived in this study is comparable with the estimates by other researchers. The computed intensity levels are used to evaluate seismic risk in CIGP. Fig. 4.5 implies that the north-western districts of Uttar Pradesh state have high seismic hazard as the main frontal thrust is passing along these regions (indicated by brown (gray in print version) color in Fig. 4.5). Districts farther from the three major fault lines of Himalayas, MBT, MCT, and MFT are seismically least critical.

Further, from Table 4.7, it can be seen that only Nath et al. (2019) predicted higher values at both Lucknow and Varanasi, compared to the works of other researchers, which may be due to the fact that the attenuation relationship was derived considering only Nepal 2015, Nepal—Bihar 1934, and Jabalpur earthquakes which are in proximity to the study region. Also, a GMPE which does not belong to India, Abrahamson and Silva (2008), was included in calculating the hazard. Hence, the site-specific hazard values computed in this research are considered for computing seismic risk at the district level in CIGP.

Seismic risk in terms of building damage and loss is computed for 54 districts (geo-units) of the CIGP region for an MCE (2% probability in 50 years). Fig. 4.6 emphasizes the average damage probability of 10 MBTs across all geo-units, for five damage states: None, Slight, Moderate, Extensive, and Complete. The figure confirms that out of 10 building types, 4 (MMB, BSR, BMS, and BCS) have a high probability of complete damage. This affirms that inventories existing in CIGP have a high probability of collapse due to the transient type of construction, which need immediate retrofitting/ replacement measures to curb financial cost and human losses. Further, MBT, MMB (Table 4.4), covering 16.5% of total households, has a maximum collapse rate of nearly 60%. Also, for each MBT, by multiplying the percentage of buildings with the collapse probability, the total buildings in CIGP that might collapse for MCE and DBE earthquakes are 36% and 11%, respectively.

Further, the geo-unit Allahabad, which falls under seismic zone II of IS 1893:2016, is identified by loose soils with average V_{s30} of 185 m/s with the highest population among other geo-units. Our results acknowledge that Allahabad has expected economic losses (Fig. 4.7) around 16 billion dollars and the maximum number of homeless (Fig. 4.8) and uninhabitable dwellings (Fig. 4.9). These findings will drive policymakers

5. Results and discussion

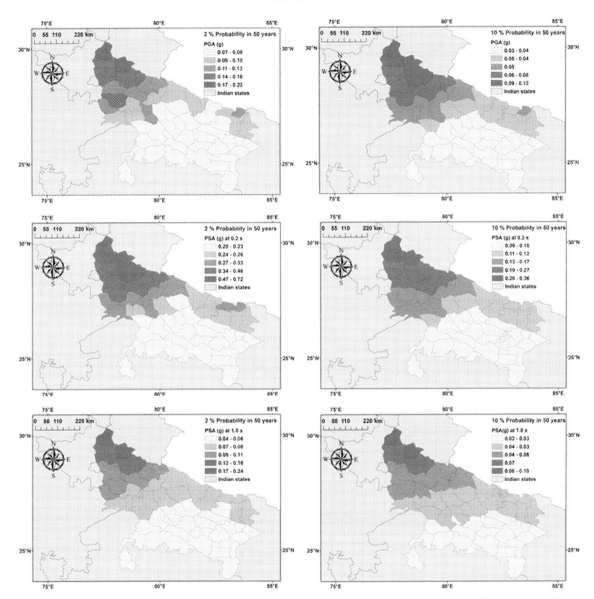

FIGURE 4.5 Seismic hazard distribution across 54 geo-units of Uttar Pradesh state in terms of PGA, PSA at 0.2 s and PSA at 1.0 s at the rock, for 475 and 2475 return periods.

and reinsurance corporations to reassess their approach and action plan.

For an MCE (2% in 50 years) and DBE (10% in 50 years), the anticipated human losses are about 12 and 4 lakhs, respectively, that are principally arising at night time (2:00 a.m.) (Fig. 4.10) and the predicted economic outlay is around 630 and 270 billion dollars, respectively.

During sensitivity analysis, the output variables monitored are economic and human

TABLE 4.7 Comparison of PGAs at 10% probability of exceedance in 50 years between different studies at six important cities.

References	PGA at 10% probability in 50 years					
	Meerut	Agra	Kanpur	Lucknow	Allahabad	Varanasi
Khattri et al. (1984)	–	–	–	0.05	–	0.05
Bhatia et al. (1999)	–	–	–	0.08	–	0.06
NDMA (2011)	0.06	0.06	0.03	0.04	0.03	0.02
Kumar et al. (2013)	–	–	–	0.04–0.07	–	–
Sitharam and Kolathayar (2013)	–	–	–	0.06–0.12	–	0.05–0.09
Nath et al. (2019)	–	–	–	0.17–0.18	–	0.09–0.11
This study	0.08	0.05	0.03	0.04	0.03	0.03

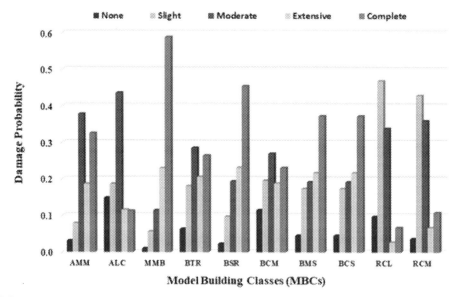

FIGURE 4.6 Histogram plot of the damage probabilities of 10 model building types adopted in this research at five damage states: None, Slight, Moderate, Extensive, and Complete.

losses. These are represented as Pie diagrams and Tornado plots. Fig. 4.11 represents a pie diagram showing the proportion of each sensitivity parameter to the total uncertainty in computing financial and human losses. The magnitude of the earthquake is highly sensitive, followed by source location.

The range of outputs for the assumed range of each input parameter can be represented by a tornado plot, which gives the two boundary

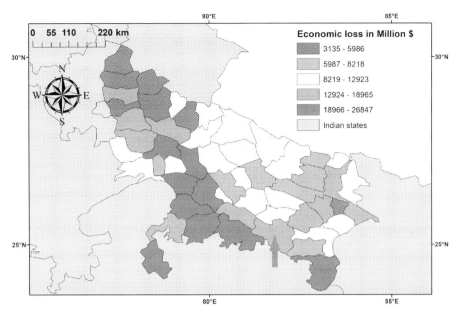

FIGURE 4.7 The anticipated economic outlay in the study area (CIGP) for MCE, with Allahabad district indicated by an arrow mark.

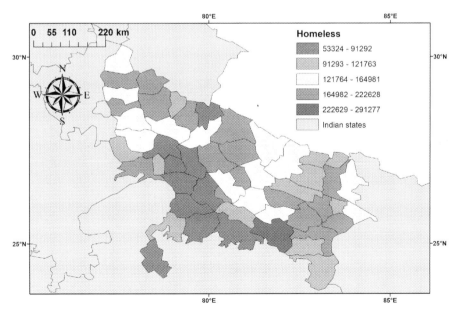

FIGURE 4.8 Allahabad district is characterized by the maximum number of Homeless in the study area (CIGP) for MCE.

68 4. Seismic risk for vernacular building classes

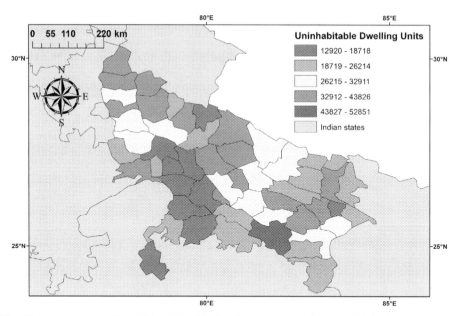

FIGURE 4.9 The maximum number of Uninhabitable dwelling units recorded in Allahabad district, in the study area (CIGP) for MCE.

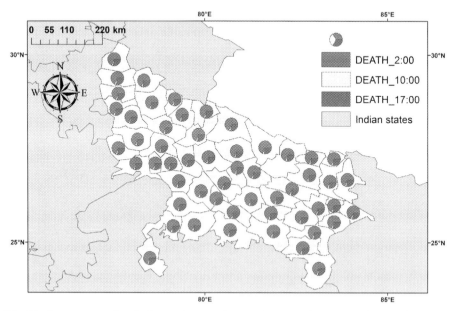

FIGURE 4.10 Pie charts representing the human losses at 2 a.m., 10 a.m., and 5 p.m. in a day, at all geo-units of the study region, dominated at 2 a.m.

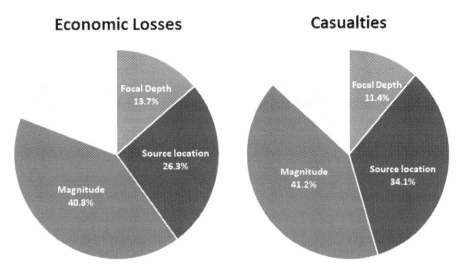

FIGURE 4.11 Pie diagram depicting the contribution of input parameters for computed economic and human losses in the study area, dominated by magnitude followed by GMPE.

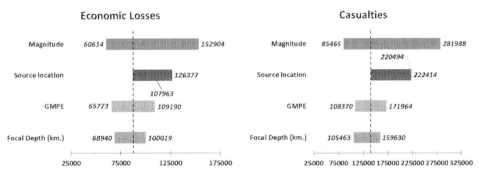

FIGURE 4.12 Tornado plots depicting the deviation in economic and human losses from the median (dotted line) for all parameters considered, dominated by magnitude followed by GMPE.

values, known as swing. The swing is denoted as the range of sensitivity of output values for the adopted range of inputs. The greatest swing was noticed for the variable magnitude, followed by source location (Fig. 4.12).

From the sensitivity study, we conclude that magnitudes followed by source location, GMPE, and focal depth, respectively, are in the order of high to low sensitivity. For the probable magnitude range of Mw 7.5–8.5, the expected economic outlay might vary between 60 and 150 billion dollars, and the human losses might vary from 0.8 to 2.8 lakhs, respectively.

6. Conclusion

Site-specific seismic hazard and earthquake loss estimation studies were computed for residential buildings in 54 geo-units (districts) of

Uttar Pradesh state in India, covering the CIGP region. The key findings for effective disaster mitigation are as follows:

- For the first time, the seismic hazard spectrum in terms of PGA and PSA (5% damping) at 25 periods for a return period of 475 and 2475 years is computed on a large scale at each district of Uttar Pradesh (54 in total).
- Earthquake loss estimation in terms of damage probability of building in entire CIGP, economic losses, and casualties in each district is reported for the first time in this study
- The results from this study reveal that MBTs and MMB (Mud Mortar Bricks with temporary roof) consisting of 16.5% of total buildings have the highest collapse probability of nearly 60%.
- Further, brick walls with stone roof (BSR) and brick walls with metal sheet roof (BMS) also have high extensive and collapse damage compared to other building groups. These three building typologies, which are located in the seismically critical region and proximity to the Himalayas, need immediate retrofitting/replacement
- Geo-unit Allahabad, even though lying in zone II as per IS 1893:2016, has the most number of homeless and uninhabitable dwellings due to the presence of poor soils. This finding will give insight to future rehabilitation and resettlement centers.
- For a future earthquake of magnitude between Mw 7.5 and 8.5, the expected economic outlay might vary from 60 to 150 billion dollars, and the human losses might vary between 0.8 and 2.8 lakhs.

The seismic risk assessment of IGPs shown has quantitatively given the related cost comparison for vernacular building classes, proving some of the types of roofing and mortar combinations as ineffective for low-cost housing initiatives, as they can be collapsed easily by earthquakes.

References

Abrahamson, N., Silva, W., 2008. Summary of the Abrahamson & Silva NGA ground-motion relations. Earthq. Spectra 24, 67–97.

ATC A 40, 1996. Seismic Evaluation and Retrofit of Concrete Buildings. Applied Technology Council, Redwood City report ATC-40.

Bagchi, S., Raghukanth, S.T.G., 2017. Seismic response of the central part of Indo-Gangetic plain. J. Earthq. Eng. 23, 183–207.

Bahadori, H., Hasheminezhad, A., Karimi, A., 2017. Development of an integrated model for seismic vulnerability assessment of residential buildings: application to Mahabad City, Iran. J. Build. Eng. 12, 118–131.

Basu, S., Nigam, N.C., 1977. Seismic risk analysis of Indian peninsula. In: Proceedings of Sixth World Conference on Earthquake Engineering, New Delhi, vol. 1, pp. 782–788.

Bhatia, S.C., Kumar, M.R., Gupta, H.K., 1999. A Probabilistic Seismic Hazard Map of India and Adjoining Regions.

Bilham, R., 2009. The seismic future of cities. Bull. Earthq. Eng. 7, 839.

BIS (Bureau of Indian Standards), 2016. Indian Standard Criteria for Earthquake Resistant Design of Structures: General Provisions and Buildings, IS 1893-Part 1, Fifth Revision. BIS, New Delhi, India.

Cattari, S., Curti, E., Giovinazzi, S., Lagomarsino, S., Parodi, S., Penna, A., 2004. A mechanical model for the vulnerability assessment of masonry buildings in urban areas. In: Proceedings of the VI Congreso Nazionale "L'ingegneria Sismica in Italia",. Genova, Italy.

Chadha, R.K., Srinagesh, D., Srinivas, D., Suresh, G., Sateesh, A., Singh, S.K., Pérez-Campos, X., Suresh, G., Koketsu, K., Masuda, T., Domen, K., 2016. CIGN, a strong-motion seismic network in central Indo-Gangetic plains, foothills of Himalayas: first results. Seismol Res. Lett. 87, 37–46.

Das, S., Gupta, I.D., Gupta, V.K., 2006. A probabilistic seismic hazard analysis of northeast India. Earthq. Spectra 22, 1–27.

Duzgun, H.S., Yucemen, M.S., Kalaycioglu, H.S., Celik, K.E., Kemec, S., Ertugay, K., Deniz, A., 2011. An integrated earthquake vulnerability assessment framework for urban areas. Nat. Hazards 59 (2), 917–947.

Federal Emergency Management Agency FEMA, 1997. NEHRP Guidelines for the Seismic Rehabilitation of Buildings, vol. 273. FEMA, Washington DC, United States.

Federal Emergency Management Agency FEMA, 2000. Commentary for the Seismic Rehabilitation of Buildings (FEMA356), vol. 7. Federal Emergency Management Agency, Washington, DC.

Federal Emergency Management Agency FEMA, 2003. HAZUS-MH. Multi-hazard Loss Estimation Methodology. Technical Manual., Washington DC, United States.

Federal Emergency Management Agency FEMA, 2005. Improvement of Nonlinear Static Seismic Analysis Procedures, vol. 440. FEMA, California, United States. Technical report, Applied Technology Council.

Freeman, S.A., 1975. Evaluations of existing buildings for seismic risk-A case study of Puget Sound Naval Shipyard. In: Proc. 1st US National Conference on Earthquake Engineering. Bremerton, Washington, pp. 113–122, 1975.

Freeman, S.A., 1978. Prediction of response of concrete buildings to severe earthquake motion. Spec. Publ. 55, 589–606.

Ghatak, C., Nath, S.K., Devaraj, N., 2017. Earthquake induced deterministic damage and economic loss estimation for Kolkata India. J. Rehabil. Civ. Eng. 5, 1–24.

Ghosh, B., Pappin, J.W., So, M.M.L., Hicyilmaz, K.M.O., 2012. Seismic hazard assessment in India. In: Proceedings of the Fifteenth World Conference on Earthquake Engineering-2012.

Jaiswal, K., Sinha, R., 2007. Probabilistic seismic-hazard estimation for peninsular India. Bull. Seismol. Soc. Am. 97, 318–330.

Kappos, A.J., Panagopoulos, G., Penelis, G.G., 2008. Development of a seismic damage and loss scenario for contemporary and historical buildings in Thessaloniki, Greece. Soil Dynam. Earthq. Eng. 28 (10–11), 836–850.

Kappos, A.J., Panagopoulos, G., 2010. Fragility curves for reinforced concrete buildings in Greece. Struct. Infrastruct. Eng. 6 (1–2), 39–53.

Khattri, K.N., Rogers, A.M., Perkins, D.M., Algermissen, S.T., 1984. A seismic hazard map of India and adjacent areas. Tectonophysics 108, 93–134.

Krawinkler, H., Seneviratna, G.D.P.K., 1998. Pros and cons of a pushover analysis of seismic performance evaluation. Eng. Struct. 20, 452–464.

Kumar, A., Anbazhagan, P., Sitharam, T.G., 2013. Seismic hazard analysis of Lucknow considering local and active seismic gaps. Nat. Hazards 69, 327–350.

Lang, D.H., Singh, Y., Prasad, J.S.R., 2012. Comparing empirical and analytical estimates of earthquake loss assessment studies for the city of Dehradun, India. Earthq. Spectra 28, 595–619.

Lang, D.H., 2013. Earthquake Damage and Loss Assessment—Predicting the Unpredictable.

Mahajan, A.K., Thakur, V.C., Sharma, M.L., Chauhan, M., 2010. Probabilistic seismic hazard map of NW Himalaya and its adjoining area, India. Nat. Hazards 53, 443–457.

McGuire, R.K., 2004. Seismic Hazard and Risk Analysis, EERI Original Monograph Series MNO-10. Earthquake Engineering Research Institute, Oakland, US, p. 221.

Molina, S., Lang, D.H., Lindholm, C.D., 2010. SELENA–An open-source tool for seismic risk and loss assessment using a logic tree computation procedure. Comput. Geosci. 36 (3), 257–269.

Molina, S., Lang, D.H., Meslem, A., Lindholm, C., 2015. SELENA v6.5-User and Technical Manual v6.5. Report No. 15-008, Kjeller (Norway) – Alicante (Spain).

Nanda, R.P., Paul, N.K., Chanu, N.M., Rout, S., 2015. Seismic risk assessment of building stocks in Indian context. Nat. Hazards 78, 2035–2051.

Nath, S.K., Adhikari, M.D., Maiti, S.K., Ghatak, C., 2019. Earthquake hazard potential of Indo-Gangetic Foredeep: its seismotectonism, hazard, and damage modeling for the cities of Patna, Lucknow, and Varanasi. J. Seismol. 1–45.

Nath, S.K., Thingbaijam, K.K.S., 2012. Probabilistic seismic hazard assessment of India. Seismol Res. Lett. 83, 136–149.

NDMA, 2011. Development of Probabilistic Seismic Hazard Map of India, Technical Report, Working Committee of Experts (WCE). National Disaster Management Authority (NDMA), New Delhi, India.

Parvez, I.A., Vaccari, F., Panza, G.F., 2003. A deterministic seismic hazard map of India and adjacent areas. Geophys. J. Int. 155, 489–508.

Petersen, M.D., Rastogi, B.K., Schweig, E.S., Harmsen, S.C., Gomberg, J.S., 2004. Sensitivity analysis of seismic hazard for the northwestern portion of the state of Gujarat, India. Tectonophysics 390, 105–115.

Porter, K., Scawthorn, C., 2007. Open-source Risk Estimation Software. Technical Report, SPA Risk. Pasadena, United States of America.

Prasad, J.S., Singh, Y., Kaynia, A.M., Lindholm, C., 2009. Socioeconomic clustering in seismic risk assessment of urban housing stock. Earthq. Spectra 25 (3), 619–641.

Prasad, J., 2010. Seismic Vulnerability and Risk Assessment of Indian Urban Housing. Ph.D Thesis. Indian Institute of Technology, Roorkee (IITR).

Raghucharan, M.C., Somala, S.N., 2018a. Generating site-specific ground motions for Delhi region for seismic vulnerability assessment of buildings—promoting disaster resilient communities. In: Urbanization Challenges in Emerging Economies: Resilience and Sustainability of Infrastructure. American Society of Civil Engineers, Reston, VA, pp. 290–299.

Raghucharan, M.C., Somala, S.N., 2018b. Seismic damage and loss estimation for central Indo-Gangetic Plains, India. Nat. Hazards 94, 883—904.

Raghucharan, M.C., Somala, S.N., Rodina, S., 2019. Seismic attenuation model using artificial neural networks. Soil Dynam. Earthq. Eng. 126, 105828.

Rong, Y., Pagani, M., Magistrale, H., Weatherill, G., 2017. Modeling seismic hazard by integrating historical earthquake, fault, and strain rate data. In: 16th International Conference on Earthquake Engineering.

Sharma, M.L., Malik, S., 2006. Probabilistic seismic hazard analysis and estimation of spectral strong ground motion on bed rock in north east India. In: 4th International Conference on Earthquake Engineering, pp. 12—13.

Sitharam, T.G., Kolathayar, S., 2013. Seismic hazard analysis of India using areal sources. J. Asian Earth Sci. 62, 647—653.

The MathWorks Inc, 2017. MATLAB R2017a - Academic Use, the Language of Technical Computing.

Verma, M., Bansal, B.K., 2013. Seismic hazard assessment and mitigation in India: an overview. Int. J. Earth Sci. 102, 1203—1218.

Wald, D.J., Allen, T.I., 2007. Topographic slope as a proxy for seismic site conditions and amplification. Bull. Seismol. Soc. Am. 97, 1379—1395.

Wyss, M., Gupta, S., Rosset, P., 2017. Casualty estimates in two up-dip complementary Himalayan earthquakes. Seismol. Res. Lett. 88 (6), 1508—1515.

Wyss, M., Gupta, S., Rosset, P., 2018. Casualty estimates in repeat Himalayan earthquakes in India. Bull. Seismol. Soc. Am. 108, 2877—2893.

Yeole, S., Curran, T.P., 2016. Investigation of post-harvest losses in the tomato supply chain in the Nashik district of India. Biosyst. Food Eng. Res. Rev. 21, 108.

Further reading

Registrar General & Census Commissioner, India, 2012. Census of India 2011. http://www.censusindia.gov.in. (Accessed 12 February 2018).

Comparative study between normal reinforced concrete and bamboo reinforced concrete

Subhojit Chattaraj

Department of Civil Engineering, MCET, Berhampore, West Bengal, India

1. Introduction

Concrete is widely used as construction material for its various advantages such as low cost, availability, fire resistance, etc., but it cannot be used alone everywhere because of its low tensile strength. So, generally, steel is used to reinforce the concrete. Though steel has a high tensile strength to complement the low tensile strength of concrete, use of steel should be limited since it is very costly and also much energy consuming in manufacturing process. Thus, a suitable substitute of this with a low cost, environmental friendly and also a less energy-consuming one is a global concern, especially for developing country. Addressing all these problems, bamboo is one of the suitable replacements of reinforcement bar in concrete for low cost constructions. Bamboo is natural, cheap, widely available, and most importantly strong in both tension and compression. The tensile strength of bamboo is relatively high and can attain 370 MPa, which makes bamboo an attractive substitute to steel in tensile loading applications. It grows very rapidly as most growth occurs during first year and becomes matured by fifth year. The strength of bamboo increases with its age and reaches to the maximum strength at 3–4 years and then starts to decline in strength. Bamboo is also an environmental friendly plant because it absorbs a lot of nitrogen and carbon dioxide in the air. Tensile property of bamboo is observed and evaluation of the use of bamboo as reinforcing bar in concrete with replacement of steel is done. The steel as a reinforcing material is a demand that is increasing day by day in most of the developing countries. There are situations when the production is not found enough to face the demand for steel. Hence, it is essential to have an alternative that is worth compared to steel. Bamboo is found in abundance, they are resilient, and hence, these can face the demand as a reinforcing material and become an ideal replacement for steel. The tensile strength property which is the main requirement of a reinforcing material is seen appreciable for bamboo, compared with other materials including steel. The structure of bamboo from its origin gives this property. The hollow tubular structure has high resistance against wind forces when it is in natural habitat. Working on the weak points of bamboo and bringing up an innovation of bamboo as a structural steel replacement would be a great alternative.

Traditionally, steel is used as reinforcement in concrete. But because of cost and availability, replacement of steel with some other suitable materials as reinforcement is now a major concern. Though bamboo has been used as a construction material, especially in developing country, until today its use as reinforcement in concrete is very limited due to various uncertainties. Since bamboo is a natural, cheap, and also readily available material, it can be a substitute of steel in reinforcing of concrete. In this chapter, aptness of bamboo as reinforcement in concrete will be evaluated. Bamboo sticks of varying cross-sections are used in this test. Also flexural strength test of bamboo reinforced beam is done to characterize the performance of bamboo as reinforcement. Comments on the selection and preparation of bamboo for reinforcing are given. Construction principles for bamboo reinforced concrete are discussed. The material used as reinforcement in concrete should show all the essential properties to make the element structurally active under load. In the case of steel, we manufacture steel to the desired proportion and test for the basic strength values as a quality check.

Similarly, the process must be done for bamboo too. Bamboo is found in nature; they have in different species. Each species differs in their characteristics, texture, thickness, and strength. Hence, it is essential to know which species is best for reinforcing and which is not.

Based on flexural strength, we will compare strength properties between normal reinforcement concrete and bamboo reinforcement concrete.

2. Methodology and materials

IS 10262:2009 has been used for the mixed design of concrete and mixing has been done by weight method.

2.1 Cement

Portland Slag Cement (PSC) has been used as cement (Tables 5.1 and 5.2).

2.2 Aggregates

These are the naturally occurring inert granular materials. They can be fine aggregates and coarse aggregates. The quantities of the fine aggregates and coarse aggregates are determined on the basis of the concrete mixes. The various tests and their respective results conducted on

TABLE 5.1 Oxide composition of PSC.

Oxide	Formula	Percentage (%)	Average (%)
Lime	CaO	44−46	45
Silica	SiO_2	26−30	27
Alumina	Al_2O_3	9.0−11.0	7.8
Iron oxide	Fe_2O_3	2.5−3.0	2.7
Magnesia	MgO	3.5−4.0	3.7
Sulfur dioxide	SO_3	2.4−2.8	2.6

TABLE 5.2 Properties of PSC

S. No.	Characteristics	Cement
1	Specific gravity	2.88
2	Initial setting time (min)	210
3	Final setting time (min)	300
4	Standard consistency (%)	50
5	Compressive strength of cement (N/mm^2)	50.42

fine aggregates and coarse aggregates as per Indian Standard Code IS 2386 are shown in Table 5.3.

2.3 Bamboo

The bamboo used in this investigation in Indian Timber Bamboo, collected in Kalyani city, Nadia. To prepare the sample, bamboo sticks of 640 mm length and around 8 mm diameter were cut and allowed to dry and season for 30 days. The thickness of the sample varies throughout its length since it is a natural material whose properties cannot be controlled strictly. The dimensions were measured at five points along the length of the sample to calculate the average dimension of the sample. During the period of season, all bamboo sticks were supported at regular interval to prevent warping.

2.4 Superplasticizer

An SNF (Sulfonated Naphthalene Formaldehyde)-based superplasticizer Sika Plast 2004 NS (A SIKA Product) was used in this investigation.

TABLE 5.3 Properties of aggregates.

Characteristics	Fine aggregates	Coarse aggregates
Specific gravity	2.45	2.50
Finenessmodulus	2.11	7.65
Water absorption (%)	1.37	3.24
Bulking (%)	24.4	—
Flakiness index (%)	—	33.53
Elongation index (%)	—	7.78
Grading	Zone II (IS 383)	—
Texture	—	Rough

The concrete mix design of the concrete is carried out as per Indian Standards code IS 10262: 2009. In this study, M30 concrete mix has been used.

Characteristic Compressive Strength at 28 days $= 30$ N/mm^2 Nominal maximum size of aggregate $= 20$ mm (Table 5.4).

3. Experimentation

In this chapter, we are mainly focusing on the comparison between normal reinforced concrete and bamboo reinforced concrete. We get different trial mixes so that we can compare the compression and tensile strength of the conventional reinforced concrete and bamboo reinforced concrete having different proportions.

Preparation of Steel Reinforcement Concrete:

- Perform shuttering work for making beam mold as per given dimension ($150 \times 150 \times 700$)
- Prepared reinforcement caging by providing 3 nos 8 mm dia bar as main reinforcement and 2 nos 8 mm dia bars as top reinforcement along with 6 mm dia bar as a stirrups@75 mm c\c.
- By making cement concrete as per given specification obtained from concrete mix design followed by 20 mm nominal size coarse aggregates, zone III fine aggregate, cement (PSC), admixtures (superplasticizer), and lab water.
- Perform casting process by using above prepared mold (Fig. 5.1).
- After completion of 72 h, we remove beam mold and dry and then place it in the curing tank (Table 5.5).

TABLE 5.4 Mix design quantity.

Cement (kg/m^3)	Sand (kg/m^3)	Coarse aggregates (kg/m^3)	Water/Cement ratio
350.22	657.43	1149.54	0.45

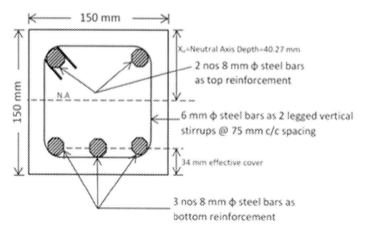

FIGURE 5.1 Steel reinforced concrete.

TABLE 5.5 Compressive strength test result for reinforced cement concrete.

Experiment No.	Trial mixes	Strength observed at 7 days (N/mm^2)	Strength observed at 14 days (N/mm^2)	Strength observed at 28 days (N/mm^2)
T1	A1	24	32	36
T2	A2	22	33	35
T3	A3	24	34	37

Preparation of Bamboo Reinforcement Concrete:

- Perform shuttering work for making beam mold as per given dimension (150×150×700)
- Prepared bamboo reinforcement caging by providing 3 nos 8 mm dia bamboo as main reinforcement and 2 nos as top reinforcement along with 6 mm dia steel bar as a stirrups @ 75 mm c\c.
- Using coal tar (alkatra) as waterproof coating over the bamboo steak. By making cement concrete as per given specification obtained from concrete mix design followed by 20 mm nominal size coarse aggregates, zone III fine aggregate, cement (PSC), admixtures (superplasticizer), and lab water.
- Perform casting process by using above prepared mold.
- After completion of 72 h, we remove beam mold and dry and then place it in the curing tank (Fig. 5.2).

4. Results

The compressive strength of reinforced cement concrete and bamboo reinforced concrete is checked in 7, 14, and 28 days. The results are given below (Table 5.6, Figs. 5.3–5.6).

The flexural strength of reinforced cement concrete and bamboo reinforced concrete is checked in 7, 14, and 28 days. The results are shown below (Tables 5.7 and 5.8).

4. Results

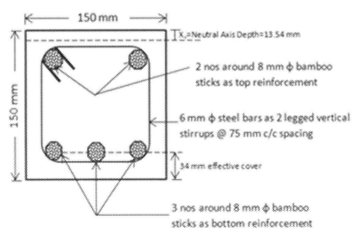

FIGURE 5.2 Bamboo reinforced concrete.

TABLE 5.6 Compressive strength test result for bamboo reinforced cement concrete.

Experiment No.	Trial mixes	Strength observed at 7 days (N/mm^2)	Strength observed at 14 days (N/mm^2)	Strength observed at 28 days (N/mm^2)
T1	B1	24	32	36
T2	B2	22	33	35
T3	B3	24	34	37

FIGURE 5.4 Steel reinforcement detailing.

FIGURE 5.3 Compressing test.

FIGURE 5.5 Bamboo reinforcement detailing.

FIGURE 5.6 Flexural strength.

TABLE 5.7 Flexural strength test result for reinforced cement concrete.

Experiment No.	Trial mixes	Strength observed at 7 days (N/mm^2)	Strength observed at 14 days (N/mm^2)	Strength observed at 28 days (N/mm^2)
T1	A1	5.33	5.97	6.23
T2	A2	5.66	5.89	6.32
T3	A3	5.12	5.62	6.75

TABLE 5.8 Compressive strength test result for bamboo reinforced cement concrete.

Experiment No.	Trial mixes	Strength observed at 7 days (N/mm^2)	Strength observed at 14 days (N/mm^2)	Strength observed at 28 days (N/mm^2)
T1	B1	2.22	2.32	2.36
T2	B2	2.25	2.37	2.49
T3	B3	2.29	2.35	2.53

5. Conclusion

- Several samples of bamboo and steel were tested and analyzed to examine their flexural properties.
- This test was carried out on mild steel including bamboo. Flexural test was carried out on 8 mm mild steel bars.
- Bamboo also was prepared to about equivalent dimension having cross-sectional area of 150 mm^2.
- The flexural test result indicates that bamboo unlike steel has a very poor flexural property and undergoes brittle failure when loaded.
- This is a huge disadvantage of using bamboo as a structural member in building construction.
- Therefore, the study concludes that due to the minimal breaking force (fb) of bamboo, it cannot be employed as a main structural member in buildings and other heavy engineering works but can be used for partition walls, ceilings, roofs, and other areas of lightweight engineering construction that is not heavy load bearing.

Further reading

Archila, H., Kaminski, S., Trujillo, D., Zea Escamilla, E., Harries, K.A., 2018. Bamboo reinforced concrete: a critical review. Mater. Struct. Constr. 51 (4). https://doi.org/10.1617/s11527-018-1228-6.

Hebel, D.E., Javadian, A., Heisel, F., Schlesier, K., Griebel, D., Wielopolski, M., 2014. Process-controlled optimization of the tensile strength of bamboo fiber composites for structural applications. Compos. B Eng. 67 (December 2017), 125–131. https://doi.org/10.1016/j.compositesb.2014.06.032.

Ingole, A., Gawande, S., Bambode, V., Khobragade, A., 2020. A review of bamboo as a reinforcement material in slab panel in modern construction. Int. J. Eng. Appl. Sci. Technol. 04 (09), 129–133. https://doi.org/10.33564/ijeast.2020.v04i09.015.

Raheem, S.B., Awogboro, O.S., Aderinto, S.J., August 2015. Investigation-into-the-Flexural-Properties-of-Bamboo-Reinforced-Concrete-Beam.docx no.

Siddika, A., Al Mamun, M.A., Siddique, M.A.B., 2017. Evaluation of bamboo reinforcements in structural concrete member. J. Constr. Eng. Proj. Manag. 7 (4), 13–19. https://doi.org/10.6106/JCEPM.2017.7.4.013.

CHAPTER 6

The power of the $GP\text{-}ARX$ model in CO_2 emission forecasting

Elham Shabani[1], Mohammad Ali Ghorbani[2], Samed Inyurt[3]

[1]Department of Agriculture Economics, Faculty of Agriculture, University of Tabriz, Tabriz, Iran; [2]Water Engineering Department, Faculty of Agriculture, University of Tabriz, Tabriz, Iran; [3]Department of Geomatics Engineering, Faculty of Engineering and Architecture, Tokat Gaziosmanpasa University, Tokat, Turkey

1. Introduction

Rapid global economic growth from $ 38.49 trillion to $ 82.44 trillion over the past two centuries and high fossil fuel energy consumption from the industrial revolution led to many environmental problems, including global warming and climate change (World Bank, 2019). However, for sustainable development, considering the environmental effects of economic growth is a critical issue. Climate change has led to adverse events such as droughts, floods, disruption of natural ecosystems, and health problems. So, one of the emerging issues of the countries is to take appropriate policies to reduce climate change and global warming (Olale et al., 2018).

Greenhouse gas emissions are known as the significant reason for global warming. The IPCC designates that GHG emissions increased 1.6% annually over the past 3 decades and predicts that GHG emissions will rise by 25%—90% in 2030 compared with those of 2000. The average global temperature has also been increased by 4.6°C over the past 100 years (IPCC, 2017). In 2013, the cost of air pollution was 5.5 million deaths in the world, and a wealth of over 5 trillion and $ 110,000 was lost in the global prosperity (World Bank, 2015). The economic consequences of global warming will be much higher than those of the World War I and World War II (Stern, 2007).

Among the greenhouse gases, carbon dioxide (CO_2), with the highest share (75%), is the most significant greenhouse gas. CO_2 emission has a critical role in global warming, reducing agricultural products and industrial productivity, immigration, and poses significant risks to living organisms and human health (Fang et al., 2018). That's why one of the most fundamental challenges facing policymakers in the world is providing appropriate solutions to reduce CO_2 emissions.

In 2018, the global CO_2 emission was 37.1 billion tons. Iran, with 672 million tons, was the seventh-largest carbon emitter among the most carbon-intensive countries worldwide

Risk, Reliability and Sustainable Remediation in the Field of Civil and Environmental Engineering
https://doi.org/10.1016/B978-0-323-85698-0.00013-7

(World Atlas of Carbon, 2018). Iran has the first rank in carbon emission between Middle Eastern and North African countries and its neighbors. In Iran, \$34 thousand billion (48.2% of GDP) is spent on the consequences of air pollution. One-tenth of Iran's deaths are due to air pollution, and 26,000 deaths attributed to air pollution are recorded annually in Iran (World Health Organization, 2017).

Ian is in an unfavorable situation in terms of carbon emissions, and making appropriate policies to reduce CO_2 emission to fight global warming is an urgent issue for Iran. To take a CO_2 emission reduction policy and evaluation its efficiency, awareness of the amounts of carbon emission before imposing any policy is essential. In this term, a prediction is knowing as the first and critical key. Knowing how the situation of CO_2 emission in the future years would a sign for a policymaker to take accurate policies. What has essential is taking an exact model to predict carbon emission, which in this term, performance metrics will help.

Due to the severe hazards of CO_2 emission and the importance of its prediction as a fundamental key to programs for global warming reduction, various studies have focused on predicting carbon emission. In these studies, various methods were applied, and the effort was to develop a prediction model with high accuracy. In this way, besides traditional models such as Autoregressive Integrated Moving Average (*ARIMA*), new Artificial Intelligence (*AI*) and Machine Learning (*ML*) models were developed. However, implementing a method will depend on the type and numbers of data. There are various research studies in terms of pollutant prediction in 2 decades.

For forecasting CO_2 emission in Taiwan, Lin et al. (2011) implemented the gray model. Pao et al. (2012) implemented the gray model, the nonlinear gray Bernoulli models (*NGBM*), and *ARIMA*. The lowest values of *MAPE, MAE,* and *RMSE* indicated the *NGBM* as the most accurate model. Lotfalipour et al. (2013) used

ARIMA and the gray model to predict CO_2 emission in Iran, and results showed that gray model with the lowest *RMSE, MAE,* and *MAPE* is the most accurate model. Samsami (2013), using Ant Colony Optimization (*ACO*) algorithm and data from 1981 to 2009 to forecast CO_2 emission for Iran, showed that the *ACO* model has high accuracy in a limited observation case. Sun et al. (2016) used Genetic Algorithm (*GA*) and backpropagation neural network (*BPNN*) and *GA-BPNN* model to predict CO_2 emission, and concluded that the combined model has high accuracy than the other examined models.

Wang and Ye (2017) implemented the *ARIMA*, gray model, and nonlinear gray multivariable model, and concluded that the latest model has higher prediction accuracy. Chen et al. (2018) to predict CO_2 developed various ML algorithms, including Gaussian processes (*GPs*), *SVM* (Support Vector Machine), *M5P*, and *BPNN*. According to performance metrics, SVM was the most accurate model. Fang et al. (2018) forecasted CO_2 emission in the United States of America, China, and Japan using improved Gaussian Process Regression based on the *PSO* algorithm. Hong et al. (2018) for forecasting the national CO_2 emissions in 2030 in South Korea implemented an optimized gene expression programming model using the metaheuristic algorithms.

Ding et al. (2020) implemented a discrete gray prediction model to predict CO_2 emission in China. Liu et al. (2020), using data from 1980 to 2015, and the *STIRPAT* model predicted production-based CO_2 emission in Beijing, China. Wen and Yuan (2020) implemented *BPNN* based on random forest and Particle Swarm Optimization (*PSO*) algorithm to forecast CO_2 emission in China's commercial department using data from 1997 to 2017. Qiao et al. (2020) to forecast CO_2 emission combined lion swarm optimizer and genetic algorithm to optimize the traditional least squares support vector machine (*LSVM*) model. They used data from 1965 to 2017. Shabani et al. (2020), to predict

CO_2 emission in the agriculture sector of Iran, implemented five different models, including Multiple Regression (*MLR*), Artificial Neural Network (*ANN*), Gaussian Process Regression (*GPR*), and Inclusive Multiple Model (*IMM*), and according to performance metrics and novel graphs concluded that IMM is the most accurate model.

The prediction has been a powerful tool for programming to reduce air pollution, and also it was a crucial option in energy policies related to low carbon. Implementing a highly accurate model to predict is the most crucial point, and the critical effort of the research studies was to introduce a most precise model. According to the literature review, newly developed *ML* methods have increased the accuracy of the models. Due to the significant carbon emission in Iran and taking emergence policies to reduce CO_2 emission, we developed a new hybrid model to predict CO_2 emission in the Iranian economy. To be a correct conclusion, we used new graphs and diagrams for choosing the most accurate model.

The rest of the study is organized as follows: Section 2 introduces the materials and methods. Section 3 shows the empirical results and Section 4 presents the conclusion.

2. Materials and method

2.1 Autoregressive Integrated Moving Average

The most famous and common model in the time series prediction is *ARIMA*, which also is called the Box–Jenkins model. *ARIMA* model is used only for stationary data, and based on the stationary statue in the various parts of a regression, there will be three different types of the ARIMA model: autoregressive model (*AR*), moving average model (*MA*), and integrated autoregressive moving average model (*ARIMA*).

Generally, an *ARIMA* model is shown as *ARIMA* (p, d, q), and for a stationary series can be shown as follows:

$$y_t = \sum_{i=1}^{p} \alpha_i y_{t-i} + \sum_{j=1}^{q} \beta_j \varepsilon_{t-j} \qquad (6.1)$$

where y_t is the parameter of the model that should be estimated, p shows the *AR* parameter, and would be determined by partial autocorrelation function (*PACF*); q denotes *MA* parameter and could be determined by autocorrelation function (*ACF*). When a series is stationary at level, d was equal to zero, and the *ARIMA* model would be changed to *ARMA* (p, q).

For an *ARIMA* model, first, the data were divided into two parts, including the modeling part (train) and the evaluation part (test). There are four steps in *ARIMA* model processing: checking the stationary of the series (determining the order of d); using *ACF* and *PACF* to determine the confidence lags order of p and q; and diagnostic checking with analysis the residuals' *ACF* and PACF. If the residuals were white nose, then we check the R^2 of the model. These steps were used for modeling parts of data. The final step is evaluation and validation. We used the built model to predict evaluation part of data and calculated the performance metrics (Yang et al., 2020).

2.2 Autoregressive Integrated Moving Average with external input

ARIMAX is usually used in signal and statistics processing. When in Eq. (6.1), y_t is also expressed as a function of exogenous variables (X); the model will be called the *ARIMAX* and can be shown as follows:

$$y_t = \sum_{i=1}^{p} \alpha_i y_{t-i} + \sum_{k=1}^{m} \theta_k X_{t-k} + \sum_{j=1}^{q} \beta_j \varepsilon_{t-j} \qquad (6.2)$$

where k denotes the confidence lag order for the exogenous variable. For a stationary series, the

ARIMAX model is shown as $ARMAX\ (p,\ q,\ m)$, and if q was equal to zero, the model could be shown as $ARX\ (p,\ m)$. In this model, in the second step, besides determining p and q, the optimal lag order for the exogenous variable should be determined. The diagnostic tests were including the residuals should be a white noses process; the highest R^2 and the lowest information criteria, including Akaike information criteria (AIC), Bayesian Information Criterion (BIC), and Schwarz information criteria (SIC). $ARMAX$ model has superiority to the $ARMA$ model, because it considers more information about the predicted variable (Yan and Chowdhury, 2013; Mei et al., 2019). For a variable such as CO_2 emission, for which there are some famous factors are affecting it, the $ARMAX$ model could be a correct model for prediction. For example, the famous hypothesis, Environmental Kuznets Curve (EKC), introduces economic growth and energy consumption as a potential factor affecting CO_2 emission (Ozjan, 2013; Kasman and Douman, 2015; Zafrious et al., 2017; Churchill et al., 2018; Adzawla et al., 2019).

2.3 Gaussian Process Regression

However, authors referred to Rasmussen and Williams (2006) as an in-depth treatment of Gaussian processes; Gaussian Process Regression (GPR) first appeared in ML research studies in Neal (1996) (Chandorkar et al., 2020). GPR is classified in the data-driven model category (Chen et al., 2013). GPR is a nonparametric full Bayesian learning algorithm, and it used to solve the supervised problems of the regression and classification (Mawloud et al., 2018). Gaussian Process (GP) models provide a probabilistic, practical, and principled approach to learning in kernel machines (Mattos et al., 2016). Compared to other ML algorithms, the advantages of the GPR model are hyperparameter estimation; easy to use; well working on small data;

considering a probability distribution over all possible values instead absolute values for each parameter in a function, and self-adaptive to enable superior parameter estimation. The output variable measurements of y can be shown as follows:

$$y = f(X(k)) + \varepsilon \qquad (6.3)$$

where X denotes input variable's measurements, ε presents noise with a Gaussian distribution and variance σ_n^2, and f denotes the unknown nonlinear function to be modeled. The space of functions has a prior probability determined as a Gaussian process owning covariance function cov $(x,\ x')$ and mean m(x) (Rasmussen and Williams, 2006).

$$f(x) \sim GP(m(x), cov(x, x')) \qquad (6.4)$$

The mean and variance of the predictive probability distribution $p\ (y_* | X,\ y,\ x_*)$ for a specified exemplar of predictor variables x_* can be shown as follows:

$$\begin{aligned} \widehat{y}_* &= m(x_*) + k_*^T \left(K + \sigma_n^2 I\right)^{-1} (y - m(x_*)) \\ \sigma_{y_*}^2 &= k_* + \sigma_n^2 - k_*^T \left(K + \sigma_n^2 I\right)^{-1} k_* \end{aligned} \qquad (6.5)$$

where K presents a covariance matrix with elements $[K]_{i,j} = $ cov $(x_i,\ x_j)$, k_* denotes a vector which has the ability to be introduced by $[k_*]_i = $ cov (x_i, x_*), $k_* = $ cov (x_i, x_*), and I shows the identity matrix. In GPR, unlike the traditional regression methods, the output of the model depends on the training dataset X, y.

By maximizing the log-likelihood function of the training datasets, the values of hyperparameters will be determined (Shabani et al., 2020):

$$\log p(y|X) = -\frac{1}{2} y^T \left(K + \sigma_n^2 I\right)^{-1} y$$

$$-\frac{1}{2} \log \left(|K + \sigma_n^2 I|\right) - \frac{n}{2} \log(2\pi) \qquad (6.6)$$

2.4 Support vector machine

The support vector machine (*SVM*) was built in 1992 by the Russian mathematician Vapnik, based on statistical learning theory. *SVM* is one of the learning methods with supervised learning, which is used to analyze data that are implemented for regression analysis and classification. The *SVM* is a classifier that is part of the kernel methods in *ML*. This learning system is used to both classify and predict the data fitness function to minimize errors in the data classification or fitness function. In linear data classification, an attempt is made to select a line that has a more reliable margin.

The purpose of the support vector machine is to determine the function $f(x)$ for the training patterns x so that it has the maximum margin of the training values y. In other words, the *SVM* is a model that fits a data-specified curve with a certain thickness so that the least error occurs in the experimental data. The two-class classification, including the classes being $y_i = +1, -1$, respectively, is considered in *SVM* (Saleh et al., 2016). In the present study, for solving the regression problems, an ε-insensitive loss function was used. In a regression, the general estimation function of *SVM* can be shown as follows:

$$f(x) = (\omega, \varphi(x)) + b \qquad (6.7)$$

where φ shows the nonlinear transformation function, and the values of ω and b are earned by minimizing risk regression:

$$R_{regression}(f) = C \sum_{i=0}^{\gamma} \Gamma\left(f(x_{i)} - y_i) + \frac{1}{2}||\omega||^2\right) \qquad (6.8)$$

where C is a constant value and determine the penalties of the estimated errors, Γ shows the cost function, and ω can be shown as follows (Kouziokas, 2020):

$$\omega = \sum_{i=0}^{\gamma} (\alpha_i - \alpha_i^*)\varphi(x_i) \qquad (6.9)$$

where α_i and α_i^* are the Lagrange multiplayer.

2.5 Autoregressive Gaussian process and autoregressive Gaussian process with exogenous variable

GPR, as a powerful nonparametric model often, is implemented for the advanced nonlinear data set. On the other side, the *ARIMA* models are extensively used in air pollution prediction research studies. When a series and the relationship between the variables are nonlinear, *ARIMA* model has no sufficient power to model these nonlinear relations. As mentioned, according to the *EKC* hypothesis, in the CO_2 emission prediction, some exogenous inputs are crucial. So, it is possible that an *ARX* model may have a better performance than AR. In the present study, the power of the *GP-ARX* model in CO_2 emission prediction will be investigated. However, in the various real and synthetic issues in the world, *GP-ARX* has better performance compared to the exiting *GPR* models (Requeima et al., 2019). If we consider an *ARX* process as below:

$$y_t = f\left(y_{t-1}, y_{t-2}, \ldots, y_{t-n}, x_t, x_{t-1}, \ldots, x_{t-m}\right) + \varepsilon$$

$$(6.10)$$

where y_t is the variable that should be predicted, t shows the time, x presents the exogenous inputs, n and m denote the optimal lags for the dependent variable and exogenous inputs, respectively, and ε denotes model's residuals and is assumed to be modeled as white Gaussian. In the *GP-ARX* model, $y_t = f(y_{t-1}, y_{t-2}, ..., y_{t-n}, x_t, x_{t-1}, ..., x_{t-m})$ will be a Gaussian Process, and the lagged values as inputs are the *ARX* section of the model.

2.6 Performance metrics

There are various performance metrics to evaluate the accuracy of the implemented models in the prediction research studies, including R-square (R^2), Root Mean Squared Error (*RMSE*), Standard Deviation (*SD*), and Mean Absolute Error (MAE). But, when the number of examined models is more, or the performance metrics are opposite, evaluation of the models to select the most accurate model will be difficult, and the probability of the incorrect decision would be more. The *Taylor* diagram, by considering three performance metrics, including Correlation Coefficient (*CC* or *R*), *RMSE*, and *SD* simultaneously, shows a specific point for each examined model. The lowest distance between the observed data and the individual point of the models determines the most accurate model (Shabani et al., 2020). The performance metrics which are used in the present study are shown below:

$$R = \sqrt{R^2} = CC$$

$$= \left(1 - \sum_{i=1}^{N} \left(\frac{CO_{2i}^{actual} - CO_{2i}^{predicted}}{CO_{2i}^{actual} - \overline{CO_2^{actual}}}\right)\right)^{0.5}$$

$$SD = \left(\frac{1}{N-1} \sum_{i=1}^{N} (error - \overline{error})\right)^{0.5}$$

$$RMSE = \left(\frac{1}{N} \sum_{i=1}^{N} error_i^2\right)^{0.5} \quad (6.11)$$

2.7 Data and study area

Iran is located in the "Middle East" and "southwestern Asia" region. Iran is the 18th biggest country, with 1.6 million square kilometers area, and with a population of 82.8 million is the 17th most populous country in the world. According to International Monetary Fund (*IMF*), in 2019, Iran with *GDP* equals $ 586,104 million was ranked 21st in the world, and according to the global carbon atlas, in 2018, it was the seventh largest carbon emitting country in the world. Some of Iran's cities, including Tehran, Tabriz, Isfahan, Mashhad, are among the most carbon emitting cities in the world. As an example, in 2019, Tehran was the second polluted province in the world.

In the present study, to predict CO_2 emission as an emerging issue in Iran, we used data including per capita CO_2 emission (shown as CO_2), per capita energy consumption (is shown as *EC*), and real per capita *GDP* (is shown as *GDP*). Required data were collected from the Iranian Statistic Center during 1961−2018. The data are divided into two train, and test parts, 80% of data is used for training, and 20% is used for the testing phase. The statistical information of the data is presented in Table 6.1.

3. Empirical results

In the present study, the *ARIMA, ARIMAX, SVM, GPR, GP-AR, and GP-ARX* models were implemented to predict CO_2 emission in Iran. The first step in implementing *ARIMA* group models is to evaluate the stationary of data. To this goal, we used Augmented Dickey-Fuller (*ADF*) and the Kwiatkowski−Phillips−Schmidt-Shin (*KPSS*) unit root test

3. Empirical results

TABLE 6.1 Statistical information of the data.

Variable	N. observation	Unit	Min	Mean	Max	Standard deviation
Per capita CO_2 emission	58	Metrics tone	1.62	4.66	8.69	2.01
Real per capita GDP	58	Billion Rial	2980.57	5636.63	10,265.71	1698.96
Per capita energy consumption	58	Million barrels equivalent of crude oil	1.87	18.81	32.29	9.41

Source: *Iranian Statistic Center (2019).*

using *Eviews11* software (Table 6.2). According to the results, variables have a unit root.

In this case, the first difference of the variables should be used to predict CO_2 emission, but to avoid missing long-run information as well as to avoid spurious regression, the cointegration between the variables was investigated via the Johansen cointegration test, and according to the results, there is at least one cointegration equation between the variables, and we could use the variables in original statue (Table 6.3).

The proper procedure of *ARIMA* and *ARIMAX* model was imposed. The structure of the models is shown in Table 6.4. Then, the models were calibrated, implementing the training data, and their training parameters are built using *Mathematica 11* software.

The Taylor diagram for the *AR, ARX, GPR, GP-AR, GP-ARX,* and *SVM* models is shown in Fig. 6.1. The Taylor diagram considering three performance metrics, including *R, RMSE,* and *SD,* gives a unique point for each model, which shows the difference between the predicted value and actual value. These unique criteria let us diagnose the most accurate model with high accuracy, and if there was any opposite in the performance metrics, do not impose our personal idea to choose the high accuracy model. According to Fig. 6.1, the *GP-ARX* model with the lowest distance from the observation (0.045) is the most accurate model

TABLE 6.2 Results of the unit root test.

Variable	ADF	KPSS	PP	Stationary statue
CO_2	I (1)	I (1)	I (1)	No stationary at level
GDP	I (1)	I (1)	I (1)	No stationary at level
EC	I (1)	I (1)	I (1)	No stationary at level

TABLE 6.3 Results of unrestricted cointegration rank test (trace).

Hypothesized No. of cointegration equation	Eigenvalue	Trace statistic	0.05 Critical value	Prob.**
None	0.251	35.832	35.192	0.042
At most 1	0.215	20.186	20.261	0.0502
At most 2	0.123	7.101	9.164	0.121

TABLE 6.4 The structure of the examined models.

Model	Input	Output
AR	$CO_{2(t-1)}$, $CO_{2(t-2)}$	$CO_{2(t)}$
ARX	$CO_{2(t-1)}$, $CO_{2(t-2)}$, $GDP_{(t)}$, $EC_{(t)}$	$CO_{2(t)}$
GPR	$GDP_{(t)}$, $EC_{(t)}$	$CO_{2(t)}$
GP-AR	$CO_{2(t-1)}$, $CO_{2(t-2)}$, $GDP_{(t)}$, $EC_{(t)}$	$CO_{2(t)}$
GP-ARX	$CO_{2(t-1)}$, $CO_{2(t-2)}$, $GDP_{(t)}$, $EC_{(t)}$	$CO_{2(t)}$
SVM	$GDP_{(t)}$, $EC_{(t)}$	$CO_{2(t)}$

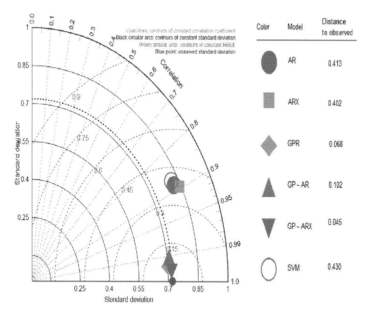

FIGURE 6.1 Taylor diagram for *the AR, ARX, GPR, GP-AR, GP-ARX,* and *SVM* models.

to predict CO_2 emission values, and it is followed *by GPR, GP-AR, ARX, AR,* and *SVM*.

Besides, for an accurate decision, we also present new colorful diagnostic diagrams in this section. Scatter diagrams for the *AR, ARX, GPR, GP-AR, GP-ARX,* and *SVM* models are presented in Fig. 6.2. This colorful novel diagram lets us to see the scatter of the predicted values best and make the correct decision. According to scatter diagrams of the models, the *GP-ARX* model with the highest correlation coefficient ($R = 0.999$) and the lowest *RMSE* (0.009) could predict CO_2 emission values with high accuracy.

The residual plots for the *AR, ARX, GPR, GP-AR, GP-ARX,* and *SVM* models are presented in Fig. 6.3. A model will predict CO_2 emission values with high accuracy if it has the highest error concentration around zero. As can be seen, in the *GP-ARX* models, the significant amounts of errors are around zero. So, the *GP-ARX* model's predicted values are nearer to the actual values, which results in high residuals around zero.

Besides, we calculated percentage of the residuals between −5 and 5 for the examined models (*LE* index). The model which has the highest errors between −5 and 5 will be the most accurate prediction model. The *LE* indices for the *AR, ARX, GPR, GP-AR, GP-ARX,* and *SVM* models are 69.23%, 70.41%, 94.64%, 92.30%, 100.00%, and 69.23%, respectively. So, the *GP-ARX* model with the highest residuals between −5 and 5 is the most accurate model for predicting CO_2 emission values.

We forecast CO_2 emission values for Iran during 2019−2024 using the *GP-ARX* model. Fig. 6.4 shows the observed and predicted values and forecasted values for the CO_2 emission in Iran using the *GP-ARX* model. High consumption of fossil fuels in economic activities, low fuel prices, and significant energy subsidies in Iran can

3. Empirical results

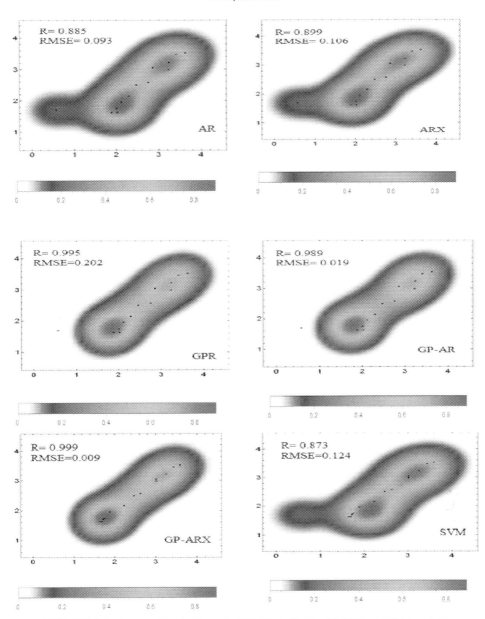

FIGURE 6.2 Scatter plots for the *AR, ARX, GPR, GP-AR, GP-ARX,* and *SVM* models.

be considered as the reasons for the upward trend in carbon emissions over the next 10 years. In the absence of new pollution reduction policies and the right decisions to reduce fossil fuel consumption, the upward trend in carbon emissions will continue.

88　　　　　　　　　　6. The power of the *GP-ARX* model in CO_2 emission forecasting

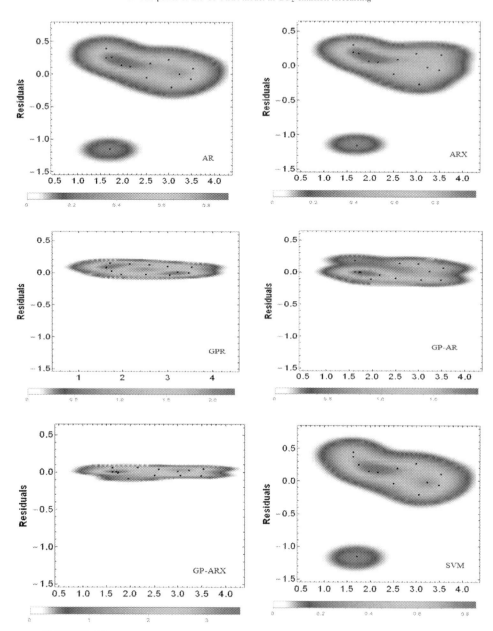

FIGURE 6.3　Residual plots for the *AR, ARX, GPR, GP-AR, GP-ARX*, and *SVM* models.

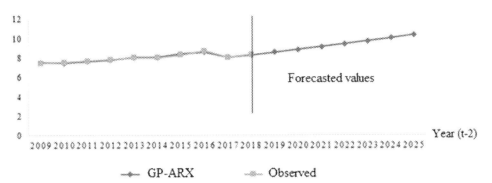

FIGURE 6.4 Observed and fitted values for CO_2 emission in Iran.

4. Conclusion

CO_2 emission, as a critical factor for global warming and climate change, has severe hazards to human health and natural ecosystems. Prediction is a powerful key for planning to reduce carbon emission, and implementing an accurate model in this term is a critical point. Due to the problematic situation of Iran in terms of CO_2 emission, proper policies to reduce carbon emission are an emerging issue. In the present paper, we examined the *GP-ARX* model's power in CO_2 emission prediction in Iran during 1961—2018 and compared it with the *AR*, *ARX*, *GPR*, and *SVM* models.

We also utilized the *Taylor* diagram, and new colorful diagnostic graphs. The performance of the *GP-ARX* model was higher than *ARX* and *GPR* models, and the developed model could increase the accuracy of the prediction. According to the Taylor diagram, the *GP-ARX* model has the lowest distance from the observation points (0.045). Besides, the *GP-ARX* model has the highest correlation coefficient (R = 0.999), the lowest *RMSE* (0.009), and the highest residuals between −5 and 5 (100%). So, the *GP-ARX* model could predict CO_2 emission values more precisely. So, it is recommended to policymakers, governments, and private and social institutes who are worried about the hazards of the CO_2 emission implemented the *GP-ARX* model to predict CO_2 emission values, to make accurate policies for mitigation its concentration. The forecasted values for CO_2 emission in Iran are on the rise, which highlights the need to further explore policies to reduce fossil fuel consumption and reduce CO_2 emission.

References

Adzawla, W., Sawane, M., Yusuf, A.M., 2019. Greenhouse gasses emission and economic growth nexus of sub-Saharan Africa. Sci. Afr. 3, e00065.

Chandorkar, M., Camporeale, E., Wing, S., 2020. Probabilistic forecasting of the disturbance storm time index: an autoregressive Gaussian process approach. Space Weather 15 (8), 1004—1019.

Chen, L., Huang, B., Liu, F., 2013. Nonlinear system identification with multiple and correlated scheduling variables. In: IFAC DYCOPS 2013 December 18—20, 2013. Mumbai, India.

Chen, S., Mihara, K., Wen, J., 2018. Time series prediction of CO_2, TVOC and HCHO based on machine learning at different sampling points. Build. Environ. 146, 238—246.

Churchill, S.A., Inekwe, J., Ivanovski, K., Smyth, R., 2018. The environmental Kuznets curve in the OECD: 1870—2014. Energy Econ. 75 (C), 389—399.

Ding, S., Xu, N., Ye, J., Zhou, W., Zhang, X., 2020. Estimating Chinese energy-related CO_2 emissions by employing a novel discrete grey prediction model. J. Clean. Prod. 259. https://doi.org/10.1016/j.jclepro.2020.120793.

Fang, D., Zhang, X., Yu, Q., Jin, T.C., Tian, L., 2018. A novel method for carbon dioxide emission forecasting based on improved Gaussian processes regression. J. Clean. Prod. 173, 143—150.

Hong, T., Joeng, K., Koo, C., 2018. An optimized gene expression programming model for forecasting the national CO_2 emissions in 2030 using the metaheuristic algorithms. Appl. Energy 228, 808–820.

IPCC, 2017. Climate Change. Synthesis Report Summary for Policymakers.

Kasman, A., Duman, Y., 2015. CO_2 emission, economic growth, energy consumption, trade and urbanization in new member and candidate countries: a panel data analysis. Econ. Model. 44, 97–103.

Kouziokas, N.G., 2020. SVM kernel based on particle swarm optimized vector and Bayesian optimized SVM in atmospheric particulate matter forecasting. Appl. Soft Comput. J. 93. https://doi.org/10.1016/j.asoc.2020.106410.

Lin, C.S., Liou, F.M., Huang, C.P., 2011. Grey forecasting model for CO_2 emissions: a Taiwan study. Appl. Energy 88 (11), 3816–3820.

Liu, Z., Wang, F., Tang, Z., Tang, J., 2020. Predictions and driving factors of production-based CO_2 emissions in Beijing, China. Sustain. Cities Soc. 53. https://doi.org/10.1016/j.scs.2019.101909.

Lotfalipour, M.R., Falahi, M.A., Bastam, M., 2013. Prediction of CO_2 emissions in Iran using Grey and ARIMA Models. Int. J. Energy Econ. Policy 3 (3), 229–237.

Mattos, C.L.C., Damianou, A., Barreto, G.A., Lawrence, N.D., 2016. Latent autoregressive Gaussian processes models for robust system identification. IFAC-papers Online 49–7 (2016), 1121–1126.

Mawloud, G., Farid, M., Danilo, C., 2018. Multi-step ahead forecasting of daily global and direct solar radiation: a review and case study of Ghardaia region. J. Clean. Prod. 201, 716–734.

Mei, L., Li, H., Zhou, Y., Wang, W., Xing, F., 2019. Substructural damage detection in shear structures via ARMAX model and optimal sub pattern assignment distance. Eng. Struct. 191, 625–639.

Neal, R.M., 1996. Bayesian Learning for Neural Networks. Springer, New York, N.Y.

Olale, E., Ochuodho, T.O., Lantz, V., El Armali, J., 2018. The environmental Kuznets curve model for greenhouse gas emissions in Canada. J. Clean. Prod. 184, 859–868.

Ozcan, B., 2013. The nexus between carbon emissions, energy consumption and economic growth in middle East countries: a panel data analysis. Energy Pol. 62, 1138–1147.

Pao, H.T., Fu, H.C., Tseng, C.L., 2012. Forecasting of CO_2 emissions, energy consumption and economic growth in China using an improved grey model. Energy 40, 400–409.

Qiao, W., Lu, H., Zhou, G., Azimi, M., Yang, Q., Tian, W., 2020. A hybrid algorithm for carbon dioxide emissions forecasting based on improved lion swarm optimizer. J. Clean. Prod. 244. https://doi.org/10.1016/j.jclepro.2019.118612.

Rasmussen, C.E., Williams, C.K.I., 2006. Gaussian Processes for Machine Learning. MIT Press, Cambridge, MA.

Requeima, J., Tebbutt, W., Bruinsma, W., Turner, R.E., 2019. The Gaussian process autoregressive regression model (GPAR). In: Proceedings of the 22nd International Conference on Artificial Intelligence and Statistics (AISTATS) 2019, Naha, Okinawa, Japan, vol. 89. PMLR.

Samsami, R., 2013. Application of Ant Colony optimization (ACO) to forecast CO_2 emission in Iran. Bull. Environ. Pharmacol. Life Sci. 2 (6), 95–99.

Shabani, E., Hayati, B., Pishbahar, E., Ghorbani, M.A., Ghahremanzadeh, M., 2020. A novel approach to predict CO_2 emission in the agriculture sector of Iran based on inclusive multiple model. J. Clean. Prod. 279. https://doi.org/10.1016/j.jclepro.2020.123708.

Shadab, A., Ahmad, S., Said, S., 2020. Spatial forecasting of solar radiation using ARIMA model. Remote Sens. Appl. Soc. Environ. 20. https://doi.org/10.1016/j.rsase.2020.100427.

Saleh, C., Dzakiyullah, N.R., Nugroho, J.B., 2016. Carbon dioxide emission prediction using support vector machine. IOP Conf. Ser. Mater. Sci. Eng. 114 (2016), 012148. https://doi.org/10.1088/1757-899X/114/1/012148.

Stern, N., 2007. The Economics of Climate Change: The Stern Review. Cambridge University Press, Cambridge.

Sun, W., Ye, M., Xu, Y., 2016. Study of carbon dioxide emissions prediction in Hebei province, China using a BPNN based on GA. J. Renew. Sustain. Energy 8 (4), 043101.

Wang, Z.X., Ye, D.J., 2017. Forecasting Chinese carbon emissions from fossil energy consumption using nonlinear grey multivariable models. J. Clean. Prod. 142, 600–612.

Wen, L., Yuan, X., 2020. Forecasting CO_2 emissions in Chinas commercial department, through BP neural network based on random forest and PSO. Sci. Total Environ. 718.

World Atlas of Carbon. http://globalcarbonatlas.org/en/content/welcome-carbon-atlas.

World Bank, 2019. World Development Indicators. Available from: http://www.worldbank.org/data/onlinedatabases/onlinedatabases.html.

World Bank, 2015. World Development Indicators, Available from: http://www.worldbank.org/data/onlinedatabases/onlinedatabases.html.

Yang, Q., Wang, J., Ma, H., Wang, X., 2020. Research on COVID-19 based on ARIMA model—taking Hubei, China as an example to see the epidemic in Italy. J. Infect. Public Health 13 (10), 1415–1418.

World Health Organization. https://www.who.int/news.

Yan, X., Chowdhury, N.A., 2013. Mid-term electricity market clearing price forecasting: a hybrid LSSVM and ARMAX approach. Electr. Power Energy Syst. 53, 20–26.

Zafeirious, E., Sofios, S., Partalidou, X., 2017. Environmental Kuznets curve for EU agriculture: empirical evidence from new entrant EU countries. Environ. Sci. Pollut. Res. 24 (18), 15510–15520. https://doi.org/10.1007/s11356-017-9090-6.

Further reading

Abdelfatah, A., Mokhtar, S.A., Sheta, A., 2013. Forecast global carbon dioxide emission using swarm intelligence. Int. J. Comput. Appl. 77 (12), 1–5.

Ahmed, K., Long, W., 2013. An empirical analysis of CO_2 emission in Pakistan using EKC hypothesis. J Int Trade Law Policy 12, 188–200.

IPCC, Climate Change, 2014. Synthesis Report Summary for Policymakers.

Tien Pao, H., Tsai, C.M., 2011. Modeling and forecasting the CO_2 emissions, energy consumption, and economic growth in Brazil. Energy Pol. 35, 58–69.

CHAPTER 7

Integrated sustainability impact assessment of trickling filter

Selvaraj Ambika[1,2]

[1]Department of Civil Engineering, Indian Institute of Technology Hyderabad, Telangana, India;
[2]Department of Climate Change, Indian Institute of Technology Hyderabad, Telangana, India

1. Introduction

The wastewater treatment is aimed to remove the pollutants before discharged back to the environment (Garg, 2012; Metcalf and Eddy, 2003). It has primary, secondary, and tertiary units working based on the physical, chemical, and biological principles (Garg, 2012; Metcalf and Eddy, 2003; Parker, 1999; Brown and Caldwell, 1980). Trickling filter is a secondary wastewater treatment unit that operates under biological especially aerobic conditions. The pretreated water is sprayed over the filter media on which the biofilm grows covering it. As the wastewater transports through the filter pores, the organic pollutants are aerobically degraded and the generated sludge will be disposed of (Garg, 2012; Metcalf and Eddy, 2003; Parker, 1999; Brown and Caldwell, 1980; Liu and Liptak, 1997; Martin and Martin, 1991; Water Environment Federation (WEF), 1996; Shammas and Wang, 2009; Wang et al., 2009; Wang et al., 1986; Eckenfelder and O'Connor, 1964; Pierce, 1978; Howland, 1958; Wang et al., 1982). The merits of the trickling filter are its resistance toward shock loadings, operable over a range of organic and hydraulic loading rates, efficient nitrification, and high-quality treatment. Industries introduce trickling filters to remove the organic pollutants from wastewater stream which required more contact area and time. It is also a cost-effective treatment resulting in alike performance compared to activated sludge process due to its natural aeration phenomenon (Garg, 2012; Metcalf and Eddy, 2003; Parker, 1999; Brown and Caldwell, 1980; Liu and Liptak, 1997; Martin and Martin, 1991; Water Environment Federation (WEF), 1996; Shammas and Wang, 2009; Wang et al., 2009; Wang et al., 1986; Eckenfelder and O'Connor, 1964; Pierce, 1978; Howland, 1958).

On one hand, the trickling filter treats the generated wastewater effectively; however, on the other hand, it impacts the environmental, economic, and socio-cultural components of sustainability positively and negatively. This impact is due to the various activities and the extensive consumption of resources and energy involved during the construction, operation, and dismantling phases of trickling filter. For

Risk, Reliability and Sustainable Remediation in the Field of Civil and Environmental Engineering
https://doi.org/10.1016/B978-0-323-85698-0.00003-4

© 2022 Elsevier Inc. All rights reserved.

instance, trickling filter treats the wastewater and avoids further damage to the environment; however, the energy consumption is directly correlated with climate change. Similarly, the presence of wastewater treatment units in any society, on the one hand, creates new jobs for the localities, on the other hand drops their land value. Knowing the positive and negative impact of trickling filter, a balance must be maintained between developmental activities and sustainability concerns (U.S. EPA, 1991). To achieve this balance, the impact assessment will work as a powerful tool (Ambika et al., 2021a; Ambika et al., 2021b). Hence, a detailed analysis is required on the impacts comprehensively to understand and quantify the impacts on sustainability. A few studies were conducted on the assessment of environmental and economic sustainability of wastewater treatment plants using a few indices (U.S. EPA, 1991; Senante et al., 2014; Cossio et al., 2020; Zhang and Ma, 2020). Briefly, the sustainability analysis of small-scale wastewater treatment was proposed by Senante et al. (2014) using a composite indicator including the social aspect which was missed in the previous studies. Along with the three aspects, and the seven scenarios different impact assessment staretegies of secondary treatment units were developed (Senante et al., 2014). Another study by Cossio et al. (2020) contextualized a set of sustainability indicators for small-scale treatment plants in low and lower middle—income countries. This study compiled the indicators from multiple literature sources and emphasized the institutional and technical dimension as a part of sustainability assessment (Cossio et al., 2020). Similarly, Zhang and Ma (2020) assessed the sustainability of a new sewage treatment plant in China. In this case study, the emergy analysis that accounts all the energy utilized in the construction and operation phases was done and accordingly the environmental sustainability was measured (Zhang and Ma, 2020). The literature review shows that scarce studies are reported aiming at a single treatment unit. Besides, to the best of the author's knowledge, no study was carried out to understand and measure the impacts of the trickling filter.

To match with the purpose and scope of having the treatment unit, the value of the spent time and monetary investment, for the future upgradation and extension, and finally for the decision-making process, the impact assessment is necessary for sustainable development. Knowing that the sustainability impact assessment (SIA) of individual treatment unit is vital to measure the impacts which act as the feedback for the further decision-making process, this chapter aims to assess the environmental, economic, and socio-cultural aspects of trickling filter considering three phases in its life cycle—construction, operation, and dismantling as shown in Fig. 7.1. This integrated SIA includes setting up of framework, analyzing the impacts in terms of qualitative and quantitative measures, and finally identifying the synergies, conflicts, and tradeoffs. The impact of the trickling filter as a secondary treatment unit of the sewage treatment plant was studied for a rural town—Hanamkonda at Warangal in the southern region of India.

2. Methodology

The tasks in this chapter include designing the trickling filter for the study area and conducting the integrated SIA. The tasks are selected based on the present stage in the SIA of trickling filter, the preciseness of the impact investigation, the depth of scrutiny, moderate time consumption, and available resources. The steps included in the integrated SIA of trickling filter are setting up of the framework, qualitative impact analysis, and impact quantification and assessment as shown in Fig. 7.2.

2.1 Designing of trickling filter

A trickling filter is designed for the rural town that has a population of 50,000 for the present study by adopting the NRC (National Research Council of US) method (Garg, 2012;

2. Methodology

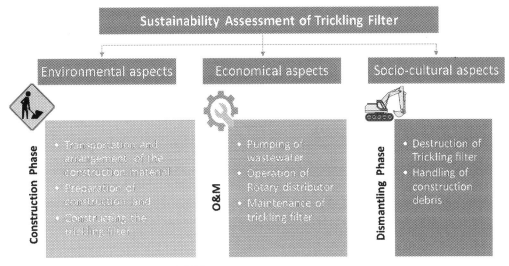

FIGURE 7.1 Life cycle of trickling filter.

FIGURE 7.2 Methodology followed in the integrated SIA of trickling filter.

Metcalf and Eddy, 2003). For the required efficiency, the trickling filter capacity can be calculated using the NRC equation as follows.

$$\eta = \frac{100}{1 + 0.0044\sqrt{\frac{Qy_i}{VF}}}$$

where η is the efficiency of trickling filter (%), Q is the flow rate of wastewater in MLD, y_i is the influent biochemical oxygen demand (BOD) in mg/L, V is trickling filter volume in ha.m, F is recirculation factor (for standard rate trickling filter, recirculation factor is 1), and Qy_i is the total BOD applied to trickling filter in kg/day.

2.2 Sustainability impact assessment

The goal of this chapter is to conduct the integrated SIA of trickling filter as the secondary treatment unit in wastewater for a rural town during construction, operation, and dismantling phases in its life cycle. Hence, the different perspectives of impact assessments are combined in the analysis and made it an integrated one. The preconditioning of the devising method was that one analysis could cover the areas that were not covered by another analysis. It was also assured that the methods in the integrated SIA were ensured to be flexible and easy to adapt to future treatment units. The SIA of trickling filter in this study included the following steps.

2.2.1 Setting up of the framework

The assessment of the impact on environmental, economic, and socio-cultural perspectives provides the details for decision-making in choosing and setting a wastewater treatment unit in a rural town. The environmental aspect considers the major three components of the environment, i.e., water, air, and land.

The economical aspect deals with the effects on the economy of the surrounding community by considering employment for the community and the cost of land of the surrounding area during its life cycle. The socio-cultural aspect is to secure the community's socio-cultural and spiritual needs equitably. It also deals with the health effects and psychological problems of the community during the life cycle of the project. The separate and partial assessments of the different aspects were compiled into an integrated SIA. Subsequently, the impact results were consolidated for the given study area to identify the intensity and direction of the potential impacts (Guidance, 2010; Warner and Preston, 1974; Gilpin, 2000).

2.2.2 Qualitative impact analysis

The possible activities under each component of sustainability were listed down during the different phases of trickling filter affecting the three pillars of sustainability. The further qualitative impact analysis was performed categorizing each activity (Ex: transport of construction materials) and criterions (air pollution) under different components. Moreover, the environmental, economic, and socio-cultural aspects of sustainability were assessed cross-cut into positive and negative, short-term and long-term, direct and indirect, and finally reversible and irreversible impacts (Warner and Preston, 1974; Gilpin, 2000, edoc). This method is useful for identifying and evaluating the interactions between various activities and environmental parameters.

2.2.3 Impact quantification and assessment

In the case of quantitative assessment, a weighing process will be applied to every scenario of activities and criterions for the deep understanding for the further decision-making process. The Leopold matrix method was adopted for quantifying the sustainability impact in this chapter. A matrix that consists of various types of activities in one dimension and sustainability components (environment, economic, and socio-cultural aspects) in another dimension was constructed for further evaluation of the interactions (Guidance, 2010). Each criterion can be given a quantitative rating or score based on its qualitative significance concerning the three aspects of sustainability. The final matrix gives efficient results as it considers both magnitude and significance. The sum of the magnitude and significance values of the environmental, economic, and socio-cultural impacts at different life cycle phases were assessed by considering 1—10 range scaling (Gilpin, 2000; edoc,; Anjaneyulu, 2007; Leopold et al., 1971).

Similarly, the sum of both acceptable and rejectable impact magnitude and significance values was calculated by considering the same factors. The impact analysis and decision-making rules in the Leopold matrix are as follows: (i) if the actual impact magnitude and significance values are within the acceptable limit, the project will be considered as having a negligible impact, (ii) in the event where the actual impact magnitude and significance values are above the acceptable limit values but less than the rejecting values, the project will be considered as causing impact; however, the impacts can be eliminated/controlled by mitigation measures. In this case, mitigation measures should be suggested to minimize the impacts, and (iii) if the actual impact magnitude and significance values are greater than the rejectable values, the project should be rejected (Gilpin, 2000, edoc,; Anjaneyulu, 2007; Leopold et al., 1971).

3. Results and discussion

The design detail of trickling filter is given in **Annexure A** and the dimensions of the trickling filter were given in Fig. 7.3. The impact was analyzed for the three life cycle phases of trickling filter which was designed for a rural town

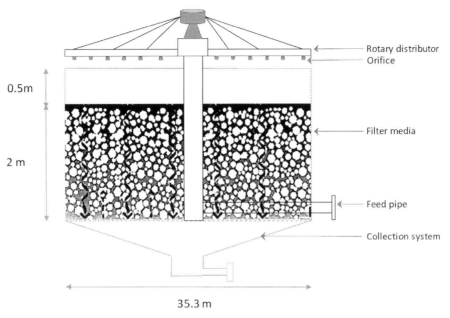

FIGURE 7.3 Trickling filter.

of India. The different activities during the construction, operation, and dismantling phases of trickling filter were analyzed for sustainability.

3.1 Setting the framework

The key sustainable aspects are subdivided into a set of activates and criterions that incorporate the basic pillars of sustainability viz a viz on environment, economy, and sociocultural aspects. At this point, the natural environment comprises soil, water, air, flora, and fauna. Economic component defines the economic condition of people who live over the surroundings of construction. The socio-cultural environment deals with the lifestyle of people and includes beauty, values, attitude, and political issues (Metcalf and Eddy, 2003).

The construction of trickling filter includes land acquisition, collection, and construction materials. The transportation of materials. These materials such as cement, fine aggregate, coarse aggregate, steel, and types of machinery are acquired to the construction site by transportation activities. This further results in disturbances of the existing ecology of that surroundings. Construction operations produce noise pollution in the neighborhood. The construction activities can destruct the flora, fauna, clean air environment, and water bodies.

The operation phase includes treatment of water, collection of sludge, and power requirements for pumping and rotary distributor. As an aerobic treatment process, trickling filter produces sludge from which odor problems may arise during its operation phase. The continuous discharge of treated water can ruin the nearby surface water systems. This subsequently results in an unaesthetic appearance to the community and ends up losing the value of the land.

The dismantling phase includes the destruction of concrete well, removal of filter media, dismantling of the rotary distributor, and transportation of raw material. Besides, the temporary employment, permanent employment, and loss of employment during the construction, operation, and dismantling phases are possibly influencing the socio-cultural aspects of sustainability.

All these impacts are analyzed qualitatively and quantitatively to understand and make the decisions to minimize the effects.

3.2 Qualitative impact analysis

The identified activities under each phase and the components of sustainability were assessed qualitatively in detail. The objectives and criteria

identified in the framework were made sure to get reflected in the assessment process within the resources, capacities, and time frame available for the exercise. The criterions were qualitatively assessed in the perspectives of positive and negative, short and long term, direct and indirect, and reversible and irreversible impacts.

3.2.1 Positive and negative impacts

The aim of this section is to compare the positive and negative impacts in the different domains and to tease out the potential conflicts. The positive and negative impacts are listed in Table 7.1. The positive impacts yield effects like

TABLE 7.1 Positive and negative impact assessment.

	Positive impacts			Negative impacts		
Phases	Environmental	Economical	Sociocultural	Environmental	Economical	Sociocultural
Construction	Not significant	Provides temporary employment	Temporary employment eases the locals life	Transportation of construction material and construction activities cause air pollution	Initial investment for construction of trickling filter	Land acquisition can cause problem
		Rise in construction material demand		Loss of vegetation over the construction site	Indirectly affects the cost of land surrounding the trickling filter	Equipment operational sound during construction
O&M phase	Improvement of wastewater quality for disposal into water bodies	Provides permanent employment	Permanent employment enhance the locals life	Continuous long-term disposal of treated wastewater would affect the ecology of receiving water	Energy consumption for operation of trickling filter	Fly nuisance
	Protecting soil from wastewater contamination	Economical compared to other treatment processes				Odor problem
	Protecting ground water from contamination	Treated wastewater can be reused				
	Safe guard aquatic life by disposing treated wastewater into water bodies					
Dismantling	Not significant	Provides temporary employment	Reduces fly and odor nuisance	Dumping of dismantled construction waste	Loss of permanent employment	Unaesthetic view causes disturbance to society

pollution reduction and improvement in the living condition of people. Wastewater produced can be treated economically and reused for irrigation purposes. Negative impacts cause adverse effects on environment, for instance, activities such as construction cause local ecological imbalance.

3.2.2 Short-term and long-term impacts

Impacts that show effects for a short period come under short-term impacts. Noise pollution and air pollution during construction phase can be categorized into short-term impacts. Long-term impacts show effects over a long period. Improper management of sludge produced during the operation phase causes odor

problems to the nearby community. The short- and long-term impacts in this study are listed in Table 7.2.

3.2.3 Direct and indirect impacts

Direct impacts are impacts that directly affect the natural conditions of soil, air, water, and socioeconomic status of the people by project activities. The transportation of construction material directly affects the surrounding ecosystem's flora and fauna. Indirect impacts are activities that are not directly due to project activities. The construction of trickling filter in an area may lead to a reduction in the cost of land in the surroundings of that area as indicated in Table 7.3.

TABLE 7.2   Short and long-term impact assessment.

Phases	Short-term impacts			Long-term impacts		
	Environmental	Economical	Sociocultural	Environmental	Economical	Sociocultural
Construction	Construction equipment causes noise pollution	Provides temporary employment	Equipment operation causes disturbance to surrounding community	Construction activities disturb the landscape	Likelihood of reduction in land cost	Not significant
	Construction activities result in air pollution			Destruction of vegetation of during treatment plant construction		
O&M phase	Not significant	Not significant	Not significant	Improvement in disposal characteristics of wastewater into water bodies	Reuse of treated wastewater for agriculture is economical	Disturbance due to lose of permanent employment
				Pollutants accumulation in receiving water due to long-term disposal which may affect aquatic life		
Dismantling	Dismantling operation causes air pollution and noise pollution	Provides temporary employment	Disturbance to surrounding community due to air and noise pollution	Not significant	Not significant	Not significant
	Unaesthetic view created by structural debris	Lose of permanent employment				

100   7. Integrated sustainability impact assessment of trickling filter

TABLE 7.3   Direct and indirect impact assessment.

	Direct impacts			Indirect impacts		
Phases	Environmental	Economical	Sociocultural	Environmental	Economical	Sociocultural
Construction	Construction activities affect the aesthetics Material transportation causes air pollution Destruction of flora and fauna during construction	New employment generation	Land acquisition	Disturbance in landscape during mining activities	Indirectly reduces the surrounding land cost	Conflicts in community due to land acquisition
O&M phase	Improvement in disposal characteristics of wastewater	Permanent employment	Fly nuisance	Not significant	Not significant	Spreading of diseases
Dismantling	It causes air pollution and noise pollution		Loss of job disturbs the locals life	Not significant	Lose of permanent employment	Psychological disturbance

3.2.4 Reversible and irreversible impacts

The impacts that are recovered by nature itself are categorized under reversible impacts. Noise pollution is generated, while construction comes under reversible impacts. The irreversible impacts are effects that cannot be recovered by the natural environment easily. The natural landscape disturbed during the construction activities cannot be recovered back easily as listed in Table 7.4.

TABLE 7.4   Reversible and irreversible impact assessment.

	Reversible impacts			Irreversible impacts		
Phases	Environmental	Economical	Sociocultural	Environmental	Economical	Sociocultural
Construction	Noise pollution due to machinery in transport of materials and construction	Not significant	Temporary psychological/ health impacts	Disturbance in landscape during mining activities Transportation of construction material causes air pollution	Decrease in cost of surrounding land	Land acquisition
O&M phase	Not significant	Not significant	Not significant	Not significant	Not significant	Odor problem Fly nuisance
Dismantling	Unaesthetic view	Not significant	Psychological nuisance	Air and noise pollution during the destruction of trickling filter	Losing the permanent unemployment	Health and psychological disturbance

In a summary, this format of qualitative analysis maintained a certain framework and the limited substitutability among different forms of available complex qualitative impact assessment tools. It indicated the relationship between different activities and criterions and the network and interrelationship among them. Besides, it searched for the best possible relationship between impacts with different components of sustainability. For instance, the economy and society are highly interconnected with each other when employment opportunity is considered as one of the criteria in all three phases of the trickling filter.

3.3 Impact quantification and assessment

The qualitative assessment based on different classifications detailed the association between the actions involved and the sustainability aspects. After the enumeration of the potential impacts of the trickling filter, it is necessary to identify major synergies, conflicts, and trade-offs across the sustainability components. Further, a more detailed analysis is required specific to the three phases of trickling filter for identifying the most significant phase. All these required the quantitative assessment that involved measurement of impact on the three pillars. Because qualitative aspects (social) are not so easily quantifiable, some argue that economic factors will be given more weight in assessments and overshadow the potential ecological and social concerns, even though these impacts may be equally or more severe than the economic impacts. Unlike economic environment impacts, it is often difficult to assign monetary values to environmental and social impacts. The complexity of the quantification analysis will largely depend on the type of tools selected and the selected factors (Guidance, 2010). To overcome the difficulty and to apply the effective weighing processes, the study considered using the Leopold matrix for impact quantification.

The impact matrices were made for all the aspects of environment by considering all the impacts during the three phases of trickling filter. The impact matrix consisted of both magnitude and significance values of a particular impact on a particular aspect of environment. The scales considered for giving these values are shown in following Tables 7.5 and 7.6.

TABLE 7.5 Scale for magnitude of impact.

Index for impact (negative)	Impact magnitude
1	Not considerable
2	Extremely low
3	Very low
4	Low
5	Moderate
6	Above moderate
7	High
8	Relatively high
9	Very high
10	Extremely high

TABLE 7.6 Scale for significance of impact.

Index for impact (negative)	Impact significance
X	Global level (massive impact)
IX	Country level
VIII	State level
VII	Regional level
VI	Zonal level
V	Town level
IV	Institution level
III	Scheme level
II	Individual level
I	Not significant

By using these scales, three categories of impact matrices were made by simple brainstorming sessions, literature reviews, and observations. The first matrix included the impact values of the actual trickling filter project (impact matrix of trickling filter). The other two matrices included both acceptable and rejecting impact values for the scenario of this particular goal. These three matrices are shown below in Tables 7.7-7.9. At this point, the environmental component is split into physical and biological means considering the pollution in the air, water, land, and noise in the former and the destruction of ecological balance, vegetation, flora fauna, and the aquatic life in the latter.

From the impact matrix of trickling filter, the total impact magnitude value (M) was found to be 131, and the total impact significance value (S) was CLXVI (166). The impact magnitude and the significance were separately quantified

TABLE 7.7 Impact matrix of trickling filter.

Rc	Aspects	Impact	Negative impact	Long-term impact	Direct impact	Irreversible impact	Magnitude (M) $\sum_{r=1}^{r=n} A_r$	Significance (S) $\sum_{r=1}^{r=n} B_r$
1	Physical	Land pollution	−4 −vi	−4 −vi	−4 −vi	−4 −vi	−16	−xvi
2		Water pollution	−1 −i	−3 −v	−1 −i	−1 −i	−6	−viii
3		Air pollution	−5 −v	−3 −v	−5 −v	−5 −v	−18	−xx
4		Noise pollution	−3 −iv	−1 −i	−3 −iv	−1 −i	−8	−x
5	Biological	Aquatic life	−2 −v	−2 −v	−2 −v	−2 −v	−8	−xx
6		Ecological balance	−2 −v	−2 −v	−1 −i	−2 −v	−7	−xvi
7		Vegetation destruction	−2 −vi	−2 −vi	−2 −vi	−2 −vi	−8	−xxiv
8	Economical	Value of land	−6 −iv	−6 −iv	−6 −iv	−6 −iv	−24	−xvi
9		Employment	−1 −i	−1 −i	−1 −i	−1 −i	−4	−iv
10	Social	Diseases	−5 −iv	−5 −iv	−5 −iv	−5 −iv	−20	−xvi
11		Conflicts	−3 −iv	−3 −iv	−3 −iv	−3 −iv	−12	−xvi
Sum of all impact magnitudes and their significances							$\sum_{C=1}^{c=11}\sum_{r=1}^{r=4} A_r$ = −131	$\sum_{C=1}^{c=11}\sum_{r=1}^{r=4} B_r$ = −CLXVI

3. Results and discussion

TABLE 7.8 Acceptable impact matrix.

Rc	Aspects	Impact	Condition of assessment				Magnitude(M_a)	Significance(S_a)
			Negative impact	Long-term impact	Direct impact	Irreversible impact	$\sum_{r=1}^{r=n} A_r$	$\sum_{r=1}^{r=n} B_r$
1)	Physical	Land pollution	−1 −i	−1 −i	−1 −i	−1 −i	−4	−iv
2)		Water pollution	−1 −i	−1 −i	−1 −i	−1 −i	−4	−iv
3)		Air pollution	−1 −i	−1 −i	−1 −i	−1 −i	−4	−iv
4)		Noise pollution	−1 −i	−1 −i	−1 −i	−1 −i	−4	−iv
5)	Biological	Aquatic life	−2 −i	−2 −i	−2 −i	−2 −i	−8	−iv
6)		Ecological balance	−1 −i	−1 −i	−1 −i	−1 −i	−4	−iv
7)		Vegetation destruction	−1 −i	−1 −i	−1 −i	−1 −i	−4	−iv
8)	Economical	Value of land	−3 −iv	−3 −iv	−3 −iv	−3 −iv	−12	−xvi
9)		Employment	−1 −i	−1 −i	−1 −i	−1 −i	−4	−iv
10)	Social	Diseases	−1 −i	−1 −i	−1 −i	−1 −i	−4	−iv
11)		Conflicts	−5 −i	−5 −i	−5 −i	−5 −i	−20	−iv
Sum of all impact magnitudes and their significances							$\sum_{C=1}^{c=11} \sum_{r=1}^{r=4} A_r$ $= -72$	$\sum_{C=1}^{c=11} \sum_{r=1}^{r=4} B_r$ $= -\mathbf{LVI}$

and found to be following the ascending order, environment (71 and 114) > socio-cultural (32 and 32) > economic (28 and 20) aspects. In the environment, impact on biological components showed more magnitude and significance during the impact assessment. This phenomenon is because the impact matrix of trickling filter was formulated by giving the maximum preferences for health issues, the value of the land, air pollution, and land pollution due to various activities involved, irreversible nature, and

associated long-term consequences. It is further reflected in the physical environment and socio-cultural aspects of the sustainability assessment.

While formulating the acceptable and rejecting matrix, the measurement of the magnitude and significance was assured to have weightage based on the depth and extent of impact due to the criterions considered. As shown in Tables 7.8 and 7.9, the acceptable (M_a) and rejecting (M_r) magnitude were measured as 72 and 299,

7. Integrated sustainability impact assessment of trickling filter

TABLE 7.9 Rejecting impact matrix.

Rc	Aspects	Impact	Condition of assessment				Magnitude(M_r) $\sum\limits_{r=1}^{r=n} A_r$	Significance(S_r) $\sum\limits_{r=1}^{r=n} B_r$
			Negative impact	Long-term impact	Direct impact	Irreversible impact		
1	Physical	Land pollution	−10 −iv	−10 −iv	−10 −iv	−10 −iv	−40	−xvi
2		Water pollution	−10 −vi	−10 −vi	−10 −vi	−10 −vi	−40	−xvi
3		Air pollution	−4 −ii	−1 −iv	−7 −iv	−1 −ii	−13	−xii
4		Noise pollution	−4 −ii	−1 −iv	−7 −iv	−1 −ii	−13	−xii
5	Biological	Aquatic life	−10 −vi	−10 −vi	−8 −vi	−5 −iv	−33	−xxii
6		Ecological balance	−10 −vi	−10 −vi	−10 −vi	−10 −vi	−40	−xxiv
7		Vegetation destruction	−7 −v	−7 −v	−7 −v	−7 −v	−28	−xx
8	Economical	Value of land	−5 −iv	−10 −iv	−5 −iv	−10 −iv	−30	−xvi
9		Employment	−5 −iv	−10 −iv	−5 −iv	−10 −iv	−30	−xvi
10	Sociocultural	Diseases	−3 −iv	−3 −iv	−3 −iv	−3 −iv	−12	−xvi
11		Conflicts	−5 −v	−5 −v	−5 −v	−5 −v	−20	−xx
Sum of all impact magnitudes and their significances							$\sum\limits_{C-1}^{c=11}\sum\limits_{r-1}^{r=4} A_r$ $= -299$	$\sum\limits_{C-1}^{c=11}\sum\limits_{r-1}^{r=4} B_r$ $= -CXC$

and acceptable (S_a) and rejecting (S_r) impact significance and as LVI (56) and CXC (190), respectively. Considering the magnitude, the sustainability components followed an order, environment > sociocultural > economy, whereas, in the view of significance, the environment occupied the topmost and had economy as the second and sociocultural component as the least in the list. In these acceptable and rejecting matrices, impact on the physical environment was higher than the biological environment.

The analysis showed that the magnitude and significance calculated from the matrix of trickling filter were in between the acceptable and rejecting values. The individual analysis of biological and physical environment, economic, and sociocultural aspects also witnessed a similar result. This reveals that the trickling filter can cause impacts; however, the impacts can be controlled or eliminated by mitigation measures.

To find the key activities and criterions, the precise analysis was done on the individual

TABLE 7.10 Decision-making on mitigation measures.

Aspects	Magnitude				Significance			
	Ma	M	Mr	Average	Ma	M	Mr	Average
Physical environment	16	48	106	61	16	54	56	36
Biological environment	16	23	101	59	12	60	66	39
Economical	16	28	60	38	20	20	52	36
Sociocultural	24	32	72	48	8	32	46	27

activates during the three phases of trickling filter and sustainability components. The average of acceptable and rejecting values was calculated for every scenario as shown in Table 7.10. The magnitude and significance of the actual matrix of trickling filter were compared with the average values. The scenarios for which the actual value is higher than the average were considered as the crucial criterion and included for further mitigation measures. Accordingly, the mitigation measures were suggested to improvise the scenarios so that they can end up reducing the magnitude and significance of the impact on the sustainability components.

3.4 Mitigation measures

The Leopold matrix analysis showed that the trickling filter can be adopted as a secondary treatment unit in the wastewater treatment plant in Hanamkonda town. Even though it is having impact values within the permissible values, there will be some negative impacts that disturb the surrounding community. To avoid such negative impacts, mitigation measures must be suggested. These mitigation measures were aimed to trade off the negative impacts (intense pollution) against the positive impacts of another criterion (income growth). It can be of (i) full compensatory method where the weak

performance of one criterion versus good outcomes of another will be compared and trade-offs will be suggested to eliminate the negative impacts, (ii) partial compensatory which sets limits to compensate, and (iii) noncompensation which does not trade off. These three methods result in strong, moderate, and weak sustainability, respectively. Generally, weak sustainability allows natural or environmental capital to be traded off against produced or manufactured capital, while strong sustainability does not allow for such substitutions (Guidance, 2010).

The findings from the detailed qualitative and quantitative analysis were (i) magnitudes of all the scenarios were less than the average and so fell in the acceptable region and (ii) significance of environment and sociocultural aspects was inclined toward the rejection values. Further numerical analysis suggested that the maximum attentiveness should be on the biological environment followed by the physical environment and socio-cultural aspects.

The destruction of the biological environment was majorly due to the destruction of vegetation and ecological balance during the construction phase and aquatic life damage during the operation phase, whereas the impact on the physical environment was influenced due to the air and land pollution during the three phases of trickling filter. In the sociocultural aspect, health issues and conflicts due to various reasons were the vital points to be included. These positive and negative feedback loops were connected to get the best possible mitigation measures. Accordingly, the trade-offs were identified between multiple scenario analyses, for instance, increase in the use of energy resources versus a decline in human capital stock (Guidance, 2010).

It was made sure that the reforms that were stipulated at the end of mitigation measures can eliminate environmentally harmful subsidies and also have positive ecological, economic, and sociocultural outcomes. Maintenance of ecological balance along with the welfare of certain labor groups and communities is an example scenario. However, the complete elimination of negative impacts on economic aspects was not possible, if the prioritization was on the safety of environment and sociocultural aspects. The example scenario is the application of dust control strategies versus community health issues like respiratory problems. It is also known that the regulations to control polluting emissions will have positive environmental effects but possibly also negative impacts on society and the economy.

Based on these assessments, the following mitigations measures were provided.

(i) The site can be selected accordingly to have minimum or no damage to the vegetation and ecology. It is better to select a site of installation farther to the community and in a barren land, so that effect on community and vegetation will be minimized

(ii) The air pollution during the construction phase can be eliminated by having dust pollution control measures like a water spraying system

(iii) The wastewater and sludge must be treated to meet the standards of its disposal to the aqueous bodies to avoid further damage

(iv) Applying the renewable energy-based treatment units

(v) Compensation for the people who lost their job

(vi) To avoid odor and flies domed trickling filter is suggested instead of a conventional trickling filter.

Installation of trickling filter may lead to some irreversible and negative impacts which could not be eliminated by using mitigation measures. So, during the impact assessment, the environmental engineers must interact with the study and design teams to find out the alternatives which may be viable for implementation and operation.

4. Conclusion

This study assessed the impacts of all the activities in the three phases of trickling filter focusing the criterions in the pillars of sustainability. After setting up the study's framework, the detailed qualitative analysis on the positive and negative impacts, short- and long-term impacts, reversible and irreversible impacts, and finally direct and indirect impacts were deeply discussed and tabulated. Using the Leopold matrix, the impact was quantified followed by the mitigation measures suggested. This integrated SIA was aimed to improve the quality of construction, operation, and dismantling process of trickling filter. The qualitative and quantitative SIA can incorporate various kinds of information expressed in different units. For instance, quantitative figures such as monetary values, and physical qualities such as pollutant emissions, and a more qualitative measure of human capital and social values. The measures of different types of impacts were standardized and ranked or rated according to their perceived degree of importance. This assessment listed out the negative impacts concerning various activities and criterions during the installation and operation of trickling filter. Besides, it reduces the uncertainties around the environment, economic and sociocultural issues, and impacts of a treatment unit. The negative impacts were discussed in detail in terms of qualitative and quantitative measures and accordingly the mitigation measures were suggested. This study helps in preparing the feasibility report of a developmental project associated with treatment plants with the concern of sustainable development. As this study is limited to adopting only a few methodologies in impact analysis, further, it can be expanded with cost–benefit analysis to find out the most economical solution. However, the details obtained from the SIA of the trickling filter and the provided mitigation measures and alternative ideas to reduce impact are useful in applying the knowledge to other treatment units.

5. Annexure A

Design of trickling filter.
Design considerations of trickling filter:
Population of Hanamkonda = 50,000 persons.
Average per capita water demand = 135 lpcd
Sewage conversion factor = 0.75.
Initial BOD present in sewage (Y_i) = 150 mg/L.
Required effluent BOD (Y_e) = 25 mg/L.
Total sewage produced (Q) = Sewage conversion factor x Population x Average per capita water demand = 50,000 × 135 x 0.75 = 5.06 MLD

Efficiency of trickling filter (η) = $\dfrac{Yi - Ye}{Yi}$ × 100

$= \dfrac{150 - 25}{150}$ × 100 = 83.33%

NRC Equation $\eta = \dfrac{100}{1+0.0044\sqrt{\dfrac{Q\,Yi}{V\,F}}}$

Designing a standard rate trickling filter, F = 1.

$$83.33 = \dfrac{100}{1 + 0.0044\sqrt{\dfrac{5.06 \times 150}{V \times 1}}}$$

V = 0.367 ha m

Organic loading rate $= \dfrac{QYi}{V} = \dfrac{5.06 * 150}{0.367}$

= 2068 (OK) (Range: 900−2200 kg/Ha-m/day).
Volume of trickling filter (V) = 0.367 Ha-m = 3670 m^3

Total depth (H) = 2m.

Surface area of trickling filter = A = $\dfrac{V}{H}$

$= 3670/2 = 1835\text{m}^2$

Provide two nos (n) = 2
Area of each trickling filter (A^1) = $\frac{A}{n}$ = 1835/2 = 917.5 m^2

Diameter of trickling filter (d) = $\sqrt{\dfrac{917.5 * 4}{\pi}}$
= 34.17m = 35m (approximately)
Provide two numbers of each 35m dia and 2m effective depth.
Overall depth = 2 m + free board = 2 + 0.5 = 2.5m.
Provide one extra unit for standby.
Therefore, two trickling filters of depth = 2.5m
diameter = 35m.

5.1 Filter media

Depth of filter media = 2m.
Volume of filter media = $\left(\dfrac{\pi}{4} * 35^2\right) * 2$
= 1923.25 m^3
Total volume of filter media required = 3*1923.25 = 5769.75 m^3
Provide filter media of 5770 m^3 of size.
Volume of concrete required for the well.
Volume of concrete required for each filter = V_c
Assume thickness of wall = 30 cm = 0.3m. Inner diameter = 35m. Outer diameter = 35.3m.
$V_c = \dfrac{\pi}{4} * (35.3^2\text{-}35^2)*2.5 = 41.41$ m^3
Total volume of concrete required = V_c = 3*41.41 = 125 m^3 of concrete.
Provide M25 concrete.
Normal mix properties for M25 = 1:1:2 (concrete: fine aggregate: coarse aggregate).
Volume of cement required = $\frac{1}{4}$ * 125 = 31.25 m^3

Volume of fine aggregate (sand) required $= \frac{1}{4}$ * $125 = 31.25 \text{ m}^3$

Volume of coarse aggregate required $= \frac{2}{4}$ * $125 = 62.5 \text{ m}^3$

5.2 Pump

Required head (H) = 2.5m

Q = 5.06 MLD = 0.0585 m^3/s

Assume efficiency of pump = 80%

Power of pump required $P = \dfrac{\gamma * Q * H}{\eta} =$

$\dfrac{9810 * 0.0585 * 2.5}{0.8} = 1795.37 \qquad w = 1.8$

Kw = 2.41 hp.

5.3 Rotary distributor

Rotary distributor is to be designed for peak discharge.

Take peak flow factor = 2

Peak discharge = 2*0.585 m^3/s = 0.117 m^3/s.

Peak discharge for each trickling filter $= \frac{0.117}{2}$ = 0.0585 m^3/s.

Assume velocity of peak flow = 2 m/s.

Area of central column = A = Q/V = 0.0585/ 2 = 0.02925 m^2

$\dfrac{\pi}{4} * d^2 = 0.02925$

d = 0.19m = 20 cm.

Diameter of central column = 20 cm. Arms of rotary distributor = 4.

Discharge per arm = 0.0585/ 4 = 0.014625 m^3/s.

Length of arm $= \dfrac{dia\ of\ filter - 3}{2} = \dfrac{35 - 3}{2}$ = 16 m.

Assume velocity through arm = 1.2 m/s

c/s area of arm = 0.014,625/ 1.2 = 0.011875 m^2

Diameter of arm = d = 0.123m = 12.3 cm.

Provide 12 cm diameter arms.

Orifices of arms.

Assume 10 mm diameter orifice.

Coefficient of discharge = c_d = 0.65.

Assume water head causing flow = h = 1.5m.

Discharge through each orifice = q = C_d A* $\sqrt{2gh}$ = 0.65*π/4*(0.01^2)* $\sqrt{2 * 9.81 * 1.5}$ = 2.769*10$^{(-4)}$ m^3/sec.

No of orifices required for each arm = $\dfrac{0.014625}{2.769 * 10(-4)}$ = 53 orifices.

Under drain system.

Assume velocity through channel = 1 m/s.

Discharge = 0.0585 m^3/s.

Area of channel = Q/V = 0.0585/ 1 = 0.0585 m^2.

Width of channel = 0.2m.

Depth of channel = 0.0585/0.2 = 0.3m.

Slope of the channel bed = S.

$$Q = \dfrac{1}{N} * A * R^{2/3} * S^{2/3}$$

N = 0.018.

A = 0.2*0.3 = 0.06 m^2.

$R = A/P = \dfrac{0.06}{0.2 + 0.3 + 0.3} = 0.075.$

After solving for slope S = 9.73*10$^{(-3)}$ = 1 in 100.

Lateral drains connected to central canal are in semicircular shape.

Let us say diameter of lateral drain (D) = 10 cm = 0.1m.

Let us assume drains run at depth = 0.3D.

For d/D = 0.3, the values of q/Q = 0.196 and a/A = 0.252

q = 0.196*Q'

$Q^1 = \dfrac{1}{N} * A* R^{2/3}*S^{1/2}$

After substituting Q^1 = 0.00815 m^3/s,

q = 0.196*Q^1 = 0.0016 m^3/s

Number of lateral drains required = Q/ q = 0.065/0.0016 = 40.

Acknowledgments

This research did not receive any specific grant from funding agencies in the public, commercial, or not-for-profit sectors.

The corresponding author would like to acknowledge the PG students—Ramesh K, Sivakumar K, Jagadeesh N, Abhishek P, and Brian N, National Institute of Technology Warangal, India, for the initiation of the work. Also, the author would like to thank Science Engineering and Research Board (SERB), India, for their support for the computing facility under Startup Research Grant (File Number: SRG/2020/000793) and Seed Grant (letter dated May 15, 2020) from Indian Institute of Technology Hyderabad, India.

References

Ambika, S., et al., 2021a. Impact of social lockdown due to COVID-19 on environmental and health risk indices in India. Environ. Res. 196 (110932). https://doi.org/10.1016/j.envres.2021.110932. https://www.sciencedirect.com/science/article/pii/S0013935121002267.

Ambika, S., et al., 2021b. Life cycle sustainability assessment of crops in India. Curr. Res. Environ. Sustainab. 3 (100074). https://doi.org/10.1016/j.crsust.2021.100074. https://www.sciencedirect.com/science/article/pii/S2666049021000505.

Anjaneyulu, V.M., 2007. Environmental Impact Assessment Methodologies, second ed. BS Publications, Hyderabad.

Brown, Caldwell, 1980. Converting Rock Trickling Filters to Plastic Media: Design and Performance. EPA-600/2-80-120. Environmental Protection Agency, Washington, D.C.

Cossio, C., Norrman, J., McConville, J., Mercado, A., Rauch, S., 2020. Indicators for sustainability assessment of small-scale wastewater treatment plants in low and lower-middle income countries. Environ. Sustainab. Ind. 6.

Eckenfelder, W.W., O'Connor, D.J., 1964. Biological Waste Treatment. Pergamon Press, NY, pp. 221—247.

edoc.hu-berlin.de.

Garg, S.K., 2012. Environmental Engineering (Vol II), twenty fourth ed. Khanna publishers, pp. 283—313Y.

Gilpin, A., 2000. Environmental Impact Assessment. Cambridge university press, The Edinburgh building, Cambridge CB2 2RU, United Kingdom.

Guidance on Sustainability Impact Assessment, 2010. Organization for economic co-operation and development.

Howland, W.E., 1958. Flow over porous media as in a trickling filter. Engineering bulletin extension series No. 94. In: Proceedings of the 12th Industrial Waste Conference, 1957, vol. 42. Purdue University, Lafayette, Ind, pp. 435—465.

Leopold, L.B., Clarke, F.E., Hanshaw, B.B., Balsley, J.R., 1971. A Procedure for Evaluating Environmental Impact. United States Department of the Interior. Geological Survey Circular 645. Washington, D.C.

Liu, Liptak, 1997. Environmental Engineering Handbook, 2d ed. The CRC Press, LLC, Boca Raton Florida.

Martin, E.J., Martin, E.T., 1991. Technologies for Small Water and Wastewater Systems, p. 122 (New York, New York).

Metcalf, Eddy, 2003. Wastewater Engineering Treatment and Reuse, fourth ed. McGraw Hill, New York.

Parker, D.S., 1999. Trickling filter mythology. J. Environ. Eng. 125 (7), 618—625.

Pierce, D.M., 1978. Upgrading Trickling Filters. US Environmental Protection Agency, Washington, DC, p. 62. June 1978.

Senante, M.M., Gómez, T., Baserba, M.G., Caballero, R., Garrido, R.S., 2014. Assessing the sustainability of small wastewater treatment systems: a composite indicator approach. Sci. Total Environ. 497—498, 607—617.

Shammas, N.K., Wang, L.K., 2009. Aerobic and anaerobic attached growth biotechnology. In: Wang, L.K., Tay, J.H., Ivanov, V., Hung, Y.T. (Eds.), Environmental Biotechnology. The Humana Press, Inc., Totowa, NJ.

U.S. EPA, 1991. Assessment of Single Stage Trickling Filter Nitrification. EPA 430/09-91-005. EPA Office of Municipal Pollution Control, Washington, D.C.

Wang, M.H.S., Wang, L.K., Poon, C.P.C., 1982. Theoretical and Empirical Developments of Attached Growth Biological Systems, Part II. US Dept. Of Commerce, National Technical Information Service. Tech. Report No. PB82-216409, p. 38. June.

Wang, L.K., Wang, M.H.S., Poon, C.P.C., 1986. In: Wang, L.K., Pereria, N.C. (Eds.), Trickling Fllters, Chapter 8 in the Handbook of Environmental Engineering, vol. 3. Humana Press, Clifton NJ, pp. 361—426.

Wang, L.K., K Shammas, N., Hung, Y.T., 2009. Advanced Biological Treatment Processes. The Humana Press, Inc., Totowa, NJ.

Warner, M.L., Preston, E.H., 1974. A Review of Environmental Impact Assessment Methodologies. Office of research and development, US Environ Protection Agency, Washington, D.C, p. 20460.

Water Environment Federation (WEF), 1996. Operation of Municipal Wastewater Treatment Plants. Manual of Practice No. 11, fifth ed., vol. 2. WEF. Alexandria, Virginia.

Zhang, J., Ma, L., 2020. Environmental Sustainability Assessment of a New Sewage Treatment Plant in China Based on Infrastructure Construction and Operation Phases Emergy Analysis. Water.

Further reading

Jebelli, J., Madramootoo, C., Tesaye Gebru, G., 2017. A simplified matrix approach to perform environmental impact assessment of small-scale irrigation schemes: an application to mekabo scheme in Tigray, Ethiopia. American J. of Environ. Eng. 7 (2), 21—34.

CHAPTER
8

Critical soil erosion prone areas identification and effect of climate change in soil erosion prioritization of Kosi river basin

Aadil Towheed, Thendiyath Roshni

Department of Civil Engineering, National Institute of Technology Patna, Patna, Bihar, India

1. Introduction

The detachment of in situ soil particles by the three natural agents (water, wind, and terrain slope) and transportation of them from one place to another is called soil erosion. Among water, wind, and slope of the terrain, water is the main agent of soil erosion in the form of precipitation and runoff. Multiple cycles of removal and deposition may occur on a single particle of the soil over its time span ranging from hours to decades or centuries. Anthropogenic activity is one of the main reasons which keep on aggravating the erosion phenomenon of the soil.

Climate change is one of the most important topics which has been discussed for decades unquestionably (Miglė and Genovaitė, 2020). Recent studies have pivoted the impacts of climate change in the Himalayan region as well as global impacts on the glacier regime (Akhtar et al., 2008; Immerzeel et al., 2012), but due to the lack of data from these remote areas and complex response of warming of glaciers, their results have significant uncertainty in the results. Changes in rainfall pattern due to climate change may cause the spatio-temporal variation in rainfall–runoff erosivity (Amanambu et al., 2019). The rate of soil erosion is liable to change with climate change and so the change in erosivity is directly linked to the climate change (Nearing, 2001).

General circulation models (GCMs) are frequently used as tools to estimate the changes in the future climate (Nahar et al., 2017). Rainfall is one of the variables of climate system in which GCMs have good skill of representation (Perkins et al., 2007). Kumar et al. (2020) utilized GCMs (CAN-ESM2, MPI-ESM-MR, CSIRO, CMCC-CMS) in the Sone command area, Bihar, and those GCMs were biased and corrected for the future period with respect to the baseline. Many studies have investigated changes in

Risk, Reliability and Sustainable Remediation in the Field of Civil and Environmental Engineering
https://doi.org/10.1016/B978-0-323-85698-0.00017-4

© 2022 Elsevier Inc. All rights reserved.

erosivity in isolation (Almagro et al., 2017; Babel and Plangoen, 2014) as well as studies based on combination of erosivity and soil erosion (Maeda et al., 2010; Segura et al., 2014) using GCMs in the different parts of the world.

There are around 82 models available for the estimation of soil loss and they are utilized according to the prevailing conditions of the climate of a particular area. Modified Universal Soil Loss Equation (MUSLE) is the modification over Universal Soil Loss Equation (USLE) (Williams, 1975) and then upgraded to the Revised Universal Soil Loss Equation (RUSLE) with the help of available data of the USDA (United States Department of Agriculture) (Renard et al., 1997). RUSLE is dominant over USLE since it has a scope for the estimation of cover management factor (C) (Merritt et al., 2003). Soil loss risk is usually evaluated using RUSLE model and can be implemented regionally as well as globally (Borrelli et al., 2016).

Change Intensity (CI) of soil erosion is a principal parameter to evaluate the variation for upcoming years. Guo et al. (2018) discussed the spatio-temporal change patterns in the soil erosion and analyzed the mechanisms. The lower Himalayan region has a major contributor of soil erosion which gets affected by climate change. Frequent floods and sediment deposition are drawbacks for the Kosi river basin due to the intense and accelerated erosion and it is essential to estimate the soil erosion and its change for the present and future period. Amanambu et al. (2019) estimated the nature of spatial-temporal variations with respect to erosivity from amount of rainfall occurrence using various GCMs. Guo et al. (2018) estimated the change intensity of soil erosion for the years 2009 and 2019 in the upper Minjiang catchment, China. No studies have been well documented regarding change intensity of soil erosion between the present and future period. Therefore, the objective of this chapter is to estimate the soil erosion for the present (2019) using observed data as well as future year (2070) using GCMs data and to evaluate the change intensity of soil erosion between these 2 years.

2. Study area and data description

Flood has been a major issue in Bihar especially due to the Kosi River being long. Bihar lists half of India's losses due to flood (https://www.indiawaterportal.org/articles/bihar-floods-causes-and-preventive-measures). Due to the frequent flood, the Kosi River is called the "Sorrow of Bihar." The Kosi River is one of the largest tributaries of the Ganga River (Mishra and Sinha, 2020). Having flown in a large area in Nepal, the Kosi River enters Bihar (India) near Hanuman Nagar (Bihar) and meets the Ganga River near Kursela in Katihar (Bihar). Fig. 8.1 shows the Kosi river basin with elevation map with highest at the north and lowest at the south. There are seven rain gauge stations according to India Meteorological Department (IMD) for the study namely Bhimnagar, Nirmali, Bahdurganj, Birpur, Murliganj, Galgalia, and Kursela. The latitude of the Kosi river basin is between $25°19'18''$N and $26°43'30''$N, whereas longitude is between $87°4'35''$E and $87°12'32''$E.

Digital elevation model (DEM) with 90 m resolution and Landsat8 were obtained from the United States Geological Survey (USGS)—based earth explorer portal (https://earthexplorer.usgs.gov/). The daily rainfall data were obtained for the seven rain gauge stations mentioned above from IMD, Pune. Fig. 8.2 shows the variation of observed average monthly rainfall for the seven rain gauge stations in which it is observed that maximum rainfall occurs in July month. Four GCMs (CanESM, CCMC, MPI, CSIRO) with the three Representative Concentration Pathways (RCPs: RCP2.6,

2. Study area and data description

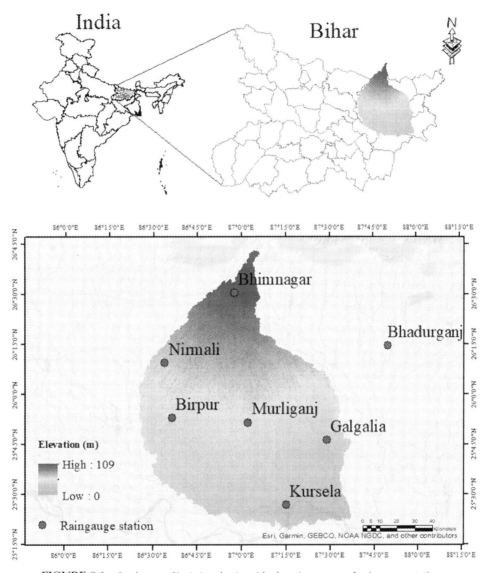

FIGURE 8.1 Study area: Kosi river basin with elevation map and rain gauge stations.

RCP4.5, and RCP8.5) were taken for the projected precipitation data, obtained from World Climate Research Program—based fifth phase of Coupled Model Intercomparison Project (CMIP5). Field data of soil were taken and analysis was done to obtain the percentage contents of sand, silt, clay, and organic matter (Meena and Jha, 2017).

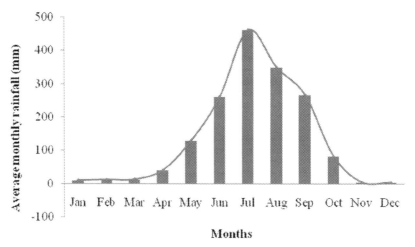

FIGURE 8.2 Variation of average monthly rainfall for the seven rain gauge stations in the study area obtained from IMD, Pune.

3. Methodology

Four Global Climate Models (GCMs) were taken from World Climate Research Program—based Coupled Intercomparison Project fifth phase (CMIP5) portal (https://esgf-node.llnl.gov/projects/cmip5/). Three GCMs (CanEMS, CMCC, and MPI) were utilized by Kumar et al. (2019) to downscale precipitation data in the Bagmati river basin, Bihar. In this study, four GCMs (CanESM, CCMC, MPI, and CSIRO) were used under three scenarios and they are RCP2.6, RCP4.5, and RCP8.5 (Kumar et al., 2020). Precipitation data taken from these GCMs were then linearly interpolated with respect to the baseline period which is taken as observed rainfall data from 1985 to 2017 from IMD, Pune. Fig. 8.3 shows the flowchart of the study.

3.1 Changes in annual soil erosion from observed period to projected year

To estimate the annual soil erosion for observed and future periods, Revised Universal Soil Loss Equation (RUSLE) was used which was proposed by Wischmeier and Smith (1978) and given as Eq. (8.1). The input factors included in the RUSLE are rainfall erosivity factor (R), soil erodibility factor (K), length and slope factor (LS), cover management factor (C), and support practice factor (P). The data required to evaluate these factors are DEM, Landsat image, rainfall data, GCM precipitation data, and soil data of the field.

$$A = R \times K \times LS \times C \times P \qquad (8.1)$$

where A represents the annual soil loss per unit area (t/ha y), R is the rainfall−runoff erosivity factor (MJ mm/ha h y), K is the soil erodibility factor (t ha h/ha MJ mm), LS is the length and slope factor, C is the crop management factor, and P is the support practice factor (Renard and Freimund, 1994).

Rainfall−runoff erosivity factor (R) is defined for the monthly or average annual rainfall (Wischmeier and Smith, 1978). Rainfall erosivity is the detachment of soil particles when the kinetic energy is transferred to them and gets eroded down the slope. R factor mainly depends

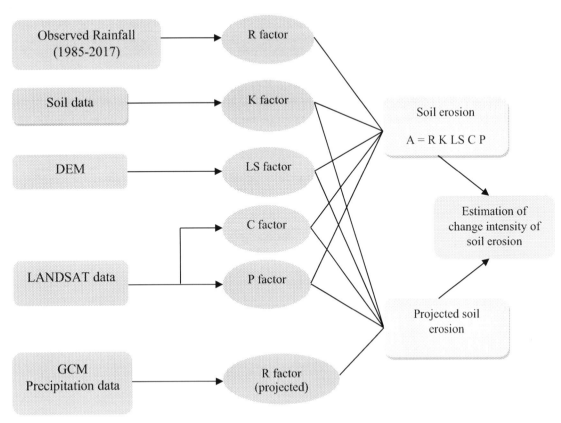

FIGURE 8.3 Detailed flowchart of the study.

on the energy of raindrops as well as the type of soil. R factor is calculated using the formula given by Renard and Freimund (1994).

$$R = 0.04830\, P^{1.610} \quad (\text{When } P < 850 \text{ mm}) \quad (8.2)$$

$$R = 587.8 - 1.219P + 0.004105P^2 \\ (\text{When } P > 850 \text{ mm}) \quad (8.3)$$

where P is average annual rainfall in mm.

The value of R factor ranges from 5470.35 to 27471.6 MJ mm/ha h y with highest in the Galgalia rain gauge station.

Soil erodibility factor (K) acts for the sediment transportability based on its erosion susceptibility. K factor predominantly depends on the amount and rate of runoff for a particular rainfall as input measured under a standard condition. Kim (2006) defined the standard condition as a unit plot, 22.6 m long, and having a 9% gradient for a continuous fallow, tilled up and down the hill slope. K factor is calculated using the following formula (Yang et al., 2003):

$$K = \frac{1}{7.6}\left\{0.2 + 0.3\exp\left[-0.0256\left(1-\frac{\text{SIL}}{100}\right)\right]\right\} \\ \left(\frac{\text{SIL}}{\text{CLA}+\text{SIL}}\right)^{0.3} \\ \left(1 - \frac{0.25\text{OM}}{\text{Org}+\exp(3.72-2.95\text{OM})}\right) \\ \left(1 - \frac{0.75\text{SN}}{\text{SN}+\exp(-5.51+22.9\text{SN})}\right) \quad (8.4)$$

where SN is given as $SN = 1.0 - SAN/100$ and CLA, SIL, SAN, and OM are represented as percentage content of clay, silt, sand, and organic matter, respectively. The value of K factor depends on the soil type (Sand, Silt, Organic matter, and Clay) within the basin.

The value of K factor ranges from 0.0285621 to 0.0420809 (t ha h/ha MJ mm).

Land and slope factor (LS) is also known as topographic factor and defined as the ratio of loss of soil in the specified condition to that of "standard" condition. "Standard" condition is defined for slope steepness of 9% with the slope length of 22.6 m (Ganasri and Ramesh, 2016). LS factor was evaluated using SRTM data obtained from USGS-based earth explorer portal (https://earthexplorer.usgs.gov/). LS factor is calculated using the formula given below:

$$LS = \left(\frac{Q_a M}{22.13}\right)^y$$
$$* \left(0.065 + 0.045 * S_g + 0.0065 * S_g^2\right) \quad (8.5)$$

where LS = Length and slope factor; S_g = Grid slope in percentage; Q_a = Flow Accumulation grid; M = Grid size; and y = dimensionless exponent which ranges from 0.2 to 0.5.

The LS factor is calculated for the baseline period and found that the value ranges from 0 to 104.018.

Crop management factor (C) is based on the vegetation cover of the study area and is the function of Normalized Difference Vegetation Index (NDVI). NDVI is derived from the Landsat data obtained from USGS-based earth explorer (https://earthexplorer.usgs.gov/). Landsat data were obtained for the months from October to December (2019). NDVI is calculated from Eq. (8.6) (Kumar and Roshni, 2019).

$$NDVI = \frac{NIR - RED}{NIR + RED} \quad (8.6)$$

C factor is computed using the equation given below:

$$C = \exp\left[-\alpha\left(\frac{NDVI}{\beta - NDVI}\right)\right] \quad (8.7)$$

The C factor ranges from 0.289469 to 1.19271.

Support practice factor (P) is defined for the soil erosion associated with the land use or the system of farming adopted. The P factor accounts for control practices that reduce the erosion capability of the runoff by their influence on drainage patterns, runoff velocity, drainage patterns, runoff concentration, and hydraulic forces actin on soil by runoff (Ganasri and Ramesh, 2016). The value of P factor varies from 0 to 1 as 0 reflects the full control on erosion and 1 reflects that there is no erosion. The P factor can either be determined with the help of image classification using remote sensing data or knowledge of experts (Panagos et al., 2015). The P factor is assigned according to the land cover and as recommended by Yang et al. (Yang et al., 2003) (Table 8.1).

3.2 Projected rainfall—runoff erosivity factor

Many researchers (Amanambu et al., 2019; Gupta and Kumar, 2017b; Panagos et al., 2017)

TABLE 8.1 Assigned P factor for different LU/LC types (Yang et al., 2003).

LU/LC types	P factor
Agriculture	0.5
Water	1
Barren	1
Built up	1
Forest	1

discussed the effects of climate change on rainfall erosivity factor and subsequently soil loss. Using bias correction technique (Kumar et al., 2020), the GCMs precipitation data were corrected with respect to the observed rainfall data for the year 2070.

3.2.1 Bias correction

Bias correction is a simple and generally preferred method to develop the relationship between modeled and observed parameters (Kumar et al., 2020). Chen et al. (2013) utilized linear bias correction for the rainfall analysis. A multiplicative factor is used in bias correction for the projected rainfall data in all GCMs that relates the observed monthly mean precipitation (1985−2017) and GCMs data (Berg et al. 2012). It is given as

$$PiBC = Pimodel (t) * f(i) \qquad (8.8)$$

$$f(i) = \frac{\overline{Piobserved}}{\overline{Pimodel}} \qquad (8.9)$$

where $PiBC$ is the bias corrected monthly precipitation (mm); $Pimodel$ is the original model precipitation (mm); $\overline{Piobserved}$ is the long-term average observed precipitation, in mm; $\overline{Pimodel}$ is the long-term average of model projections of precipitation (mm); t is the time step (annual steps in this case); i is an index of the month; and f is the multiplicative factor for each month i.

4. Results and discussions

4.1 Estimation of soil erosion for the years 2019 and 2070

With the help of remote sensing data and GIS, priority level maps for soil erosion were developed for the year 2019 and the projected year 2070. The maps were prioritized into five priority levels for both the years and shown in Fig. 8.4. The five priority levels are based on the severity of soil erosion, first priority level (>20 t/ha/y), second priority level (20−10.31 t/ha/y), third priority level (10.30−5.16 t/ha/y), fourth priority level (5.15−1.51 t/ha/y), and fifth priority level (<1.50 t/ha/y). It was observed from Fig. 8.4A and B that the eastern area of the Kosi river basin lies in the first priority level where the Galgalia rain gauge station is situated. In the year 2019 (Fig. 8.4A), the eastern area comes under third priority level which is more than that of in the year 2070 (Fig. 8.4B). The southern area is in the fifth priority level in both the years as it affected due to soil erosion.

4.2 Change intensity in annual soil erosion from 2019 to 2070

To monitor the spatial-temporal changes in soil erosion effectively, the change intensity of soil erosion has been estimated between the years 2019 and 2070. CI has been classified into seven levels based on the standard deviation and histogram distribution of map: intensive decrease (CI < −63.29 t/ha/y), moderate decrease (−63.28 t/ha/y < CI < −43.29 t/ha/y), mild decrease (−43.28 t/ha/y < CI < −23.27 t/ha/y), stable (−23.26 t/ha/y < CI < −3.26 t/ha/y), mild increase (−3.25 t/ha/y < CI < 16.76 t/ha/y), moderate increase (16.77 t/ha/y < CI < 36.78 t/ha/y), and intensive increase (CI > 36.79 t/ha/y). Fig. 8.5 shows the change intensity map soil erosion between 2019 and 2070. It was observed from Table 8.2 that majority of area (43.642%) of CI is under "stable" zone

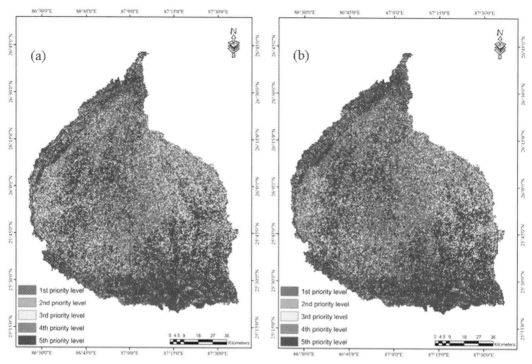

FIGURE 8.4 Priority level maps for erosion in the years (A) 2019 and (B) 2070.

followed by mild increase zone (38.361%), mild decease (11.595%), moderate decrease (3.682%), intensive decrease (2.469%), moderate increase (0.240%), and least in intensive increase (0.010%). Some area near Galgalia rain gauge station is covered under "intensive increase" where the Galgalia rain gauge station is situated and the highest observed annual average rainfall occurred during the baseline (1985–2017). Majority of central area of the Kosi river basin covered in the "stable" zone.

Based on the above analyses of soil loss prioritization and CI, it may be generalized that the eastern area of the Kosi river basin is severely affected with the soil erosion in both the years (2019 and 2070) and majority of area of soil erosion in the "stable" zone as per the CI of soil erosion. According to the CI, the possible remedial measures may be taken in the more affected area.

4. Results and discussions

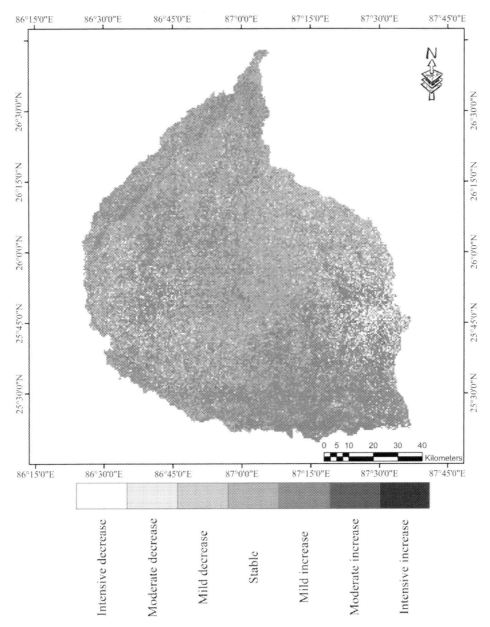

FIGURE 8.5 Change intensity of soil erosion map from 2019 to 2070.

TABLE 8.2 Area covered under different change intensities.

Change intensity (CI)	Intervals (t/ha/y)	Area (*100 ha)	Percentage area
Intensive decrease	CI < −63.29	281.74	2.469
Moderate decrease	−63.28 < CI < −43.29	420.10	3.682
Mild decrease	−43.28 < CI < −23.27	1322.98	11.595
Stable	−23.26 < CI < −3.26	4779.57	43.642
Mild increase	−3.25 < CI < 16.76	4377.00	38.361
Moderate increase	16.77 < CI < 36.78	27.44	0.240
Intensive increase	CI > 36.79	1.18	0.010

5. Conclusions

The Kosi river basin is very much prone to frequent flood and sediment deposition has been a big issue in the basin. Hence, annual soil erosion due to climate change has been estimated for the current as well as for the projected year. From the above results, it was observed that the first priority level of soil erosion (>20 t/ha/y) affects the eastern portion of the Kosi river basin where Galgalia rain gauge station is situated and this portion is severely affected in the year 2019 as compared to the projected year 2070.

The CI has also been estimated between the years 2019 and 2070 and is classified into seven zones and they are intensive decrease, moderate decrease, mild decrease, stable, mild increase, moderate increase, and intensive increase. The reason for the estimation of CI is to observe which portion of the Kosi river basin is stable. It was observed that majority of area is under stable zone in the Kosi river basin which is 43.462%. The least percentage (0.01%) of CI is in the intensive increase which is nearby the Galgalia rain gauge station which is also having the highest observed average annual rainfall.

This study is based on observed as well as remotely sensed data and the spatial variation of soil erosion is analyzed with the help of RUSLE model for the year 2019 and projected year 2070. The results of the present study would help to recognize the portion of severely affected in the Kosi river basin for the future period. The observations are crucial for the upcoming studies since this research would give a viewpoint on the relationships between rainfall and soil loss for the present as well as the future period.

References

Akhtar, M., Ahmad, N., Booij, M.J., 2008. The impact of climate change on the water resources of Hindukush-Karakorum-Himalaya region under different glacier coverage scenarios. J. Hydrol. 355 (1−4), 148−163. https://doi.org/10.1016/j.jhydrol.2008.03.015.

Almagro, A., Oliveira, P.T.S., Nearing, M.A., Hagemann, S., 2017. Projected climate change impacts in rainfall erosivity over Brazil. Sci. Rep. 1−12. https://doi.org/10.1038/s41598-017-08298-y.

Amanambu, A.C., Li, L., Egbinola, C.N., Obarein, O.A., Mupenzi, C., Chen, D., 2019. Spatio-temporal variation in rainfall-runoff erosivity due to climate change in the Lower Niger Basin, West Africa. Catena. https://doi.org/10.1016/j.catena.2018.09.003.

Babel, M.S., Plangoen, P., 2014. Projected rainfall erosivity changes under future climate in the upper nan watershed, Thailand. J. Earth Sci. Climatic Change 05 (10). https://doi.org/10.4172/2157-7617.1000242.

Berg, P., Feldmann, H., Panitz, H.J., 2012. Bias correction of high resolution regional climate model data. J. Hydrol. 448−449, 80−92.

Borrelli, P., Diodato, N., Panagos, P., 2016. Rainfall erosivity in Italy: a national scale spatio- temporal assessment rainfall erosivity in Italy: a national scale spatio-temporal. Int. J. Digital Earth. https://doi.org/10.1080/17538947.2016.1148203.

Chen, J., Brissette, F.P., Chaumont, D., Braun, M., 2013. Finding appropriate bias correction methods in downscaling precipitation for hydrologic impact studies over North America. Water Resour. Res. 49 (7), 4187–4205.

Ganasri, B.P., Ramesh, H., 2016. Assessment of soil erosion by RUSLE model using remote sensing and GIS - a case study of Nethravathi Basin. Geosci. Front. 7 (6), 953–961. https://doi.org/10.1016/j.gsf.2015.10.007.

Guo, B., Yang, G., Zhang, F., Han, F., Liu, C., 2018. Dynamic monitoring of soil erosion in the upper Minjiang catchment using an improved soil loss equation based on remote sensing and geographic information system. Land Degrad. Dev. 29 (3), 521–533. https://doi.org/10.1002/ldr.2882.

Gupta, S., Kumar, S., 2017. Simulating climate change impact on soil erosion using RUSLE model–A case study in a watershed of mid-Himalayan landscape. J. Earth Syst. Sci. https://doi.org/10.1007/s12040-017-0823-1.

Immerzeel, W.W., van Beek, L.P.H., Konz, M., Shrestha, A.B., Bierkens, M.F.P., 2012. Hydrological response to climate change in a glacierized catchment in the Himalayas. Climatic Change 110 (3–4), 721–736. https://doi.org/10.1007/s10584-011-0143-4.

Kim, H.S., 2006. Soil Erosion Modeling Using RUSLE and GIS on the IMHA Watershed, South Korea. Doctoral dissertation. Colorado State University, USA.

Kumar, K., Singh, V., Roshni, T., 2019. Efficacy of the hybrid neural network in statistical downscaling of precipitation of the Bagmati river basin. J. Water Climate change 1–21. https://doi.org/10.2166/wcc.2019.259.

Kumar, S., Roshni, T., 2019. NDVI-rainfall correlation and irrigation water requirement of different crops in the Sone river-command, Bihar. Mausam 2, 339–346.

Kumar, S., Roshni, T., Kahya, E., Ghorbani, M.A., 2020. Climate change projections of rainfall and its impact on the cropland suitability for rice and wheat crops in the Sone river command, Bihar. Theor. Appl. Climatol. 142 (1–2), 433–451. https://doi.org/10.1007/s00704-020-03319-9.

Maeda, E.E., Pellikka, P.K.E., Siljander, M., Clark, B.J.F., 2010. Potential impacts of agricultural expansion and climate change on soil erosion in the Eastern Arc Mountains of Kenya. Geomorphology 123 (3–4), 279–289. https://doi.org/10.1016/j.geomorph.2010.07.019.

Meena, R.S., Jha, R., 2017. Approximating soil physical properties using geo-statistical models in lower Kosi basin of Ganga river system, India prone to flood inundation. Int. J. Civ. Eng. Technol. 8 (5), 1445–1459.

Merritt, W.S., Letcher, R.A., Jakeman, A.J., 2003. A review of erosion and sediment transport models. Environ. Model. Soft. 18, 761–799. https://doi.org/10.1016/S1364-8152(03)00078-1.

Miglė, J., Genovaitė, L., 2020. Climate change concern, Personal responsibility and actions related to climate change mitigation in EU countries: Cross-cultural analysis. J. Clean. Prod. 281. https://doi.org/10.1016/j.jclepro.2020.125189.

Mishra, K., Sinha, R., 2020. Geomorphology Flood risk assessment in the Kosi megafan using multi-criteria decision analysis: a hydro-geomorphic approach. Geomorphology 350, 106861. https://doi.org/10.1016/j.geomorph.2019.106861.

Nahar, J., Johnson, F., Sharma, A., 2017. Assessing the extent of non-stationary biases in GCMs. J. Hydrol. https://doi.org/10.1016/j.jhydrol.2017.03.045.

Nearing, M.A., 2001. Potential changes in rainfall erosivity in the U.S. with climate change during the 21st century. J. Soil Water Conserv. 56 (3), 229–232.

Panagos, P., Ballabio, C., Borrelli, P., Meusburger, K., Klik, A., Rousseva, S., et al., 2015. Rainfall erosivity in Europe. Sci. Total Environ. 511, 801–814. https://doi.org/10.1016/j.scitotenv.2015.01.008.

Panagos, P., Ballabio, C., Meusburger, K., Spinoni, J., Alewell, C., Borrelli, P., 2017. Towards estimates of future rainfall erosivity in Europe based on REDES and World-Clim datasets. J. Hydrol. 548, 251–262.

Perkins, S.E., Pitman, A.J., Holbrook, N.J., McAneney, J., 2007. Evaluation of the AR4 climate models' simulated daily maximum temperature, minimum temperature, and precipitation over Australia using probability density functions'. J. Clim. 20 (17), 4356–4376. https://doi.org/10.1175/JCLI4253.1.

Renard, K.G., Freimund, J.R., 1994. Using monthly precipitation data to estimate the R-factor in the revised USLE. J. Hydrol. 157 (1–4), 287–306.

Segura, C., Sun, G., McNulty, S., Zhang, Y., 2014. Potential impacts of climate change on soil erosion vulnerability across the conterminous United States. J. Soil Water Conserv. 69 (2), 171–181. https://doi.org/10.2489/jswc.69.2.171.

Williams, J.R., 1975. Sediment-yield prediction with universal equation using runoff energy factor. In, present and prospective technology for predicting sediment yield and sources. Proceedings of the Sediment Yield Workshop. USDA Sedimentation Lab., Oxford, MS, pp. 244–252. ARS-S-40, 28-30 November 1972.

Wischmeier, W.H., Smith, D.D., 1978. Predicting rainfall erosion losses: A Guide to Conservation Planning. The USDA Agricultural Handbook No. 537.

Yang, D., Kanae, S., Oki, T., Koike, T., Musiake, K., 2003. Global potential soil erosion with reference to land use and climate changes. Hydrol. Process. 2928 (May 2002), 2913–2928. https://doi.org/10.1002/hyp.1441.

Further reading

Uddin, K., Murthy, M.S.R., Wahid, S.M., Matin, M.A., 2016. Estimation of soil erosion dynamics in the Koshi basin using GIS and remote sensing to assess priority areas for conservation. PLoS One. https://doi.org/10.1371/journal.pone.0150494.

CHAPTER 9

Adaptive Kriging Monte Carlo Simulations for cost-effective flexible pavement designs

Deepthi Mary Dilip, Aleena Joy, Anamika Venu

BITS Pilani Dubai Campus, United Arab Emirates

1. Introduction

The need for sustainable infrastructure is ever increasing and a paradigm shift in design and construction is needed to achieve them. Owing to the relatively inexpensive and easily available binder materials, around 80% of the road surfaces, across the globe, are flexible (asphalt) pavements. Although cheaper to construct than their rigid (concrete) counterparts, the more frequent maintenance requirements often result in less sustainable design solutions. Despite the simple structure, the flexible pavement behavior is quite complex because the materials used for construction are often highly nonlinear and nonhomogeneous in their behavior under traffic loading, and because their properties change, sometimes drastically, with time. The Mechanistic-Empirical pavement design is widely accepted in the pavement engineering community because it connects pavement performance observed in the field with the structural responses under wheel loads. Moreover, the M-E framework allows the effect of the material variabilities on the pavement performance to be assessed rationally and captured systematically in the flexible pavement analysis, ensuring that the structure remains safe during its design life (Dilip and Babu, 2013; Mittal and Swamy, 2020). The performance of flexible pavements is expressed in terms of distresses exhibited during their service life, which include rutting, fatigue cracking, and longitudinal and transverse cracking among others. As the most prevalent form of structural distress in flexible pavements is fatigue cracking (Dong and Huang, 2012), and rutting along with road roughness (Ogwang et al., 2019), the two distress of fatigue cracking and deep structural rutting have been considered in this study. The fatigue failure criterion is based on limiting the tensile strain at the bottom of the asphalt layer, and is assessed using the fatigue life, defined as the cumulative load repetitions before which certain percentage of cracked surface area (usually taken as 20%) is observed. Similarly, the rutting failure is based on limiting the subgrade compressive strain, and is evaluated using the rutting life, defined as the cumulative load repetitions which produces a specified rut-depth (assumed to be around 0.25-inch) (Luo et al., 2019) in flexible pavements.

With heavier trucks plying the roads today, newly constructed pavements are failing well before their design life (Abu Abdo and Jung, 2020). When the uncertainties in materials, loads, and environmental parameters are not duly accounted for in the design process itself, the expected performance of asphalt pavements is seen to be quite different from that observed in the field (Mittal and Swamy, 2020). The presence of uncertainties in flexible pavement design cannot be fully avoided and is quite pervasive, arising from a number of sources including the material properties and strength, traffic loads, drainage conditions, construction and compaction procedures, and climatic factors such as temperature and rainfall. It is essential to recognize the presence of all the major sources that contribute to pavement design life uncertainty, as the more the sources are identified, quantified, and factored into the design process, the higher the level of reliability. The uncertainties can be broadly classified into two sources: aleatory uncertainties and epistemic uncertainties. The aleatory uncertainty manifests as pavement distresses due to the natural (inherent) variability of the pavement materials along the length of the pavement structure and the highly uncertain deterioration of the structure under wheel loads with each passing vehicle (and time). Moreover, the pavement performance is adversely influenced by the high variability exhibited by the materials used in various layers (Noureldin et al., 1994; Timm and Birgisson, 1998) that results in the uncertainty of the resilient moduli within each pavement layer. The epistemic uncertainty arises due to our inability to model the real world perfectly as well as due to the limited information/data available to build these models. The epistemic uncertainty can arise either due to the inaccuracies in the estimation of the design (input) parameters from

data (i.e., design parameter uncertainty) as well as due to the model uncertainty. To tackle these uncertainties, numerous efforts have made to move from the deterministic pavement design to a reliability-based design approach (ARA Inc, 2004; Austroads, 2004).

In a reliability-based design setting, the probabilistic techniques such as the first and second-order reliability method (FORM and SORM) which are approximate techniques and the simulation-based Monte Carlo Simulation (MCS) approach are adopted to compute the failure probability under the expected design traffic. Regardless of the approach to compute the structural reliability, the probabilistic analysis requires repeated calls to the pavement analysis models for the computation of the critical responses under loading. To introduce computational efficiency in the reliability-based design of flexible pavements, these models are replaced by surrogate models such as the response surface methods (Dilip et al., 2013; Luo et al., 2018) to compute the critical pavement responses. For instance, in the M-E framework, the critical pavement responses under traffic loads can be determined from simple linear-elastic pavement analysis software such as KENPAVE or the finite-element based software such as ABAQUS. These models that surrogate the real-world pavement structure, and introduce an epistemic uncertainty due to our lack of understanding of the exact pavement behavior. As the Kriging metamodel allows the quantification of the error induced by substituting the pavement analysis software by the metamodel, the Adaptive Kriging Monte Carlo Simulation approach is proposed in this study to design structures that meet target levels of reliability.

1.1 Reliability-based design of flexible pavements

The pavement design reliability is defined as the probability that a pavement as designed will withstand the design number of load applications during the desired service life while maintaining both structural and functional integrity (AASHTO, 1993). To incorporate reliability analysis in the M-E pavement structural design process, and ensure that the failure probability is within acceptable limits, the uncertainties contributing to the load repetition variability need to be first identified. In a deterministic setting, the general form of the distress (allowable load repetition) models for fatigue cracking and rutting for pavements takes the following forms (Asphalt Institute, 1991; IRC: 37, 2018):

$$N_f = k_1 \left(\frac{1}{\epsilon_t}\right)^{k_2} \left(\frac{1}{E_1}\right)^{k_3} \quad (9.1)$$

$$N_r = k_4 \left(\frac{1}{\epsilon_z}\right)^{k_5} \quad (9.2)$$

where N_f and N_r are the allowable fatigue and rutting load repetitions, respectively, ϵ_t is critical tensile strain (in microstrain), ϵ_z is the vertical subgrade strain (in microstrain), E_1 is the modulus of elasticity of bituminous surfacing (MPa), and $k_1 - k_5$ are the regression coefficients. According to the Mechanistic-Empirical Pavement Design Guide (MEPDG), the rutting life (N_r) and fatigue life (N_{f-HMA}) of the HMA layer in the AASHTO design guide (AASHTO, 2008) are quantified as follows:

$$N_r = \left(\frac{\Delta_{p(HMA)}}{\beta_{r1} k_z \varepsilon_{r(HMA)} 10^{k_{r1}} T^{k_{r3}}}\right)^{1/k_{r2}\beta_{r2}} \quad (9.3)$$

$\Delta_{p(HMA)}$ = accumulated permanent distortion in HMA layer (in), $\varepsilon_{r(HMA)}$ = elastic strain calculated by the structural response model at mid-depth of HMA layer (in/in), T = pavement or mix temperature ($^\circ$F), $k_{r1,r2,r3}$ = global field calibration parameters, $\beta_{r1,r2,r3}$ = local or mixture

field calibration constants, and k_z = depth confinement factor

$$N_{f-HMA} = k_{f1} C_H \beta_{f1} (\varepsilon_t)^{k_{f2}\beta_{f2}} (E_{HMA})^{k_{f3}\beta_{f3}} \quad (9.4)$$

N_{f-HMA} = allowable number of axle-load applications for flexible pavement, ε_t = tensile strain at critical locations calculated by the structural response model (in/in), E_{HMA} = HMA modulus (psi), $k_{f1,r2,r3}$ = global field calibration parameters, $\beta_{f1,r2,r3}$ = local or mixture field calibration constants, and C_H = thickness correction term, dependent on the type of cracking.

The allowable load repetitions for any pavement structure depend on the critical pavement responses under loading, which in turn are a function of the pavement design parameters. The uncertainties in the input parameters that have a significant impact of the pavement reliability are the layer thicknesses H and the resilient moduli E. The uncertainties of these parameters can be either aleatory or epistemic or both, and can be represented as random variables with defined mean and standard deviation or its complete probability distribution. Based on the premise that the pavement is considered to have failed when the traffic demand exceeds the pavement life, the limit state or performance function for fatigue may then be formulated for fatigue failure as follows:

$$P_f^f = P\left(N_f < N_d\right) = P\left(g_f(\boldsymbol{H}, \boldsymbol{E}) < 0\right) \quad (9.5)$$

$g_f = \frac{N_f}{N_d} - 1$ where N_d is the actual or expected number of load applications (demand), and in a similar manner for rutting failure as

$$P_f^r = P(N_r < N_d) = P\left(g_r(\boldsymbol{H}, \boldsymbol{E}) < 0\right) \quad (9.6)$$

1.1.1 Estimation of failure probability

In the reliability-based analysis of flexible pavements, the FORM and SORM are frequently adopted to compute the failure probabilities P_f based on approximating the limit state function locally (with a linear or quadratic Taylor expansion), as these methods are very efficient and only a relatively small number of model evaluations are needed to calculate P_f (Luo et al., 2019). However, when nonlinear limit state functions are adopted to replace the pavement analysis models, these approximate techniques further contribute to the epistemic uncertainties. Although the Monte Carlo simulation, a direct sample-based estimation of the expectation of the failure probability, is more accurate than the approximate techniques, the high computational cost of the MCS approach is often the deterrent in reliability-based analysis of structures. In this study, the evaluation of the limit state function $g_k(H, E)$ is carried out by computing the critical pavement responses corresponding to fatigue and rutting failure using the Multi-Layer Elastic Analysis software (Al-Rumaithi, 2021) in MATLAB (2016) that finds the stresses, strains, and deflections of pavement or soil layers subjected to circular loading. As the responses are computed directly in MATLAB, crude MCS can be carried out, without introducing further model uncertainty due to the use of surrogate models. Nevertheless, the evaluation of the limit state function is still costly, and the Monte Carlo simulation approach become intractable when large number of limit state function evaluations are required to compute the pavement reliability at different levels of traffic and uncertainty. To achieve a balance between computational efficiency and accuracy, the Adaptive Kriging Monte Carlo Simulation is adopted in this study.

1.1.2 Adaptive Kriging Monte Carlo Simulations

Adaptive Kriging Monte Carlo Simulation (AK-MCS) combines Monte Carlo simulation with Kriging (i.e., Gaussian process modeling) metamodels that are adaptively built around the limit state functions. The AK-MCS approach is based on the Kriging theory which considers the correlation between two different samples (i.e., real model M and metamodel $\widehat{M}$) depending on the distance between input variables. When surrogate models, such as the Kriging metamodel $(\widehat{M})$, can be built to replace the computationally expensive pavement analysis programs like Multi-Layer Elastic Analysis software (M), adopting MCS to compute the failure probability can be easily justified, particularly when the target reliability levels are quite conservative (order of magnitude ranging from 99% to 80%) in the case of flexible pavements. The experimental design samples for building the metamodels can be achieved through Latin Hypercube Sampling (LHS) scheme to ensure that the entire parameter space is sampled from. However, while these metamodels can accurately predict the critical pavement responses in the vicinity of the experimental design samples, the LHS samples are generally not optimal to estimate the failure probability. When such metamodels are used to evaluate the structural reliability, an adaptive experimental design algorithm increases the accuracy of the surrogate model in the vicinity the limit state function, and thereby the reliability estimate.

To build the adaptive experimental design, an initial approximate Kriging metamodel is first constructed based on a small number of Latin Hypercube samples computed using the pavement analysis model M. To this experimental design, carefully selected samples are added based on the current estimate of the limit state surface $g_{\downarrow k}, (k \in (f, r))$ (Marelli and Sudret, 2014). For this a learning function, which is a measure of the attractiveness of a candidate sample x with respect to improving the estimate of the failure probability when it is added to the experimental design, needs to be defined. The objective of the learning function is to choose samples that are close to the limit state surface. Because the computation of the failure probability requires only the sign of the performance function values, the U-function (Echard et al., 2011) is adopted to identify the misclassification that happens when the sign of the surrogate model and the sign of the underlying limit state function do not match. The U-function is for a Gaussian process is defined as

$$U(x) = \frac{\left|\mu_g(x)\right|}{\sigma(x)} \tag{9.7}$$

where $\mu_g(x)$ and $\sigma(x)$ are the prediction mean and standard deviation of $\sigma(x)$. The corresponding probability of misclassification is then

$$\Phi(-U(x)) \tag{9.8}$$

where Φ is the CDF of a standard Gaussian variable. Thus, the next candidate sample from the set is chosen as the one that maximizes the probability of misclassification (Dubourg, 2011).

2. Numerical examples

To illustrate the advantage of the Adaptive Kriging MCS, a probabilistic analysis is carried out for a three-layer and four-layer flexible pavement with an asphalt surface course and a granular course(s) laid over a semiinfinite subgrade. For the three-layer pavement structure, the allowable load repetitions were computed using the Asphalt Institute distress models, and an expected traffic level of 1 msa was assumed. In the case of the four-layer structure, the MEPDG design equation was adopted for an expected traffic level of 25 msa. The design parameters adopted for the study are presented in Table 9.1. The layer thicknesses were assumed to be normally distributed with the Coefficient of Variation (COV) values adopted as 15%, 20%, and 20% respectively, while the resilient moduli were assumed to be lognormally distributed with COV values of 10%, 30%, 20%, and 20%, respectively (Saride et al., 2019).

TABLE 9.1 Design parameters.

No	Layer	Thickness (mm)		Layer modulus (MPa)	
		Section 1	Section 2	Section 1	Section 2
1	Asphalt surface	150	150	2000	2000
2	Granular base	350	250	350	350
3	Granular subbase	–	300	–	250
4	Subgrade	Semiinfinite		80	80

2.1 Validation of adaptive Kriging metamodels with crude MCS

For both the three-layer and four-layer pavement structures, the Kriging predictor for the critical tensile and compressive strains is built using a Gaussian autocorrelation function. Studies have shown that arbitrarily selecting the model (Kriging trend and correlation) can lead to poor probability of failure estimates for complex systems (Sundar and Michael, 2019). In order to validate the metamodel, the failure probability was computed by running crude Monte Carlo Simulations. The results, presented in Fig. 9.1, indicate that the Kriging model with a quadratic trend function provides a higher level of accuracy when compared with the linear trend, and is adopted for further analysis. However, the MCS requires around 10^5 simulations to keep the COV within 2%, and as the time taken to run one simulation using the Multi-Layer Elastic Analysis software around 0.2 s, the crude MCS can become a tedious process. On the other hand, the failure probability values estimated by the AK-MCS were found to be accurate, as presented in Table 9.2, and needed only 100 calls to the Multi-Layer Elastic Analysis model. This clearly illustrates the benefit of using the AK-MCS approach instead of the crude MCS methodology.

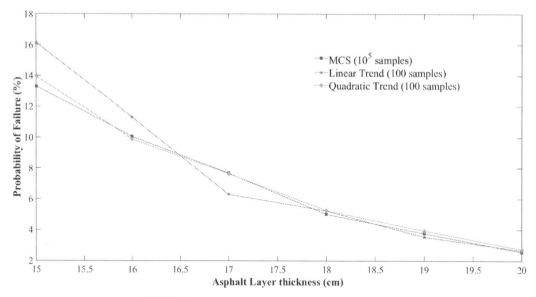

FIGURE 9.1 Validation of the Kriging Model trend.

TABLE 9.2 Validation of failure probability estimation by AK-MCS

	3-Layer section		4-Layer section	
Layer	MCS	AK-MCS	MCS	AK-MCS
Fatigue failure				
Probability of failure (%)	90.6	90.51	90.7	90.3
Rutting failure				
Probability of failure (%)	81.3	81.4	99.7	99.1
No of calls	10^5	100	10^5	100

3. Cost-effective flexible pavement structures

While there are a number of different ways to incorporate sustainable practices in pavement design, the initiative by transportation agencies to combine the traditional pavement materials with recycled plastic waste is of great significance today, as it also addresses the global problem of plastic waste disposal. Although a number of studies using waste plastic in the bituminous mixtures have reported cost savings particularly when plastic was used to replace the bituminous binder, the cost of synthesizing polymer-modified bitumen using recycled plastic may exceed the cost of producing unmodified bitumen, at an industrial scale. While incorporating recycled plastic waste will lessen the burden on the landfills, the cost effectiveness of such sustainable pavements also needs to be investigated thoroughly, before they can be implemented in practice.

The economic feasibility of incorporating recycled plastic into the bituminous mixture is evaluated by the cost of construction per square yard ($/sy) and keeping the costs of all other layers other than the asphalt surface layer a constant. Based on the assumed density of HMA of 0.055 ton/inch, the required cost for constructing the asphalt layer can be computed as a function of its thickness. The cost of the unmodified asphalt surface course is adopted as $65.37 per ton of HMA and the cost of the polymer-modified bitumen is taken as $65.52 per ton of HMA per ton of HMA (Luo et al., 2019). In a deterministic analysis, the cost effectiveness of the unmodified and polymer-modified asphalt mixture may be compared by defining the cost effectiveness as the ratio of the expected performance (say fatigue life) with the unit mixture cost (FHWA, 2017) to evaluate how the savings due to increased pavement life could offset the initial costs. The incorporation of plastic in flexible pavement construction is expected to improve the pavement stiffness and durability, thereby increasing the pavement life. Hence, in a reliability-based design, the cost effectiveness of the alternative structures can be compared at the same levels of pavement failure of the conventional pavement. For the purpose of this comparative study, the resilient modulus values of the unmodified and modified asphalt layers were adopted from literature (Souliman et al., 2016).

To illustrate this concept, the pavement reliability is estimated using the AK-MCS technique at different pavement thickness levels. The comparative analysis is carried out by keeping the uncertainties (i.e., the COV level) at the same level for all the design parameters, and the results are presented in Fig. 9.2. For the pavement section considered in the study, the alternative pavement section was found to be cost effective at all levels of reliability and for both fatigue and rutting failure. However, as can be observed from Fig. 9.2, for reliability levels below 90%, the decrease in costs for the alternative pavement section is negligible.

3.1 Effect of asphalt layer COV on the cost effectiveness

In the reliability-based design, the flexible pavement structures are designed to meet specific target levels of reliability. Thus, when alternative materials are used to replace the conventional ones in the design of sustainable flexible pavements, the cost effectiveness is also compared at the specified reliability levels of 90%, 95%, and 98% as illustrated in Fig. 9.3. As the COV level of the resilient modulus modified asphalt layer has not been documented in literature, the COV levels were assumed to be the same as that of the conventional bitumen layers. However, when waste plastic from different sources are used to increase the pavement layer

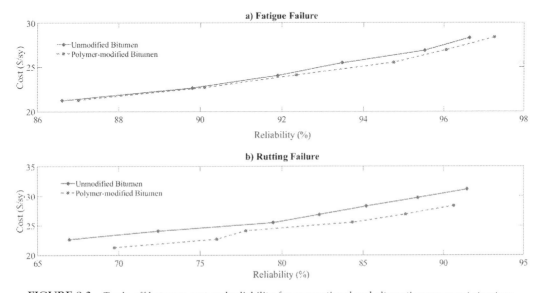

FIGURE 9.2 Trade-off between cost and reliability for conventional and alternative pavement structure.

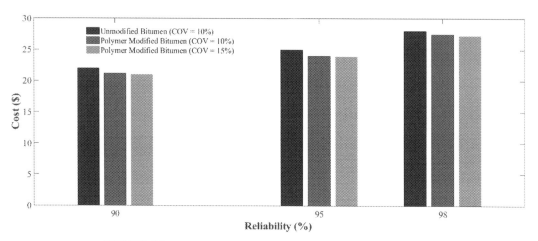

FIGURE 9.3 Cost-effectiveness at specified levels of target reliability.

stiffness, it can be reasonably expected that the uncertainties in the layer modulus may increase. Therefore, the effect of the uncertainties on the cost effectiveness of the alternative designs is studied by considering two levels of COV, i.e., 10% and 15%, where the COV of conventional (asphalt) layer modulus is assumed to be 10%. The four-layer pavement structure is considered in this study, and the distress model proposed by Mechanistic-Empirical Pavement Design Guide (MEPDG) was adopted for the estimation of system reliability. The results indicate that even in the presence of higher uncertainties in the (modified) asphalt layer, the use of polymer-modified bitumen is more cost effective that the conventional design.

4. Conclusions

The deterioration of asphalt (flexible) pavements under heavy traffic loads is aggravated under uncertainties, and the need to build cost-effective pavement structures while preserving natural reserves led to the use of alternative materials like recycled plastics in the asphalt pavement industry. Although sustainable pavements are still an aspirational goal, efficient construction practices and the use of recycled/waste materials can go a long way toward reducing environmental impacts (Usman and Masirin, 2019). As a result, many researchers have focused on determining the optimum quantities of plastic that will increase the pavement stiffness and thereby result in longer lasting and more reliable roads. However, the effect of incorporating uncertainties in the design of such alternative pavement structures has not been adequately addressed. In this chapter, the probabilistic analysis of flexible pavement structure within the M-E design is carried out, to address the premature failure of flexible pavements under heavy loads, predominantly caused by rutting, fatigue cracking among other pavement distresses.

The Adaptive Kriging Monte Carlo Simulation (AK-MCS) algorithm is implemented to analyze the cost effectiveness of adopting alternative pavement structures, in the presence of uncertainty. Metamodels are widely used for the propagation of uncertainty when the computational expense of the probabilistic modeling is high, particularly when Monte Carlo Simulations are adopted. As the use of such surrogate models further introduces uncertainty in the analysis, the Kriging model (or Gaussian process predictor) is adopted to predict the pavement responses, to quantify the epistemic uncertainty. For higher efficiency as well as accurate reliability estimates, the Kriging method is integrated with Monte Carlo simulation (AK-MCS) to replace the original limit-state function through MCS. The accuracy of the AK-MCS reliability estimate as well as the Kriging model adopted (trend and correlation function) can be verified through crude MCS; although this is a time-consuming process, the direct computation of the critical pavement responses is enabled through the use of the Multi-Layer Elastic Analysis software. The AK-MCS is shown to be very efficient as the probability of failure, computed with only a small number of calls to the true pavement analysis model, is highly accurate. This level of accuracy is particularly significant when alternative design sections are compared in a probabilistic analysis and the reliability estimates are fairly close, as in the case of the sections considered in the study. Using the AK-MCS, it was shown that the pavement structures modified with plastic are found to be more cost effective than the conventional design alternatives, even in the presence of higher uncertainties.

References

AASHTO, 1993. "Guide for Design of Pavement structures." American Association of State Highway and Transportation Officials, Washington, DC.

AASHTO, 2008. Mechanistic- Empirical Pavement Design Guide: A Manual of Practice, AASHTO Designation MEPDG-1, American Association of State Highway and Transportation Officials, Vol. 7(4). Design of Structures & Machines, Washington, DC, pp. 453–472.

Abu Abdo, A.M., Jung, S.J., 2020. Investigation of Reinforcing Flexible Pavements with Waste Plastic Fibers in Ras Al Khaimah, UAE, Road Materials and Pavement Design. Taylor and Francis.

Al-Rumaithi, A., 2021. Multi-layer Elastic Analysis, MATLAB Central File Exchange. https://www.mathworks.com/matlabcentral/fileexchange/69465-multi-layer-elastic-analysis.

ARA Inc, 2004. ERES Consultants Division. Guide for Mechanistic-Empirical Design of New and Rehabilitated Pavement Structures. Final Report, NCHRP Project 1-37A. Transportation Research Board of the National Academies, Washington, DC.

Asphalt Institute, 1991. Thickness Design-Asphalt Pavements for Highways and Streets, Manual Series No. 1. Asphalt Institute, Lexington, KY.

Austroads, 2004. Pavement Design. Austroads, Sydney, Australia.

Dilip, D.M., Babu, G.L.S., 2013. Methodology for pavement design reliability and back analysis using Markov chain Monte Carlo simulation. J. Transport. Eng. 139 (1), 65–74.

Dilip, D.M., Ravi, P., Babu, G.L.S., 2013. System reliability analysis of flexible pavements. J. Transport. Eng. 139 (10), 1001–1009.

Dong, Q., Huang, B., 2012. Evaluation of influence factors on crack initiation of LTPP resurfaced-asphalt pavements using parametric survival analysis. J. Perform. Constr. Facil. 28 (2), 412–421.

Dubourg, V., 2011. Adaptive Surrogate Models for Reliability Analysis and Reliability-Based Design Optimization. Ph. D. thesis. Université Blaise Pascal, Clermont-Ferrand, France.

Echard, B., Gayton, N., Lemaire, M., 2011. AK-MCS: an active learning reliability method combining Kriging and Monte Carlo simulation. Struct. Saf. 33 (2), 145–154.

FHWA, 2017. Towards Sustainable Pavement Systems : A Reference Document. https://www.fhwa.dot.gov/pavement/sustainability/ref_doc.cfm.

IRC:37, 2018. Guidelines for the Design of Flexible Pavements. Indian Road Congress, New Delhi.

Luo, Z., Karki, A., Pan, E., Abbas, A.R., Arefin, M.S., Hu, B., 2018. Effect of uncertain material property on system reliability in mechanistic-empirical pavement design. Construct. Build. Mater. 172, 488–498.

Luo, Z., Hu, B., Pan, E., 2019. Robust design approach for flexible pavements to minimize the influence of material property uncertainty. Construct. Build. Mater. 225, 332–339.

Marelli, S., Sudret, B., 2014. UQLab: a framework for uncertainty quantification in Matlab. In: Proceedings of the 2nd International Conference On Vulnerability, Risk Analysis and Management (ICVRAM2014), Liverpool (United Kingdom), pp. 2554–2563.

MATLAB, 2016. MATLAB Version 7.10.0 (R2010a). The MathWorks Inc., Natick, Massachusetts.

Mittal, A., Swamy, A., 2020. The Effect of Model Uncertainty on the Reliability of Asphalt Pavements, Transportation Research. Springer, pp. 771–780.

Noureldin, A., Sharaf, E., Arafah, A., Al-Sugair, F., 1994. Estimation of standard deviation of predicted performance of flexible pavements using AASHTO model. Transport. Res. Rec. 1449, 46–56.

Ogwang, A., Madanat, S., Horvath, A., 2019. Optimal cracking threshold resurfacing policies in asphalt pavement management to minimize costs and emissions. J. Infrastruct. Syst. 25 (2), 04019003.

Saride, S., Peddinti, P.R., Basha, M.B., 2019. Reliability perspective on optimum design of flexible pavements for fatigue and rutting performance. J. Transp. Eng. B Pavements 145 (2), 04019008.

Souliman, M.I., Mamlouk, M., Eifert, A., 2016. Cost-effectiveness of rubber and polymer modified asphalt mixtures as related to sustainable fatigue performance. Procedia Eng. 145, 404–411.

Sundar, V.S., Michael, D.S., 2019. Reliability analysis using adaptive kriging surrogates with multimodel inference. ASCE-ASME J. Risk Uncertain. Eng. Syst. A Civ. Eng. 5 (2), 04019004.

Timm, D., Birgisson, B., Newcomb, D., 1998. Development of mechanistic-empirical pavement design in Minnesota. Transport. Res. Rec. 1629 (1), 181–188.

Usman, N., Masirin, M.I.M., 2019. Performance of Asphalt Concrete with Plastic Fibres, Use of Recycled Plastics in Eco-Efficient Concrete. Woodhead Publishing, pp. 427–440.

CHAPTER 10

Aggregating risks from aquifer contamination and subsidence by inclusive multiple modeling practices

Maryam Gharekhani[1], Rahman Khatibi[2], Ata Allah Nadiri[1,3,5], Sina Sadeghfam[4]

[1]Department of Earth Sciences, Faculty of Natural Sciences, University of Tabriz, Tabriz, East Azerbaijan, Iran; [2]GTEV-ReX Limited, Swindon, United Kingdom; [3]Traditional Medicine and Hydrotherapy Research Center, Ardabil University of Medical Sciences, Ardabil, Iran; [4]Department of Civil Engineering, Faculty of Engineering, University of Maragheh, Maragheh, East Azerbaijan, Iran; [5]Medical Geology and Environmental Research Center, University of Tabriz, Tabriz, Iran

1. Introduction

A risk aggregation problem is presented in this chapter for the management of plains and their underlying aquifers exposed to risks from contamination and subsidence. This chapter builds on the authors' ongoing research to transform commonly used vulnerability indices to risk indices on both contamination and subsidence, which are traditionally studied separately in terms of their vulnerability, each using a set of data layers accounting for their local variability. For instance, DRASTIC is a set of seven data layers invoking collectively aquifer vulnerability to anthropogenic contamination; likewise, ALPRIFT is a set of seven different data layers invoking collectively that of a plain to subsidence. This chapter treats vulnerability as local property but risk as mathematical products of local and system-wide properties. Thus, a new set of data layers are introduced to represent system-wide impacts through an appropriate algorithm for both components. The algorithm for each risk is facilitated by using the framework of Origin, Source, Pathways, Receptors, and Consequences (OSPRC) and risk cells, as introduced by the authors recently, Sadeghfam et al. (2020a,b), Nadiri et al. (2018c). Various modeling strategies are discussed to carry out the risk aggregation problem.

Risk-based decision-making is now all pervasive, see Chapter 1 and some of the methodologies for calculating risk are as follows: (i) probabilistic techniques to quantify risk as products of consequences of hazard and their likelihood, as reviewed by Khatibi (2011); (ii) techniques that combine load and resistance, as reviewed by Tung et al. (2005) and Sadeghfam et al. (2018a), but in reality, these techniques cater for the reliability of operational systems; and (iii) indexing risk as products of vulnerability to measure local properties (referred to by the authors as Passive Vulnerability Indices, PVIs) multiplied by measures of impacts, where some refer to these as hazard (as reviewed in Living with risk, 2004) but Sadeghfam et al. (2020a,b) refer to them as Active Vulnerability Indices (AVIs). The probabilistic techniques render absolute risk values often known as risk quantification techniques, but they require extensive amount of data and are precluded in this chapter. For sparse data, risk indexing techniques are used and this chapter transforms vulnerability indexing by DRASTIC and ALPRIFT into risk indexing problems.

This chapter takes one step toward formulating appropriate risk analysis tools for aquifers under sparse data to serve risk analysis processes, which consist of risk assessment, risk management, and risk communication. Some of the research on aquifer risks may broadly be classified as follows: (i) control of water-related health risks, e.g., US EPA (2001), UNICEF (2012), WHO (2009), and Krishna et al. (2019); (ii) groundwater remediation, e.g., US EPA (2001), Li et al. (2016), and Qian et al. (2020); and (iii) the risk of contamination or subsidence assessed in terms of hazard and vulnerability in GIS environment, e.g., Huang et al. (2012), Wu et al. (2010), Bilgot and Parriaux (2009), Matzeu et al. (2017), Li et al. (2012), Wang et al. (2012), and Nadiri et al. (2018a). This chapter is focused on the latter category and builds risk mapping using the DRASTIC and ALPRIFT vulnerability maps and extending their concepts for risk mapping.

The authors refer to both DRASTIC and ALPRIFT as frameworks, which comprise consensually selected data layers for seeking a new sense out of the combined data. Each of the frameworks pools together seven data layers with a scoring system of prescribed rates to account for local variations and prescribed weights to account for their relative importance. Prescribing the values makes the procedure susceptible to subjectivity and a requirement for checking the information content of the results. This gives rise to a topical research to reduce impacts of prescribed values, see Nadiri et al. (2018b, 2017a,b,c) and Sadeghfam et al. (2016). The seven data layers for both DRASTIC and ALPRIFT are outlined later in this chapter, which are used widely and these are used as PVIs. This chapter uses a data layer for the AVI as follows: (i) Velocity of groundwater (Ve) is used as AVI similar to Nadiri et al. (2018c) for risk mapping of contaminants and (ii) Density of abstraction wells (De) is used as AVI for risk mapping of subsidence.

The aggregation of risks from contamination and subsidence is explored by using Ardabil plain, Ardabil province, northwest Iran, for which vulnerability studies have already been published, for DRASTIC vulnerability indices, see Nadiri et al. (2017b) and for ALPRIFT vulnerability indices, see Nadiri et al. (2020). These results are reproduced here without repeating them, which use Inclusive Multiple Models (IMMs), as defined by Khatibi et al. (2020) and Khatibi and Nadiri (2020). They are implemented at two levels: **Level 1** for both DRASTIC and ALPRIFT vulnerability indices, three variations of fuzzy models (Sugeno FL (SFL), Mamdani FL (MFL) and Larsen FL (LFL)) are used to reduce the inherent subjectivity in their weights by learning their values from both the data and target values; **Level 2** combines these models by another model to combine the outputs of the models at Level 1, which serves as a supervised learning process to learn from the Level 1 results and from the target values.

The risk aggregation problem is made possible by dividing a study area into as many risk cells as necessary, within each one of which the risk from one origin is studied and this chapter considers two risk cells: (i) Risk Cell 1 accounts for contamination; and (ii) Risk Cell 2 accounts for subsidence. Each risk cell is treated idiosyncratically with no restriction on selecting the modeling strategy or on the number of risk cells. The implementation of risk aggregation problems also uses the OSPRC framework, which breaks down a study area into pixels or grid cells, see Khatibi (2008), Khatibi (2011), and Nadiri et al. (2018c). The pixels in each risk cell are identified with one of these processes and in this way risks from different origins can be treated in an appropriate manner, e.g., contamination and subsidence. Notably, Khatibi (2008) argues that the SPRC framework is widely known to have emerged in the decade before 2000 but without the origin dimension and suggests its possible uses for integrating risks with different idiosyncrasies and Nadiri et al. (2018c) make a successful use of the OSPRC.

Risk aggregation problems are investigated by a modularized modeling strategy, in which one module is used for each risk cell and each delivering a risk index value at each pixel. These need to be aggregated and this chapter investigates three strategies: (i) the risk indices at each pixels from Risk Cells 1 and 2 are simply added; (ii) a data fusion technique developed to learn a representative value from the values of each risk cell; and (iii) the same data fusion scheme is used for the normalized risk indices. The simple algorithm used in this chapter for data fusion is based on catastrophe theory as applied in the decision theory given by Cheng et al. (1996). Data fusion is used widely and the best example is the human mind capable of combining and making sense from different senses. The strategies are tested in Ardabil plain, renowned for its agricultural activities, but since 1980s, fertilizers and pumpage have accelerated risk of contamination of the aquifer with nitrate and its concentration exceeding the permissible value of 10 mg/L (as per United State Environmental Protection Agency - USEPA 2012) is considerable; as well as to subsidence.

2. Study area

2.1 Physical system

The study area is described by the authors elsewhere, see Nadiri et al. (2017a,b) and outlined below. Ardabil plain with approx. 990 km^2 (see Fig. 10.1) is located in central parts of Ardabil province and is part of the Qarasu (blackwater) basin, which drains three major mountain ranges: (i) Mount Savalan (Sabalan) at its west, which is 4811 m AMSL with a permafrost crater at the peak; (ii) Qushchu mountains ranges at its southern side; and (iii) Baghri (also known as Talesh) mountains at its east. The aquifer is the outcome of geomorphological processes drained by Qarasu and its tributaries including Balikhli Chay, Qara Chay, Hire Chay, Naqiz Chay, Namin Chay, Nuran Chay, and Sulaya Chay.

2.2 Geological context

Following the expansion of the Laramide orogeny phase in the study area, volcanic activities in the Eocene period have given rise to the formation of Mount Savalan and Baghri. Its geological formations comprise: (i) limestone formation mostly occurred in the Jurassic period located in the northern part, (ii) volcanic formations composed from Basalt (Cretaceous) to Andesite and Trachyandesite (Quaternary) in the eastern part as the dominant formation of the basin; (iii) tuff formations, marl, and sandstone mostly occurred in Neogene and located in the northwest of the area; (iv) conglomerates during the Cretaceous epoch and volcanic conglomerate, tuff conglomerate during Neogene and Quaternary; and (v) traces of recent alluvium related to Quaternary mostly at the north of the study area (Nadiri et al., 2017a).

Fractures are likely at the west and related to igneous and pyroclastic rocks originating from

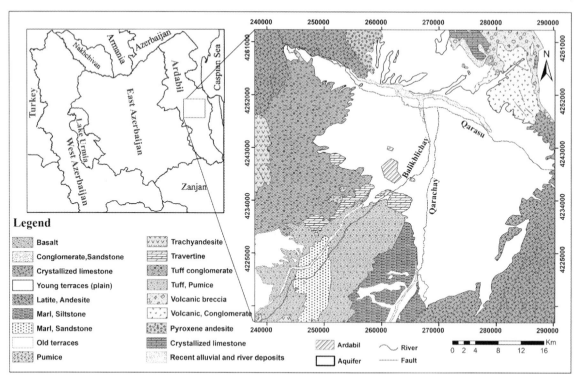

FIGURE 10.1 Geological map and location of study area.

volcanic activity of the Mount Savalan, which serve as the main source to recharge the alluvial aquifer. Noncarbonate rocks of Lahar and conglomerate outcrops in the western part of the plain are likely to feed the aquifer and covers approx. 430 km^2. The plain comprises alluvial deposits of sand, silt, and clay, a composition susceptible to subsidence in agricultural regions.

The region is exposed to the activity of the main faults at its southern parts, which run in the southwest to northeast direction related to the upper Eocene volcanic activities. The Savalan volcanic activities in the Quaternary period became the primary feature and following tertiary tectonic processes. The faults in the area include (i) Savalan faults are outcomes of volcanic activities, but subsequent lava flows include fractures and sometimes deep faults in different directions near the Mount Savalan and give rise to hot water mineral springs; (ii) East Ardabil Fault along the direction of south to north, parallel to the Astara fault and continues toward the Masouleh fault; and (iii) the Talesh fault runs for approx. 400 km along in many series largely in the direction of northeast to southwest.

2.3 Hydrogeology of study area

The subbasin relating to Ardabil plain is approx. 4000 km^2, of which 25% is unconfined and under the plain but its 75% of the area expands to mountainous areas. Geoelectric studies show that changes in alluvium thicknesses at the margins of the plain are often low but gradually increase toward its middle so that higher thicknesses are noted in the middle parts of the plain and at eastern parts (Nadiri et al., 2017a). The sources recharging the aquifer comprise (i) groundwater flows though the joints and fractures at the mountainous upper catchment; (ii) rainfall directly falls on the plain; and (iii) the aquifer and the watercourses interact with net losses and gains varying through time. Groundwater in the aquifer is withdrawn through 3669 deep and semideep withdrawal wells, 391 natural springs, and 48 qanats.

2.4 Land use

The major city in the study area is the historic city of Ardabil with a population of more than half a million and with an additional one million living in the plain. The main economic activities are agriculture for its fertility and some processing industries around Ardabil including drink and food packing, producing electronic devices, producing paper, and factories for producing plastic tools.

3. Methodology

A prototype for aggregating risks to aquifers is modularized as follows: (i) examine a study area, its spatial layout, data availability with a view to identify risk origins and subdivide the study area into pixels and a number of risk cells, (ii) identify origins of risk; and allocate one module for each risk origin; (iii) formulate a modeling strategy for each module to ascertain its risk indices; and (iv) a further module is developed to aggregate the risks, which may just comprise algorithms to aggregate risks. The modeling strategy in this chapter is carried out by identifying two risk cells and therefore three modules are needed, in which Module 1 is used for risk

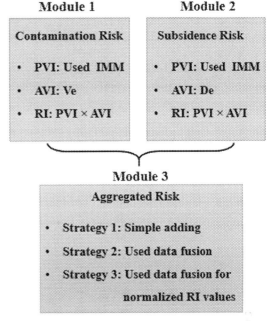

FIGURE 10.2 Three Modules to mapping aggregated risk.

indexing from nitrate-N contamination, Module 2 for subsidence, and Module 3 for their aggregation (Fig. 10.2). These are outlined below.

3.1 Preliminary concept

3.1.1 Risk cells and OSPRC framework

The research in this chapter employs the concept of risk cells presented by Nadiri et al. (2018c) and Sadeghfam et al. (2018a,b). It transforms a study area into as many risk cells as it required, where one risk cell is assumed not to interact with others at the level of lower properties. The only interaction is the aggregation of risk values. For this purpose, the study area is subdivided to grids, which are identical for each and all risk cells. The risk at each risk cell is studied through the OSPRC framework. The addition of the origin dimension, "O," makes it possible to be specific in the way the risk emerges. To this end, the origin dimension makes it possible to discern such variations as anthropogenic from geogenic activities, diffuse loads from point loads (point sources), diversity of pollutants, and reactant or conservative pollutants. Origins of source express potentials and are different than Source, defined as the triggered activities. Pathways focus on the way risks are propagated system wide, but Receptors and Consequences normally extend to the social dimension, but these are not addressed in this chapter owing to the lack of appropriate data.

3.1.2 DRASTIC framework for mapping vulnerability to anthropogenic contamination

This chapter formulates specific vulnerability using the DRASTIC framework in response to anthropogenic impacts. The formulation was presented by the United State Environmental Protection Agency (U.S. EPA), which provides a general indication of groundwater vulnerability potentials to surface contaminations (Aller et al., 1987). It considers seven hydrogeological data layers, which comprise the following: Depth to water table (D), net Recharge (R), Aquifer media (A), Soil media (S), Topography or slope (T), Impact of the vadose zone (I), and hydraulic Conductivity (C) of the aquifer. Each data layer is assigned with rates to account for its local variation within each pixel and weights to account for its relative importance through the following steps: (i) discretize the study area in GIS into pixels of an appropriate size; (ii) reorganize the data as an array of seven raw data layers for each pixel, each with prescribed values; (iii) assign the values of the rates at each pixel to the raw values, as recommended by Aller et al. (1987); and (iv) assign weights to each data layer, as recommended by Aller et al. (1987). The authors regard the subsequent vulnerability index as PVI expressed, as follows:

$$PVI = D_r D_w + R_r R_w + A_r A_w + S_r S_w + T_r T_w + I_r I_w + C_r C_w$$

(10.1)

where D, R, A, S, T, I, and C are the seven data layers and the subscripts "r" and "w" refer to the rates and weights, respectively.

3.1.3 ALPRIFT framework and InSAR data for mapping vulnerability to subsidence

The ALPRIFT framework, introduced by Nadiri et al. (2018a), maps Subsidence Vulnerability Indices (SVIs) of any aquifer/plain system, which pools together seven data layers. The data layers are assumed to be independent of one another and comprise Aquifer media (A), Land use (L), Pumping of groundwater (P), Recharge (R), Impacts instigated by aquifer thickness (I), Fault distance (F), and water Table decline (T). Each data layer is assigned with rates to account for local variation and weights to account for its relative importance through the following steps: (i) discretize the study area in GIS into pixels of an appropriate size; (ii) reorganize the data as an array of seven raw data layers for each pixel, each with prescribed values; (iii) assign the values of the rates at each pixel to the raw values; and (iv) assign weights to each data layer. These render PVI, expressed as follows:

$$PVI = A_r A_w + L_r L_w + P_r P_w + R_r R_w + I_r I_w + F_r F_w + T_r T_w$$

(10.2)

where PVI uses the seven data layers, n, in which subscript "r" denotes rates and "w" that of weights, the values of which are prescribed by Nadiri et al. (2018a). Notably, minimum and maximum of PVI indices are 24 and 240, respectively.

The remote sensing data for the research in this chapter uses satellite images to assess land subsidence and to detect land subsidence from Sentinel-1A (Interferometric Synthetic Aperture Radar - InSAR). Open access InSAR images can be downloaded from https://scihub.copernicus.eu/, which have a vertical resolution of 10 m, and as such they are unsuitable for a direct assessment of subsidence and need some ground-truthing. The period investigated is from August 2015 to August 2016. For more background information and alternative approaches, see Nadiri et al. (2018a, 2020). The subsequent data processing is through the following steps: (i) measure a set of values at a set of observation wells identifiable by their geo-references; (ii) use GIS to lay out the InSAR image over the pixels at the study area and transfer their values to each pixel; (iii) apply the ENVI ground-truth modules to adjust automatically InSAR values at each pixel using a set of sparsely measured subsidence values; (iv) carry out ground-truthing and normalize the subsidence data layer (i.e., calculate $\frac{SGT_i}{SGT_{max}}$); (v) condition the output, as follows:

$$CPVI_i = \frac{SGT_i}{SGT_{max}} \times PVI_{max} \qquad (10.3)$$

where $CPVI$ = Conditioned PVI at pixel i; PVI_{max} = the maximum PVI calculated from the ALPRIFT framework; SGT_i = subsidence value (in cm) at grid cell i, SGT_{max} = maximum SGT values.

3.1.4 Risk mapping

The domain of the influence of each data layer for contamination and subsidence can be local or system wide. Similar to Sadeghfam et al. (2020a), the chapter argues that a local domain of influence would refer to the correlation among a number of pixels as a measure of PVI of the DRASTIC data layers and ALPRIFT data layers. These are suitable to serve as PVI.

A further data layer is introduced to extract information for the system-wide propagation or as a measure of system-wide impacts. This comprises a data layer in terms of groundwater Velocity (Ve), to be referred as Ve data layer; see Table 10.1 for more information. The rate of the data layer at each pixel takes the value of the velocity at that pixel subject to normalization. This is sufficient to calculate its AVI values for risk indexing as follows:

Active Vulnerability Index:
$$AVI_{Contamination} = Ve_w Ve_r \qquad (10.4)$$

Ve_w is set to 1 and the value of Ve_r is the actual velocity at each pixel. Similarly, a further data layer is introduced for risk mapping of subsidence to extract information for system-wide propagation or as a measure of system-wide impacts. This comprises a data layer in term of Density of abstraction wells (De), to be referred as De data layer; see Table 10.1 for more information. The rate of the data layer at each pixel takes the value of the density of abstraction wells at

TABLE 10.1 Definition of AVI data layers.

Data layers	Description	Data layers	Method	Weight	Reference map
Velocity of groundwater (Ve)	Expresses groundwater movement in aquifers based on horizontal hydraulic conductivity and hydraulic gradient. The higher velocity, the greater is the transfer of contaminants in the aquifer.	Used 55 OWs.	1. Prepare groundwater level by ordinary kriging, 2. Calculate hydraulic gradient, 3. Use the hydraulic conductivity data layer (C) 4. Multiply (2) by (3) above.	1	Fig. 10.3B1
Density of abstraction wells (De)	Increasing the density of abstraction wells leads to increased pumpage; natural withdrawal of groundwater would strain soils and induce subsidence.	Used the location of 3669 available abstraction wells.	Processed for each pixel using the kernel density method in GIS platform.	1	Fig. 10.3B2

that pixel subject to normalization. This data layer indicates a potential for propagating system-wide impacts and accounts for activating hydrostatic pressure through withdrawing from any control volume. Withdrawing hydrostatic pressure triggers active subsidence processes and hence Active Vulnerability Index (AVI), expressed as follows:

Active Vulnerability Index:
$$AVI_{Subsidence} = De_w De_r \qquad (10.5)$$

De_w is set to 1 and the value of De_r is the actual density of abstraction wells at each pixel. The authors have published a proof of concept for the two quantities of PVI and AVI; see Sadeghfam et al. (2020a,b). The similarities are striking between the collective of PVI and AVI with the concept of risk in terms of the collective of hazard and likelihood, as in the classical definition of risk, expressed as follows (Almoussawi and Christian 2005; Khatibi 2011):

$$Quantified\ Risk = Consequences\ of\ Hazards \times Likelihood \qquad (10.6a)$$

The gist of Eq. (10.6a) is that hazard is a local property but can prevail at any location, whereas likelihood is a statistical measure of impacts within the system. Nadiri et al. (2018c) and Sadeghfam et al. (2020a) introduce the terms passive and active as the key for expressing risk indices in the same way as Eq. (10.6a) as follows:

$$Risk\ Index = PVI \times AVI \qquad (10.6b)$$

3.1.5 Inclusive multiple models

Aquifer risk mapping for nitrate-N by Module 1 and for subsidence by Module 2 needs to be sufficiently accurate before they can be aggregated over a plain. Khatibi et al. (2020) and Khatibi and Nadiri (2020) present a view of modeling practices and classify them as Exclusionary Multiple Models (EMMs) and Inclusive Multiple Modeling (IMM) practices. EMM

practices normally give rise to developing innovative models, but they criticize them for using inappropriate superlative expressions such as selecting a "superior" model from multiple models and thereby rejecting the others. In reality, EMM results are unlikely to be superior but may be regarded as fit-for-purpose. However, they argue that IMM practices unify many modeling strategies by putting emphasis on learning from multiple models to enhance accuracies. In particular, they suggest a greater learning from error residuals to extract as much information as possible from the data and target values through supervised learning techniques to become defensible models.

Supervised learning techniques require the availability of target values, but this is not always available. Without any learning, the parametric values are prescribed on the basis of expert judgment, e.g., Eqs. (10.4) and (10.5), or by using learning only from the input data. Both are the cases here for risk mapping and risk aggregation problems. One of the techniques used for learning from the input data alone is data fusion. A variety of techniques supports data fusion and they include statistical matching, Bayesian inference (as reviewed by Chen and Han 2016), gray relational analysis (Bai et al., 2016), and moving average filters (Rodriguez et al., 2015). The data fusion technique used in the chapter is based on catastrophe theory as it is rooted in the decision theory by Cheng et al. (1996), which is detailed below.

3.1.6 Risk aggregation problem

Transforming vulnerability indices into risk indices and aggregating their values for aquifers are not topical yet due to lacking the appropriate methodology. Those reported by the authors comprise the following: Sadeghfam et al. (2020a) transformed the SVIs into Risk Indices (RIs) by breaking down the ALPRIFT into ALRIF (characterizing PVI) and water-driven PT data layers (characterizing AVI). They applied data fusion to combining PVI, AVI, or RI with InSAR results to produce more consistent results, and then predicted the PVI, AVI, and RI maps by Artificial Neural Network (ANN). Therefore, they applied supervised learning to produce PVI, AVI, and RI maps, whereas in the chapter the authors applied supervised learning to produce PVI but explore an unsupervised learning to produce to the risk aggregation problem.

The chapter investigates risk aggregations from two origins of contamination and subsidence, for each of which two risk cells are delineated—Risk Cell 1 and Risk Cell 2. Three strategies are formulated to aggregated risk indices from contamination and subsidence origins, as follows:

Strategy 1: This uses the simple algorithm of adding the risk indices at each pixel without any learning from the input data.

Strategy 2: This uses a data fusion technique based on catastrophe theory rooted in the decision theory by Cheng et al. (1996). It produces a new estimate of Risk Index, RI_{agg} from RI_{cont} and from RI_{subs}, as follows:

$$RI_{agg} = \left(\sqrt[3]{RI_{cont}} + \sqrt[2]{RI_{subs}} \right) \Big/ 2 \qquad (10.7)$$

This is similar to Sadeghfam et al. (2020a), which also presents more details. The approach is tantamount to data fusion, which aims to extract new representative information from a set of different strands of existing information.

Strategy 3: This is similar to Strategy 2, but uses normalized RI data in Eq. (10.7).

3.2 Modeling strategy

3.2.1 Modules 1 and 2—vulnerability indices

The authors have already published defensible vulnerability mapping results using IMM for both nitrate-N contamination and subsidence as a result of overabstraction. The results are reproduced for a further research in the chapter and therefore their modeling strategies are only outlined below without any detail. The modeling strategies for both Modules 1 and 2 were carried out at three levels, as follows:

Level 0: Data layers are abstracted from raw data for both DRASTIC and ALPRIFT frameworks, as well as observed data and InSAR data. Additionally, the data layers for AVI are also processed at this stage, which comprise Ve data layer for Module 1 and De data layer for Module 2.

Level 1: Both modules at Level 1 were constructed using Fuzzy Logic (FL) formulated by the procedures given by: Sugeno FL (SFL), Mamdani FL (MFL), and Larsen FL (LFL), the procedure for which is given by Nadiri et al. (2017b, 2020). These models enable supervised learning from both input data and target values.

Level 2: Module 1 uses ANN to combine the models at Level 1 and Module 2 uses Genetic Expression Programming (GEP), both of which also incorporate their target values. These are detailed by Nadiri et al. (2017b, 2020). The inputs to these models were taken from the outputs of their models at Level 1 and both models are supervised and learn from the same target values at Level 1. Further details are given by Nadiri et al. (2017b, 2020).

Level 3: Under normal circumstances, the modeling strategy would involve a Level 3 phase, for which the vulnerability maps for both Modules 1 and 2 would be regarded as respective PVI values and a further step would be taken to estimate (i) Ve and De data layer and (ii) estimate RI for both Modules 1 and 2, as per Eq. (10.6b). All these steps were implemented for the results presented in the chapter.

3.2.2 Module 3

Module 3 estimates the aggregated risk indices using three strategies as presented in Section 3.1.6.

3.3 Best practice in preparing dataset

The best practice procedure for the DRASTIC and ALPRIFT data layers is presented by

Nadiri et al. (2017b, 2020) and these are not repeated here. The new data layers introduced in the chapter are summarized in Table 10.1, which also outlines the various decisions made for the mapping of Ve and De data layers.

The study area is divided into 3760 number of pixels of size 500×500 m with a provision of 55 observation wells to record groundwater levels; 3669 number of abstraction wells; and 34 number of wells with geological logs. The tedious data preparation involves a set of decisions, which are presented in detail by Nadiri et al. (2020, 2017a,b) and they are reproduced in Appendix I.

3.4 The performance metrics

The performance of PVI, AVI, and RI is measured by Receiver Operating Characteristics (ROCs). ROC curves are a spatial goodness-of-fit metric for spatially distributed variables. Since its first use by electrical engineers, it is now used widely and includes hydrology, groundwater problems and psychology, data mining, machine learning, and model performance assessment. ROC curves comprise a plot of True Positive Rates (TPRs) against False Positive Rates (FPRs) at various threshold settings. ROC is also referred to as relative operating characteristic, as it compares two operating characteristics (TPR and FPR). The shape of the ROC curve is information on its own as it indicates visually on the performance of the modeled variable. The Area Under Curve (AUC) of ROC is also a performance metric, where an AUC value of 1 signifies a perfect fit but that of 0.5 signifies a strong presence of noises.

Some of the mapping results are banded using Jenks' optimization method to identify the optimum arrangement of different classes, in which the minimum average deviation is found from the class mean in terms of maximizing the deviation of each class from that of other classes. The key feature of the Jenks optimization method is the reduction of the variance within the classes

is in terms of maximizing the class ranges (Jenks, 1967). The paper uses five bands as follows: Band 1 (low), Band 2 (relatively low), Band 3 (moderate), Band 4 (relatively high), and Band 5 (high). Using this technique reduces the subjectivity associated with indices classification by expert judgment.

The results not using the Jenk's' method are also banded but in the range of five equal bands and for normalized indices, these are (band 1: 0−0.2; band 2: 0.2−0.4; band 3: 0.4−0.6; band 4: 0.6−0.8; band 5: 0.8−1).

4. Results

This section reproduces the PVI maps for both DRASTIC and ALPRIFT from authors' published works; AVI maps are produced by this research and the products of PVI and AVI are mapped to produce risk indices. These values are worked out at each pixel and the results are banded within five ranges for the presentation of the results. The calculated risk values are further aggregated by investigating three strategies.

4.1 Risk mapping against contamination and subsidence—Modules 1 and 2

The vulnerability mapping for the DRASTIC and ALPRIFT data layers is presented by Nadiri et al. (2017b, 2020) and these are reproduced in this chapter as PVI maps in Fig. 10.3A1 for nitrate-N contamination and Fig. 10.3A2 for subsidence. Further results are produced by the research for both Ve and De data layers and these are presented in Fig. 10.3B1 and 10.3B2 for both cases. Risk mapping against contamination and subsidence is processed as per Eq. (10.6b) and these are presented in Fig. 10.3C1 and 10.3C2.

A visual comparison between values in Fig. 10.3A1 and 10.3B1 (PVI, AVI for contamination), as well as Fig. 10.3A2 and 10.3B2 (PVI, AVI for subsidence), indicates that PVI and AVI are quite

4. Results 143

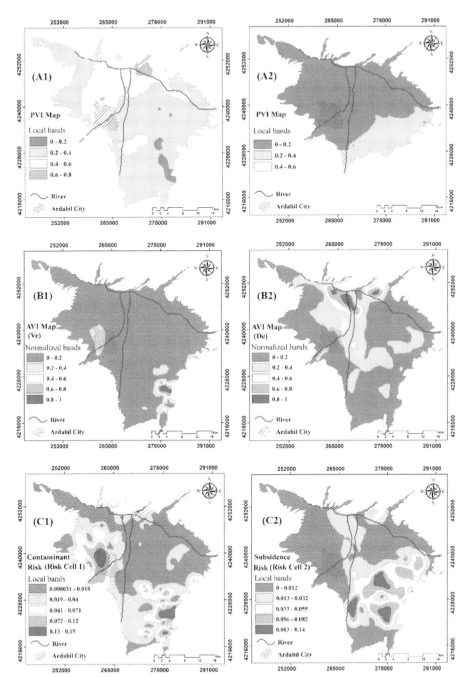

FIGURE 10.3 Mapping results: (A) PVI maps (reproduced from authors' published works); (B) AVI maps; (C) Risk Indices (RI) maps. Note: 1 denoted nitrate-N contamination, e.g., (3A1) and 2 denotes subsidence, e.g., (3A2).

independent but their products presented in Fig. 10.3C1 and 10.3C2 reflect influences from both PVI and AVI, as well as hotspots stemming from their products.

There is a problem with interpreting relative values of risk indices as the products of two relative values of PVI and AVI between "0" and "1" will naturally be less than their individual values. A drop in the risk index value is not considered to be a problem and therefore the banding is carried out between the maximum and minimum values.

The risk bands due to contamination in Fig. 10.3C1 reveal two zones of hotspots, at northwest and southeast of the study area. At the northwest, PVI values are high but AVI is high at southeast. Similarly, the risk bands due to subsidence in Fig. 10.3C2 reveal one zone of hotspots, which is seemingly influenced by high PVI values. Relatively high AVI values at north of Fig. 10.3B2 do not seemingly influence the risk map. These results reveal possible problems with normalizing and this is discussed in more detail in due course.

Observed nitrate-N contamination and subsidence maps are displayed in Fig. 10.4A1 and 10.4A2, which make up the basis for target values both in Modules 1 and 2 for the mapping of PVI (vulnerability) values. Their performances measured by ROC/AUC are reproduced in Fig. 10.4B1 and 10.4B2, which provide evidence of their IMM modeling strategy being capable of producing defensible results. However, no learning is associated with risk indexing and this is discussed below.

The results presented in Fig. 10.4B1 and 10.4B2 display ROC/AUC performance metrics for AVI and risk indices. Although AVI involves no learning, the signals for its statistical agreement with the respective observed results are quite strong (AUC = 0.91 for the contamination AVI mapping and AUC = 0.86 for subsidence AVI mapping). The signals for risk mapping are even stronger (AUC = 0.93 for contamination RI mapping and AUC = 0.90 for subsidence RI mapping). The derived risk indices do not maximize the extracted signals from the data and therefore may not be considered as defensible. However, they are certainly fit-for-purpose and if their features do not match with observed values on a one-to-one basis, this is because they employ algorithms without learning from the target values. Notably, risk maps discover features that are not readily visible from observed values, PVI and AVI. The results make case for an exploration of the risk aggregation problem.

4.2 Risk aggregation—Module 3

Three strategies are investigated for the risk aggregation problems and these are used to investigate the collective exposure of the study area to risk from nitrate-N contamination and subsidence. Strategy 1 uses the simple strategy of adding up risk indices at each pixel; Strategy 2 uses a simple scheme based on decision theory; and Strategy 3 normalizes the risk for each risk cell and then uses the same scheme as in Strategy 2. The results of Strategies 1—3 are presented in Fig. 10.5A—C. Each of these figures agrees broadly with one another but they differ from each other in degree of relative risk.

The results presented in the chapter on subsidence risk as Module 2 reveals one zone of hotspots at the south of the plain, which is seemingly influenced by high PVI values. Relatively high AVI values do not seemingly influence the subsidence risk map.

The results presented in the chapter on aggregating risks as Module 3 by three strategies to collective exposure of the study area to risk from nitrate-N contamination and subsidence show that the southeast of the aquifer is a hotspot area and is exposed to contamination and subsidence risks.

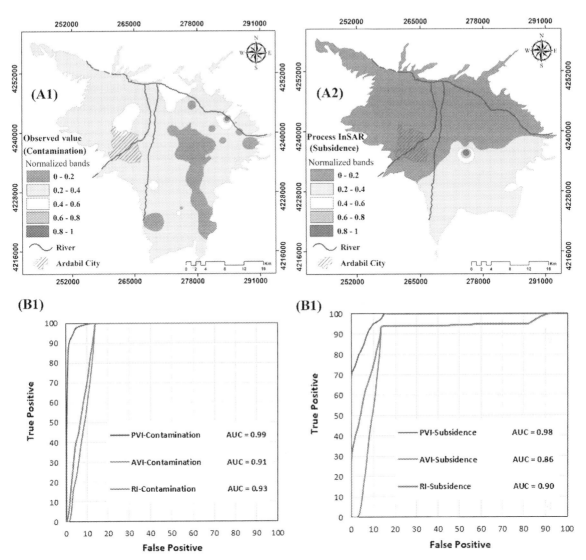

FIGURE 10.4 Observed values and measuring performance of PVI, AVI, and RI and (4A1) measured nitrate-N contamination, (4A2) InSAR-based subsidence; (4B1) ROC/AUC of PVI, AVI, and RIs for nitrate-N contamination and (4B2) ROC/AUC of PVI, AVI, and RIs for subsidence.

5. Discussion

To the best knowledge of the authors, the risk aggregation problem from multiple origin and multiple sources is not topical in research owing to the absence of any flexible technique. The research presented in the chapter builds on the authors past works, e.g., Nadiri et al. (2018c) and Sadeghfam et al. (2020a,b).

A proof of concept for risk aggregation problems is presented Nadiri et al. (2018c) in using the DRASTIC framework for anthropogenic contaminant of nitrate-N and the SPECTR framework for contaminants from geogenic

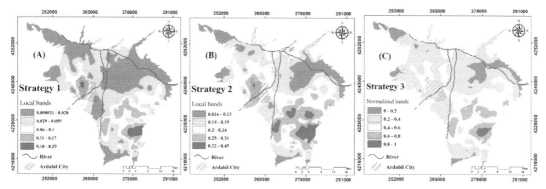

FIGURE 10.5 Aggregated risk maps: (A) Strategy 1 (used Jenks' method); (B) Strategy 2 (used Jenk's' method); (C) Strategy 3.

origin of arsenic. Using the OSPRC (Origins, Source, Pathways, Receptors, Consequences) framework, they employed two risk cell, one for each origin and captured the idiosyncrasies of each contaminant. Similarly, a proof of concept for subsidence risk indexing is given by Sadeghfam et al. (2020a,b), who identify appropriate data layers for PVI and AVI. They use a scheme based on catastrophe theory to carry out data fusion and to study PVI, AVI, or RI with respect to InSAR results to produce more consistent results.

Modeling studies using multiple models at one level are used widely, but these researchers often claim to produce superior results, which are little more than fit for purpose. This chapter is not focused on them, but it is pointed out that they normally focus on one contaminant, although Nadiri et al. (2018c) published vulnerability mapping results using Support Vector Machine at one level but with the aim of studying both nitrate-N and arsenic contamination PVI mapping. The results presented in the chapter for Module 1 on contamination studies and in Module 2 on subsidence make use of two-levels IMM strategies. The actual models used together help to harden the extracted information from the data toward producing defensible results. In some cases, the authors use a three-level modeling strategy for obtaining defensible modeling results, e.g., Nadiri et al. (2021).

The chapter uses AVI to transform PVI into RI mapping. Data layer for AVI used Velocity of groundwater (Ve) for mapping contaminants and Density of abstraction wells (De) is used for mapping subsidence. The PVI (DRASTIC and ALPRIFT) extracts information (nitrate contamination and subsidence), which is of local significance, whereas AVI (Ve and De) are of system-wide significance. In practice, other data layers may also be suitable for AVI, but these are planned for the future research studies on the subject. The chapter is mainly focused to explore the underlying ideas and demonstrate their proof of concept for further research.

Although ROC/AUC of PVI and AVI indicates that signals are quite strong, but the derived RI mappings do not maximize the extracted signals from the data and therefore may not be considered as defensible. This is because there is no observed aggregated risk value. One of the aims of Sedghi et al. (2022) (Chapter 11 of this book) is to explore the techniques for deriving a surrogate for aggregated risks. They apply data fusion scheme to 10 groundwater indices and show that the emerging surrogate index is a defensible representation of the individual indices. The authors plan to transfer the learning to an IMM risk aggregation problem with two levels of learning. Even simple risk aggregation strategies presented in this chapter are certainly fit for purpose and if their features do not match with observed values on a one-to-one basis, this is because they employ algorithms without learning from the target values.

The chapter presents evidence on the proof of concept for the risk aggregation problem for plains over aquifers, where there are two origins of risks. In reality, the methodology does not restrict the number of origins or contaminants, as it is a very capable approach and can cope evidently with complex problems. The only constraint is the absence of observed aggregate risk values, but further works are ongoing to employ data fusion algorithms and thereby measure the performances of aggregated RI mappings.

The study area was under rural and pristine conditions prior to the 1980s, but since then, the use of nitrate-based fertilizers took off and these had adverse chain effects on water consumptions. Thus, suddenly, pumpage increased in the study area in a background of no or little planning systems using decision-making by participation. As remarked by Llamas and Martínez-Santos (2005), "Intensive groundwater use is not a panacea, and it will not necessarily solve the world's water problems. In fact, should the prevailing anarchy continue, serious problems may appear in the mid- or long-term (two to three generations)." With no established system for equitable water use and no control on water quality, the sustainability of the study area is now questionable in the long run. The absence of equitable policies for water usage and unsustainable pumpage is quite typical in Iran but with local variations.

6. Conclusion

The chapter presents risk aggregation for the management of plains/aquifers exposed to risks from contamination and subsidence. This problem is solved by a modeling strategy using three module, as follows: Module 1: mapping Contaminating Risk for a study area by two-levels IMM strategies; (ii) Module 2: mapping Subsidence Risk for a study area by two-levels IMM strategies; and (iii) Module 3:

mapping aggregated risks related to contamination and subsidence by combining the results from Modules 1 and 2. Three strategies are investigated in Module 3. Strategy 1 uses the simple strategy of adding up risk indices at each pixel; Strategy 2 uses a simple scheme based on decision theory; and Strategy 3 normalizes the risk for each risk cell and then uses the same scheme as in Strategy 2.

The chapter uses AVI to transform PVI into Risk Index (RI) mapping in Modules 1 and 2. The data layer for AVI used Velocity of groundwater (Ve) for mapping contaminants and Density of abstraction wells (De) is used for mapping subsidence. The performance of PVI, AVI, and RI results is investigated by using ROC/AUC metrics and they show that the ROC/AUC of PVI and AVI signals is quite strong, but due to the lack of any measured aggregated risk, there is no technique yet to derive RI mappings by learning from the data to maximize the extracted signals; hence, RI maps may be considered as fit for purpose but not defensible.

The strength of this research is on aggregating risks from different origins of risk and its weaknesses is the absence of observed aggregate risk values for measuring the performances of aggregated risk indices mappings. A possible solution is discussed to be by using data fusion techniques.

Appendix I

The set of seven data layers were processed using the best practice procedure given in Table 10.A.1. These produced the rated DRASTIC data layers as given in Fig. 10.A.1A—G.

The set of seven raster layers were processed using the best practice procedure given in Table A.10.2. These produced the rated DRASTIC data layers as given in Fig. 10.A.2A—G.

148   10. Aggregating risks from aquifer contamination and subsidence by inclusive multiple modeling practices

TABLE 10.A.1   Definition of DRASTIC data layers.

Data layers	Description	Data layers	Interpolation method	Weight	Reference map
Depth to water (D)	The depth of unsaturated layer measured from surface to the water table. The less the depth, the more vulnerable the aquifer to pollution.	Used the 55 observation wells available.	Ordinary kriging (OK)	5	Fig. 10.A.1A
Net recharge (R)	Net recharge is the total quantity of water, which is infiltrated in the ground and reaches the water table.	The relationship for recharge given by Scanlon et al. (2002) is modified by adding the pumped water volume in a year.	—	4	Fig. 10.A.1B
Aquifer media (A)	It refers to the aquifer capacity to attenuate the pollutants. The larger grain size, higher the permeability and aquifer vulnerability.	Used the 34 geological logs available.	Ordinary kriging (OK)	3	Fig. 10.A.1C
Soil media (S)	It is between ground surface and unsaturated soil and is the upper weathered zone of the earth.	This was obtained from the soil cover map of Ardabil plain, published by the Ardabil Regional water Authority.	—	2	Fig. 10.A.1D
Topography (T)	It refers to the slope of the land surface. Low topography (slope) encourage retention of water for longer periods and thus the higher the rate of infiltration and vulnerability	The ASTER DEM data were used with 28 m spatial resolution to estimate the slope over the study area.	—	1	Fig. 10.A.1E
Impact of the vadose zone (I)	It is defined as the unsaturated zone material that is the zone above the water table. If the vadose zone contains the less permeable layer such as clay, the lower rating is assigned.	Used the 34 geological logs available.	Ordinary kriging (OK)	5	Fig. 10.A.1F
Hydraulic conductivity (C)	Hydraulic conductivity is the ability of aquifer to transmit water for a given hydraulic gradient. The higher the conductivity, the more vulnerable the aquifer.	The conductivity parameter was estimated by using the pumping test data from the analysis of 50 pumping tests.	Ordinary kriging (OK)	3	Fig. 10.A.1G

Appendix I

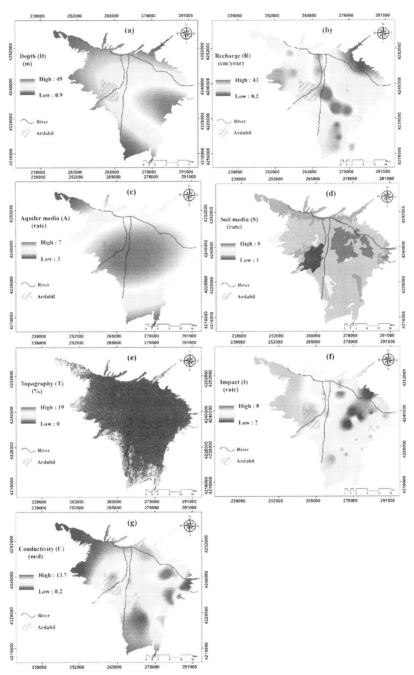

FIGURE 10.A.1 DRASTIC parameters (A) Depth to Water, (B) Net Recharge, (C) Aquifer Media, (D) Soil Media, (E) Topography, (F) Impact of the Vadose Zone and (G) Hydraulic Conductivity.

TABLE 10.A.2 Definition of ALPRIFT data layers.

Data layers	Description	Data layers	Interpolation method	Weight	Reference map
Aquifer media (A)	Expresses soil texture and structure.	Used the 34 geological logs available.	Inverse Distance Weighting (IDW)	5	Fig. 10.A.2A
Land use (L)	Land use considers anthropogenic activities and natural processes, which would alter soil texture and trigger subsidence	Land use map was obtained from Iran National Cartographic Centre	—	4	Fig. 10.A.2B
Pumping of groundwater (P)	Pumpage and natural withdrawal of groundwater would strain soils and induce subsidence.	The measured monthly discharge data from 3669 available abstraction wells (October 2015—October 2016)	Inverse Distance Weighting (IDW)	4	Fig. 10.A.2C
Recharge (R)	Recharge is the infiltrating surface water from topsoil, which percolates to groundwater.	The relationship for recharge given by Scanlon et al. (2002) is modified by adding the pumped water volume in a year.		3	Fig. 10.A.2D
Impact of aquifer thickness (I)	Impacts focused on aggregated thickness of saturated/unsaturated media from land surface to bedrock. The higher the thickness, the greater sensitivity to subsidence	This parameter was obtained by the geo-electric surveys.	Inverse Distance Weighting (IDW)	2	Fig. 10.A.2E
Fault distance (F)	Faults are natural factors affecting subsidence accounting for tectonic movements	Fault distance was calculated by the Euclidean distance method in a GIS using the map by the Geological Survey and mineral Exploration of Iran	—	1	Fig. 10.A.2F
Water table decline (T)	Groundwater level decline reduces fluid pressure and increases the effective stress and encourages subsidence. It was considered for 1-year period.	The water table declination at 34 observation wells was calculated by subtracting the water table for 1-year period (October 2015—October 2016)	Inverse Distance Weighting (IDW)	5	Fig. 10.A.2G

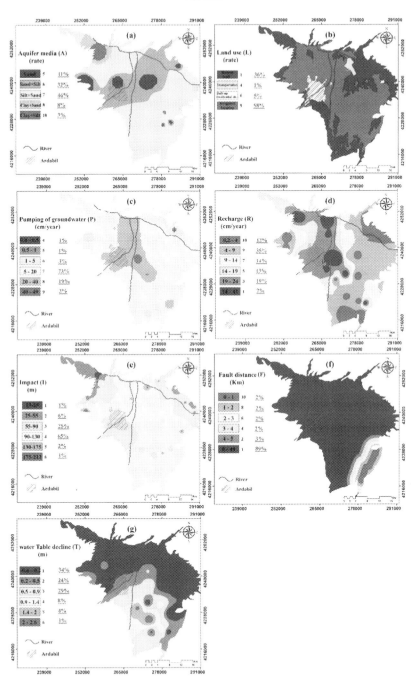

FIGURE 10.A.2 Raster BAF data layers—rated data; (A) Aquifer media; (B) Land use; (C) Pumping of groundwater; (D) Recharge; (E) Impacts in terms of thickness; (F) Fault distance; (G) water Table decline. Note: The A, P, I, and T data layers use the IDW interpolation technique.

References

Aller, L., Bennett, T., Lehr, J.H., Petty, R.J., Hackett, G., 1987. DRASTIC: A Standardized System for Evaluating Ground Water Pollution Potential Using Hydrogeologic Settings. EPA 600/2-87-035. U.S. Environmental Protection Agency, Ada, Oklahoma.

Almoussawi, R., Christian, C., 2005. Fundamentals of quantitative risk analysis. J. Hydroinf. 7 (2), 61−77.

Bai, Y., Xie, J., Wang, X., Li, C., 2016. Model fusion approach for monthly reservoir inflow forecasting. J. Hydroinf. 18 (4), 634−650.

Bilgot, S., Parriaux, A., 2009. Using geotypes for landslide hazard assessment and mapping: a coupled field and GIS-based method. Geophys Res 11.

Chen, Y., Han, D., 2016. Big data and hydroinformatics. J. Hydroinf. 18 (4), 599−614.

Cheng, C.H., Liu, Y.H., Lin, Y., 1996. Evaluating a weapon system using catastrophe series based on fuzzy scales. In: Proceedings of Soft Computing in Intelligent Systems and Information. IEEE, pp. 212−217. https://doi.org/10.1109/AFSS.1996.583593. In: Proceedings of soft computing in intelligent systems and information.

Huang, B., Shu, L., Yang, Y.S., 2012. Groundwater overexploitation causing land subsidence: hazard risk assessment using field observation and spatial modelling. Water Resour. Manag. 26 (14), 4225−4239.

Jenks, G.F., 1967. The data model concept in statistical mapping. Int. Yearb. Cartogr. 186−190.

Khatibi, R., 2008. Systemic nature of, and diversification in systems exposed to, flood risk. In: Brebbia, C.A. (Ed.), Proceedings of Flood Recovery, Innovation and Response (FRIAR2). WIT Press (July).

Khatibi, R., 2011. Evolutionary systemic modelling of practices on flood risk. J. Hydrol. 401 (1−2), 36−52.

Khatibi, R., Ghorbani, M.A., Naghshara, S., Aydin, H., Karimi, V., 2020. A framework for 'inclusive multiple modelling' with critical views on modelling practices− applications to modelling water levels of Caspian Sea and Lakes Urmia and van. J. Hydrol. 587, 124923.

Khatibi, R., Nadiri, A.A., 2020. Inclusive Multiple Models (IMM) for predicting groundwater levels and treating heterogeneity. Geosci. Front. 12 (2). https://doi.org/10.1016/j.gsf.2020.07.011.

Krishna, A.K., Rama Mohan, K., Dasaram, B., 2019. Assessment of groundwater quality, toxicity and health risk in an industrial area using multivariate statistical methods. Environ. Syst. Res. 8, 26.

Li, C.N., Lo, C.W., Su, W.C., Lai, T.Y., Hsieh, T.K., 2016. A study on location-based priority of soil and groundwater pollution remediation. Sustainability 8 (4), 377.

Li, Y., Li, J., Chen, S., Diao, W., 2012. Establishing indices for groundwater contamination risk assessment in the vicinity of hazardous waste landfills in China. Environ. Pollut. 165, 77−90.

Living with Risk, 2004. A global review of disaster reduction initiatives. Inter-Agency Secretariat of the International Strategy for Disaster Reduction (UN/ISDR) 429 p.

Llamas, R., Martínez-Santos, P., 2005. Intensive groundwater use: silent revolution and potential source of social conflicts M. J. Water Resour. Plan. Manag. 131 (5).

Matzeu, A., Secci, R., Uras, G., 2017. Methodological approach to assessment of groundwater contamination risk in an agricultural area. Agric. Water Manag. 184, 46−58.

Nadiri, A.A., Khatibi, R., Khalifi, P., Feizizadeh, B., 2020. A study of subsidence hotspots by mapping vulnerability indices through innovatory 'ALPRIFT' using artificial intelligence at two levels. Bull. Eng. Geol. Environ. 1−15.

Nadiri, A.A., Gharekhani, M., Khatibi, R., 2018b. Mapping aquifer vulnerability indices using artificial intelligence-running multiple frameworks (AIMF) with supervised and unsupervised learning. Water Resour. Manag. 32, 3023−3040. doi.org/10.1007/s11269-018-1971-z.

Nadiri, A.A., Gharekhani, M., Khatibi, R., Asgari Moghaddam, A., 2017b. Assessment of groundwater vulnerability using supervised committee to combine fuzzy logic models. Environ. Sci. Pollut. Res. 24 (9), 8562−8577.

Nadiri, A.A., Gharekhani, M., Khatibi, R., Sadeghfam, S., Asgari Moghaddam, A., 2017a. Groundwater vulnerability indices conditioned by supervised intelligence committee machine (SICM). Sci. Total Environ. 574, 691−706.

Nadiri, A.A., Razzag, S., Khatibi, R., Sedghi, Z., 2021. Predictive groundwater levels modelling by Inclusive Multiple Modelling (IMM) at multiple levels. Earth Sci. India 14 (2), 749−763. https://doi.org/10.1007/s12145-021-00572-y.

Nadiri, A.A., Sadeghfam, S., Gharekhani, M., Khatibi, R., Akbari, E., 2018c. Introducing the risk aggregation problem to aquifers exposed to impacts of anthropogenic and geogenic origins on a modular basis using 'risk cells. J. Environ. Manag. 217, 654−667.

Nadiri, A.A., Taheri, Z., Khatibi, R., Barzegari, G., Dideban, K., 2018a. Introducing a new framework for mapping subsidence vulnerability indices (SVIs). Sci. Total Environ. 628, 1043−1057.

Nadiri, A.A., Sedghi, Z., Khatibi, R., Gharekhani, M., 2017c. Mapping vulnerability of multiple aquifers using multiple models and fuzzy logic to objectively derive model structures. Sci. Total Environ. 593, 75−90.

Qian, H., Chen, J., Howard, K.W.F., 2020. Assessing groundwater pollution and potential remediation processes in a multi-layer aquifer system. Environ. Pollut. 263, 114669.

Rodriguez, A., Bermudez, M., Rabunal, J.R., Puertas, J., 2015. Fish tracking in vertical slot fishways using computer vision techniques. J. Hydroinf. 17 (2), 275–292.

Sadeghfam, S., Ehsanitabar, A., Khatibi, R., Daneshfaraz, R., 2018b. Investigating 'risk'of groundwater drought occurrences by using reliability analysis. Ecol. Indicat. 94, 170–184.

Sadeghfam, S., Hassanzadeh, Y., Khatibi, R., Moazamnia, M., Nadiri, A.A., 2018a. Introducing a risk aggregation rationale for mapping risks to aquifers from point-and diffuse-sources—proof-of-concept using contamination data from industrial lagoons. Environ. Impact Assess. Rev. 72, 88–98.

Sadeghfam, S., Hassanzadeh, Y., Nadiri, A.A., Zarghami, M., 2016. Localization of groundwater vulnerability assessment using catastrophe theory. Water Resour. Manag. 30 (13), 4585–4601.

Sadeghfam, S., Khatibi, R., Dadash, S., Nadiri, A.A., 2020b. Transforming subsidence vulnerability indexing based on ALPRIFT into risk indexing using a new fuzzy-catastrophe scheme. Environ. Impact Assess. Rev. 82, 106352.

Sadeghfam, S., Nourbakhsh Khiyabani, F., Khatibi, R., Daneshfaraz, R., 2020a. A study of land subsidence problems by ALPRIFT for vulnerability indexing and risk indexing and treating subjectivity by strategy at two levels. J. Hydroinf. 22 (6), 1640–1662.

Scanlon, B., Healy, R., Cook, P., 2002. Choosing appropriate techniques for quantifying groundwater recharge. Hydrol. J. 10 (1), 18–39.

Sedghi, Z., Rostami, A., Khatibi, R., Nadiri, A.A., Sadeghfam, S., Abdoallahi, A., (submitted as Chapter 22 of This Book) 2022. Mapping Aggregated Index of Groundwater Quality Indices for Aquifer Management Using Inclusive Multiple Modeling (IMM) Practices; Risk, Reliability and Sustainability. In: Roshni, T., Samui, P., Bui, G., Kim, D. and Khatibi R. (Eds.).

Tung, Y.K., Yen, B.C., Melching, C.S., 2005. Hydrosystems Engineering Reliability Assessment and Risk Analysis. McGraw-Hill Professional, New York, USA.

UNICEF, 2012. State of the World's Children.

USEPA, 2001. Sampling Procedures for the 2001 National Sewage Sludge Survey. Office of Science and Technology, Washington, D.C.

USEPA, 2012. National Primary DrinkingWater Regulations. US Environmental Protection Agency. EPA816-F- 09-004.

Wang, J., He, J., Chen, H., 2012. Assessment of groundwater contamination risk using hazard quantification, a modified DRASTIC model and groundwater value, Beijing Plain, China. Sci. Total Environ. 432, 216–226.

WHO, 2009. Guidelines for Drinking-Water Quality. Word Health Organization.

Wu, Y.P., Jiang, W., Ye, H., 2010. Karst collapse hazard assessment system of wuhan city based on GIS. In: 2010 International Symposium in Pacific Rim, Taipei, Taiwan, April 26–30.

CHAPTER 11

Mapping and aggregating groundwater quality indices for aquifer management using Inclusive Multiple Modeling practices

Zahra Sedghi[1], Ali Asghar Rostami[2], Rahman Khatibi[3], Ata Allah Nadiri[4,7,8], Sina Sadeghfam[5], Alireza Abdoallahi[6]

[1]Department of Earth Sciences, Faculty of Natural Sciences, University of Tabriz, Tabriz, East Azerbaijan, Iran; [2]Department of Water Engineering, University of Tabriz, Tabriz, East Azerbaijan, Iran; [3]GTEV-ReX Limited, Swindon, United Kingdom; [4]Traditional Medicine and Hydrotherapy Research Center, Ardabil University of Medical Sciences, Ardabil, Iran; [5]Department of Civil Engineering, Faculty of Engineering, University of Maragheh, Maragheh, East Azerbaijan, Iran; [6]Department of Water Engineering, Bu-Ali Sina University, Hamedan, Iran; [7]Medical Geology and Environmental Research Center, University of Tabriz, Tabriz, Iran; [8]Department of Earth Sciences, Faculty of Natural Sciences, Institute of Environment, University of Tabriz, Tabriz, East Azerbaijan, Iran

1. Introduction

A set of 10 indices is studied in this chapter to understand the water quality of aquifers and risks to health, which are impacted by geogenic and/or anthropogenic activities often stemming from poor or nonexistent planning systems, reflected by sparse data availability. The indices studied in the chapter comprise Groundwater Quality Index (GQI), GroundWater Quality Index (GWQI), and eight Risk Indices (RIs) on carcinogenic and noncarcinogenic (Dermal and Oral pathways for Adult and Child). The measured data comprise a set of sampled data from Observation Wells (OWs) but the *aggregated index* provides an insight into the spatial health of a study area. In countries with underdeveloped planning traditions and practices, the capacity for decision-makers and planners is not often commensurate with risk-based decision-making practices and not so with decision-making by participation. To help the buildup of competencies for safeguarding environmental systems in any meaningful way, the chapter explores capturing the information content of multiple indices in terms of a single index and therefore the research presented in the chapter introduces two innovatory features into aquifer management practices: (i) deriving an aggregated index to capture the signals within the multiple indices; (ii) the Inclusive Multiple Modeling (IMM) is applied as a strategy to data fusions problems with unsupervised learning at Level 1 and supervised learning at Level 2.

A study of aquifer water quality is carried out by using GQI, originally introduced by Babiker et al. (2007) and GWQI, originally introduced by Ribeiro et al. (2002). These are objective procedures, where the indexes are calculated from the field data and the basic mathematical technique of an objective scoring system involving rates in terms of their ratios to prescribed standards; and weights in terms of the relative importance of the particular water quality constituent. Notably, both indices range normally between 1 and 100, but the GWQI values closer to 100 signify poor quality, whereas the GQI values closer to 1 signify poor quality. This research compares GQI and GWQI indices results and evaluates these frameworks and consequently shows that their main disadvantage is the uncertainty of the rate and weight assessment of these frameworks. The fuzzification of these frameworks increases the accuracy of results by overcoming the disadvantage. The chapter also uses a set of eight indices through the well-established procedure of Human Health Risk Assessment (HHRA), developed by USEPA (1989), which assesses health risk to the human body exposed to contaminated water. They evaluate possible adverse harms to human health impacting on adults and children receiving contaminants through dermal and oral pathways in the formal carcinogenic and noncarcinogenic risks. Risk exposures stem from oral ingestion and dermal absorption of contaminants (USEPA, 1989, 2004; Xiao et al., 2019; Bodrud-Doza et al., 2019). Trace elements enter the human body system through contaminated drinking water, air, and nutrients; gradually settle as sediment; and

accumulate in human adipose tissue, muscle, bone, and joints (Saçmaci et al., 2012). Exposures to trace elements (Cd, As, Cr, Pb) even at low concentrations interfere with the normal functioning of bodies; and their higher-concentration intakes affect adversely on human health (Valavanidis et al., 2010).

Topical research works on these 10 indices are wide and include Adimalla and Qian. (2019), Lal and Datta (2018), Khadra and Stuyfzand (2018). Generic features of these techniques are classified in Table 11.1. However, the review of these indices or their alternatives approaches is outside the scope of the chapter, as these available capabilities are used here as established tools. The research aspects of the chapter are discussed below.

The research in the chapter is focused on investigating the aggregation of the above 10 indices into a single index by using IMM practices introduced by Khatibi et al. (2020), Nadiri and Khatibi (2021), who contrast it with Exclusionary Multiple Modeling (EMM) practices characterizing existing practices. They criticize EMM practices for being focused on selecting "the superior model" from multiple models under a study. They capture the scientific thinking in IMM practices by pooling four dimensions in a framework: Model Reuse (MR), Hierarchical Recursion (HR), Elastic Learning Environment (ELE), and Goal Orientation (GO). They present the proof of concept for various disciplines of modeling, e.g., see Khatibi et al. (2020), Nadiri and Khatibi (2021), Nadiri et al. (2017, 2018), Sadeghfam et al. (2018), and Rostami et al. (2020). These published works provide some proof of concept that modeling practices can maximize the information extracted from the data, and as such, they are defensible, whereas the results by EMM practices are only fit for purpose.

The use of IMM requires a modeling strategy to be formulated in terms of activities at two levels and the strategy presented in the chapter comprises the following levels. **At Level 0**, recorded sample data are used to derive

TABLE 11.1 Groundwater Quality Indices developed and used in different parts of the world.

Name of index	Reference	Constituents				Parameters					Models	Key contribution
		Major ions	Minor ions	Properties	Trace element	WQI	GWQI	GQI	HRA	Data fusion		
GWQI/ FGWQI	Vadiati et al. (2016)	#	#	#		*	*	*			Fuzzy	Used a hybrid fuzzy-based GWQI (FGWQI) to evaluate groundwater quality; concluded that hybrid FGWQI is significantly more accurate than traditional WQI.
RBFNNWQI	Hameed et al. (2017)	#		#		*					ANN	Assessed water quality by radial basis function neural network model and quality index.
GIS Base	Adimalla and Qian (2019)	#	#	#			*		*		GIS	Studied health risk from nitrate in drinking groundwater using the WQI index.
RFHRA	Wu et al. (2020)	#	#	#	#				*		Random Forest	A classification of groundwater quality and evaluation of the amount of arsenic using health index.
GIS, statistical techniques	Ravindra et al. (2019)	#	#	#			*		*		GIS	Evaluated health index, groundwater quality, and chemical analysis using a multivariate technique.
GQI/FGQI	Jha et al. (2020)	#	#					*			GIS, Fuzzy logic	A novel hybrid framework integrating Fuzzy Logic with the GIS-based GQI is proposed in this study for assessing groundwater quality and its spatial variability.
ANN-PMI	Singha et al. (2020)				#				*		ANN	Assessed groundwater heavy metal pollution indices studies by deep learning.
HPI/HEI	Long et al. (2020)				#				*		GIS	The water pollution situation was assessed using heavy metal contents and evaluation indices, and human health risks were evaluated on the basis of both carcinogenic and noncarcinogenic aspects.

Continued

TABLE 11.1 Groundwater Quality Indices developed and used in different parts of the world.—cont'd

Name of index	Reference	Constituents				Parameters					Models	Key contribution
		Major ions	Minor ions	Properties	Trace element	WQI	GWQI	GQI	HRA	Data fusion		
HHR/WQI	Zhang et al. (2020)	#	#	#	#				*		GIS	Studied heavy metal health risk in drinking groundwater using the WQI index.
FHI	Mohanta et al. (2020)	#	#	#					*		Fuzzy logic	Assessed health risk index of fluoride concentration in drinking groundwater using fuzzy logic.
Data-F used 10 indices	Present study	#	#	#	#	*	*	*	*	GIS, IMM	Groundwater quality, Hazard Index, health risk mapping, using an IMM modeling strategy.

10 indices (GQI, GWQI, CRI-DA, CRI-OA, CRI-DC, CRI-OC, NRI-DA, NRI-OA, NRI-DC, and NRI-OC). **At Level 1**: an unsupervised data fusion technique is used to aggregate the 10 indices together for the derivation of a single representative index capturing the risk features. **At Level 2**, an Artificial Neural Network (ANN) is formulated to extract a new "aggregated index" by using the same input data of 10 neurons as in Level 1 but the output of Level 1 is used as its target values. This is one strategy at Level 1 but another strategy is presented in due course. Therefore, the strategy uses both unsupervised and supervised learning techniques. The models or algorithms selected in the chapter are a deliberately simple or most commonly used scheme for two reasons: (i) the models are means to an end and not necessarily the aim or the best tool in the trade; and (ii) in this way, the derived conclusions are outcomes of the strategy and the procedure but not of a property of the particular model selection.

The study area relates to a subbasin of Qizil Ozen, which flows through Zanjan province, northwest Iran, where there is a limited number of sample measurements and probably the only samples for study the area. Normally, equitable planning practices through decision-making by participation are not strong in the country and this is often reflected by the sparsity of the available data. Therefore, the use of 10 maps can be particularly challenging to decision-makers. The research presented in the chapter is driven by the proposition that the above planning situation is quite common in countries with poor or nonexisting planning practices. The driver of this research is an academic initiative and its innovative features include the following: (i) an investigation into aggregating multiple indices into a single representative index; (ii) an application of the IMM practices into a new field of research; and (iii) the application of the IMM strategy itself includes an innovation to test the aggregated index by two different strategies, as presented in due course.

2. Study area and data availability

2.1 Physical description of the study area

Gultepe-Zarinabad subbasin is the study area presented in the chapter, which is one of the upper subbasins of Qizil Ozen (Golden River or Red River) in the southwest of Zanjan province, adjacent to Hamadan province, northwest Iran. The river is 670 km long and flows into the Caspian Sea by draining Qaflanti Dagi (Mount Qaflanti or Qaflankuh) at its upper reaches, where the study area is located. Its basin area is 5124 km^2, of which 38% is plain and 62% is highland. Agricultural activities are predominant in a large part of the study area (Taheri Tizro et al., 2014), but there are also many mining activities in recent decades.

The study area is characterized by its temperate climate and cold winters followed by moderate summers with the average minimum temperature recorded in January ($-3.6°C$) and the maximum temperature in July ($21.8°C$), at the climate station in Zanjan. Its climate is classified as the semiarid as per De Martin approach. Over the recent 20-year period (1985−2005), the range of annual rainfall was recorded to be 272−373 mm. Rainfall mainly occurs during the wet period, with its maximum in November and December. The lowest rainfall normally occurs in July and August, which correspond to the dry period (Taheri Tizro et al., 2014).

2.2 Geological

The region is located in the western part of a central tectonic zone, where the oldest geological series of rocks, the Gultepe-Zarinabad subbasin, comprises slate and schistose rocks as partly metamorphosed relating to the Upper Triassic−Jurassic sequences. Crystallized limestones are observed between these rocks. Rocks with low metamorphic formations are Nayeen and Shemshak formations followed by outcrops in the northeast and southeast parts of the plain. Conglomerate and sandstone with a limestone base of Cretaceous rocks relate to the Albian along with andesite-basalt, Albian shales, and shales related to the Upper Cretaceous, which are exposed in the southern parts and metamorphosed. Cretaceous exposures form slopes against Jurassic rocks. Tertiary rocks lie in the older rocks with their unconformities composed of conglomerate and red sandstone formations, known as Fajan formations, which are of the Eocene age.

Volcanic rocks, such as andesite—dacite—rhyolite, are manifest in the southern parts along with green marl and sandstone. Miocene rocks (Upper Red Formation) of clastic sediments, including marls, red sandstone, and conglomerates, are mixed with gypsum and salt layers and manifest in northern parts. Quaternary sediments consist of clay and sand with their horizontal beddings deposited in the central and western parts of the plain. Quaternary deposits (recent alluvium) dominate in a large part of the plain. The main aquifer system of high transmissivity was developed within these sediments at the study area. Pliocene sediments include clay and sand with horizontal beddings deposited in the central and western parts of the basin, which lie at the Upper Red Formation (Taheri Tizro et al., 2014).

2.3 Hydrogeology

The spinal watercourse of Qizil Ozen runs along the south to the northerly direction and collects tributaries from highlands and the study area forms one of its headwater subbasins. The river receives tributaries and the watercourses flowing into the main river within the study area (Fig. 11.1) are more than 10. The river in the study area is a permanent one but its tributary watercourses are not. Therefore, the interaction of the unconfined Gultepe-Zarinabad aquifer with the watercourses is rather complex. Groundwater withdrawal from the aquifer is through wells, springs and qanats (Fig. 11.2).

The unconfined aquifer in the study area is an outcome of the alluvial deposits of Gultepe-Zarinabad plain. The depth of the water table is highly variable; it is relatively shallow (6 m below ground surface) in the central part near Zarinabad River, while the depth is more than 60 m below ground surface in the eastern regions. The water table has the highest level in April or May, and the lowest level in September or October. Groundwater flow originates from the recharge area of the upper hills of east and northeast and slopes toward the central part of the plain and ultimately discharges to Zarinabad River.

As concluded from the pumping test analysis, the variability of transmissivity in the study area is from 30 to 900 m^2/day. The lower transmissivity occurs in the northeast and the higher values of transmissivity are determined in the north and north-east. Estimates for abstraction rates at the boreholes range from about 30 m^3/h to 120 m^3/h. Due to the limited availability of lithological data, it was difficult to study the subsurface geology and this posed severe constraints to compile a holistic picture. These shortcomings overcame by using geophysical data. The maximum illuviation thickness in Gultepe-Zarinabad plain is equal to 181 m (Taheri Tizro et al., 2014).

2.4 Data availability to define the pollution baseline

The samples, taken in 2019 for this study, were analysed in the hydrogeological laboratory at the University of Hamadan, which included major ions and a set of minor and trace ions, as well as pH, Electric Conductivity (EC) by standard methods (APHA, 2005). The measurement comprise 28 samples, the analytical precision of which were calculated by the Normalised Inorganic Charge Balance. The Charge Balance Error (%CBE) of the samples was determined to be less than ±5, deemed acceptable for groundwater studies (Hounslow, 1995). Table 11.2 comprises the statistical summary of the 22 groundwater quality parameters.

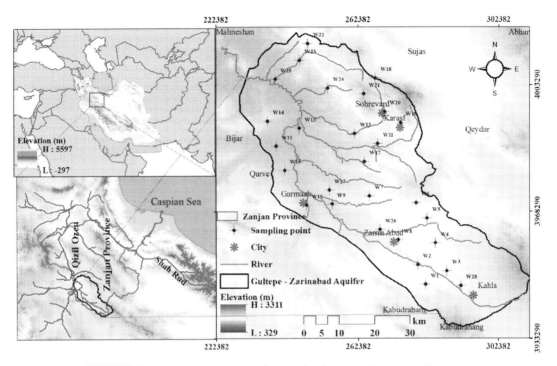

FIGURE 11.1 Location map of study area showing groundwater sampling sites.

It covers major ions: SO_4^{-2}, Na^+, Ca^{+2}, Mg^{+2}, K^+, HCO_3, and Cl^- and the properties: TH, TDS, EC, pH, and trace elements of the following elements: Al, As, Cd, Cu, Cr, Fe, Mn, Ni, Pb, Co, and Zn. Table 11.2 also shows the allowable limits, which are exceeded in the case of EC, TDS, as well as of TH and As, Fe, Pb, and Mn elements. Notably, the allowable levels of WHO (2011) are used and these are displayed in Fig. 11.3.

3. Methodology

This section presents the methodology for each of the 10 indices comprising the following: GQI, GWQI, carcinogenic, and noncarcinogenic Risk Indices (Oral and Dermal mode for Adults and Children). The modeling strategy is based on IMM practices at two levels as laid out in this section and illustrated in Fig. 11.4. The modeling strategy includes activities at Level 0 to carry out the calculation of the 10 indices, which are used for modeling the aggregated index by data fusion techniques at Levels 1 and 2. The chapter sets a strategy based on the proposition that the challenge of coping with multiple indices in countries with little track records on planning can be helped with by an "aggregated index." The strategy seeks to capture significant signals of multiple indices in the single aggregated index.

3.1 Basic techniques

The research presented in the chapter uses several basic techniques to produce indices. These basic techniques are outlined below.

3.1.1 Spatial modeling

All the indices covered in the chapter use a spatial distribution technique to be outlined at this outset. There are two approaches for spatially distributing index values: (i) The sample readings are spatially distributed over a

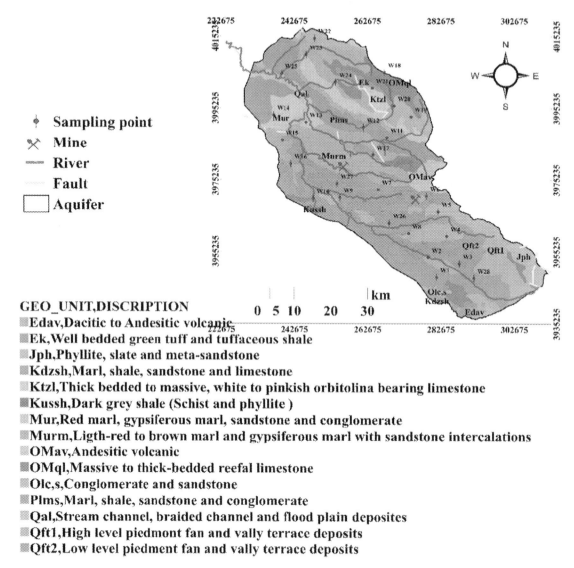

FIGURE 11.2 Geological map of the study area.

grid and the interpolated values of each contaminant are calculated for each pixel. The calculations are then carried out at each pixel giving a spatial view of the contaminants. (ii) The indices are calculated at each OW and their spatial distributions are mapped out using a spatial interpolation technique. The study employs the second approach. The spatially distributed values are then used by the IMM modeling strategy presented in Section 3.2.

The chapter uses the Kriging interpolation method and unlike other point interpolation methods (nearest point or moving average), the Kriging technique is a statistical technique. It performs a weighted averaging on point values (e.g., those at OWs), where the output at a pixel or OW is estimated as equal to the sum of the product of point values and weight and then divided by the sum of weights. The weight factors in the Kriging technique are determined by

3. Methodology

163

TABLE 11.2 Statistics of the sampled data.

Parameters	Units	Max	Min	Mean	SD	$CV_{(0/0)}$	WHO standards	Exceeding OW
pH		8.98	4.85	7.15	0.83	12	6.5 - 8.5	
EC	*(μS/cm)*	**17440**	295	1939	3184	164	1000	W9,W10,W12,W15,W16,W24,W25,W27
TDS	*(mg/L)*	**14073**	193	1727	2668	155	500	W9,W10,W12,W15,W16,W24,W25,W27
TH	*(mg/L.CaCo₃)*	**2090**	168	491	419	85	100	
Al	*(μg/L)*	29	1.18	7.24	6.44	89	100	
As	*(μg/L)*	862	1.04	36.48	159	436	10	W5, W15, W20, W22, W24
Co	*(μg/L)*	1.57	0.05	0.28	0.33	118	50	
Cr	*(μg/L)*	16	1.03	4.04	3.29	81	50	
Cu	*(μg/L)*	17.16	0.16	2.83	3.89	137	1000	
Fe	*(μg/L)*	**7668**	47	563	1403	249	300	W15, W24, W26, W27
Mn	*(μg/L)*	**543**	0.28	27	100	368	400	W24
Ni	*(μg/L)*	11	0.42	2.65	2.26	85	70	
Pb	*(μg/L)*	29	0.95	6.71	6.66	99	10	W7, W10, W15, W24, W27
Zn	*(μg/L)*	10	0.59	3.72	2.35	63	3000	
Cd	*(μg/L)*	0.38	0.05	0.12	0.06	51	3	
K⁺	*(mg/L)*	0.831	0.007	0.11	0.16	148	10	
Na⁺	*(mg/L)*	38	0.52	7.54	7.66	102	200	
Mg²⁺	*(mg/L)*	22	1.33	4.47	4.29	96	30	
Ca²⁺	*(mg/L)*	28	1.99	5.36	4.86	91	75	
SO₄⁻²	*(mg/L)*	34	0.43	6.02	7.31	121	200	
CI⁻	*(mg/L)*	39	0.165	6.58	9.24	140	250	
HCO₃⁻	*(mg/L C₃CO₃)*	9	2.08	4.05	1.63	40	*No guideline*	
Color Code		*Major Ions*		*Trace Element*		*Properties*		**Bold** value exceeding standards

Note 1: Standard is from World Health Organization (WHO) standard (World Health Organization, 2011)

Note 2: All the fractional differences between the total cations and total anions (Edmond et al., 1995) were <5% and hence reliability of their analysis.

Note 1: Standard is from World Health Organization (WHO) standard (World Health Organization, 2011).

Note 2: All the fractional differences between the total cations and total anions (Edmond et al., 1995) were <5% and hence reliability of their analysis.

using a user-defined semivariogram model through the output of their spatial correlation operation and the pattern analysis (Isaaks and Srivastava, 1990).

3.1.2 Groundwater quality index

Groundwater Quality Index (GQI) is an objective scoring scheme used widely by practitioners to assess aquifer water suitability for drinking, and its mathematical formulation bears some similarity with such schemes as DRASTIC as given by Aller et al. (1987). GQI was given by

Babiker et al. (2007), which synthesizes different available water quality data by indexing them relative to the standards of the World Health Organization (WHO, 2011), see Rufino et al. (2019). The objectivity of the approach is due to using a procedure for calculating its parameters (rates and weights) from observed sample data and published standards, as opposed to prescribed values in DRASTIC. The index is an attempt to the challenge for expressing the overall water quality condition given the spatial variability for each and often of multiple contaminants

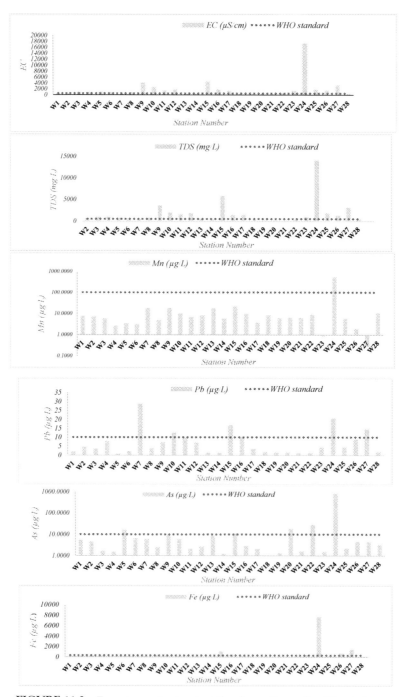

FIGURE 11.3 Excessive contaminants at the observation wells of the study area.

3. Methodology

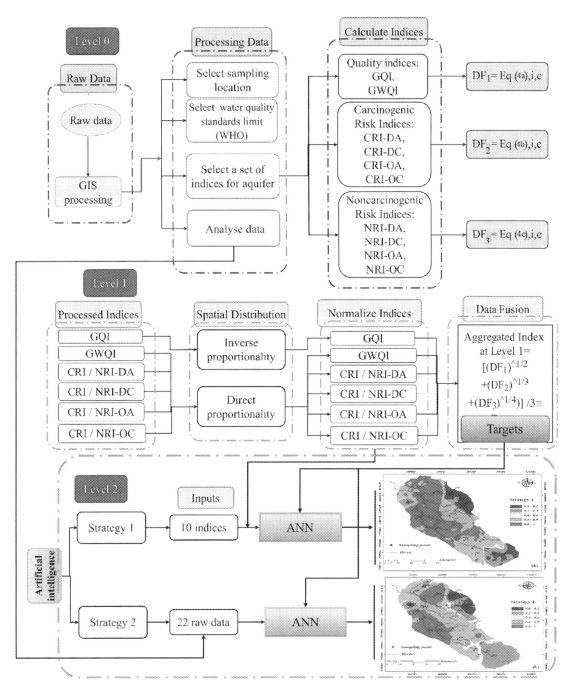

FIGURE 11.4 Flowchart of the methodology.

under a wide range of physical, chemical (e.g., ions of Cl^-, Na^+, Ca^{2+}), and biological factors relative to the standards set by the World Health Organization, WHO (2011).

The process is in four steps: (i) calculate the C index based on the measured concentrations and drinking water standard (WHO), see Eq. 11.1; (ii) calculate rating values *r-values* between 1 and 10 as expressed by Eq. 11.2; and (iii) calculate weight values, w, see Eq. 11.3; (iv) calculate GQI using ranks r and relative weights w of every variable and data layer, see Eq. 11.4. These are expressed as follows:

$$C = (X_r - X)/(X_r + X) \qquad (11.1)$$

$$r = 0.5 * C^2 + 4.5 * C + 5 (r \text{ is between } 1-10) \qquad (11.2)$$

$$w = \text{the mean rating value, } r \text{ (of each layer)}$$

$$w \text{ (for non} - \text{carcinogenic elements}(r \leq 8)$$
$$= \text{the mean rating value } r + 2(r \leq 8) \qquad (11.3)$$

$$GQI = 100 - \left(\sum_i^n r_i w_i / N \right) \qquad (11.4)$$

where X_r is the measured concentration, at each observation well; X is maximum permissible level of drinking water constituent as per WHO (2011). The r-value is the rate of contaminant $(1-10)$ and is calculated by Eq. 11.2; and w is the relative weight of each index (Eq. 11.3). If the weight value for noncarcinogenic elements is ≤ 8, the calculated weight must be added by 2. In Eq. 11.4, N is the total number of parameters used in the suitability analyses. Notably, the value of C in Eq. 11.1 uses the rate value of 1 to indicate a minimum impact on groundwater quality; the rate 10 is used to indicate the maximum impact; and the median level is set to 5. Overall, GQI varies between 1 and 100, although it is known to exceed 100, as well. The index value closer to 1 indicates poor but closer to 100 indicates good quality water.

3.1.3 Ground water quality index

The Ground Water Quality Index (GWQI) is another water quality index based on the WHO drinking standard, introduced first by Ribeiro et al. (2002). GWQI shows a different approach to water quality related to the suitability of water for drinking purposes but more or less with a similar rationale. In general, water quality indices are primarily devised to determine the suitability of the groundwater for drinking purposes (Tiwari and Mishra 1985), and like GQI, it is calculated in four steps, using a weighted arithmetic index method (Horton 1965; Brown et al., 1972; Ramakrishnaiah et al., 2009; Adimalla et al., 2018), as follows:

1. Assign weights, w_i, to each element (i) of drinking water constituents; its values are assigned based on the expert opinion from the lowest value of 1 to the highest value of 5. See Table 11.3 for the values used in this study.
2. Calculate the relative weight, W_i, for n, *the number of elements*, as expressed by the following:

$$W_i = \frac{w_i}{\sum_i^n w_i} \qquad (11.5)$$

3. Define "quality rate" Q_i for the ith constituents expressed in terms of its concentration, C_i is concentration of each constituent at each observation well, and s_i, is given by the WHO guidelines for drinking water, and also by multiplying by 100, expressed as follows:

$$q_i = 100 \times \frac{C_i}{S_i} \qquad (11.6)$$

4. Finally, calculate the GWQI through the subindex, SI_i, which is expressed as follows:

$$SI_i = W_i \times q_i \qquad (11.7)$$

$$GWQI_i = \sum_i^n SI_i \qquad (11.8)$$

3. Methodology

TABLE 11.3 Relative weight of chemical parameters.

Parameter	Unit	WHO standards	Weight (w_i)	Weight for GQI	Weight for GWQI
pH	—	6.5–8.5	4	4.87	0.05
EC	(μS/cm)	1000	5	6.5	0.06
TDS	(mg/L)	500	5	6.5	0.06
Al	(μg/L)	100	2	1.24	0.02
As	(μg/L)	10	5	6	0.06
Co	(μg/L)	50	1	3.03	0.01
Cr	(μg/L)	50	5	3.51	0.06
Cu	(μg/L)	1000	2	1.01	0.02
Fe	(μg/L)	300	4	4.47	0.05
Mn	(μg/L)	400	5	1.75	0.06
Ni	(μg/L)	70	1	3.81	0.01
Pb	(μg/L)	10	5	5.6	0.06
Zn	(μg/L)	3000	1	1.02	0.01
Cd	*(μg/L)*	3	5	3.16	0.06
TH	(mg/L.CaCo$_3$)	100	4	7.71	0.05
K^+	(mg/L)	10	2	1.07	0.02
Na^+	(mg/L)	200	2	1.25	0.02
Mg^{2+}	(mg/L)	30	2	1.87	0.02
Ca^{2+}	(mg/L)	75	2	1.47	0.02
SO_4^{-2}	(mg/L)	250	4	1.13	0.05
Cl^-	(mg/L)	250	3	1.17	0.04
HCO_3	*(mg/L CaCO$_3$)*	*No guideline*	3	1.18	0.04
Total weight			72		1

Note 1: the weights for TDS, As, Cr, Mn, Cd, and EC are the highest.
Note 2: the weights for Co, Ni, and Zn are the least.
Note 3: The weights for the remaining have values are in between.

Likewise, GWQI is another objective index, which displays similar features to DRASTIC vulnerability indices for being a scoring scheme of rates and weights. It employs prescribed standard values S_i and these are specified in Table 11.3. Overall, GWQI varies between 1 and 100, although it is known to exceed 100, as well. The index value closer to 1 indicates good but closer to 100 indicates poor quality water.

3.1.4 Human Health Risk Assessment

The research in the chapter also employs HHRA, given by USEPA (1989), which brings together the concepts of human health and risk toward studying the degree of harm to human exposures against carcinogenic or noncarcinogenic ions through dermal pathways and inhalation through the oral pathways. The procedure comprises eight indices in terms of risks being carcinogenic and noncarcinogenic health risks (USEPA, 2009) and both are calculated for adults and children exposed to dermal and oral pathways. The excess intake of contaminants through drinking water can cause serious health hazards in human beings. The amounts of contaminants in the human body depend on its actual concentration in water and the intensity of drinking $day^{-1} kg^{-1}$ of body weight. The procedure employs four steps as follows:

Hazard identification during which excessive contaminants are identified with respect to their allowable levels, both carcinogenic and noncarcinogenic consequences.

Dose−response assessment establishes a relationship between the degree of exposure and adverse health responses. For both carcinogenic and noncarcinogenic impacts, a dose−response relationship is used, which is described in terms of Cancer Slope Factor (CSF) for carcinogenic impacts and Reference Dose (RfD) for noncarcinogenic impacts and their prescribed values for different ions are given by USEPA (2004).

Exposure assessment for human exposure to a contaminant is considered in terms of intensity, time, and frequency, which are carried out for oral and dermal pathways. Exposures by oral pathways use Chronic Daily Intake (CDI) and those of dermal ones in terms of Dermal Absorbed Dose (DAD), both of which require a set of prescribed parameters as detailed in USEPA (1991, 2004). CDI values are expressed as follows:

$$CDI_{Oral} = \frac{(CW \times IR \times EF \times ED)}{(BW \times AT)} \quad (11.9)$$

$$CDI_{Dermal} = \frac{(CW \times CA \times K_p \times ET \times EF \times ED \times CF)}{(BW \times AT)}$$

$$(11.10)$$

where AT was the average time for noncarcinogenic (days); BW represents the average body weight (kg); CF is the volumetric conversion factor for groundwater; CSF is Cancer Slope Factor; CW represents the concentration of the contaminant in water (mg/L); ED is the exposure duration (year); EF is the frequency of exposure (days/year); ET was the exposure time (h/day); IR is the ingestion rate of water (L/day); K_p is dermal permeability coefficient in samples (cm/h); and SA is the exposed skin area (cm^2) whose values are shown in Table 11.A.1; RfD and CSF values are shown in Table 11.A.2.

Risk characterization expresses the probability of harmful impacts on humans exposed to a characteristic contaminant for both carcinogenic and noncarcinogenic risks (USEPA, 1989):

$$Risk_{Dermal} = DAD \times CSF, Risk_{Oral}$$
$$= CDI \times CSF, Risk = Risk_{Dermal} + Risk_{Oral}$$
$$(11.11)$$

$$NCHQ_{Dermal} = \frac{DAD}{RfD}, NCHQ_{Oral}$$
$$= \frac{CDI}{RfD}, NCHQ = NCHQ_{Dermal} + NCHQ_{Oral}$$
$$(11.12)$$

3.1.5 Data fusion

The modeling strategy in the chapter employs a data fusion technique. It is an "umbrella" term for those approaches that combine data or variables aiming to improve quality, to reduce uncertainty, or to discover new knowledge/features. See and Abrahart (2001) define data fusion as the amalgamation of information from different data sources, which is widely

3. Methodology

practiced in electrical engineering. A decision making model for ship collision avoidance is developed in this paper by use of the information fusion methods (Liu et al., 2008), each with its own definition and procedures. Abdelgawad and Bayoumi (2011) present another categorization as low-level fusion, medium-level fusion, high-level fusion, and multilevel fusion.

A variety of techniques supports data fusion and they include statistical matching, Bayesian inference (as reviewed by Endres and Augustin, 2016), moving average filters (Rodriguez et al., 2015), and gray relational analysis (Huang and Chu, 1999). In this study, data fusion is carried out to combine index values by indicators (GQI, GWQI) and (CRI-DA, CRI-OA, CRI-DC, CRI-OC, NRI-DA, NRI-OA, NRI-DC, and NRI-OC) to produce a single output, a scheme as depicted in Fig. 11.4 and outlined below.

$$DF_1 = \left[\frac{\sqrt[3]{GWQI} + \sqrt[2]{GQI}}{2} \right] \qquad (11.13)$$

$$DF_2 = \left[\frac{\sqrt[5]{CRI - DA} + \sqrt[4]{CRI - OA} + \sqrt[3]{CRI - DC} + \sqrt[2]{CRI - OC}}{4} \right] \qquad (11.14)$$

$$DF_3 = \left[\frac{\sqrt[5]{NRI - DA} + \sqrt[4]{NRI - OA} + \sqrt[3]{NRI - DC} + \sqrt[2]{NRI - OC}}{4} \right] \qquad (11.15)$$

$$DF_{Total} = \left[\frac{\sqrt[4]{DF_3} + \sqrt[3]{DF_2} + \sqrt[2]{DF_1}}{3} \right]$$
$$= \text{Aggregated Index} \qquad (11.16)$$

The results would depend on the justification for selecting each and any of these radicals. A more objective way is to use the procedure embedded in the decision theory given by Hansson (2005) and further discussed by Sadeghfam et al. (2021). However, the chapter uses a more pragmatic approach as outlined in the due course.

3.1.6 Artificial Neural Networks

The research uses an artificial intelligence technique to improve the signal content of the aggregate index. This is not to seek the optimum technique but to use a standard technique as a means to an end. Therefore, ANN is used for its wide application and is specified below.

ANNs are parallel information processing techniques, which employ a set of neurons to transform input data into output signals, where the neurons in a layer are not connected to each other but each one to all the neurons of the next layer. It emulates the working of the brain but mathematically it employs polynomials for model fitting by using a set of weights known as activation functions. The chapter uses a Multilayer feedforward Perceptron (MLP) network, where its topology comprises an input layer, a hidden layer, and an output layer. Learning the values of the weights between the input–hidden and hidden–output layers in

this study is based on testing various alternatives through a trial-and-error procedure. The study found the performance of the hyperbolic tangent sigmoid and pure linear functions to produce the optimum activation functions for the input layer, and the hidden layer, respectively.

3.1.7 Normalization of indices

Each index used in data fusion is normalized by a scheme to convert the index values to a corresponding array of indices in the range of "0" to "1," so that the highest value reflects the highest risk. Attention is given to further transform the

GWQI indices as by definition their worst values occur at the lower index values. Thus, the following normalization schemes are used:

$$X_r = \frac{X_i - X_{min}}{X_{max} - X_{min}} \quad (11.17)$$

$$X_r = \frac{X_{max} - X_i}{X_{max} - X_{min}} \quad (11.18)$$

where i is the number of the data points; X_{min} and X_{max} are minimum and maximum index values, respectively; X_r is the normalized index and is considered as a variable.

3.2 Modeling strategy

The IMM modeling strategy, shown in Fig. 11.4 and presented in the chapter, comprises the activities at two levels, which additionally involves those at Level 0.

3.2.1 Level 0: calculating the index values

The calculation of the 10 indices (GQI, GWQI, CRI-DA, CRI-OA, CRI-DC, CRI-OC, NRI-DA, NRI-OA, NRI-DC, and NRI-OC) is routine and is used to discover of the health of aquifers or their inherent risks.

3.2.2 Level 1: data fusion

The 10 indices are transformed into a single aggregated index at Level 1 using Eqs. 11.13–11.16 to produce the aggregated index at Level 1. However, this requires decisions to be made on the order of the index used. Data fusion is applied in four steps, as follows: (i) DF1 (Eq. 11.13) is applied to combine values of GQI and GWQI indices; (ii) DF2 (Eq. 11.14) is applied to aggregate values of the carcinogenic risks indices (CRI-DA, CRI-OA, CRI-DC, CRI-OC); (iii) DF3 (Eq. 11.15) is applied to aggregate values of the noncarcinogenic risks indices (NRI-DA, NRI-OA, NRI-DC, and NRI-OC); (iv) DF$_{Total}$ (Eq. 11.16) is applied to aggregating the outputs of

DF1, DF2, and DF3 to produce their "Aggregated Index."

The weights of each index, in terms of the degree of the radical in Eqs. 11.13–11.16, are assigned based on the expert opinion. Due to the importance of the carcinogenicity index, greater weight was assigned to this index. Then lower weights are assigned to noncarcinogenicity and water quality indices, respectively. In the calculation of indices, due to the importance of indices associated with the oral and dermal pathways for children and adults, first, a higher weight is given to adults than to children due to their susceptibility to carcinogenicity and noncarcinogenicity. The main differences between GQI and GWQI indices are related to the manner of finding weight for each data layer. In the GQI index, the weights are calculated using average rates in each data layer (see Eq. 1c) and in the GWQI index, the weight is assigned by expert opinion based on knowledge for the particular case study.

The water quality indices and carcinogenic and noncarcinogenic health indices are calculated at Level 0, and then at Level 1, they are normalized in the range of 0–1, to reduce impacts of scales. The normalization uses Eq. 11.17 for all the indices, except for the GQI, which uses Eq. 11.18.

3.2.3 Level 2: further fusion strategies

Two alternative approaches are investigated at Level 2 for further data fusion but using supervised learning by ANN, as follows:

Strategy 1: This comprises an ANN model, the inputs for which are the 10 indices obtained at Level 0 and its target values are the output of data fusion at Level 1.

Strategy 2: This also comprises an ANN model, the inputs for which are the 22 arrays of raw data used for calculating the indices at Level 0 and its target values are the output of data fusion at Level 1.

3.3 Performance metrics and banding spatial results

No performance metrics are required for calculating the indices at Level 0. However, the performance of the representative index at Levels 1 and 2 can be studied by using ROC/AUC. This is now the industry norm to study the spatial behavior of variables in 2D.

Receiver Operating Characteristic (ROC) is often used for spatial goodness-of-fit. It is an analysis tool for radar images and is known as "Signal Detection Theory" to differentiate between the blips from a friendly ship, an enemy target, or noise. The signal detection theory is now used in wider modeling practices. The ROC curve accuracy is measured by the Area Under Curve (AUC) and its value of 1 represents a perfect test but 0.5 reflects the existence of strong noise.

The ANN models were constructed using best practice procedures, which set the number of neurons at the hidden layer by a trial-and-error procedure from 1 to a higher value. For each run, penalty functions are used in terms of Root Mean Square Errors (RMSE) and goodness-of-fit is used in terms of Correlation Coefficient (CC) or R^2. The model selection is based on choosing the neurons with the least RMSE value. In the ANN model, the data are subdivided into two parts: 80% of the data points are selected randomly for the training phase and the remaining 20% for testing.

The models at Level 1 are examined for being fit-for-purpose and those at Level 2 for defensibility. The following performance metrics are used to assess the performance at Level 2 using CC and RMSE. The CC values vary between 1 and -1, in which a value near zero indicates poor correlations but that approaching 1 indicates a near "perfect" performance. RMSE values range from 0 (the "perfect model") to any higher real numbers and is a global measure of the model fit between the observed values and modeled values.

The results are nominally banded by using the normalized indices and dividing them into five equal bands as follows: Band $0 - 0.2$ (low risk), Band $0.2 - 0.4$ (relatively low risk), Band $0.4 - 0.6$ (moderate risk), Band $0.6 - 0.8$ (relatively high risk), and Band $0.8 - 1$ (high risk).

The ROC/AUC metrics for the aggregated indices at Levels 1 and 2 require the comparison of two spatially distributed values of the modeled indices and observed (target) indices for the assessment of False Negative and True Positive values. There is no single value for their comparisons, but some exploration was carried out to select an appropriate value by testing 0.2, 0.4, 0.6, and 0.8. This is discussed further in due course.

4. Results

4.1 Results at level 0

4.1.1 GQI and GWQI

The GQI and GWQI indices are calculated in a spreadsheet platform as per the procedure given in Sections 3.1.2 and 3.1.3 when using sampled data and prescribed parametric values in Table 11.3; their results are given in Fig. 11.5 and Table 11.4, which also includes prescribed values using the allowable limit (SI) recommended by WHO (2011).

The results presented in Fig. 11.5 show that approximately 17% of the wells in the study area expose consumers to risks when the aquifer water is used for human consumption. The remaining 83% of the wells may be in good condition, but this can rapidly change due to advection/diffusion processes taking place in aqueous environments. The table shows that there are As and Pb contaminations in relation to the wells in interactions with watercourses. For the spatial distribution of As and Pb in the catchment, high contamination sites have perceptually been known to local residents, referring to as *Qotur Su*, which means scabby water.

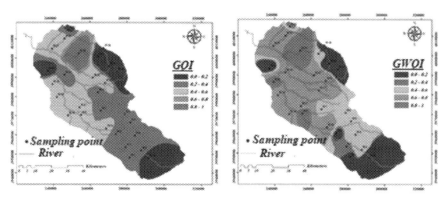

FIGURE 11.5 Drinking water quality category maps of the study area.

TABLE 11.4 Performance metrics of Level 0, 1, 2, and 3 for 10 indices and strategies.

	Area under the curve					Model performance			
				Asymptotic 95% Confidence Interval		R^2		RMSE	
Test Result Variable(s)	Area	Std. Error[a]	Asymptotic Sig.[b]	Lower Bound	Upper Bound	Training	Testing	Training	Testing
GQI	0.87	0.007	0.00	0.86	0.89	—		—	
GWQI	0.86	0.006	0.00	0.88	0.90				
CRI-DA	0.86	0.007	0.00	0.84	0.87				
CRI-OA	0.86	0.007	0.00	0.84	0.87				
CRI-DC	0.86	0.007	0.00	0.84	0.87				
CRI-OC	0.86	0.007	0.00	0.84	0.87				
NRI-DA	0.78	0.009	0.00	0.78	0.82				
NRI-OA	0.89	0.006	0.00	0.88	0.91				
NRI-DC	0.84	0.007	0.00	0.84	0.87				
NRI-OC	0.89	0.006	0.00	0.88	0.91				
Strategy 1	0.92	0.003	0.00	0.96	0.97	0.92	0.9	0.04	0.05
Strategy 2	0.94	0.001	0.00	0.98	0.98	0.96	0.94	0.01	0.02

The test result variable(s): GQI, GWQI, CRI-DA, CRI-OA, CRI-DC, CRI-OC, NRI-OA, NRI-DC, NRI-OC have at least one tie between the positive actual state group and the negative actual state group. Statistics may be biased.
Strategy 1: three neurons in the hidden and one neuron in the output layers; number of training epoch: 500; AUC = 0.92.
Strategy 2: three neurons in the hidden and one neuron in the output layers; number of training epoch = 100; AUC = 0.94.
[a] *Under the nonparametric assumption.*
[b] *Null hypothesis: true area = 0.6.*

The results in Fig. 11.5 and Table 11.4 further identify possible hotspots at areas with poor water quality for drinking and the areas with good water quality, where the conditions remain within their pristine conditions at the upper catchment to the south of the study area and patches at northeast and northwest. However, deterioration is evident along the floodplains, where agriculture is practiced, with possible impacts from the mining industry. Although the role of the mining industry may be grave, there is not enough information yet for a direct study of their impacts. The results in Fig. 11.5 set a pattern that good water quality still prevails at upper catchments of the south and northeast and northwest of the study area, but poor water quality may occur along the floodplains of Qizil Ozen. There are some divergences between the bands for GQI and GWQI.

As per spatial variability of the trace elements (As, Pb, Fe, Mn), high contamination areas match with downstream and the dense watercourses areas. The risk of trace elements pollution increases in these areas. Spatial distribution of GQI and GWQI identifies visually poor and good quality zones and both results converge together by and large. The divergences between the respective results are significant and driven by the principle of precaution using both indices are advantageous. Areas with poor quality for drinking can be more than 20% of the study area.

4.1.2 USEPA indices for human health risk

Human health risk by chronic exposures to the contaminated groundwater via dermal and oral pathways is calculated according to Sections 3.1 and 3.2 for adults and children. The results for carcinogenic risks are shown in Fig. 11.6a1—d1 and for noncarcinogenic risks in Fig. 11.6a2—d2.

The spatial distribution of risk indices along the study area indicates that good water quality for drinking prevails at areas where the conditions still remain within their pristine conditions at the upper catchment to the south and patches at the northeast and northwest. However,

deterioration is evident in the floodplains with agriculture practices and significant mining.

The convergence between the four CRIs is remarkable, but some divergence is also observed, particularly for CRI-OC. These convergence and divergences are also mirrored for the NRIs, although the divergence between CRI and NRI values seems significant.

4.2 Level 1: results of data fusion by radicals

The results at Level 1 are obtained by a data fusion technique specified by Eqs. 11.13—11.16, which is an unsupervised process of learning from the data. Its output is the aggregated index distributed over the study area. The results are investigated by ROC/AUC in Fig. 11.7A. However, ROCs require setting a threshold value to compare False Positive and True Negative values. This was carried out by a trial-and-error procedure and testing the AUC values for 0.2, 0.4, 0.6, and 0.8 values but that of 0.6 was found to produce the optimum threshold for setting FN and TP comparisons.

The results shown in Fig. 11.7A show that the aggregated index extracts a significant amount of information from each index and each and all of them are fit-for-purpose. The results in Fig. 11.7B display the spatial distribution of the aggregated index, which extracts significant amount of information. It brings to light the pattern observed that at the upper catchments (south and northeast and northwest) water quality is expected to be good but along the floodplains water quality seems to deteriorate.

4.3 Level 2: results of strategies 1 and 2

Strategy 1 uses ANN, the inputs for which are the 10 indices, and the target values in terms of the aggregated index as the output of the data fusion at Level 1. Strategy 2 uses ANN, the inputs for which are the 22 arrays of raw data and the target values are the aggregated index as the output of the data fusion at Level 1 (similar to Strategy 1). To train the ANN model

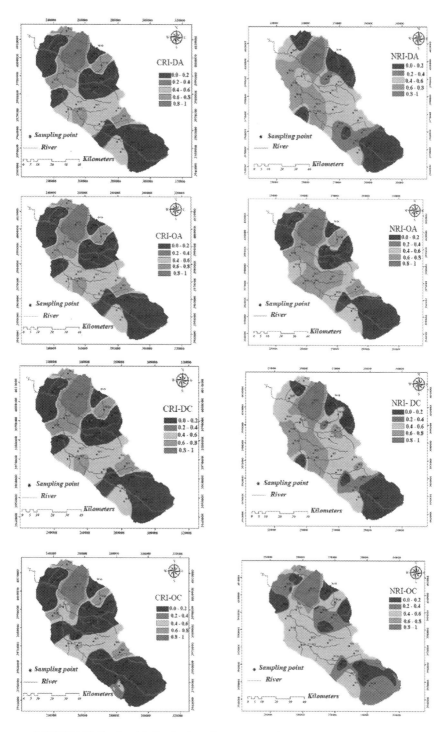

FIGURE 11.6 Spatial distribution of carcinogenic/noncarcinogenic risk indices.

4. Results

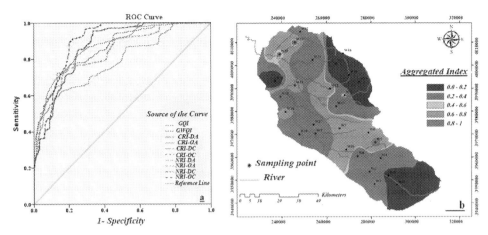

FIGURE 11.7 Modeling results at Level 1: (A) ROC/AUC of Level 1, (B) spatial distribution of the aggregated index.

in Strategies 1 and 2, a three-layer perceptron algorithm network (MLP-LM) was used, in which the optimum transformer function was found to be Tansig in the second layer and Purlin in the third layer. Table 11.4 presents the performance of both ANN1 and ANN2 models and the various default parameters used for the implementation of both models. Overall, both models are evidently fit-for-purpose.

The results for the performance metrics of ROC/AUC for both strategies are shown in Fig. 11.8A, according to which Strategy 1 compares well with Strategy 2. Strategy 1 is supervised learning from the data and Level 1 results and may seem tautological, i.e., the individual indices are processed at Level 1 using one algorithm and at Level 2 using another algorithm. Therefore, it is logical to be skeptical of its real value. Strategy 2 is in reality a check on the results of Strategy 1 and Fig. 11.8A shows that the degree of information in both strategies is just as good, and the better performance of Strategy 2 is marginal. If Strategy 2 performs marginally better than Strategy 1, why should one employ IMM at two levels and why not resort to Strategy 2 alone? The reason for this is clear that Strategy 2 cannot be performed by the aggregated index at Level 1, which delivers target values from Level 1. Therefore, the overall conclusion is that the results of Level 1 are defensible.

Strategy 2: The spatial distribution of the aggregated risk by Strategy 1 is shown in Fig. 11.8B and that by Strategy 2 in Fig. 11.8C.

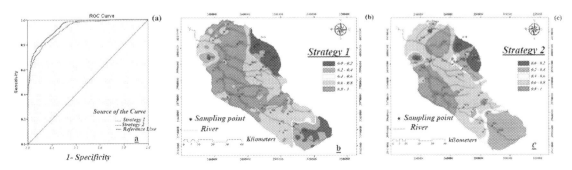

FIGURE 11.8 Level 2 results: (A) Performance metrics; (B) mapping Strategy 1; and (C) mapping Strategy 2.

There is good convergence between the respected bands but marginal divergences are also significant.

4.4 Overview of the results

The basic premise of the chapter is to aggregate multiple indices for aquifer management and derive one single index. Although there are various data fusion techniques for this type of problem, they are all deemed to be fit-for-purpose if successful, as there is no observed aggregated value to measure the performance of the aggregated index. However, the results of the IMM strategy formulated in the chapter provide progressive improvements measured by the performance metrics of ROC/AUC. A visual comparison of the results shows that (i) overall, the indices are indicative of low bands (good quality) at headwaters/highlands but these deteriorate toward lower catchment, where there are agricultural practices and mining activities. The aggregated index indeed captures the information at Level 1 using an unsupervised learning scheme, which is demonstrably capable of extracting the information on adverse water quality and risks to human health. The results presented in Figs. 11.7A and 11.8A indicate that the AUC values improve from $AUC_{Level\ 1} = 0.87, 0.86, 0.86, 0.86, 0.86, 0.86, 0.78, 0.89, 0.84, 0.89$ to $AUC_{Strategy\ 1} = 0.92$ and $AUC_{Strategy\ 2} = 0.94$. The investigation at Level 2 is also interesting as Strategy 1 is checked independently by Strategy 2. There is not much difference between Strategy 1 and Strategy 2 and this indicates that the ability of ANN is quite strong to learn the aggregated index at Level 1. Overall, the results provide a proof of concept for aggregated indices.

5. Discussion

The chapter investigates the aggregation of 10 water quality and risk to human health indices to obtain an aggregated risk using an IMM strategy. The results show that the aggregated index is indeed capable of capturing the appropriate signals in the base indices as a single aggregated index with improved performance metrics. These results are sufficient to treat them as a proof of concept for the modeling strategy. The choices of the techniques in implementing the modeling strategy were based on a deliberate choice of less sophisticated techniques or low-grade techniques rather than selecting the state-of-the-art research techniques. This may be regarded as a safety factor for the rational of the modeling strategy presented in the chapter. Further improvements on modeling techniques are likely and they would add to the defensibility of the results.

GQI and GWQI water quality indices are commonly used to evaluate the quality of groundwater for drinking with wide applications and research (Vadiati et al., 2016; Hameed et al., 2017; Adimalla and Qian., 2019; Wu et al., 2020; Ravindra et al., 2019; Jha et al., 2020; Singha et al., 2020; Zhang et al., 2020) topical research on these indices is still strong on reducing their inherent subjectivity. Although the chapter is not concerned directly with reducing subjectivity, IMM is an approach well suited for this. The remaining eight risk indices (CRI-DA, CRI-DC, CRI-OA, CRI-OC, NRI-DA, NRI-DC, NRI-OA, and NRI-OC) also suffer from subjectivity due to prescribed parametric values, but reducing inherent uncertainty is topical to reduce uncertainties, e.g., see Sadeghfam et al. (2021). The chapter is concerned with both reducing subjectivity and uncertainty but indirectly and the modeling strategy takes an effective step to ensure the defensibility of the results.

The banding of the indices presented in the chapter is nominal and not risk based. Ideally, a risk-based approach is the most appropriate approach. However, there are no such techniques yet available. Therefore, the authors use a nominal banding system and are reluctant to associate the bands with natural language descriptions of very good, good, moderate, poor, and very poor.

Khatibi (2011) reviews the emergence of risk and argues that risk may have two or three dimensions of (i) adverse effects often in terms of local adverse effects; (ii) system-wide actuating effects, e.g., probability of occurrence or water advection/diffusion contaminants in an aquifer; and (iii) responsibility. The term aggregation may have various connotations including heaps without taking account of the inherent dimensionality of the variables. Therefore, questions may be asked if the 10 indices are of similar dimensionalities before they qualify for aggregation. Currently, the use of the term aggregation justifies the modeling strategy and the rationale of the research in the chapter, but further investigations are planned to understand the dimensional homogeneity of the indices to be aggregated.

To the best knowledge of the authors, there is no similar work by other researchers on aggregating indices together to simplify the tasks of managers. Therefore, no comparative study is necessary.

To the best knowledge of the authors, there is no historic record of aquifer water quality data for the study area other than those produced by the authors. So, there is no objective way to benchmark the results with established baselines. However, both agriculture using modern fertilizers and mining activities at the study area are of a recent origin and only go back to post-1990s, prior to which the baseline conditions are deemed to be pristine, sustainable, and safe. The results presented in the chapter provide evidence that this upper catchment is in problem and it is unlikely the problem to disappear. In fact, if anything, it is likely to worsen. The underlying issue is that the problem is readily attributable to the absence of any planning system or to a poor planning practice both in the study area and in the country as a whole. The study area is a portrayal of similar situations within the country befitting the tragedy of the commons stemming from the absence of a planning system. Arguably, a planning system based on risk-based decision-making and decision-making by participation is the steppingstone

for solving some of the problems. These are further elaborated in Khatibi (2022) by the author.

6. Conclusions

Impacts of trace elements contaminating aquifers are investigated by mapping their water quality of a study area using the following 10 indices: GQI, GWQI, CRI-DA, CRI-DC, CRI-OA, CRI-OC, NRI-DA, NRI-DC, NRI-OA, and NRI-OC. The first indices are quite popular and measure the quality of water for drinking and the remaining eight are used to assess risk to human health stemming from trace elements. The research in the chapter is a pursuit of a simple rationale that if a planning system in a country is yet to be established, should risk indices not be fused together in one single aggregated signal, which extracts the relevant signals? The chapter provides a proof-of-evidence for the aggregated index.

A pilot study is presented at the upper catchment of Qizil Ozen, a significant tributary of the Caspian Sea, where modern fertilizers have become available since the 1990s and there are additional mining activities. An IMM practice is used to formulate a modeling strategy to aggregate the above 10 indices to produce an aggregated index distributed over the study area. The modeling strategy in this study compensates for the lack of information by presenting the aggregated index at the study area through data fusion and data extraction at two levels, and using observation data, it is possible to check its performance. However, the scope of the study did include hydrogeochemical studies of the contaminants, and therefore, the contamination origins remain unknown, although out-of-hand, they are attributed to poor planning practices stemming from modern agricultural fertilizers and mining practices.

The results are found to be defensible by using appropriate performance metrics and therefore the IMM strategy makes it possible to aggregate the 10 indices into one single aggregated index. The new index extracts appropriate signals from each of the individual indices, and as such, it becomes a representative of each and all the 10 indices. This is particularly suitable for countries, where the tradition of planning practices is yet to develop. This procedure is quite innovative but still has certain aspects that need to be investigated further. The ultimate aim is that the results must be defensible by not being a product of the particular assumptions of modeling procedures, in which the results are deemed as fit-for-purpose. The results have an indirection implication for the need for risk-based decision-making and a planning system responsive to participatory practices.

Appendix I

TABLE 11.A.1 Definitions, symbols, units, and values associated with equations used for health risk assessment.

P	Meaning	Unit	Oral values (Adult)	(Child)	Dermal values (Adult)	(Child)	References
AT	Average exposure time for ingestion	Days	25,550	3650	Noncarcinogenic effects $= ED \times 365 = 10,950$ (Adult). Carcinogenic effects $AT = 70 \times 365 = 25,550$	2190 (Child). Carcinogenic effects $AT = 70 \times 365 = 25,550$	USEPA (2004), Xiao et al. (2019), Bodrud-Doza et al. (2019)
BW	Average body Weight of a population group	Kg	70	25	70	25	USEPA (2004), Giri and Singh (2015), Xiao et al. (2019), Bodrud-Doza et al. (2019)
CF	Conversion factor	L/cm^3			1.1000		Bodrud-Doza et al. (2019), Yang et al. (2012)
CDI	Chronic daily intake	μg/kg/day	—	—	—	—	Yang et al. (2012), Xiao et al. (2019), Bodrud-Doza et al. (2019)
CW	Concentration in water	μg/L	—	—	—	—	Study data
ED	Exposure Duration through ingestion	Year	70	10	30	6	USEPA (2004), Bortey-Sam et al. (2015), Xiao et al. (2019), Bodrud-Doza et al. (2019)
EF	Dermal exposure frequency	days/year	365	350	USEPA (2004), Rahman et al. (2017), Xiao et al. (2019), Bodrud-Doza et al. (2019)		
ET	Exposure time in the shower	h/event	—	—	0.58	1	USEPA (2004), Xiao et al. (2019), Bodrud-Doza et al. (2019)
IR	Daily groundwater ingestion rate	L/day	2.2	1	—	—	Giri and Singh (2015), Rahman et al. (2017), Xiao et al. (2019), Bodrud-Doza et al. (2019)

TABLE 11.A.1 Definitions, symbols, units, and values associated with equations used for health risk assessment.—cont'd

P	Meaning	Unit	Oral values (Adult)	Oral values (Child)	Dermal values (Adult)	Dermal values (Child)	References
Kp	Dermal permeability coefficient	cm/hr			Al(0.001), As(0.001), Cr(0.003), Cu(0.001), Fe(0.001), Mn(0.001), Ni(0.004), Pb(0.001), Zn(0.0006), Cd(0.001)	USEPA (2004), Xiao et al. (2019), Bodrud-Doza et al. (2019)	
SA	Exposed skin area during bathing	cm^2	—	—	18,000	6600	USEPA (2004), Xiao et al. (2019), Bodrud-Doza et al. (2019)

TABLE 11.A.2 Dermal permeability coefficient, reference dose, slope factor, and gastrointestinal absorption coefficient for each element.

Elements	Units	Noncarcinogen Oral RfD (µg/Kg/day)	Noncarcinogen Dermal RfD (µg/Kg/day)	Carcinogen SF (kg × day/mg)
Al	(µg/L)	1000 (USEPA, 2020)	200 (USEPA, 2004)	—
As	(µg/L)	0.3 (USEPA, 2020)	0.285 (USEPA, 2004)	1.5
Cr	(µg/L)	3 (USEPA, 2020)	0.075 (USEPA, 2004)	—
Cu	(µg/L)	40 (USEPA, 2020)	12 (USEPA, 2004)	—
Fe	(µg/L)	700 (USEPA, 2020)	140 (USEPA, 2004)	—
Mn	(µg/L)	24 (USEPA, 2020)	0.96 (USEPA, 2004)	—
Ni	(µg/L)	20 (USEPA, 2020)	0.8 (USEPA, 2004)	—
Pb	(µg/L)	1.4 (Saleem et al., 2019)	0.42 (USEPA, 2004)	0.042
Zn	(µg/L)	300 (USEPA, 2020)	60 (USEPA, 2004)	—
Cd	(µg/L)	0.5 (USEPA, 2020)	0.025 (USEPA, 2004)	—

References

Abdelgawad, A., Bayoumi, M., February 2011. Sand monitoring in pipelines using Distributed Data Fusion algorithm. In: 2011 IEEE Sensors Applications Symposium. IEEE, pp. 217–220.

Adimalla, N., Qian, H., 2019. Groundwater quality evaluation using water quality index (WQI) for drinking purposes and human health risk (HHR) assessment in an agricultural region of Nanganur, south India. Ecotoxicol. Environ. Saf. 176, 153–161. https://doi.org/10.1080/10807039.2018.1460579.

Adimalla, N., Li, P., Qian, H., 2018. Evaluation of groundwater contamination for fluoride and nitrate in semi-arid region of Nirmal Province, South India: a special emphasis on human health risk assessment (HHRA). Human and Ecological Risk Assessment. Int. J. https://doi.org/10.1080/10807039.2018.1460579.

Aller, L., Bennett, T., Lehr, J., Petty, R., et al., 1987. US EPA/Robert S. Kerr Environmental Research Laboratory EPA. https://doi.org/EPA/600/2-87/035.

APHA, 2005. Standard Methods for the Examination of Water and Wastewater. American Public Health Association/American Water Works Association/Water Environment Federation, Washington DC.

Babiker, I.S., Mohamed, M.A., Hiyama, T., 2007. Assessing groundwater quality using GIS. Water Resour. Manag. 21 (4), 699–715. https://doi.org/10.1007/s11269-006-9059-6.

Bodrud-Doza, M., Islam, S.D.U., Hasan, M.T., Alam, F., Haque, M.M., Rakib, M.A., Asad, M.A., Rahman, M.A., 2019. Groundwater pollution by trace metals and human health risk assessment in central west part of Bangladesh. Groundw. Sustain. Dev. 9, 100219. https://doi.org/10.1016/j.gsd.2019.100219.

Bortey-Sam, N., Nakayama, S.M.M., Ikenaka, Y., Akoto, O., Baidoo, E., Mizukawa, H., Ishizuka, M., 2015. Health risk assessment of heavy metals and metalloid in drinking water from communities near gold mines in Tarkwa, Ghana. Environ. Monit. Assess. 187, 397.

Brown, R.M., McClelland, N.I., Deininger, R.A., O'Connor, M.F., 1972. A Water Quality Index—Crashing the Psychological Barrier. Indicators of Environmental Quality. Springer, pp. 173—182.

Edmond, J.M., Palmer, M.R., Measures, C.I., Grant, B., Stallard, R.F., et al., 1995. The fluvial geochemistry and denudation rate of the Guayana Shield in Venezuela, Colombia, and Brazil. Geochim. Cosmochim. Acta 59 (16), 301—325 doi:10.1016/0016-7037(95)00128-M.

Endres, E., Augustin, T., 2016. Statistical matching of discrete data by Bayesian networks. Workshop and Conference Proceedings 52, 159—170.

Giri, S., Singh, A.K., 2015. Human health risk assessment via drinking water pathway due to metal contamination in the groundwater of Subarnarekha River Basin, India. Environ. Monit. Assess. 187, 63. https://doi.org/10.1007/s10661-015-4265-4.

Hameed, M., Sharqi, S.S., Yaseen, Z.M., Afan, H.A., Hussain, A., Elshafie, A., 2017. Application of artificial intelligence (AI) techniques in water quality index prediction: a case study in tropical region, Malaysia. Neural Comput. Appl. 28 (1), 893—905. https://doi.org/10.1007/s00521-016-2404-7.

Hansson, S.O., 2005. Decision Theory, A Brief Introduction. Royal Institute of Technology (KTH).

Horton, R.K., 1965. An index number system for rating water quality. J. Water Pollut. Control Fed. 37, 300—306. http://www.scirp.org/(S(i43dyn45teexjx455qlt3d2q))/reference/Ref.

Hounslow, A.W., 1995. Water Quality Data: Analysis and interpretation. Lewis Publisher 397. http://www.archaeology.ws/2004-11-29.htm.

Huang, Y.P., Chu, H.C., 1999. Simplifying fuzzy modeling by both gray relational analysis and data transformation methods. Fuzzy Sets Syst. 104 (2), 183—197. https://doi.org/10.1016/S0165-0114(97)00212-1.

Isaaks, E.H., Srivastava, R.M., 1990. An Introduction to Applied Geostatistics Illustrated Edition. Oxford University Press, p. 595.

Jha, M.K., Shekhar, A., Jenifer, M.A., 2020. Assessing groundwater quality for drinking water supply using hybrid fuzzy-GIS-based water quality index. Water Res. 179, 115867. https://doi.org/10.1016/j.watres.2020.115867.

Khadra, W.M., Stuyfzand, P.J., 2018. Simulation of saltwater intrusion in a poorly karstified coastal aquifer in Lebanon (Eastern Mediterranean). Hydrogeol. J. 26, 1—18. https://doi.org/10.1007/s10040-018-1752-z.

Khatibi, R., 2011. Evolutionary systemic modelling of practices on flood risk. J. Hydrol. 401, 36—52. https://doi.org/10.1016/j.jhydrol.2011.02.006.

Khatibi, R., 2022. (submitted as Chapter 11 of this book) A basic framework to overarch sustainability, risk and reliability—A critical review. In: Roshni, T., Samui, P., Bui, G., Kim, D., Khatibi, R. (Eds.), Risk, Reliability and Sustainability.

Khatibi, R., Ghorbani, A., Naghshara, S., Aydin, H., Karimi, V., 2020. Introducing a framework for 'Inclusive Multiple Modelling' with critical views on modelling practices - applications to modelling water levels of Caspian Sea and Lakes Urmia and Van. J. Hydrol. 587, 124923. https;//DOI:10.1016/j.jhydrol.2020.124923.

Lal, A., Datta, B., 2018. Modelling saltwater intrusion processes and development of a multi-objective strategy for management of coastal aquifers utilizing planned artificial freshwater recharge. Model. Earth Syst. Environ. 4 (1), 111—126. https://link.springer.com/article/10.1007/s40808-017-0405-x.

Liu, Y., Wang, S., Du, X., et al., 2008. A multi-agent information fusion model for ship collision avoidance. IEEE Int. Conf. Mach. Learn. Cybern. 6—11. https://doi.org/10.1109/ICMLC.2008.4620369.

Long, X., Liu, F., Zhou, X., Pi, J., Yin, W., Li, F., Huang, S., Ma, F., 2020. Estimation of spatial distribution and health risk by arsenic and heavy metals in shallow groundwater around Dongting Lake plain using GIS mapping. Chemosphere 269 (128698). https://doi.org/10.1016/j.chemosphere.2020.128698.

Mohanta, V.L., Singh, S., Mishra, B.K., 2020. Human health risk assessment of fluoride-rich groundwater using fuzzy-analytical process over the conventional technique. Groundw. Sustain. Dev. 10 (100291). https://doi.org/10.1016/j.gsd.2019.100291.

Nadiri, A.A., Khatibi, R., 2021. Inclusive Multiple Models (IMM) for predicting groundwater levels and treating heterogeneity. Geosci. Front. 12 (2021), 713—724. https://doi.org/10.1016/j.gsf.2020.07.011.

Nadiri, A.A., Sedghi, Z., Khatibi, R., Gharekhani, M., 2017. Mapping vulnerability of multiple aquifers using multiple models and fuzzy logic to objectively derive model structures. Sci. Total Environ. 593, 75—90.

Nadiri, A.A., Sedghi, Z., Khatibi, R., Sadeghfam, S., 2018. Mapping specific vulnerability of multiple confined and unconfined aquifers by using artificial intelligence to learn from multiple DRASTIC frameworks. J. Environ. Manag. 227, 415—428. https://doi.org/10.1016/j.jenvman.2018.08.019.

Rahman, M.M., Islam, M.A., Bodrud-Doza, M., Muhib, M.I., Zahid, A., Shammi, M., Tareq, S.M., Kurasaki, M., 2017. Spatio-temporal assessment of groundwater quality and human health risk: a case study in Gopalganj, Bangladesh. Expo. Health. https://doi.org/10.1007/s12403-017-0253-y.

Ramakrishnaiah, C.R., Sadashivaiah, C., Ranganna, G., 2009. Assessment of water quality index for the groundwater in Tumkur Taluk, Karnataka State, India. E-J. Chem. 6 (2), 523–530. https://doi.org/10.1155/2009/757424.

Ravindra, K., Thind, P.S., Mor, S., Singh, T., Mor, S., 2019. Evaluation of groundwater contamination in Chandigarh: source identification and health risk assessment. Environ. Pollut. 255, 113062. https://doi.org/10.1016/j.envpol.2019.113062.

Ribeiro, L., Paralta, E., Nascimento, J., Amaro, S., Oliveira, E., Salgueiro, R., 2002. A agricultura a delimitac ao das zonas vulnera'veis aos nitratosdeorigem agrı'cola segundo a Directiva 91/676/CE. In: Proc. III Congreso Ibe'rico sobre Gestio'n e Planificacio'n del Agua. Universidad de Sevilla, Spain, pp. 508–513.

Rodrígueza, S., De Paza, J.F., Villarrubiaa, G., Zato, C., Bajo, J., Corchado, J.M., 2015. Multi-Agent Information Fusion System to manage data from a WSN in a residential home. Inf. Fusion 23, 43–57. https://doi.org/10.1016/j.inffus.2014.03.003.

Rostami, A.A., Karimi, V., Khatibi, R., Pradhan, B., 2020. An investigation into seasonal variations of groundwater nitrate by spatial modelling strategies at two levels by kriging and co-kriging models. J. Environ. Manag. 270 (15), 110843. https://doi.org/10.1016/j.jenvman.2020.110843.

Rufino, F., Busico, G., Cuoco, E., Darrah, T.H., Tedesco, D., 2019. Evaluating the suitability of urban groundwater resources for drinking water and irrigation purposes: an integrated approach in the Agro-Aversano area of Southern Italy. Environ. Monit. Assess. 191 (12), 768.

Saçmaci, Ş., Kartal, Ş., Sacmaci, M., 2012. Determination of Cr (III), Fe (III), Ni (II), Pb (II) and Zn (II) ions by FAAS in environmental samples after separation and preconcentration by solvent extraction using a triketone reagent. Fresenius Environ. Bull. 21 (6), 1563–1570.

Sadeghfam, S., Hassanzadeh, Y., Khatibi, R., Moazamnia, M., Nadiri, A.A., et al., 2018. Introducing a risk aggregation rationale for mapping risks to aquifers from point- and diffuse-sources—proof-of-concept using contamination data from industrial lagoons. Environ. Impact Assess. Rev. 72, 88–98. https://doi.org/10.1016/j.eiar.2018.05.008.

Sadeghfam, S., Khatibi, R., Nadiri, A.A., Ghodsi, K., et al., 2021. Next stages in aquifer vulnerability studies by integrating risk indexing with understanding uncertainties by using generalised likelihood uncertainty estimation. Expos. Health 13 (4), 1–15. https://doi.org/10.1007/s12403-021-00389-6.

Saleem, M., Iqbal, J., Shah, M.H., et al., 2019. Seasonal variations, risk assessment and multivariate analysis of trace metals in the freshwater reservoirs of Pakistan. Chemosphere 216, 715–724. https://doi.org/10.1016/j.chemosphere.2018.10.173.

See, L., Abrahart, R., 2001. Multi-model data fusion for hydrological forecasting. Comput. Geosci. 27 (8), 987–994. https://doi.org/10.1016/S0098-3004(00)00136-9.

Singha, S., Pasupuleti, S., Singha, S.S., Kumar, S., 2020. Effectiveness of groundwater heavy metal pollution indices studies by deep-learning. J. Contam. Hydrol. 235, 103718.

Taheri Tizro, A., Voudouris, K., Vahedi, S., 2014. Spatial variation of groundwater quality parameters: a case study from a semiarid region of Iran. Int. Bull. Water Resour. Dev. 1, 3.

Tiwari, T.N., Mishra, M., 1985. A preliminary assignment of water quality index of major Indian rivers. Indian J. Environ. Protect. 5 (4), 276–279.

USEPA, 1989. Risk Assessment Guidance for Superfund Volume I Human Health Evaluation Manual (Part A).

USEPA, 1991. Human health evaluation manual, supplemental guidance: "standard default exposure factors". OSWER Directive 9285, 6-03.

USEPA, 2004. Risk Assessment Guidance for Superfund Volume I: Human Health Evaluation Manual (Part E). http://www.epa.gov/oswer/riskassessment/ragse/pdf/introduction.pdf.

USEPA (US Environmental Protection Agency), 2009. Baseline Human Health Risk Assessment Vasquez Boulevard and I-70 Superfund Site. Denver CO. http://www.epa.gov/region8/superfund/sites/VB-170-Risk.pdf. (Accessed 20 January 2011).

USEPA, 2020. Regional Screening Level (RSL) Summary Table (TR 1E-06 THQ 1.0). https://semspub.epa.gov/work/HQ/197414.pdf. (Accessed 8 January 2020).

Vadiati, M., Asghari-Moghaddam, A., Nakhaei, M., Adamowski, J., Akbarzadeh, A.H., 2016. A fuzzy logic based decision-making approach for identification of groundwater quality based on groundwater quality indices. J. Environ. Manag. 184, 255–270. https://doi.org/10.1016/j.jenvman.2016.09.082.

Valavanidis, A., Vlachogianni, T., 2010. Metal Pollution in Ecosystems. Ecotoxicology Studies and Risk Assessment in the Marine Environment. University of Athens, Greece.

Water Quality Data: Analysis and interpretation. Lewis Publisher, 1995 397. http://www.archaeology.ws/2004-11-29.htm.

WHO, 2011. Guidelines for drinking-water quality. In: Recommendations, third ed., vol. 1. WHO, Geneva.

Wu, C., Fang, C., Wu, X., Zhu, G., 2020. Health-risk assessment of arsenic and groundwater quality classification using random Forest in the Yanchi region of Northwest China. Expos. Health 12 (4), 761–774.

Xiao, J., Wang, L., Deng, L., Jin, Z., 2019. Characteristics, sources, water quality and health risk assessment of trace elements in river water and well water in the Chinese Loess Plateau. Sci. Total Environ. 650, 2004–2012.

Yang, M., Fei, Y., Ju, Y., Ma, Z., Li, H., 2012. Health risk assessment of groundwater pollution—a case study of typical city in North China Plain. J. Earth Sci. 23 (3), 335–348. https://link.springer.com/article/10.1007/s12583-012-0260-7.

Zhang, Q., Xu, P., Qian, H., 2020. Groundwater quality assessment using improved water quality index (WQI) and human health risk (HHR) evaluation in a semi-arid region of northwest China. Expos. Health 12, 487–500. https://link.springer.com/article/10.1007/s12403-020-00345-w.

Further reading

Chau, K.W., 2006. A review on integration of artificial intelligence into water quality modelling. Mar. Pollut. Bull. 52, 726–733.

USEPA, (US Environmental Protection Agency), 2001. Baseline Human Health Risk Assessment Vasquez Boulevard and I-70 Superfund Site. Denver, CO. http://www.epa.gov/region8/super fund/sites/VB-170-Risk.pdf.

Zeng, X.X., Liu, Y.G., You, S.H., Zeng, G.M., Tan, X.F., Hu, X.J., Hu, X., Huang, L., Li, F., 2015. Spatial distribution, health risk assessment and statistical source identification of the trace elements in surface water from the Xiangjiang River, China. Environ. Sci. Pollut. Res. 22, 9400–9412.

CHAPTER 12

Liquefaction hazard mitigation using computational model considering sustainable development

Sufyan Ghani, Sunita Kumari

Department of Civil Engineering, National Institute of Technology Patna, Patna, Bihar, India

1. Introduction

Engineering structures are exposed to various natural calamities resulting in different forms of hazards. These primely designed and well-built structures are designed to withstand such hazard to utmost strengths, but their responses are most vulnerable when it comes to withstanding ground shaking due to earthquakes. Liquefaction is one of the most disastrous phenomena that arise due to earthquake. The destructive nature and random occurrence of liquefaction phenomenon have attracted the attention of many geotechnical engineers. Under the influence of cyclic loading imparted by earthquake, development of excess pore water pressure in saturated soil leads to the loss of strength and stiffness which is the major criteria of liquefaction. Soil failure due to liquefaction causes severe lifeline and structural damage in affected areas. Therefore, there is a crucial need to evaluate the liquefaction prone locations for reduction of seismic risks. Well-established geologic and geotechnical data for an area help in mapping of liquefaction susceptibility. Seed and Idriss (1971) provided the first simplified method for assessing liquefaction hazards which were based on the concept of effective stress. Further, various researchers have introduced numerous methods that are based on effective stress and energy dissipation concepts and have been found to be an effective tool for evaluating liquefaction potential of soil. India has always been an epicenter of high intensity earthquakes like Latur earthquake 1993, Jabalpur earthquake 1997, Bhuj earthquake 2001, and Koyna earthquake 1967. These high-intensity earthquakes have established that seismically no region of India can be considered stable and safe. These high-intensity earthquakes have caused serious damages to life and structures leading to huge loss of capital and resources. Even today, safe guarding a structure against liquefaction seems to be a challenging and cumbersome task for geotechnical engineers. Therefore, assessing liquefaction potential of a soil deposit before designing and construction of a structure is a common practice. Presently, these applicable methods for evaluating liquefaction are equation-based empirical approaches. These empirical methods require large data set based on extensive laboratory and in situ testing for proper evaluation of liquefaction which involves high-end calculations. The dependency of empirical methods on experimental results and extensive calculation makes them a bit resource and time consuming. Also a skilled supervision is required for estimating the liquefaction behavior of soil using these empirical methods. At the present time, computational modeling is being used frequently for finding the solutions to various engineering problems including assessment of liquefaction behavior of soil deposits. Artificial Neural Network (ANN) is a widely accepted technique applied by numerous researchers to evaluate the liquefaction behavior of soil (Wang and Rahman, 1999; Juang and Chen, 1999; Goh, 2002; Omer et al., 2009; Farrokhzad et al., 2010; Samui and Sitharam, 2011; Kumar et al., 2012; Ghani and Kumari, 2021a,b; Ghani et al., 2021a,b). The accuracy and robustness of ANN model can be significantly enhanced by choosing suitable inputs and providing large datasets for training and testing of the model. A powerful ANN model provides more precise and realistic results as compared to empirical methods. In this chapter, Idriss and Boulanger (2006) empirical approach is used to evaluated liquefaction potential of soil, and based on the obtained factor of safety (FOS), an ANN model has been developed which predicts the factor of safety against liquefaction (FL) which attempts to overcome the major drawback of empirical methods.

The computational and empirical methods adopted in this chapter are a robust and appropriate method to evaluate liquefaction susceptibility of soil, but there are many possible factors that may cause uncertainty in determining liquefaction. Therefore, accurate and precise prediction of liquefaction is highly significant for designing structures, especially when the study area has seismic active zones. Proper forecasting of soil liquefaction guarantees the safety and serviceability of such engineering projects. Thus, in the process of liquefaction evaluation, it must be assured that the factor of safety of the soil deposits is measured with appropriate level of accuracy. To obtain the most reliable method among the two aforementioned methods, a reliability analysis has been performed. Reliability-based design methods have been active for most of the civil engineering works such as building structure, dam, analysis of slope, and settlement of foundation (Juang et al., 1999; Cao et al., 2017; Janalizade et al., 2015; Jha and Suzuki 2009; Umar et al., 2018; Hui et al., 2021; Pei et al., 2021; Rajeswari and Sarkar, 2021; Dudzik and Potrzeszcz-Sut, 2021; Ghani and Kumari, 2021c). Considering these facts, this chapter aims to simplify the process of evaluating soil's liquefaction behavior in a broader domain involving the least experimental datasets. Advanced first-order second-moment (AFOSM) reliability analysis has been performed to obtain reliability indices (β) and probability of liquefaction (P_L) for better understanding and clarity of the results.

2. Study area and data collection

Over the past few decades, it has been well established that several parts of India have been vulnerably exposed to high-intensity earthquakes followed by liquefaction. Many portions of the nation propose threat to liquefaction. Considering the intensity of these threats, this chapter proposes to analyze liquefaction susceptibility of Muzaffarpur district (26.1197° N and 85.3910° E) of Bihar, India. Muzaffarpur district is located in seismically active zones that fall close to the Gangetic planes. The considered study area portrays one of the most vulnerable conditions to undergo liquefaction. The majority of the area has alluvial soil deposits with high percentages of plastic and nonplastic fines. Due to the presence of river belts, the proposed study area is prone to floods every monsoon making soil fully saturated. The overall ground water table is nearly 4–5 m below the ground level, as well as the soil at surface level is loose and distorted. Therefore, an alluvial soil deposit with saturated soil condition or high water table level and loose soil on surface clearly makes an ideal case for triggering the phenomenon of liquefaction during an earthquake.

Bihar state is among the highly populated states in the country. Closely packed urban structures with high population density and lesser mitigation techniques to withstand liquefaction phenomenon raise an alarm for high risk and fatality. Therefore, for this chapter, the data obtained from an investigation site mentioned above and tested in NIT Patna Laboratory are used to analyze the liquefaction behavior of soil deposits having clay particles. Table 12.1 highlights the statical details of the projected datasets obtained from the study area and Fig. 12.1 demonstrates the variation of $(N_1)_{60}$ and FC along their respective depth.

3. Theoretical details of empirical and computational model

The following sections present the details of empirical and computational models used in this chapter. Idriss and Boulanger (2006) semi-empirical approach has been adopted to analyze

the liquefaction potential of the site. Furthermore, ANNs have been employed for the same.

3.1 Idriss and Boulanger (2006) empirical approach

For liquefaction assessment of soil deposits, the Cyclic Stress Ratio (CSR) that illustrates the loading that is imposed by earthquakes on the soil has been proposed as follows:

$$(CSR)_{7.5} = 0.65 \left(\frac{\sigma_{vo} a_{max}}{\sigma'_{vo}} \right) \frac{r_d}{MSF} \frac{1}{K_\sigma} \quad (12.1)$$

where σ_{vo} and σ'_{vo} are the total and effective vertical overburden stress at some specified depth z; a_{max} is the peak horizontal ground acceleration in $g's$; r_d is a stress reduction factor, MSF represents the magnitude scaling factor which is applied to alter the induced CSR to the reference earthquake magnitude of 7.5; and K_σ presents the correction factor for effective overburden.

Stress reduction coefficient (r_d) is calculated using analytical procedures and is adequately expressed as a function of depth (z) and earthquake magnitude (M). Following expression for r_d is applicable to a depth where $z \leq 34$ m, in which z is depth in meters and M is earthquake magnitude.

$$Ln(r_d) = \alpha(z) + \beta(z)M \quad (12.2)$$

$$\alpha(z) = -1.012 - 1.126 \sin\left(\frac{z}{11.73} + 5.133\right) \quad (12.3)$$

TABLE 12.1 Statical details of the dataset.

Variables	Unit	Maximum	Minimum	1st quartile	Median	3rd quartile	Mean
Wc/LL	—	1.23	0.62	0.74	0.89	1.15	0.95
$(N_1)_{60}$	—	35.00	2.00	20.00	26.50	28.10	16.25
FC	%	96.00	34.33	38.75	46.75	81.20	58.37
PGA	—	0.36	0.24	0.24	0.36	0.36	0.30
M_w	—	7.5	6.0	6.0	7.0	7.0	6.8

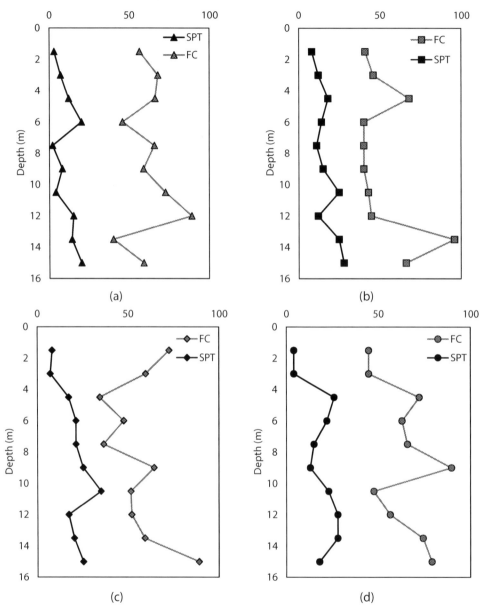

FIGURE 12.1   The variation of SPT value and fine content along the depth for (A) site A (B) site B (C) site C, and (D) site D.

$$\beta(z) = 0.106 + 0.118 \sin\left(\frac{z}{11.28} + 5.142\right) \quad (12.4)$$

whereas the following expression is applicable for $z > 34$ m;

$$r_d = 0.12 \exp(0.22M) \quad (12.5)$$

Equations used to determine cyclic resistance ratio (CRR) and corresponding factor of safety (FOS) generated from the corrected blow count $(N_1)_{60}$ has been listed below:

$$CRR = \exp\left[\frac{(N_1)_{60cs}}{14.1} + \left\{\frac{(N_1)_{60cs}}{126}\right\}^2 - \left\{\frac{(N_1)_{60cs}}{23.6}\right\}^3 + \left\{\frac{(N_1)_{60cs}}{25.4}\right\}^4 - 2.8\right]$$

$$(12.6)$$

Where term $(N_1)_{60cs}$ depends on $(N_1)_{60}$ and given as follows:

$$(N_1)_{60CS} = (N_1)_{60} + \Delta(N_1)_{60} \quad (12.7)$$

The following equation is used to evaluate factor of safety (FOS) against liquefaction, which is described as follows:

$$FOS = \frac{CRR}{CSR} \quad (12.8)$$

3.2 Artificial neural network

An ANN is one of the widely acknowledged advanced soft computing approaches which is stimulated by the functionality of the human brain (Goh, 1996; Maier et al., 2010; Wambua et al., 2016; Mokhtarzad et al., 2017; Yang and Wang, 2020; Hasson et al., 2020; Khoo et al., 2021; Ghani et al., 2021a,b). ANN generates a multifaceted network mapping between input and output variables which can approximate nonlinear functions. Among the numerous methods available to develop an ANN model, multilayer perceptron (MLP) neural network is one of the most commonly applied methods which can solve the complex mathematical

problems that require nonlinear equations by defining proper weights. The typical MLP consists of at least three layers. The first layer is termed as an input layer; last layer is termed as output layer, whereas the remaining layers present between the input and output layers are called hidden layers. Coulibaly et al. (2000) suggested that one hidden layer is adequate for an ANN model to approximate multifaceted nonlinear functions for given datasets. Observations drawn from this chapter also suggest that one hidden layer was suitable to estimate the relationship between the principal components and factor of safety against liquefaction of the soil deposits. Therefore, this chapter uses an ANN model which comprises one input layer, one output layer, and one hidden layer. Once the network is properly trained with adequate datasets, it can be validated, and further the trained network can be used to make predictions for a new set of data that it has never been introduced. Due to its efficiency and wide applicability, it is one of the most suitable computational models applied in the field of liquefaction assessment and has been used by various researchers (Farookhzad et al., 2010; Kamatchi et al., 2010; Samui and Sitharam, 2011; Kumar et al., 2012; Prabakaran et al., 2015; Ramezani et al., 2018; Ghani and Kumari, 2021a,b; Ghani et al., 2021a,b; Kamura et al., 2021; Babacan and Ceylan, 2021). Fig. 12.2 shows the typical representation of ANN architecture.

The developed computational model uses data cases divided into training and testing data obtained from different sites in Muzaffarpur district, India. The developed models consider geotechnical parameters which significantly affect the liquefaction resistance of the soil deposits as well as parameters that are related to the intensity of earthquake and ground acceleration as input parameters. The five input parameters are selected as per the literature observations which include SPT blow count $(N_1)_{60}$, ratio of water content and liquid limit

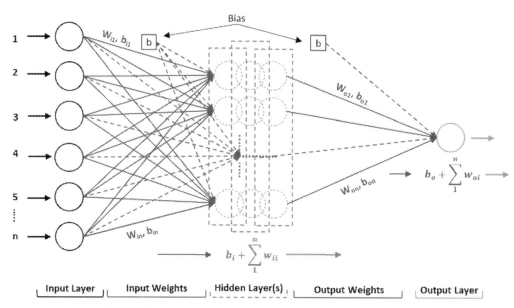

FIGURE 12.2   A typical architecture of artificial neural network.

(wc/LL), fine content (F.C), peak ground acceleration (a_{max}/g), and magnitude of earthquake (M_w). The computational model developed is one of the effective approaches to study liquefaction of soil deposits. For the ANN model, 1 hidden layer with 10 hidden neurons was considered, and because of the randomness present in ANN, the numbers of hidden neurons are finalized based on trial and error run. The linear function (*trainlm*) and tangent sigmoid (*tansig*) function are used as training function and transfer function, respectively. Furthermore, the performance of the developed ANN model was determined using several statical performance parameters mentioned in the following section.

4. Advanced first order second moment reliability method

Application of reliability analysis in geotechnical engineering problems has been widely practiced by numerous researchers to overcome the uncertainty and misleading results (Juang et al., 1999; Cao et al., 2017; Janalizade et al., 2015; Jha and Suzuki, 2009). The methods adopted to evaluate liquefaction are strongly based on theory with independent soil parameters and high correlation between soil properties which leads to certain complexity during the evaluation of reliability index and failure probability. The complexity and uncertainty in estimations often lead to highly inaccurate results. Therefore the use of highly effective and advance reliability techniques has been adopted for the present study. AFOSM techniques specifically, the Hasofer–Lind reliability index method is computed as follows (Ditlevsen, 1981):

$$\beta = \min_{x \in F} \sqrt{(X - m)^T C^{-1} (X - m)} \quad (12.9)$$

where X represents vector of random variables in a function given by $G(X) = 0$; m represents vector of mean values; and C represents covariance matrix. The minimization in Eq. (12.1) is performed over the failure domain F corresponding to the region $G(\mathbf{X}) < 0$.

$$\beta^2 \min_{G(X)=0} \left\{ \left[\frac{(X_1 - m_1)^2}{\sigma_1^2} \right] + \left[\frac{(X_2 - m_2)^2}{\sigma_2^2} \right] \right.$$
$$+ \left[\frac{(X_3 - m_3)^2}{\sigma_3^2} \right]$$
$$+ \left[\frac{(X_4 - m_4)^2}{\sigma_4^2} - 2\frac{(X_4 - m_4)(X_5 - m_5)\rho_{45}}{\sigma_4 \sigma_5} \right.$$
$$\left. + \frac{(X_5 - m_5)^2}{\sigma_5^2} \right] * \left(\frac{1}{1 - \rho_{45}^2} \right) \right\}$$

(12.10)

where x_1 represents w_c/LL; x_2 represents $(N_1)_{60}$; x_3 represents fine content (FC); x_4 represents peak ground acceleration (PGA); and x_5 represents magnitude of earthquake (M_w); m_i signifies mean value of input variables ($i = 1$ to 5); σ_i represents standard deviation of input variable x_i; and ρ_{45} is correlation coefficients between input variable x_4 and x_5.

Reliability index (β) is determined using Eq. (12.10) and the determined values of reliability indices are grouped in Table 12.2 for all the sites.

5. Data processing and analysis

For the application of computational analysis, normalization of the datasets is considered as one of the most crucial stage. Normalization is performed in the preprocessing stage in order to cancel out the dimensionality effect of the variables by creating new values that maintain the general distribution and ratios in the source data, while keeping values within a scale applied across all numeric columns used in the model. Therefore, before developing any model, the input and output variables have been normalized between 0 and 1 using the following expression:

$$x_{NORMALISED} = \left(\frac{x - x_{\min}}{x_{\max} - x_{\min}} \right)$$

(12.11)

In which $x_{\min}$ and $x_{\max}$ represent the minimum and maximum value of parameter (x) under consideration, respectively. This approach is called "min-max" normalization technique. After the normalization has been performed, the dataset is then divided into training and testing subsets. The major significance of dividing dataset in Training and Testing phases is to validate the robustness of the developed model. By dividing dataset in training and testing parts, one can minimize the effects of data discrepancies and obtain a better understanding about the characteristics of the developed model. The first subset is used to fit/train the model and therefore referred as training dataset. The second subset is not used to train the model; instead, the input element of the dataset is provided to the developed model and then predictions are made and compared to the expected values. Because the data in the testing set already contain known values for the attribute that you want to predict, it is easy to determine whether the model's predictions are correct. Therefore, the second dataset is referred to as the test dataset. Therefore, 70% of the whole dataset is extracted at random to build the training subset, while the balance 30% data is used as testing subset.

The performance of the computational model is generally investigated by various statical performance parameters enlisted below (Asteris et al., 2021; Kardani et al., 2021a,b; Kumar et al., 2021). Formulas are used to calculate mentioned parameters, i.e., R, R^2, MAE, and

TABLE 12.2 Statical performance parameters for the developed ANN model.

ANN model	R^2	RMSE	MAE
Training	1.00	0.03	0.009
Testing	0.994	0.017	0.013

RMSE and are presented in Eqs. (12.1)–(12.4), respectively.

1. Correlation coefficient (R),
2. Coefficient of Determination or Model Fit Value (R^2),
3. Maximum absolute error (MAE),
4. Root mean square error (RMSE)

$$RMSE = \sqrt{\frac{1}{n}\sum_{i=1}^{n}(y_i - x_i)^2} \quad (12.12)$$

$$R^2 = \left[\frac{\sum_{i=1}^{n}(x_i - \bar{x})(y_i - \bar{y})}{\sqrt{\sum_{I=1}^{N}(x_i - \bar{x})^2 \sum_{I=1}^{N}(y_i - \bar{y})^2}}\right]^2 \quad (12.13)$$

$$MAE = \frac{1}{n}\sum_{i=1}^{n}|y_i - x_i| \quad (12.14)$$

R^2 ranges from 0 to 1. The closer it is to 1, the better the fit. If R^2 is equal to 1, it means perfect linear relationship exists between the dependent variable and independent variables, while R^2 equal to 0 indicates that independent variables have no impact on the dependent variable.

MSE and RMSE are the estimates of standard error; smaller the values, the better the fit. These measurements provide an excellent indication of the quality of the fit when the prediction is important for the model. The values of R^2, MSE, and RMSE determine the goodness of the model.

6. Results and discussion

This section outlines the results determined from the empirical and developed computational model. ANN-based computational model can predict the liquefaction behavior of soil with utmost ease and accuracy. A comparative study for the same has been performed to determine the robustness such models and their respective results have been presented in this section below. Fig. 12.3 clearly highlights that the developed ANN model has similar predictions as per the empirical equations. The F_L determined using ANN model can be as a prominent parameter for defining soils liquefaction behavior. The major drawback of these empirical methods discussed in aforementioned sections

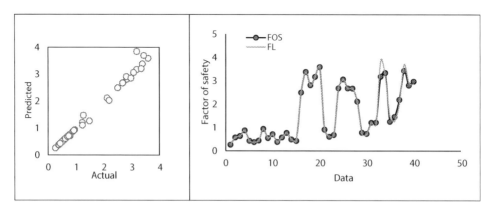

FIGURE 12.3 Trend of FOS and F_L as per the data frequency.

has been significantly addressed with the use of computational method. Dependency of empirical equations on high-end extensive experimental investigation and cumbersome calculation for evaluating liquefaction behavior of soil deposit is reduced up to a considerable extent while using ANN model. The developed ANN model uses artificial intelligence to train itself and becomes a high-end liquefaction prediction model. It uses a set of input data's to predict the desired output with utmost ease. Once a model is trained and tested, it can serve as a robust tool for prediction liquefaction behavior of soil for new dataset as well.

Fig. 12.4 describes the frequency plot of factor of safety determined using empirical and computational approach. It has been observed that the range of factor of safety is similar for both the methods. Furthermore, AFOSM reliability analysis has been performed on the obtained F_L values from developed ANN model to justify its reliability and applicability as an effective tool for measuring liquefaction response of soil deposits. Reliability analysis provides a clear insight about the accuracy of the proposed model. Using Eq. (12.10), reliability indices (β) have been estimated for the whole site which in turn help in evaluating probability of liquefaction. Instances in the past prove that for soil deposits with factor of safety against liquefaction greater than unity also liquefy up to a considerable extent; therefore, describing liquefaction behavior of soil in terms of probability helps engineers and designers while taking crucial mitigation decisions. Fig. 12.5 presents the trend of probability of liquefaction according to depth for all the four sites. It has been observed in some cases that the liquefaction probability reduces up to a considerable extent. To obtain a clear insight on the response of probability of liquefaction and factor of safety against depth, a 3D surface plot has been presented in Fig. 12.6. It demonstrates the variation and disparity among the two variables as per the variation in depth for all the four boreholes. The finding suggests that as the depth increases, the probability of soil to undergo liquefaction tends to increase or as depth increases, FL tends

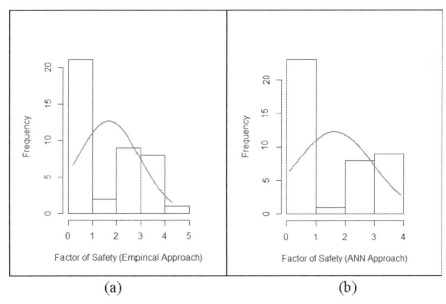

FIGURE 12.4 Histogram and frequency plot for highlighting the factor of safety determined using (A) empirical approach and (B) computational approach.

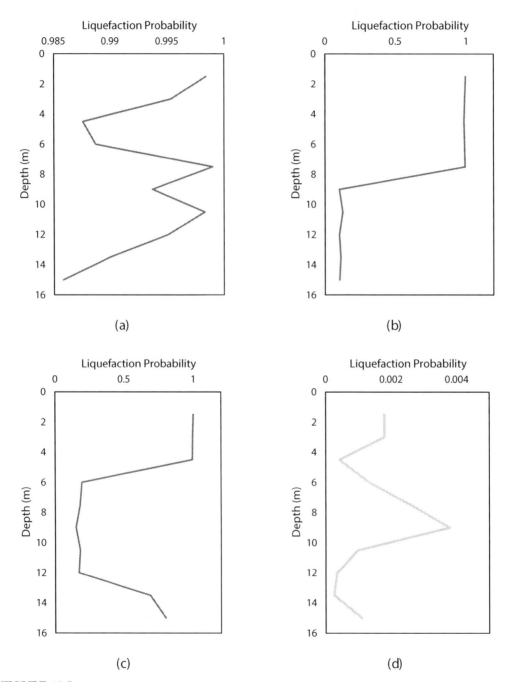

FIGURE 12.5 Depth wise variation of probability of liquefaction (A) site A (B) site B (C) site C and (D) site D.

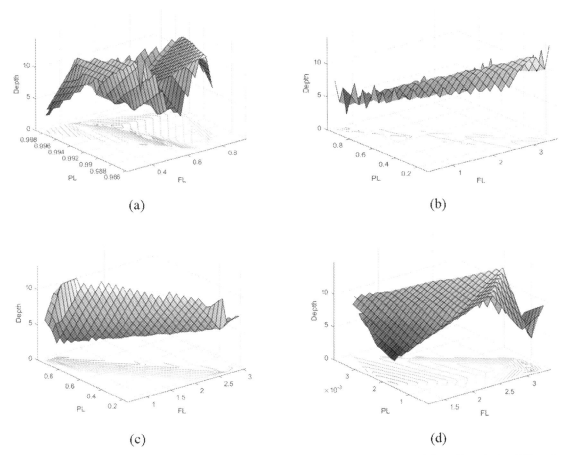

FIGURE 12.6 A 3D surface plot predicting the variation of factor of safety (FL) and probability of liquefaction (PL) against depth of (A) site A, (B) site B, (C) site C, and (D) site D.

to be on the safer side, i.e., above unity. Site-A exhibited a higher probability of liquefaction as compared to the other sites. Contrary to site-A, site-D exhibited insignificant liquefaction probability, whereas for site-B and site-C probability of liquefaction varied from high to low as per the depth. Soil layer closer to ground surface has less overburden pressure and is more vulnerable to liquefaction under high seismic motion, whereas soil deposits deep below the ground layer have high overburden pressure which restricts soil movement and considerably reduces its tendency to liquefy. Figs. 12.5 and 12.6 exhibited the similar trend of liquefaction probability against depth, which satisfies the determined results.

This chapter focuses on the response of fine content against liquefaction susceptibility of soil. According to several researchers, fine content increases soil resistance capability against liquefaction up to considerable extent thus resulting in lower liquefaction probability. Due to dilative nature of fine-grained soil, it influences the soil's ability to develop excess pore pressures which is one of the governing criteria for liquefaction. Presence of fine in sandy soil

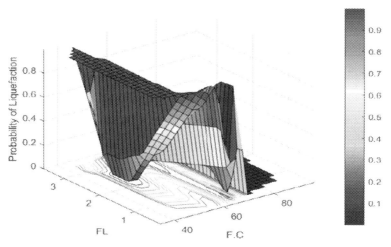

FIGURE 12.7 3D surface plot predicting the variation of factor of safety (FL) and probability of liquefaction (PL) as per fine content (FC).

decreases the penetration resistance up to a great extent as fines have lower permeability as which results in increased excess pore pressures on the penetration device thus resulting in lower effective stresses and lower penetration resistance. Fig. 12.7 shows the trend of FL and PL against fine content. It is evident from the drawn observation that with an increase in fine content, there is an orderly increase in FL and systematic decrease in probability of liquefaction of the soil deposit.

7. Conclusion and summary

This chapter focuses on liquefaction assessment of soil deposit using computational methods to provide an alternative and reliable tool for predicting the response of soil against liquefaction. The observations made from the present work confirm that the use of artificial intelligence for hazard mitigation can save us from incurring massive damages caused due to hazards like liquefaction. The following conclusions are drawn from the study:

1. The artificial neural network (ANN) model was trained to predict factor of safety against liquefaction (FL) using basic geotechnical and seismic parameters such as wc/LL, FC, $(N_1)_{60}$, PGA, and M_w.
2. Statical parameters suggest that the developed model shows high correlation with globally accepted Idriss and Boulanger (2006) semiempirical method.
3. AFOSM reliability analysis of the developed ANN model suggests that the proposed model is a reliable and efficient tool in predicting liquefaction probability of soil deposits.
4. A reliable and competent computational tool which overcomes the major limitations of empirical approaches will contribute immensely in resource efficiency. This method may be categorized as a sustainable method for evaluating and predicting risk against any seismic hazards due to its cost effectiveness and quick prediction ability.
5. This chapter also promotes the use of computational tools for predicting liquefaction probability of seismically active

zones at preliminary stage of construction to avoid any risk and threat of structure against liquefaction.

References

Asteris, P.G., Skentou, A.D., Bardhan, A., Samui, P., Pilakoutas, K., 2021. Predicting concrete compressive strength using hybrid ensembling of surrogate machine learning models. Cement Concr. Res. 145, 106449.

Babacan, A.E., Ceylan, S., 2021. Evaluation of soil liquefaction potential with a holistic approach: a case study from Araklı (Trabzon, Turkey). Boll. Geofis. Teor. Appl. 62 (1).

Cao, Z., Wang, Y., Li, D., 2017. Probabilistic Approaches for Geotechnical Site Characterization and Slope Stability Analysis. Springer Berlin Heidelberg.

Coulibaly, P., Anctil, F., Bobée, B., 2000. Daily reservoir inflow forecasting using artificial neural networks with stopped training approach. J. Hydrol. 230 (3–4), 244–257.

Ditlevsen, O., 1981. Uncertainty Modeling with Applications to Multidimensional Civil Engineering Systems. McGraw-Hill International Book Co.

Dudzik, A., Potrzeszcz-Sut, B., 2021. Hybrid approach to the first order reliability method in the reliability analysis of a spatial structure. Appl. Sci. 11 (2), 648.

Farrokhzad, F., Choobbasti, A.J., Barari, A., 2010. Artificial neural network model for prediction of liquefaction potential in soil deposits. In: International Conferences on Recent Advances in Geotechnical Earthquake Engineering and Soil Dynamics, 4.

Ghani, S., Kumari, S. Liquefaction susceptibility of high seismic region of Bihar considering fine content. In Basics of Computational Geophysics. Elsevier, pp. 105–120.

Ghani, S., Kumari, S., 2021. Liquefaction study of fine-grained soil using computational model. Innov. Infrastruct. Solut. 6 (2), 1–17.

Ghani, S., Kumari, S., 2021. Sustainable development of prediction model for seismic hazard analysis. In: Sustainable Development through Engineering Innovations: Select Proceedings of SDEI 2020. Springer Singapore, pp. 701–716.

Ghani, S., Kumari, S., Bardhan, A., 2021. A novel liquefaction study for fine-grained soil using PCA-based hybrid soft computing models. Sādhanā 46 (3), 1–17.

Ghani, S., Kumari, S., Choudhary, A.K., Jha, J.N., 2021. Experimental and computational response of strip footing resting on prestressed geotextile-reinforced industrial waste. Innov. Infrastruct. Solut. 6 (2), 1–15.

Goh, A.T.C., 1996. Neural-Network modeling of CPT seismic liquefaction data. J. Geotech. Eng. ASCE 122 (1), 70–73.

Goh, A.T.C., 2002. Probabilistic neural network for evaluating seismic liquefaction potential. Can. Geotech. J. 39, 219–232.

Hasson, U., Nastase, S.A., Goldstein, A., 2020. Direct fit to nature: an evolutionary perspective on biological and artificial neural networks. Neuron 105 (3), 416–434.

Hui Ma, C., Yang, J., Cheng, L., Ran, L., 2021. Research on slope reliability analysis using multi-kernel relevance vector machine and advanced first-order second-moment method. Eng. Comput. 1–12.

Idriss, I.M., Boulanger, R., 2006. Semi-empirical procedures for evaluating liquefaction potential during earthquakes. J. Soil Dyn. Earthq. Eng. 26 (2), 115–130.

Janalizade, A., Naghizadehrokni, M., Naghizaderokni, M., May 2015. Reliability-based method for assessing liquefaction potential of soils. In: 5th ECCOMAS Thematic Conference on Computational Methods in Structural Dynamics and Earthquake Engineering.

Jha, S.K., Suzuki, K., 2009. Reliability analysis of soil liquefaction based on standard penetration test. Comput. Geotech. 36 (4), 589–596.

Juang, C.H., Chen, C.J., 1999. Cpt-based liquefaction evaluation using artificial neural networks. Comput. Aided Civ. Infrastruct. Eng. 14 (3), 221–229.

Juang, C.H., Rosowsky, D.V., Tang, W.H., 1999. Reliability-based method for assessing liquefaction potential of soils. J. Geotech. Geoenviron. Eng. 125 (8), 684–689.

Kamatchi, P., Rajasankar, J., Ramana, G.V., et al., 2010. A neural network based methodology to predict site-specific spectral acceleration values. Earthq. Eng. Eng. Vib. 9, 459–472.

Kamura, A., Kurihara, G., Mori, T., Kazama, M., Kwon, Y., Kim, J., Han, J.T., 2021. Exploring the Possibility of Assessing the Damage Degree of Liquefaction Based Only on Seismic Records by Artificial Neural Networks. Soils and Foundations.

Kardani, N., Bardhan, A., Kim, D., Samui, P., Zhou, A., 2021. Modelling the energy performance of residential buildings using advanced computational frameworks based on RVM, GMDH, ANFIS-BBO and ANFIS-IPSO. J. Build. Eng. 35, 102105.

Kardani, N., Bardhan, A., Samui, P., Nazem, M., Zhou, A., Armaghani, D.J., 2021. A novel technique based on the improved firefly algorithm coupled with extreme learning machine (ELM-IFF) for predicting the thermal conductivity of soil. Eng. Comput. 1–20.

Khoo, Y., Lu, J., Ying, L., 2021. Solving parametric PDE problems with artificial neural networks. Eur. J. Appl. Math. 32 (3), 421–435.

Kumar, V., Venkatesh, K., Tiwari, R.P., Kumar, Y., 2012. Application of ANN to predict liquefaction potential. Int. J. Comput. Eng. Res. 2 (2), 379–389. ISSN: 2250–3005.

Kumar, M., Bardhan, A., Samui, P., Hu, J.W., R Kaloop, M., 2021. Reliability analysis of pile foundation using soft computing techniques: a comparative study. Processes 9 (3), 486.

Maier, O., Wiethoff, C.M., 2010. N-terminal α-helix-independent membrane interactions facilitate adenovirus protein VI induction of membrane tubule formation. Virology 408 (1), 31–38.

Mokhtarzad, M., Eskandari, F., Vanjani, N.J., Arabasadi, A., 2017. Drought forecasting by ANN, ANFIS, and SVM and comparison of the models. Environ. Earth Sci. 76 (21), 729.

Omer, M., Bani-Hani, K., Safieh, B., 2009. Liquefaction assessment by artificial neural networks based on CPT. Int. J. Geotech. Eng. 3 (2), 289–302. https://doi.org/10.3328/IJGE.2009.03.02.289-302.

Pei, J., Han, X., Tao, Y., Feng, S., 2021. Lubrication reliability analysis of spur gear systems based on random dynamics. Tribol. Int. 153, 106606.

Prabakaran, K., Kumar, A., Thakkar, S.K., 2015. Comparison of Eigen sensitivity and ANN based methods in model updating of an eight-story building. Earthq. Eng. Eng. Vib. 14, 453–464. https://doi.org/10.1007/s11803-015-0036-z.

Rajeswari, J.S., Sarkar, R., 2021. Reliability analysis of single pile in lateral spreading ground: a three-dimensional investigation. In: Geohazards. Springer, Singapore, pp. 383–398.

Ramezani, M., Bathaei, A., Ghorbani-Tanha, A.K., 2018. Application of artificial neural networks in optimal tuning of tuned mass dampers implemented in high-rise buildings subjected to wind load. Earthq. Eng. Eng. Vib. 17, 903–915. https://doi.org/10.1007/s11803-018-0483-4.

Samui, P., Sitharam, T.G., 2011. Machine learning modelling for predicting soil liquefaction susceptibility. Nat. Hazards Earth Syst. Sci. 11 (1–9), 2011.

Seed, H.B., Idriss, I.M., 1971. Simplified procedure for evaluating soil liquefaction potential. J. Soil Mech. Found. Div. ASCE 97, 1249–1274. SM8.

Umar, S.K., Samui, P., Kumari, S., 2018. Deterministic and Probabilistic Analysis of Liquefaction for Different Regions of Bihar, Geotechnical and Geological Engineering, vol. 36, pp. 3311–3321. Springer, Published online.

Wambua, R.M., Mutua, B.M., Raude, J.M., 2016. Prediction of missing hydro-meteorological data series using artificial neural networks (ANN) for Upper Tana River Basin, Kenya. Am. J. Water Resour. 4 (2), 35–43.

Wang, J., Rahman, M.S., 1999. A neural network model for liquefaction-induced horizontal ground displacement. Soil Dynam. Earthq. Eng. 18 (8), 555–568.

Yang, G.R., Wang, X.J., 2020. Artificial neural networks for neuroscientists: a primer. Neuron 107 (6), 1048–1070.

CHAPTER 13

Probabilistic risk factor—based approach for sustainable design of retaining structures

Anasua GuhaRay

Department of Civil Engineering, BITS-Pilani Hyderabad Campus, Telangana, India

1. Introduction

Soil is a naturally available heterogeneous material, formed through complex processes and hence shows a wide range of variabilities in its index and engineering properties. This unpredictability in geotechnical engineering can result from spatial variability of soil, collection of limited number of samples, error in laboratory testing, faulty equipment, human error, and the limitations of the scientific models used to relate the laboratory or field properties with bearing capacity and deformation behavior of soil. Geotechnical practitioners perceived that, due to profitable hindrance, the probability cannot be nullified. Hence, the foundations of infrastructures need to be designed considering the reasonably least probability. Unless all the sources of uncertainty are encompassed in the designs judiciously, it is not feasible to categorically assess the risk intrigued in geotechnical systems.

The increasing frequency of catastrophic failures of geotechnical structures has compelled the geotechnical scientists to focus on the urgency of reliability-based approaches for design of geotechnical structures. Traditional design of any structure is established on the postulation of Factor of Safety (FS). However, in this approach, it is difficult to include unconditionally all the sources of precariousness in determining soil parameters. This method reflects a reliable result only when the input geotechnical index and engineering properties can be accurately assessed. Proper estimation of the range of shear strength and consolidation properties of soil is a real challenge for geotechnical structures. The mathematical approach of reliability analysis takes into account these uncertainties of field variables.

The foremost approach of reliability based analysis was proposed in Terzaghi Lecture (Casagrande, 1965) on "calculated risk." Over the past decade, Harr (1984), Kulhawy (1992), Lacasse and Nadim (1997), and Duncan (2000) developed approaches for determining reliability index (β) and probability of failure (P_f). Different approaches for calculating reliability index like

First-Order Second-Moment Methods (Baecher and Christian, 2003; Duncan, 2000; Hassan and Wolff, 1999; Hasofer and Lind, 1974), Point Estimate Methods (Nyugen and Chowdhury, 1984, 1985; Rosenblueth, 1975), Response Surface Methodology (Becker, 1996a,b; Cornell, 1990; Orr, 2000), and Monte Carlo Simulation have been proposed.

Ang and Tang (1984) and Tang (1996) assessed the gross critical reliability index of a cantilever wall for noncorrelated and perfectly correlated failure modes considering overturning and sliding failures. Zevgolis and Bourdeau (2010) enumerated the external stability of a reinforced cantilever retaining wall for static case and overall failure system was addressed as a series of correlated failure modes. They concluded that all the failure modes were positively correlated to each other and the margin of failure was affected by the degree of correlation. Zevgolis and Bourdeau (2010) considered beta distribution for the random variables, since upper bound of lognormal distribution and both upper and lower bounds of normal distribution go to infinity. The authors carried out probabilistic analysis by Monte Carlo Simulation and observed that the First-Order Reliability Methods may often lead to uneconomic design. Castillo et al. (2004) applied both deterministic and probability-based approaches for the retaining wall designs. They suggested an approach which minimizes the cost or optimizes an objective function, determines the failure probabilities based on their upper bounds, and allows both P_f and FS to coexist. The authors also carried out a sensitivity analysis by transforming the data parameters with reference to a cantilever retaining wall.

Low (2005) demonstrated different reliability-based design approaches (Hasofer–Lind Method and FORM) for retaining walls and pointed out the dissimilarities between reliability-based and partial safety factor designs. The authors also recommended the use of Hasofer–Lind Method and FORM instead of Monte Carlo Simulation since the results obtained by the two methods are in very good agreement and the former methods are less

time consuming. Babu and Basha (2006, 2008a) used a target reliability index of 3 to propose an optimum design of cantilever retaining walls. Sensitivity analysis (by local parameter differential sensitivity analysis method) was carried out for different wall proportions and soil properties for all modes of failure and it was concluded that friction angles of backfill soil and cohesion of foundation soil were the most sensitive parameters. Goh et al. (2009) used Monte Carlo simulation and concluded that the same partial factor can have different degrees of risk based on the extent of uncertainty of the soil parameters. Daryani and Mohamad (2014) carried out system reliability analysis of cantilever retaining walls driven in cohesionless soils. The authors concluded that although the reliability index increased with the wall cross-section, the rotational failure mode was unaffected by the same.

Hoeg and Murarka (1974) underlined the need for probabilistic approach for optimum design of a gravity retaining wall. They illustrated that higher values of FS may not always warrant a safe structure and neglecting variability of engineering properties of soil may lead to high failure probabilities. Blazquez and Der Kiureghian (1987) considered the failure modes to be either noncorrelated or perfectly correlated and evaluated the reliability index of the retaining wall subjected to seismic load. Srivastava and Babu (2010) used RSM for developing approximate and simple functional relationships between input and output variables. They concluded that much less computational effort was required for RSM compared to the other popular methods of reliability analysis. Low et al. (2011) suggested a simpler and short code for obtaining the system failure probability of a retaining wall against overturning and sliding modes of failure. Researchers like Hornberger and Spear (1981), Sobol (1993), Andres (1997), Saltelli et al. (1999), and Oakley and O'Hagan (2004) proposed different perspectives for estimating sensitivity of geotechnical variables. Babu and Basha, 2006, 2008b applied the differential analysis method to gauge the same.

A thorough review of the past research works shows the recommendation of using partial safety factors compared to a lumped factor of safety for different geotechnical variables. However, considering the randomness of soil properties, even design with partial safety factor approach may lead to uneconomic design. To the best of author's knowledge, there is very limited perspective to contemplate this "partial factor of safety" based on variations of different geotechnical parameters. Past literature highlights different methodologies for calculating sensitivity and failure probability (P_f). However, design approach assimilating the combined effects is yet to be adequately addressed and incorporated. Hence, this chapter aims to develop a novel approach by combining both reliability and sensitivity of random variables on failure probability, which will lead to an economic design. The proposed approach is applied to a cantilever and a gravity retaining wall. Design charts are also proposed for these structures, for different ranges of random variables.

2. Articulation of probabilistic risk factor

This chapter aims to develop a technology to recognize the prospective failure modes of a system and to analyze the consequences of these failures on the system along with mitigation of these effects. In risk analysis, risk is defined as the amalgamation of severity of damage and the probability of its occurrence. The Risk Priority Number (RPN) used to study risk is defined as the product of severity, occurrence, and detection (Rausand et al., 2004; Stamatis, 1995). This concept of RPN is extended to propose a novel factor called Probabilistic Risk Factor (R_f) for each random variable which considers the sensitivity S_i of each variable and the P_f of each failure mode. R_f determined as the product of the probability of failure (P_f) and sensitivity (S) and is given by

$$R_f(i) = 1 + \sum_{j=1}^{n} P_f(j) \times S(i) \qquad (13.1)$$

where i and j are random variables and failure modes, respectively.

The original values of the random variables are modified by these R_f values to get the amended values with variations incorporated into them. The structure is then reanalyzed with the improved soil properties. The redesigned structure satisfies all the stability requirements.

3. Cantilever retaining wall

A cantilever earth retaining wall (Fig. 13.1) of height H has a backfill inclined at an angle β with the horizontal, subjected to a surcharge q per unit area. The backfill and the foundation soil are considered to be homogenous throughout and independent of spatial variability. The mean values and variation of both backfill and foundation soil properties, provided in Table 13.1, are considered from literature due to unavailability of any specific site data (Duncan, 2000; Harr, 1984; Kulhawy, 1992). Table 13.2 summarizes the deterministic FS values as determined by conventional limit equilibrium equations for different failure modes of the retaining wall.

To determine P_f of the failure modes, each performance function is interpreted as $f_i(x) = (FS)_i - 1$, where i represents different failure modes. The structure fails when $f_i(x) < 0$.

Monte Carlo Simulation is used to calculate P_f for different modes. 30,000 random data points are simulated in MATLAB for the mean and COV of random variables as mentioned in Table 13.1 (Zevgolis and Bourdeau, 2010). After generating the histograms, each is fitted to an appropriate statistical distribution and the goodness of fit is examined by Anderson Darling A^2 test (Anderson and Darling 1952). The P_f for all modes of failure is calculated for $f_i(x) < 0$ (Table 13.3). From Table 13.3, it can be observed that the cantilever retaining wall fails in sliding and eccentricity modes of failure.

13. Probabilistic risk factor—based approach for sustainable design of retaining structures

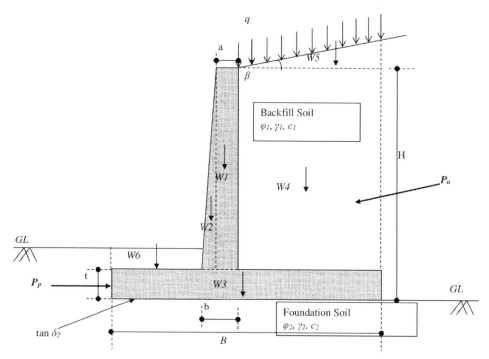

FIGURE 13.1 Prototype cantilever retaining wall.

TABLE 13.1 Statistical measures of input variables.

Random parameters	Mean (μ)	Coefficient of variation (%)	Statistical distribution
γ_1	18 kN/m^3	7	Gaussian
φ_1	35 degrees	13	Log-normal
γ_2	19 kN/m^3	7	Gaussian
φ_2	22 degrees	13	Log-normal
c_2	30 kN/m^2	50	Log-normal
Q	30 kN/m^2	—	—
B	15 degrees	—	—

TABLE 13.2 Factor of safety from deterministic equations for cantilever retaining walls.

Sliding (FS_{sli})	Overturning (FS_{ot})	Eccentricity (FS_{ecc})	Bearing (FS_{bc})
$(FS)_{sli} = \dfrac{\tan \delta_2 \sum F_V + P_{ph}}{\sum F_H}$	$(FS)_{ot} = \dfrac{\sum M_R}{\sum M_o}$	$(FS)_{ecc} = \dfrac{B}{6e}$	$(FS)_{bc} = \dfrac{q_{ult}}{q_{max}}$
1.50	2.67	1.65	2.34

TABLE 13.3 Failure Probability for different failure modes of cantilever retaining wall.

	Sliding	Overturning	Eccentricity	Bearing
P_f	0.101	0.000	0.127	0.020
β	1.276	–	1.142	2.052

The Analysis of Variance F-test is used in the present analysis to determine the relative sensitivity of the input random variables on the failure modes. In this variance-based method, 1σ, i.e., 68.27% variation of the values falls within 1 unit of standard deviation (σ) of the mean μ. The relative sensitivity of the random variables on four modes of failure is shown in Fig. 13.2.

Figs. 13.2A–D show that φ_1 is the most significant parameter and its variation has pronounced effect on overturning and eccentricity modes of failure. It is also observed that the unit weights of both backfill and foundation soil have negligible effect on all the modes of failure. As is obvious from the limit equilibrium equations, the internal angle of friction of foundation soil has a noteworthy impact on bearing and sliding modes of failure. Cohesion of foundation soil has a paramount effect on bearing mode of failure. The random variables are then divided by the calculated R_f values to yield the modified values. The deviations of the random variables are captured in these modified values (Table 13.4).

Fig. 13.3 shows the range of R_f of geotechnical variables for φ_1, for constant values of other variables. With increase in COV of φ_1 from 0% to 20%, R_f for φ_1 improves from 1.15 to 1.85. The

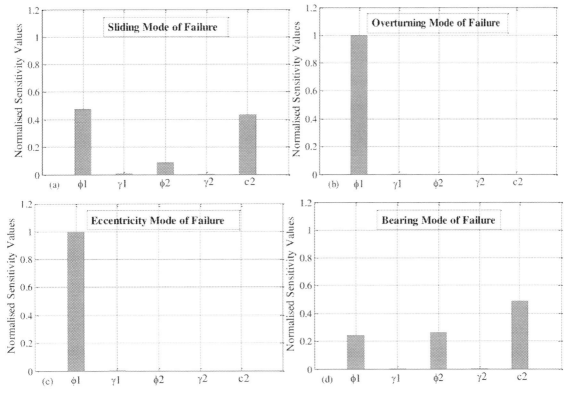

FIGURE 13.2 Sensitivity analysis by ANOVA F-Statistics for (a) Sliding, (b) Overturning, (c) Eccentricity and (d) Bearing modes of failure.

TABLE 13.4 Probabilistic risk factors for different random variables for cantilever retaining wall.

Random parameters	Original values	Normalized $S \times P_f$	(R_f)	Corrected values
φ_1	35 degrees	0.72471685	1.75	20 degrees
γ_1	18 kN/m^3	0.00220017	1.00	18 kN/m^3
φ_2	22 degrees	0.05585550	1.10	20 degrees
γ_2	19 kN/m^3	0.00026821	1.00	19 kN/m^3
c_2	30 kN/m^2	0.21695926	1.25	24 kN/m^2

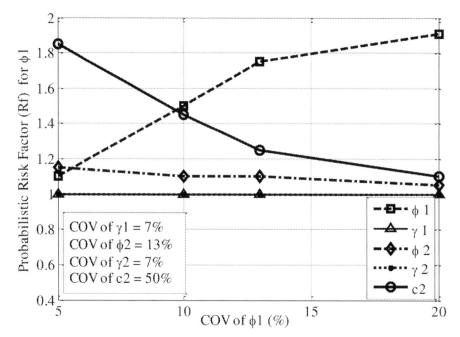

FIGURE 13.3 Variation of Probabilistic Risk Factor (Rf) with φ1.

R_f of γ_1 and γ_2 remains unaltered at unity throughout. R_f of φ_2 does not change with variations of other variables except φ_1 (1.15–1.05). For lesser COV of φ_1, R_f for φ_1 is minimum and that for c_2 is high, while the contrary situation occurs for lesser variations of c_2, i.e., R_f for c_2 is small and that for φ_1 is high. Hence, it may be stated that lesser variation of soil properties does not require a global factor of safety of 1.5 to be applied to the entire structure. On the other hand, partial factor of safety, in the form of R_f, can be assigned depending on variations of soil properties, which may lead to an economic design.

The generalized recommendations for safe and economic design of cantilever retaining wall are as follows:

- Width at top of stem $(a) = \geq 0.3$ m
- Bottom width of stem $(b) = 0$ for H ≤ 5 m

$b = 0.12$H for H > 5 m for $COV_{\varphi 1} \leq 5\%$ and $COV_{\varphi 1} = 5\%–10\%$

- Toe length (L_t) = 0.18H
- Heel length (L_h) = 0.3H for H ≤ 5 m,
  L_h = 0.35H for H > 5 m for $COV_{\varphi 1} \leq 5\%$

  L_h = 0.3H for H ≤ 5 m, L_h = 0.4H for H > 5 m for $COV_{\varphi 1} = 10\%$

- Base thickness, $t = 0.09H$

4. Gravity retaining wall

A prototype gravity retaining wall is analyzed for different modes of failure (Fig. 13.4). The limit equilibrium equations for analysis of different failure modes are similar to that of cantilever retaining wall, with the exception that passive pressure is not considered in the present analysis for conservative design. The deterministic FS (Table 13.5) shows that the wall does not yield to any modes of failure. The least cross-sectional area required to achieve this FS is 6.75 m².

Table 13.6 demonstrates the probability of failure for different modes for COV of φ_1 and $\varphi_2 = 13\%$, γ_1 and $\gamma_2 = 7\%$, and $c_2 = 50\%$. The maximum values of COV are used in illustration. It can be seen from Table 13.6 that sliding mode of failure is the dominating failure mode, contributing maximum to the total probability of failure. Fig. 13.5 reflects the sensitivity effect

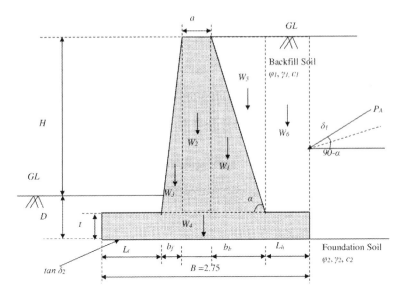

FIGURE 13.4 Prototype gravity retaining wall.

TABLE 13.5 Factor of safety from deterministic equations for gravity retaining walls.

Sliding (FS_{sli})	Overturning (FS_{ot})	Eccentricity (FS_{ecc})	Bearing (FS_{bc})
$\dfrac{\tan \delta_2 \sum F_V + \frac{2}{3} B c_2}{\sum F_H}$	$\dfrac{\sum M_R}{\sum M_o}$	$\dfrac{B}{6e}$	$\dfrac{q_{ult}}{q_{max}}$
1.49	3.25	2.33	3.32

TABLE 13.6 Failure Probability for different failure modes of gravity retaining wall.

Modes of failure	Sliding	Overturning	Eccentricity	Bearing
P_f	0.0165	0.001	0.001	0.003

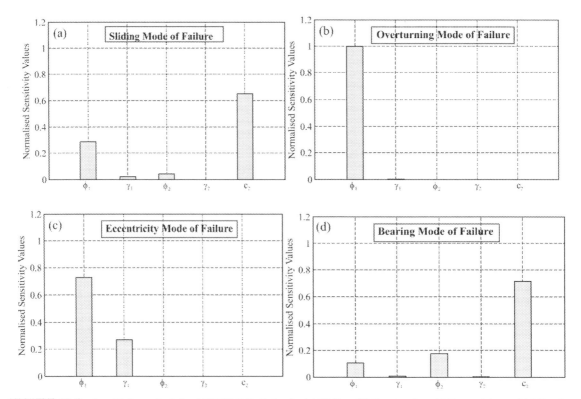

FIGURE 13.5 Sensitivity analysis by ANOVA F-Statistics for (a) Sliding, (b) Overturning, (c) Eccentricity and (d) Bearing modes of failure.

of the random variables as per ANOVA F-statistic results.

Fig. 13.5 reflects that φ_1 has a significant effect on primarily overturning and eccentricity modes of failure. While γ_1 and γ_2 have trivial impact on all failure modes, φ_2 is very sensitive to bearing and sliding modes of failure. c_2 has a noteworthy effect on sliding and bearing modes of failure. Table 13.7 shows the probabilistic risk factors for different random variables. Fig. 13.6 reflects the set of R_f for geotechnical variables with variation of φ_1, with constant values of other variables.

It is observed that for small variations of φ_1, R_f for φ_1 is low and that for c_2 is high; on the other hand, it is opposite for small variations of c_2, i.e., R_f for c_2 is small and that for φ_1 is high. The pertinent dimensions of the retaining wall, as obtained by the present method of analysis, and the reduction (in %) of cross-sectional area, based on a range of φ_1, are outlined in Table 13.8.

4. Gravity retaining wall

TABLE 13.7 Probabilistic risk factors for different random variables for gravity retaining wall.

Random variables	Original values	Normalized $S \times P_f$	R_f	Corrected values
φ_1	35 degrees	0.279674939	**1.30**	26.93 degrees
γ_1	18 kN/m^3	0.016958802	**1.05**	17.15 kN/m^3
φ_2	22 degrees	0.04873781	**1.05**	20.95 degrees
γ_2	19 kN/m^3	4.3946E-05	**1.00**	19.00 kN/m^3
c_2	30 kN/m^2	0.654584502	**1.70**	17.65 kN/m^2

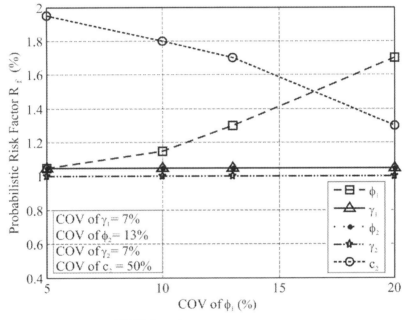

FIGURE 13.6 Variation of Rf with φ1.

TABLE 13.8 Modification of the structure.

			COV of φ_1		
Dimensions (m)	Deterministic	5%	10%	13%	20%
H	6.0	6.0	6.0	6.0	6.0
L_t	0.30	0.30	0.30	0.30	0.30
L_h	0.75	0.75	0.75	0.75	0.85
A	0.3	0.30	0.35	0.45	0.50
T	0.75	0.75	0.75	0.75	0.75
Area (m^2)	6.75	6.12	7.04	7.90	9.60
% Savings	—	+9.25%	−4.58%	−17.45%	−41.70%

The generalized design recommendations for gravity retaining walls (as per Fig. 13.4) are as follows:

- $a \geq 0.3$ m
- $t = 0.12$H
- $L_t \geq 0.06$H
- $L_h \geq 0.14$H
- $B = 0.4-0.48$H for $COV_\varphi \leq 5\%$ and $B = 0.5-0.6$H for $COV_\varphi = 5-10\%$

5. Conclusion

The disastrous effects of unpredictability at various levels of geotechnical design have led the researchers to recognize the importance of the role of reliability-based perspective in the design of geotechnical structures. A detailed study of the past research works highlights different methodologies for determining the probability of failure (P_f) of different modes of failure and sensitivity analysis of random variables on these failure modes. This chapter couples these two approaches mathematically and proposes a probabilistic risk factor—based approach, which is expected to produce more profitable design. Probabilistic Risk Factors (R_f) are proposed for different geotechnical earth structures corresponding to different variations of random variables. The probabilistic risk factor—based design approach is applied to prototype cantilever and gravity retaining walls. The backfill and foundation soil properties are considered as random variables and risk factors are presented for each random variable corresponding to a range of COV of φ_1.

In the present analysis of earth retaining walls, the internal angle of friction (φ_1) of the backfill soil emerges as the most sensitive parameter. The study also identifies sliding and eccentricity as the major failure modes which affect the gross failure of the system. The analysis of the retaining wall shows that the overturning failure mode accorded the least to the global P_f, although it was affected by perturbation of φ_1 to the maximum extent. Different R_f values for range of variations of φ_1 are identified based on the suggested approach. The study concludes that pertinent risk factors need to be designated to different geotechnical input variables depending on their scatter as per the in situ conditions.

Overall, the proposed probabilistic risk factor—based design technique can be applied for the investigation and design of a wide variety of geotechnical earth structures. It proposes to incorporate different types and degrees of uncertainties in design to ensure reliable and economic design of different earth structures.

References

Anderson, T.W., Darling, D.A., 1952. Asymptotic theory of certain "goodness-of-fit" criteria based on stochastic processes. Ann. Math. Stat. 23, 193–212.

Andres, T.H., 1997. Sampling method and sensitivity analysis for large parameter sets. J. Stat. Comput. Simulat. 57, 77–110.

Ang, A.H.S., Tang, W.H., 1984. Probability Concepts in Engineering Planning and Design, Decision, Risk and Reliability, vol. 2. John Wiley & Sons.

Babu, G.L.S., Basha, B.M., 2006. In: Inverse Reliability Based Design Optimisation of Cantilever Retaining Walls, 3rd International ASRANET Collquium, 10-12th July, Glasgow, UK.

Babu, G.L.S., Basha, B.M., 2008a. Optimum design of cantilever sheet pile walls using inverse reliability approach. Comput. Geotech. 35, 134–143.

Babu, G.L.S., Basha, B.M., 2008b. Optimum design of cantilever retaining walls using target reliability approach. Int. J. GeoMech. 8 (4), 240–252.

Baecher, G.B., Christian, J.T., 2003. Reliability and Statistics in Geotechnical Engineering. Wiley, New York.

Becker, D.E., 1996a. Eighteenth Canadian geotechnical colloquium: limit states design for foundations, Part 1, an overview of the foundation design process. Can. Geotech. J. 33, 956–983.

Becker, D.E., 1996b. Limit state design for foundations. Part II: development for national building code of Canada. Can. Geotech. J. 33 (6), 984–1007.

Blazquez, R., Der Kiureghian, A., 1987. Seismic Reliability of Retaining Walls, 5th International Conference on Application of Statistics and Probability in Soil and Structural

Engineering (ICASP 5), BC (Canada). Vancouver, pp. 1149–1156.

Castillo, E., Munguez, R., Teran, A.R., Canteli, A.F., 2004. Design and Sensitivity Analysis using the probability-safety-factor method: an application to retaining walls. Struct. Saf. 26, 156–179.

Cornell, J.A., 1990. How to Apply Response Surface Methodology?, the ASQC Basic References in Quality Control: Statistical Techniques, 8. ASQC, Wisconsin.

Daryani, K.H., Mohamad, H., 2014. System reliability-based analysis of cantilever retaining walls embedded in granular soils. Georisk 8, 3.

Duncan, J.M., 2000. Factors of safety and reliability in geotechnical engineering. J. Geotech. Geoenviron. Eng. ASCE 126 (4), 307–316.

Goh, A., Phoon, K., Kulhawy, F., 2009. Reliability analysis of partial safety factor design method for cantilever retaining walls in granular soils. J. Geotech. Geoenviron. Eng. 135 (5), 616–622.

Harr, M.E., 1984. Reliability-based Design in Civil Engineering, 1984. Henry M. Shaw Lecture, Dept. of Civil Engineering, North Carolina State University, Raleigh, N.C.

Hasofer, A.M., Lind, N.C., 1974. A extract and invariant first order reliability format. J. Engg. Mech. ASCE 100 (EM-1), 111–121.

Hassan, A.M., Wolff, T.F., 1999. Search algorithm for minimum reliability index of earth slopes. J. Geotech. Geoenviron. Eng. 125 (4), 301–308.

Hoeg, K., Murarka, R., 1974. Probabilistic analysis and design of a retaining wall. J. Geotech. Eng. Div. ASCE 100 (3), 349–366.

Hornberger, G., Spear, R., 1981. An approach to the preliminary analysis of environmental systems. J. Environ. Manag. 12, 7–18.

Kulhawy, F.H., 1992. On the Evaluation of Soil Properties. ASCE Geotech (Spec. Publ. No. 31), 95–115.

Lacasse, S., Nadim, F., 1997. Uncertainties in Characterizing Soil Properties, Publ. No. 201. Norwegian Geotechnical Institute, Oslo, Norway, pp. 49–75.

Low, B.K., 2005. Reliability-based design applied to retaining walls. Geotechnique 55 (1), 63–75.

Low, B.K., Zhang, J., Tang, W.H., 2011. Efficient system reliability analysis illustrated for a retaining wall and a soil slope. Comput. Geotech. 38, 196–204.

Nyugen, V.U., Chowdhury, R., 1984. Probabilistic study of soil-Pile stability in strip coal mines-two techniques compared. Int. J. Rock Mech. Min. Sci. Geomech. Abstr. 21 (6), 303–312.

Nyugen, V.U., Chowdhury, R., 1985. Simulation for Risk with correlated variables. Geotechnique 35 (1), 47–58.

Oakley, J., O'Hagan, A., 2004. Probabilistic sensitivity analysis of complex models: a Bayesian approach. J. Roy. Stat. Soc. B Stat. Methodol. 66, 751–769.

Rosenblueth, E., 1975. Point Estimates for Probability Moments, vol. 72. Proc National Academy of Sciences of United States of America, pp. 3812–3814.

Orr, T.L.L., 2000. Selection of characteristic values and partial factors in geotechnical designs to Eurocode 7. Comput. Geotech. 26, 263–279.

Rausand, M., Hoylan, A., 2004. System Reliability Theorie, Models, Statistical Methods, and Applications, second ed. Wiley Series in Probability and Statistics.

Srivastava, A., Babu, G.L.S., 2010. Reliability analysis of gravity retaining wall system using response surface methodology. Indian Geotech. J. 40 (2), 124–128.

Stamatis, D.H., 1995. Failure Mode and Effect Analysis: FMEA from Theory to Execution. American Society for Quality (ASQ), Milwaukee, Wisconsin.

Saltelli, A., Tarantola, S., Chan, K.P.S., 1999. A quantitative model independent method for global sensitivity analysis of model output. Technometrics 41, 39–56.

Sobol', I., 1993. Sensitivity estimates for nonlinear mathematical models. Math. Model Civ. Eng. 1, 407–417.

Tang, W.H., 1996. Correlation, multiple random variables, and system reliability. In: Fenton, G.A. (Ed.), Workshop Presented at GeoLogan 97 Conference: Probabilistic Methods in Geotechnical Engineering, July 15, Logan, Utah, USA, pp. 39–50.

Zevgolis, I.E., Bourdeau, P.L., 2010. Probabilistic analysis of retaining walls. Comput. Geotech. 37, 359–373.

CHAPTER 14

Blast-induced flyrock: risk evaluation and management

Avtar K. Raina[1], Ramesh Murlidhar Bhatawdekar[2,3]

[1]CSIR-Central Institute of Mining and Fuel Research & AcSIR, Nagpur, Maharashtra, India; [2]Department of Mining Engineering, Indian Institute of Technology, Kharagpur, West Bengal, India; [3]Geotropik-Centre of Tropical Geoengineering, Department of Civil Engineering, Universiti Teknologi Malaysia, Johor Bahru, Johor, Malaysia

1. Introduction

Despite the issues related to storage, transportation, and use of explosives and ensuing side effects, rock breakage by blasting has been in vogue for few centuries now (Buffington, 2000). Blasting is expected to continue to be the major excavation method owing to its economics, least itinerary, and ease of deployment despite the need of shift to mechanical excavations (Sobko et al., 2019). Moreover, the cost of mechanical excavations is still high with spatial constraints in maneuvering and hence limited applications.

Blasting has a wide application in civil and mining projects of all sorts that may be open or underground excavations with subclasses of varied nature (Fig. 14.1). Civil engineering projects can be classified as tunneling, infrastructure projects such as highways, airports, bridges, etc. (Kang and Paulson, 1998; Lam, 1999; Ricketts et al., 2004). Excavation required for such projects is smaller in quantity as compared to large opencast mines, thus few holes per blast for excavation of civil engineering projects (Hendron, 1978; Ricketts et al., 2004; Tripathy et al., 2016; Zhou et al., 2019).

On the other hand, large opencast mines require large quantity of excavated rock on sustainable basis (Persson et al., 1994; Jimeno et al., 1997; Prakash et al., 2013; Esposito et al., 2017). Several holes per blast with larger diameter drilling are planned in opencast or surface mines for different requirements of the mining operations (Beyglou et al., 2017; Janković and Valery, 2002; Scott et al., 2002; Singh et al., 2016). During blasting only 20%–25% explosives energy is utilized for breaking rock into smaller fragments, throw of rock, and formation of blasted rock muck pile (Berta, 1990; Liu and Katsabanis, 1997; Jhanwar et al., 2000; Singh and Xavier, 2005; Raina et al., 2014; Calnan, 2015). The residual explosives energy causes environmental effect like ground vibration, air overpressure, flyrock, dust and fumes, and noise (Bhandari, 1997; Pal Roy, 2005; Hajihassani et al., 2014; Roy et al., 2016).

Risk, Reliability and Sustainable Remediation in the Field of Civil and Environmental Engineering
https://doi.org/10.1016/B978-0-323-85698-0.00016-2
© 2022 Elsevier Inc. All rights reserved.

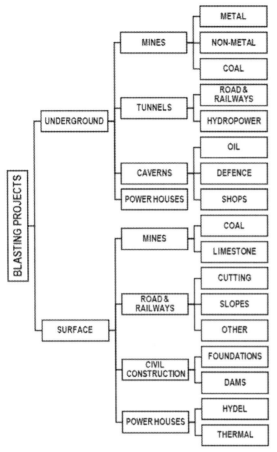

FIGURE 14.1 Types of excavations in civil and mining where blasting is used as rock breaking method.

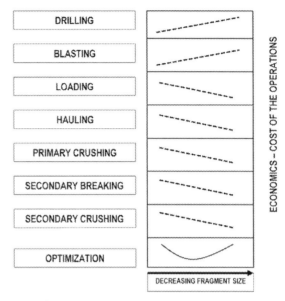

FIGURE 14.2 A schematic method of mine-mill fragmentation system and optimization.

Blasting is an essential unit operation of Mine-Mill Fragmentation System (Hustrulid, 1999) and the cost economics of various unit operations have conflicting relationships with fragmentation (Mackenzie, 1966). Hence, a case for optimization is imperative (Fig. 14.2) A production pattern is thus an outcome of optimization in which the fragmentation is optimized and other unwanted outcomes like flyrock are minimized to a level where these do not inflict damage to anything in their path. This in physical terms means effective use of explosive energy. Thus, in mining as well as civil excavations, the objective of blasting is to fracture the in situ rock and displace it to a proper distance forming good profile and optimum run of mine (RoM) for achieving higher overall efficiency in excavation, transport, crushing, and grinding and thus obtain maximum profitability (Persson et al., 1994; Kanchibotla et al., 1999; Grundstrom et al., 2001; Ouchterlony, 2005). Occurence of flyrock however, constraints the fragmentation system.

Before discussing flyrock in detail, it will not be out of place to define the basic terminology of blasting for a better understanding of the subject. Three major components are involved in blasting viz. the rock or rockmass, the explosive, and blasthole diameter (d). The rockmass and explosives can be expressed in terms of their strength or other properties, e.g., Compressive and Tensile strength (σ_c, σ_t), density (ρ_r), s- and p-wave velocities (v_s, v_p) of rock, and detonation velocity (c_d) and density (ρ_e) of explosive. These properties dictate the basic blast design that is defined in terms of Burden (B), Spacing (S), Blasthole Depth (l_{bh}) in relation to a given Bench Height (H_b), Stemming length (L_s), Delay in firing from row-to-row (t_{rr}), and hole-to-hole (t_{hh}) as shown in Fig. 14.3.

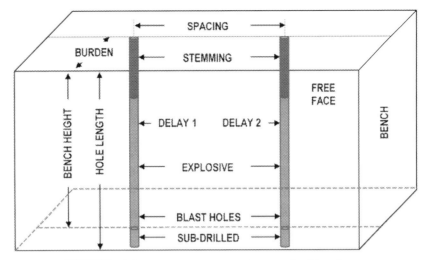

FIGURE 14.3 Basic blasting terminology used in blast design.

Drill diameter is the fundamental operational factor that is decided by the production demand, cost economics, and bench height. Burden is defined by the drill diameter and the rockmass properties. Rest of the variables have direct relationship with the burden (Ash, 1973; Konya and Walter, 1991; Hustrulid, 1999). Once charged with explosives and fired, the blast manifests in fragmentation (Mean Fragment Size, k_{50}; Uniformity Index, n), Throw (R_T), Ground Vibrations (measured in terms of peak particle velocity, v_{max}), air overpressure (measured in terms of sound pressure, p_{oa}), heat, noxious gases, and flyrock (generally expressed a flyrock distance R_f). Out of these, the fragmentation and throw are desirable, whereas the rest of the outcomes are undesirable.

2. Flyrock definition and causes

Explosive energy, in addition to fragmentation, throw, ground vibration, air overpressure, etc., causes throw of rock fragment beyond the stipulated blast area known as Flyrock (Fig. 14.4) and is the most unacceptable occurrence in the operation of blasting (IME, 1997).

Several other definitions of flyrock available in literature are as follows:

▶ Flyrock is essentially the product of uncontrolled venting of the gases that are developed when a charge is initiated (Davies, 1995).
▶ Flyrock is the rock that is propelled through the air during mine blasting (Dick et al., 1983; Fletcher and D'Andrea, 1990).
▶ Flyrock can be defined as an undesirable throw of the material (Siskind and Kopp, 1995).
▶ Flyrock, also called rock throw, is uncontrolled propelling of rock fragments produced in blasting and constitutes one of the main sources of material damage and harm to people (Jimeno et al., 1997).
▶ Rock fragments thrown unpredictably from a blasting site by the force of explosion (Persson et al., 1994).
▶ Flyrock is a rock fragment that travels excessive and unwanted distances from a blast face in surface blasting under the impulse of explosive gases (Raina and Murthy, 2016a)

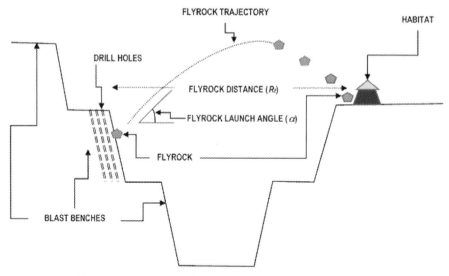

FIGURE 14.4 Definition of flyrock with respect to the pit Configuration and habitat.

Despite the above explicit definitions, confusion in terminology of flyrock exists in published domain, which must be explained in the interest of science, to avoid misrepresentation in future and to improve search and traceability of the literature. Flyrock is the rock fragment emanating from a blast face due to explosive gas pressure that is projected beyond desired throw of the fragmented rockmass. Flyrock distance is the horizontal distance traveled by a flyrock after its ejection from blast face to the final place of landing that must include the distance traveled due to rebound (Fig. 14.5), should the case be. Flyrock trajectory is the path on which the flyrock has moved and landed at a particular place. Flyrock angle is the angle with respect to horizontal at which flyrock was ejected from the face.

The mechanisms of flyrock occurrence include face burst, rifling, and cratering (Richards and Moore, 2004). Raina et al. (2014) provided a complete set of situations in which flyrock can occur. Any imbalance of the explosive energy, variation in mechanical strength of rockmass, and alteration in confinement can result in flyrock (Bajpayee et al., 2004). Though it is estimated that only 1% of explosive energy is transformed in to flyrock (Berta, 1990), it is sufficient to be categorized as a dangerous consequence, resulting in serious bodily injuries or fatalities and also may cause severe damage to property (Adhikari, 1999; Raina et al., 2015). It will be appropriate to state that flyrock has a potential to kill and damage anything that falls in its way. That makes the subject interesting than ground vibrations that have received tremendous attention so far.

The classical works (Lundborg, 1974; Roth, 1979) initiated the investigations on flyrock. Hustrulid (1999), however, pointed to the shortcoming of such works and thus paved the way for further research. Various studies have reported rare occurrence of flyrock (Baliktsis and Baliktsis,2004; Bhandari, 1997; Bhowmik et al., 2004; Ghasemi et al., 2012; Persson et al., 1994; Raina et al., 2011; Venkatesh et al., 1999).

Despite the initial efforts (Livingston, 1956; Lundborg, 1974; Lundborg et al., 1975; Davies, 1995), the research on flyrock had been practically shelved till 2002, when CSIR-CIMFR (India) initiated a comprehensive study on the subject (Raina, 2006) and rekindled the research

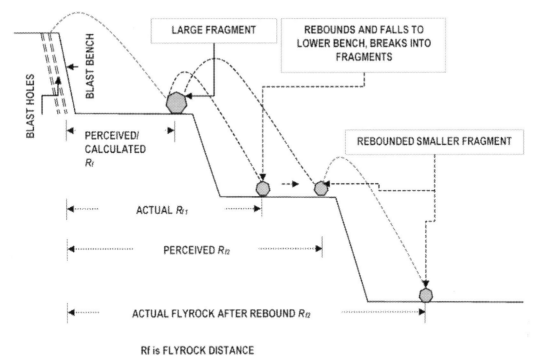

FIGURE 14.5 Flyrock distance definitions with respect to general launch and rebound.

efforts on flyrock. Subsequently, a few major research efforts followed in terms of Ph.D. thesis (Stojadinović, 2013; Raina, 2014; Trivedi, 2015) that have contributed significantly to the knowhow on flyrock. Various researchers have reviewed findings on flyrock by other researchers, which provide comprehensive view of causes and methodology in prediction of flyrock due to blasting (Raina et al., 2014; Bhatawdekar et al., 2018; Murlidhar et al., 2020).

There are umpteen reports of incidents of flyrock accidents like those in Johor Bahru, Malaysia, involving fatalities and damages to property (Mohamad et al., 2013b, 2018), etc., which are a cause of concern to mine management in general and blasters in particular. Further, several studies carried out into investigation and statistics of accidents (Table 14.1) caused by blasting (Siskind and Kopp, 1995; Bajpayee et al., 2002, 2004; Kecojevic and Radomsky, 2005; Mohamad et al., 2013a, 2016, 2018; Raina et al., 2014) favor that the subject should be investigated in detail.

Incidents of flyrock occur due to two types of variables viz. controllable and uncontrollable (Table 14.2). Blasting engineer can control some of the variables during blast design and charging of holes and hence known as controllable parameters. Inadequate burden or stemming, deviated drilling, excessive specific charge, and selection of improper delays are some of the causes of flyrock related to controllable parameters. On the other hand, natural properties of rockmass being blasted and associated structures and nonhomogeneity and geology which cannot be changed are known as uncontrollable variables.

Input parameters are based on blast design parameters such as hole diameter, burden, spacing, bench height, stemming length, powder factor, and maximum charge per delay. Various

214
14. Blast-induced flyrock: risk evaluation and management

TABLE 14.1 Accident statistics logged by various researchers (Raina, 2014).

References	Period	Blasting injuries	% of flyrock injuries in blasting-related accidents
Mishra and Mallick (2013)	1996–2011	30	24.19%
Verakis (2011)	2010–11	18	38.00%
Bajpayee et al. (2004)	1978–98	281	40.57%
Verakis and Lobb (2007)	1994–2005	168	19.05%
Little (2007)	1978–98	412	68.20%
Kecojevic and Radomsky (2005)	1978–2001	195	27.69%
Adhikari (1999)	–	–	20.00%

TABLE 14.2 Controllable and noncontrollable variables in blasting vis-a-vis the output variables (Hustrulid, 1999).

Input (independent)		Output (dependent)
Controllable	**Uncontrollable**	**Output (dependent)**
1. Hole diameter	1. Density of rock mass	Group A
2. Hole depth	2. Compressive strength (intact rock)	1. Fragment size
3. Subdrill	3. Tensile strength (intact rock)	2. Muck profile (swell, heave)
4. Bench height	4. In situ P-wave velocity (dynamic)	3. Throw
5. Hole inclination	5. Modulus of elasticity (dynamic)	4. Ground vibration
6. Burden	6. Poisson's ratio	
7. Spacing	7. Number of joints	Group B
8. Stemming length	8. Joint Aperture	1. Excess throw and flyrock
9. Type of stemming material	9. Joint spacing	2. Overbreak
10. Drilling pattern	10. Joint orientation	
11. Explosive type	11. Joint condition	
12. Explosive density	12. Joint roughness	
13. Initiation system	13. Water (sometimes controllable)	
14. Initiation sequence	14. Degree of weathering	
15. Loading method	15. Unknown rock conditions (random)	
16. Number of free faces		
17. Presence of water		
18. Blast size and configuration		
19. Blasting direction		

researchers use also ratios of blast design parameters instead of single parameter. Burden spacing ratio, burden to hole diameter, bench height to hole diameter, and stemming length to burden ratio are some of the input parameters that have been used by recent researchers. Input parameter may consist of one of rockmass property such as Rock Mass Rating (*RMR*), rock density, Rock Quality Designation (*RQD*), or Geological Strength Index (*GSI*).

Findings from these investigations show that primary causes of flyrock occurrence are as follows:

- Inadequate stemming or burden,
- Inappropriate blast design, layout,

- Excessive specific charge, high explosive charge concentration,
- Abnormal delay timing between holes or rows,
- Large variation in rockmass properties and geology, and
- Presence of incompetent strata within competent rockmass, voids, and fissures, damage of the blast face due to previous blasts.

The operational causes of accidents due to flyrock due to blasting include

- Lack of competency in blasting team or supervisor to identify risk of flyrock,
- Lack of efforts to warn, inspect, and evacuate employees or nearby personnel from blasting zone,
- Inappropriate blasting shelter arrangements for personnel within blasting zone,
- Negligence in utilizing blast shelter by humans within blast danger zone,
- Insufficient number of guards posting to prevent persons or vehicle entering blasting area,
- Weakness in security arrangement for blasting area, and
- Inadequate communication between guards and blasting team.

Several studies have been carried out for correlating rare occurrence of flyrock and significant parameters. In order to control the flyrock, it is important to understand the blast design (Monjezi et al., 2013). Also, flyrock can be controlled by site study and building information model related to drilling and blasting (Venkatesh et al., 1999; Bhatawdekar et al., 2019). The adverse impact of flyrock due to such factors cannot be easily avoided. Some of the effects such as excessive back break from the previous blast can result in flyrock (Gate et al., 2005; Karami et al., 2006; Sanchidrián et al., 2007; Monjezi et al., 2014; Ebrahimi et al., 2016).

3. Brief analysis of data in literature

An analysis of 23 published works was conducted to have a broad understanding of the research over a range of analyses with the independent blast design, rock variables, and dependent variable like flyrock distance (Table 14.3). The total data amounts to around 3000 events of flyrock and is a significant development. This is interesting as the initial approach around 2004 when the author started working on flyrock, the response from the community, was not in favor of the research.

The drill diameters that have been reported in such studies have a wide range (Fig. 14.6) from 75 to 165 mm with a gap in 115−145 mm. Accordingly, the variables like burden (Fig. 14.7) and stemming length (Fig. 14.8) vary with gaps in the data.

Bench height (Fig. 14.9) and the density of rock (Fig. 14.10) used has a better representation with rock types varying from granite, cooper ore, and limestone with similar trend of data in flyrock distance (Fig. 14.11). Unfortunately, the data of coal mines, that form the major mining arena, are lacking. Similarly, the specific charge varies over a broad spectrum with data lacking in the range of 0.5−0.6 and 0.7−0.9 kg/m^3 (Fig. 14.12). Though the trend is in tune with the findings of Lundborg (1974), the lacunae in prediction and risk analysis are evident' (Fig. 14.13). Some of the data points in Fig. 14.13 at a specific charge of 0.2−0.35 kg/m^3 are doubtful as at such low specific charges it is difficult for flyrock to travel to distances of over 200−300 m unless and until unscientific mining has been practiced. Moreover, the variables like burden and stemming length are dependent on the charge diameter that has a relationship with the blasthole diameter and are ingrained in the specific charge. Such use of dependent and highly correlated variables results in aliasing in analysis as is evident from the exploratory analysis provided in Table 14.4.

TABLE 14.3 Summary of flyrock data obtained from published literature.

Sl. No.	References	Rock type	Number of data sets	Hole diameter (mm) Min	Max	Burden (m) Min	Max	Stemming length (m) Min	Max	Bench height (m) Min	Max	Specific charge (kg/m³) Min	Max	Rock Mass property Name, unit	Min	Max	Flyrock distance (m) Min	Max	Remarks
																			Spacing 2.65 to 3.9 m
1.	Armaghani et al. (2014)	Granite	44	89	150	1.9	3.2	1.7	3.6			0.67	1.05	ρ_r	2.3	2.8	60	405	
2.	Rezaei et al. (2011)	Iron ore	490			2.0	6.5	2.0	10.0	5.0	17.5	0.13	0.35	ρ_r	1.9	4.9	10	70	
3.	Monjezi et al. (2011b)	Copper	195	75	115	2.5	5.0	2.0	4.5	6.0	16.0	0.11	0.95	RMR	36.0	47.0	20	100	B to S ratio
4.	Marto et al. (2014)	Granite	113					1.5	3.5	7.5	22.0	0.31	0.96	ρ_r	2.2	2.9	43	206	SHRN = 15,44
5.	Monjezi et al. (2010b)	Iron ore	250	76	115	3.0		3.5		7.0	18.0	0.13	0.35	ρ_r	2.4	4.4	26	186	
6.	Ghasemi et al. (2012)	Copper	150	89	152	2.5	5.0	2.0	4.5			0.15	0.40				30	85	
7.	Koopialipoor et al. (2019)	Granite																	B/S Ratio; DNA
8.	Monjezi et al. (2011a)	Iron ore		75	115			2.0	10.0	7.0	18.0	0.13	0.18	BI	59.5	84	23	196	SMR = 35,60
9.	Amini et al. (2012)	Copper	245	115	152	3.0	5.0	2.8	4.5	9.5	15.0	0.21	0.93				20	100	
10.	Trivedi et al. (2014)	Limestone	95											RQD = ?					UCS = ?; DNA
11.	Ghasemi et al. (2014)	Copper	230			2.0	5.0	2.0	4.5	9.6	16.0	0.18	0.93				20	100	
12.	Trivedi et al. (2015)	Limestone	125	115	165														Data for 25 sets only
13.	Faradonbeh et al. (2016a)	Iron ore	97			2.1	3.0	1.6	2.0	8.3	10.4	0.21	0.26				210	310	

No.	Reference	Rock type															Remarks
14.	Hasanipanah et al. (2018)	Granite	62	90	90	1.5	3.4	1.0	7.1			0.14	0.37	RMR			
15.	Adhikari (1999)	Limestone	54	100	150	2.4	6.0	0.5	3.6			0.30	0.98		25	300	B/D = 21 to 40, ST/D = 4 to 36, Inappropriate delay
16.	Zhou et al. (2020a)	Granite	67	75	150			1.4	4.5	10.0	29.0	0.45	1.14		67	354	B to S ratio = 0.4 to 0.95
17.	Monjezi et al. (2013)	Copper	76	76	152	2.0	5.0	1.6	4.0			0.10	1.29		10	100	RMR = 37 to 47
18.	Armaghani et al. (2016)	Granite	62			1.5	3.4	1.9	3.7			0.14	0.37		72	128	S = 2.2 to 4.1, Q_{max} = 93.5 to 245.2 kg, RMR = 44 to 73
19.	Faradonbeh et al. (2016b)	Granite	62	75	150			1.4	4.5	10.0	29.0	0.45	1.14		67	354	BS Ratio 0.4 to 0.95; Q_{max} = 48 to 594 kg
20.	Faradonbeh et al. (2018)	Granite	76	75	150	1.5	3.2	1.7	3.6			0.67	1.05		60	405	B = 2.65 to 4 m
21.	Guo et al. (2021)	Granite	240				3.4	1.9	3.7	10.0	23.0	0.34	1.09		72	328	
22.	Raina et al. (2011)	7 Rock Types															
23.	Zhou et al. (2020b)	Granite	62			1.5	3.4	1.9	3.7	10.0	23.0	0.34	1.08		72	328	

Q_{max}, maximum charge per delay; SHRN, Schmidt Hammer Rebound Number; RMR, Rock Mass Rating; RQD, Rock Quality Designation; ρ_r, Rock Density.

218　　14. Blast-induced flyrock: risk evaluation and management

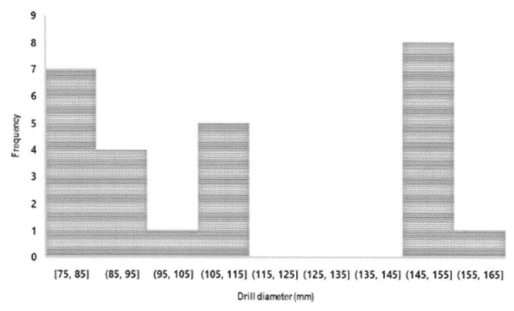

FIGURE 14.6　Frequency of drill diameters in published literature pertaining to flyrock.

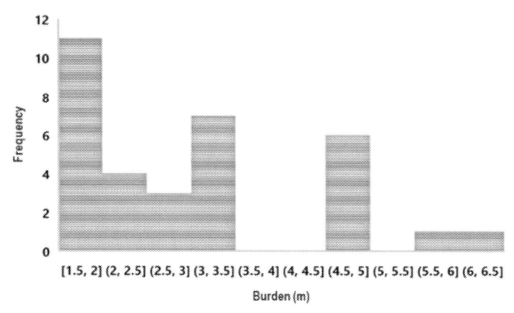

FIGURE 14.7　Frequency of burden in published literature pertaining to flyrock.

3. Brief analysis of data in literature

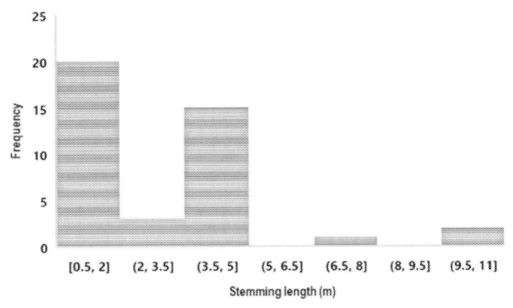

FIGURE 14.8 Frequency of stemming length in published literature pertaining to flyrock.

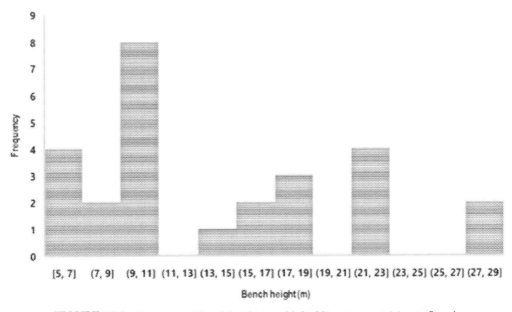

FIGURE 14.9 Frequency of bench height in published literature pertaining to flyrock.

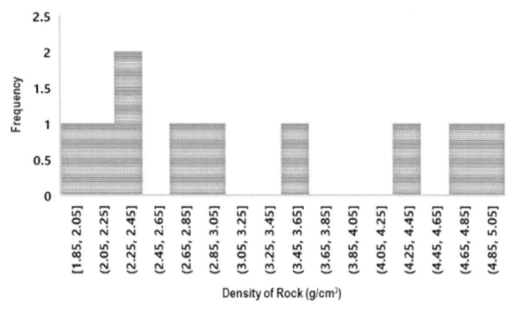

FIGURE 14.10 Frequency of rock density in published literature pertaining to flyrock.

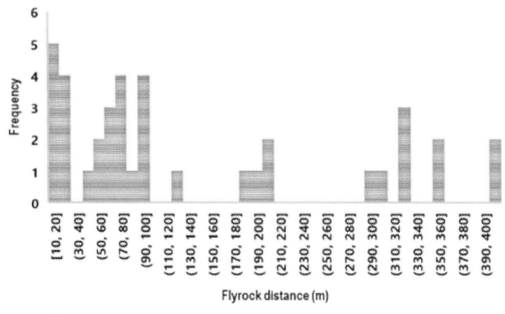

FIGURE 14.11 Frequency of flyrock distance in published literature pertaining to flyrock.

In addition, one outcome which is conspicuously absent from the predictions, is the size of flyrock, that has been ignored by most of the researchers.

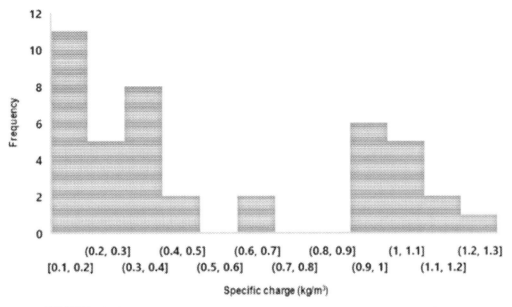

FIGURE 14.12 Frequency of specific charge in published literature pertaining to flyrock.

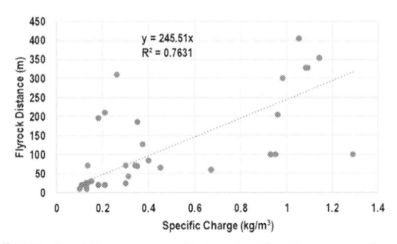

FIGURE 14.13 Flyrock Distance versus specific charge in published literature pertaining to flyrock.

4. Impact of geology on flyrock and associated risk

Geological factors play a crucial role in inducing flyrock during blasting. Geological discontinuities may consist of faults, folds, joints, mud seams, voids, and localized fissures to extended cracks. A void within blast hole could result in localized overcharging and causing flyrock from blown out hole. Such conditions include incompetent strata in a competent rockmass (Fig. 14.14), presence of voids that results in high concentration of explosive (Fig. 14.15), too high an explosive column (Fig. 14.16),

TABLE 14.4 Statistical analysis using ANOVA of the published data by various authors as presented in Table 14.3.

Source	Sum of squares	Df	Mean square	F-value	P-value	Result
Model	1038.48	5	207.70	31.19	<0.0001	Significant
A-Drill diameter	95.44	1	95.44	14.33	0.0006	Significant
B-Burden	8.84	1	8.84	1.33	0.2569	Insignificant
C-Stemming Length	3.99	1	3.99	0.5987	0.4441	Insignificant
D-Bench ht.	2.38	1	2.38	0.3571	0.5539	Insignificant
E-Specific charge	229.75	1	229.75	34.50	<0.0001	Significant

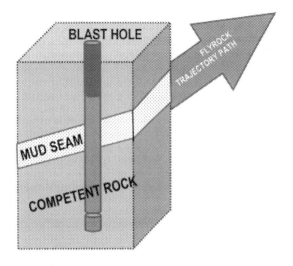

FIGURE 14.14 Presence of incompetent strata in a competent rockmass.

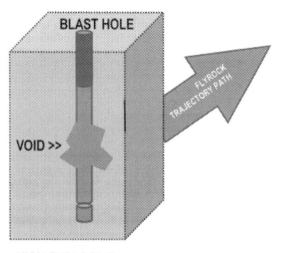

FIGURE 14.15 Presence of voids in rock.

damaged front row burden (Fig. 14.17), too less stemming (Fig. 14.18), and damaged front row with uneven burden (Fig. 14.19).

Geological irregularities can be identified with geological mapping in highwall face. With technological development, drills are equipped with rate of penetration measurement system. Drill logs from such machines provide useful information for blast design purpose in geologically disturbed areas such as incompetent rocks, voids, and mud seams. In one of the fatal accidents due to blasting, investigation showed that sandstone overburden covered with clay was a crucial factor causing flyrock (Shea and Clark, 1998) where stemming became ineffective due to clay layer on the top portion of the blasthole. Based on local geology at individual blast hole, explosive charging should be modified. Geology can dictate powder factor and desired fragmentation and even flyrock (Wallace, 2001).

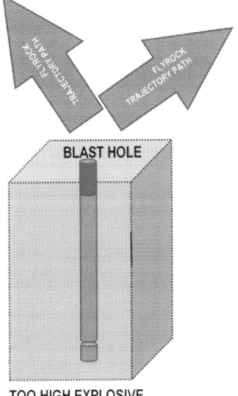

FIGURE 14.16 Too high an explosive column.

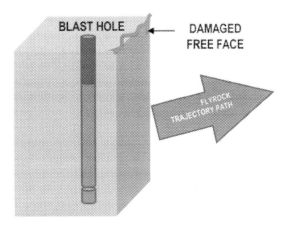

FIGURE 14.17 Front row burden damaged.

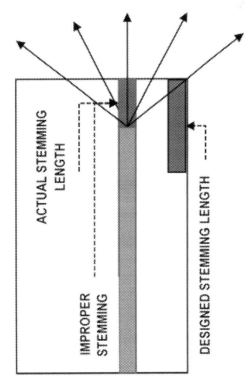

FIGURE 14.18 Too less stemming.

5. Models for flyrock distance prediction

The available literature can be classified as follows:

- Flyrock distance as a function of shot conditions/blast design variation
- Flyrock distance as a function of rock properties and inhomogeneities in rockmass
- Empirical methods of prediction of flyrock distance—explicit models
- Advanced analytical methods to predict flyrock distance—implicit models

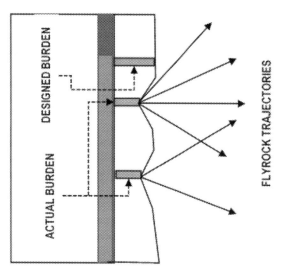

FIGURE 14.19 Damaged front row.

- Flyrock risk and management
- Flyrock Danger Zone, also known as security area demarcation

Such classifications and review have been extensively dealt by Raina et al. (2014) with impetus on development of a new predictive regime. However, there are certain shortcomings in addressing flyrock that include the following:

1. Flyrock is a chance arising out of single or different combinations of variables entering into the analytical domain,
2. Flyrock cannot be produced and hence there are severe constraints on the design of experiments. This is more prominent if distances of more than 100 m of flyrock are to be visualized,
3. Flyrock control being a statutory compliance; the reporting of flyrock is minimal that obscures the data availability,
4. The data on flyrock is not appropriate or represented in perfect scientific terms. Though, the initial studies present an excellent approach, the subsequent research efforts have used variables that are aliased in nature, and
5. Two major domains viz. the blasting operation and postejection phenomenon introduce variables that are extremely difficult to control or beyond control.

The above constraints, however, do not preclude the possibilities of research on flyrock. Accordingly, some of the major contributions are listed for the inquisitive reader. As mentioned earlier, there have been several attempts to model the flyrock distance with various methods. These methods can be further classified as follows:

1. Kinematic-based models,
2. Empirical models employing various rock and blast design variables,
3. Application of implicit methods, generally known as intelligent methods like Artificial Neural Networking etc., that have been presented later, and
4. Risk analysis using empirical or advanced analytical and simulation techniques.

A compilation of the kinematic and empirical models for prediction of flyrock distance has been made in Table 14.5.

Such models given in Table 14.5 that use rock, explosive properties, and charge distribution in the blasthole can be used to have an estimate of the flyrock distance provided all factors are known. The performance of the models with aliasing variables without specific mention of rock properties and special conditions for flyrock will need to be validated. However, the postrelease phenomenon will need to be taken care of in all models as discussed below, unless otherwise specified.

5.1 Postrelease behavior of flyrock

The fragment of rock once released from a blast face is subject of projectile motion in air.

TABLE 14.5 Kinematic and empirical models for prediction of flyrock distance as developed by various researchers.

Sl. No.	Model type	Author(s)	Model
1.	Empirical	Raina and Murthy (2016a)	$\ln(R_f) = 2.22 + 5.63 \times 10^{-8}Z_r + 7.63 \times 10^{-4}\rho_{ee} - 1.8 \times 10^{-3}(B \times S) + 2.14\frac{l_c}{l_{bh}}$
2.	Empirical	Khandelwal and Monjezi (2013)	$R_f = 88.1311 - 2.8214 l_{bh} - 0.1134S - 2.8338B + 2.6665 l_s + 52.7774q - 4.789 b_{sd}$
3.	Empirical	Ghasemi et al. (2012)	$R_f = 6946.547\left[B^{-0.796}S^{0.783}l_s^{1.994}H_b^{1.649}d^{1.766}\left(\frac{q}{Q_h}\right)^{1.465}\right]$
4.	Empirical	Aghajani-Bazzazi et al. (2010)	$R_f = \ln(1.770 + 0.320q - 0.259B - 0.90S)$
5.	Semiempirical	McKenzie (2009)	$B_{sd} = \frac{l_s + 0.0005 l_c d}{0.00923\left(l_c d^3 \rho_e\right)^{0.333}}$; $v_e = K_v\left[\frac{d}{k_x}\right]\left[\frac{2.6}{\rho_r}\right]$; $K_v = 0.0728 \times B_{sd}^{-3.251}$; $b_d = \frac{L_f}{k_x \rho_r}$ ρ_e (explosive density, g/cm^3), b_d (correction factor for air drag), L_f is the shape factor of flyrock fragment (1.1–1.3 for most of the flyrock fragments), k_x (flyrock diameter, mm), ρ_r (rock density, g/cm^3)
6.	Empirical	Richards and Moore (2004)	Face burst: $R_f = \frac{k^2}{g}\frac{\sqrt{q_l}^{2.6}}{B}$; Cratering: $R_f = \frac{k^2}{g}\frac{\sqrt{q_l}^{2.6}}{l_s}$ Defined face burst, cratering, and rifling mechanisms of flyrock. Only two are shown as rifling is not considered important. Air drag, etc., not incorporated
7.	Mathematical	St. George and Gibson (2001)	$v_e = \frac{3\rho_e c_d^2 \Delta t}{32 k_x \rho_r}$ ρ_e (explosive density g/cm^3), ρ_r (rock density g/cm^3), k_x (flyrock diameter cm)
8.	Empirical	Gupta (1990)	$\frac{l_s}{B} = 155.2 R_f^{-1.37}$ observed reduction in the distance of flyrock not appreciable for l_s/B ratio >0.6
9.	Mathematical	(Roth J.A. 1979)	$v_e = (2E')^{0.5}\left(\frac{q_l}{m_l}\right)$; $R_f = v_e \frac{2\sin 2\,\theta}{g}$ $\frac{c_d}{3}$ for (2E')$^{0.5}$ for different types of explosives suggested, e.g., for ANFO the value is 0.44. Second Eq. is also used by (Chiapetta et al., 1983)
10.	Semiempirical	Lundborg (1974)	$R_f = 260 d^{\frac{2}{3}}$; $k_x = 0.1 \times d^{\frac{2}{3}}$ The coefficient 260 was reduced to 40 for normal blasting

Legend: R_f is the flyrock distance in m, ρ_r is the density of rock in kg/m^3, d is the diameter of the drill hole in inches, and k_x is the size of rock fragment in m traveling maximum distance; (2E′)$^{0.5}$, modified Gurney's constant is smaller than (2E)$^{0.5}$ as the direction of detonation is tangent to the rock and v_e is the exit velocity of the flyrock in m/s. v_e is the exit velocity of fragment in m/s, θ is angle of flyrock projection in degrees and g is acceleration due to gravity in m/s^2, l_s is length of stemming in m, B is burden in m, α_{bi} is blasthole angle in degree, q_l is explosive mass/m in kg/m, B_{sd} is the scale depth of burial, l_c is the charge length expressed as a multiple of hole diameter, with a maximum value of eight for hole diameter (d) less than 100 mm and a maximum value of 10 for diameters greater than 100 mm, d is drill diameter in mm, c_d is velocity of detonation of explosive in m/s, Δt is length of impulse time, q is the overall specific charge (kg/m^3) of the blast, Q_h is the charge per hole in kg, H_b is bench height (m), l_{bh} is the blasthole length (m), S is spacing (m), B is burden (m), l_s is stemming length (m), q is specific charge (kg/m^3), and b_{sd} is the specific drilling (1/m^2).

The effect of air drag is dominant as it is directly related with the velocity of a fragment in motion. McKenzie (2009) pointed that kinematic equations overpredict the flyrock distance, as air drag is ignored in calculations and is a basic error in most of such models. A fundamental model that can be used to estimate the air drag (μ_{air}) or the retardation due to movement of an object or flyrock in air is given by Chernigovskii (1986):

$$\mu_{air} = b_d v_x^2$$

$$b_d = \frac{c_x \rho_{air} L_x}{2m}$$

The drag factor (b_d) depends on the shape (L_x) and mass of fragments (m) in kg and the density of air (ρ_{air}) in kg/m^3. v_x is the velocity of fragment at a given time in m/s. The drag coefficient c_x for fragments of different shapes varies between 1.2 and 1.8 with the conditions:

If $\mu_{air} < 0.3$ ..., no correction is needed; if $\mu_{air} > 0.3$ ..., correction for drag is needed.

Stojadinović et al. (2013) provided a method to work out the drag coefficient for use considering ballistic trajectory equations for flyrock for which the terminal velocity of the flyrock fragment is required. Terminal velocity is the velocity of a fragment immediately before it hits the ground. However, the terminal velocity of flyrock is difficult to ascertain since exact position of the landing of a flyrock is not known. The effect of air drag has received its due attention from researchers and still continues to be investigated for obvious reasons of nonlinear retardation of flyrock fragment in air. However, the influence of other factors viz. the Magnus Effect and the wind velocity on the travel of flyrock fragment in air cannot be ignored. Also, in some cases, the flyrock fragment rebounds after its initial landing on surface (Raina et al., 2013), thereby affecting the actual flyrock distance. It is also assumed that the empirical equations developed presume the final distance of travel of the flyrock.

6. Use of intelligent techniques in flyrock prediction

During the last decade, various computational techniques have been used for prediction of flyrock as presented in Table 14.6. Such models are implicit in nature and may have been developed for specific conditions and variables that may not hold good for other mining conditions and will need to be thoroughly examined.

There are umpteen instances of intelligent methods as summarized above that report the R^2 of +0.90 which means that, at sites where these models were developed, the flyrock incidences should be practically negligible or totally under control. However, such instances need to be seen yet.

7. Flyrock risk and management measures

Ground vibrations and air overpressure due to blasting have received significant attention from the researchers and the regulations and guidelines are more or less finalized. Despite the certain amount of subjectivity, the rules for ground vibrations are largely accepted. However, flyrock, in contrast, cannot be subjective as no one "can be hit" by an imaginary flyrock (Rosenthal and Morlock, 1987). There can be small incidences to fatal accidents due to flyrock as explained by Verakis (2011) and shown in Fig. 14.20 (after Bird and Germain, 1996).

The earlier risk methods in blasting can be traced to Bandyopadhyay et al. (2003) who deployed fuzzy set theory in evaluating risk using linguistic variables/values and Yager's methodology for ordinal multiobjective decision based on fuzzy sets to evaluate risk due to environmental factors of blasting in mines including flyrock. Many such attempts to model the flyrock risk have been compiled in Table 14.7 with the methods used and comments thereof.

TABLE 14.6 Different computational and intelligent methods used by various authors for flyrock distance prediction.

Sl. No.	Reference	Models used for prediction of flyrock	Comments
1.	Fattahi and Hasanipanah (2021)	ANFIS-GOA, ANFIS-CA	$R^2 = 0.974$ $R^2 = 0.953$
2.	Li et al. (2021)	RMSE values of (0.1991, 0.179, 0.1994, 0.2137, and 0.1787) and (0.2351, 0.1819, 0.1937, 0.1929, and 0.1679) for GA-ANN, PSO-ANN, ICA-ANN, ABC-ANN, and FA-ANN FA-ANN as the best selected model	Best portion data selected using FDM
3.	Ye et al. (2021)	GP RF Inputs from GP and RF models were used to MCS	$R^2 = 0.8360$ $R^2 = 0.8266$
4.	Nguyen et al. (2021)	WOA−SVM−RBF (RMSE = 5.241) WOA−SVM−L (RMSE = 9.080) [ANFIS, GBM, RF, CART, and ANN] RMSE from 5.804 to 6.567] WOA−SVM−P (Lower performance) WOA−SVM−HT (Lower performance)	$R^2 = 0.977$ $R^2 = 0.937$ $R^2 = 0.965$ to 0.973
5.	Dehghani et al. (2021)	GEP (RMSE = 5.85) COA used review initial blast design and maximum flyrock value reduced by 43.6%	$R^2 = 0.91$
6.	Masir et al. (2021)	Risk assessment of flyrock using FFTA-MCDM	
7.	Zhou et al. (2020)	MRA, ANN. Maximum charge per delay has highest impact on flyrock	
8.	Han et al. (2020)	BN - Hole diameter most influential parameter for prediction of flyrock	
9.	Lu et al. (2020)	ORLEM ELM ANN	$R^2 = 0.958$ $R^2 = 0.955$ $R^2 = 0.912$
10.	Kalaivaani et al. (2020)	RFNN-PSO (NS = 0.921, MABE = 13.86, and RMSE = 15.79)	$R^2 = 0.933$
11.	Zhou et al. (2020b)	ANN Train R2 = 0.900 and Test R2 = 0.906 PSO algorithm used for minimization of flyrock	$R^2 = 0.900$
12.	Murlidhar et al. (2020b)	BBO-ELM(COP = .95, RMSE = 18.84) PSO-ELM(COP = .93, RMSE = 21.51) ELM (COP = .85, RMSE = 32.29)	$R^2 = 0.94$ $R^2 = 0.93$ $R^2 = 0.79$

Continued

TABLE 14.6 Different computational and intelligent methods used by various authors for flyrock distance prediction.—cont'd

Sl. No.	Reference	Models used for prediction of flyrock	Comments
13.	Murlidhar et al. (2020a)	Review of environmental issues including prediction of flyrock by other researchers ANN, ANFIS, ANN- GA, ANN-ICA, ANN-PSO, FIS, MVRA, SVM	
14.	Koopialipoor et al. (2019)	ICA-ANN PSO-ANN GA-ANN	$R^2 = 0.94$ $R^2 = 0.93$ $R^2 = 0.79$
15.	Guo et al. (2019a)	SVR (RMSE = 7.239) SVR-GLMNET (RMSE = 3.737)	$R^2 = 0.972$ $R^2 = 0.993$
16.	Guo et al. (2019b)	DNN (Train R2 = 0.9829, Test R2 = 0.9781, RMSE = 8.2690 and 9.1119) ANN (Train R2 = 0.9093, Test R2 = 0.8539, RMSE = 19.0795 and 25.05120) WOA used for minimization of flyrock after DNN which was 13.5% higher than actual measured value	$R^2 = 0.9829$ $R^2 = 0.9093$
17.	Hudaverdi, and Akyildiz (2019)	MDA. New classification of flyrock into three groups 1. Low flyrock throw, 2. high flyrock throw 3. Excessive flyrock throw	
18.	Kumar et al. (2018)	PSO-ANN, ICA-ANN, BP-ANN, (MRA). PSO-ANN was the best model.	
19.	Asl et al. (2018)	ANN-FA	$R^2 = 0.93$
20.	Mohamad et al. (2018)	Blast design has influence of 69% accuracy in prediction of flyrock. Balance parameters based on geological conditions.	
21.	Rad et al. (2018)	LS-SVM (MSE = 16.25) SVR (MSE of 31.58)	$R^2 = 0.969$ $R^2 = 0.945$
22.	Hasanipanah et al. (2018)	RES (MEDAE = 11.39, VAF = 89.42, RMSE = 14.35) MLR (MEDAE = 11.39, VAF = 64.15, RMSE = 26.42)	$R^2 = 0.894$ $R^2 = 0.641$
23.	Faradonbeh et al. (2018)	GEP (Train R2 = 0.920, Test R2 = 0.924) FA - reduction in flyrock distance by 34%	$R^2 = 0.920$
24.	Ghasemi et al. (2018)	M5P (VAF = 92, RMSE = 3.9).	$R^2 = 0.920$
25.	Bhatawdekar et al. (2018)	Review of models used by other researchers for prediction of flyrock ANN, PSO-ANN, ICA-ANN, SVM	

7. Flyrock risk and management measures

No.	Reference	Method / Remarks	Results
26.	Adikari and Hmtga (2018)	Environmental issues due to the blasting slope stability, geological conditions	
27.	Hasanipanah et al. (2017b)	PSO (RMSE = 12.01) MLR (RMSE = 18.64)	$R^2 = 0.966$ $R^2 = 0.886$
28.	Hasanipanah et al. (2017a)	RT (MEDAE = 0.872) MLR (MEDAE = 0.860)	$R^2 = 0.947$ $R^2 = 0.772$
29.	Dehghani and Shafaghi (2017)	DE = Differential evaluation algorithm DA = dimensional analysis algorithm.	$R^2 = 0.91$ $R^2 = 0.752$
30.	Bakhtavar et al. (2017)	H-DAFIS; Crucial parameters for fly rock are dependent upon S x RMR x HL- B x HL x ST	
31.	Yari et al. (2017)	DEA, LA HL = 12.1 m, B = 3.5 m, S = 4.5 m, and ST = 3.8 most suitable blast pattern for minimizing flyrock, back break at copper mine	
32.	Jahed Armaghani et al. (2016)	ANN ANFIS	$R^2 = 0.925$ $R^2 = 0.964$
33.	Armaghai et al. (2016)	MR MCS The simulated flyrock by MCS was 236.3 m as compared to actual flyrock of 238.6 m	
34.	Raina and Murthy (2016a)	Surface response analysis Input parameters determined using ANN	$R^2 = 0.82$
35.	Raina and Murthy (2016b)	ANN	
36.	Yari et al. (2016)	BPNN	
37.	Stojadinović et al. (2016)	ANN for prediction of flyrock launch velocity	
38.	Trivedi et al. (2016)	BPNN and MVRA for prediction of flyrock and fragment size	
39.	Faradonbeh et al. (2016b)	GP (RMSE = 17.638 and VAF = 89.917) NLMR (RMSE = 26.194 and VAF = 81.041)	$R^2 = 0.908$ $R^2 = 0.816$
40.	Faradonbeh et al. (2016a)	GEP Train RMSE 2,119, Test RMSE 5.788 GP Train RMSE 2.119, Test 10.062	
41.	Saghatforoush et al. (2016)	ANN ABC improved efficiency of prediction by 61%.	

Continued

TABLE 14.6 Different computational and intelligent methods used by various authors for flyrock distance prediction.—cont'd

Sl. No.	Reference	Models used for prediction of flyrock	Comments
42.	Armaghani et al. (2015)	ANN ANFIS	$R^2 = 0.86$ $R^2 = 0.95$
43.	Raina et al. (2015)	Review of empirical and semiempirical methods	
44.	Armaghani et al. (2014)	BP-ANN PSO-ANN	
45.	Marto et al. (2014)	ANN-ICA	$R^2 = 0.95$
46.	Ghasemi et al. (2014)	ANN FIS	$R^2 = 0.94$ $R^2 = 0.95$
47.	Trivedi et al. (2014)	ANN	$R^2 = 0.98$
48.	Monjezi et al. (2013)	ANN	$R^2 = 0.98$
49.	Raina et al. (2013a)	ANN	-
50.	Mohamad et al. (2013a)	ANN	$R^2 = 0.97$
51.	Monjezi et al. (2013)	SVM, ML MVRA	$R^2 = 0.95$ $R^2 = 0.44$
52.	Amini et al. (2012)	SVM ANN	$R^2 = 0.97$ $R^2 = 0.92$
53.	Raina et al. (2011)	Dynamic flyrock danger zone, risk analysis, factor of safety	
54.	Rezaei et al. (2011)	FIS	$R^2 = 0.98$
55.	Monjezi et al. (2011a)	ANN	$R^2 = 0.97$
56.	Monjezi et al. (2010a)	ANN	$R^2 = 0.98$

7. Flyrock risk and management measures 231

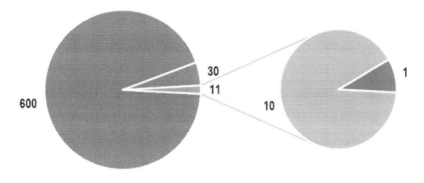

■ Incidents ■ Accidents ■ Serious Accidents ■ Fatal Accidents

FIGURE 14.20 Safety in accidents. *Modified after Bird, F.E., Germain, G.L. 1996. Loss Control Management: Practical Loss Control Leadership. Det Norske Veritas Inc., USA, pp. 1–5.*

TABLE 14.7 Flyrock risk analysis by various authors as found in published literature.

Sl. No.	Year	Reference	Methods in short
1.	2020	Hasanipanah and Bakhshandeh Amnieh (2020)	Fuzzy rock engineering system; claims that the method is successful in evaluating the parameters that affect flyrock, which facilitate decisions to be made under uncertainties.
2.	2020	Zhou et al. (2020a)	Nonlinear models and Monte Carlo (MC) simulation. 260 data points from Malaysia. Q_{max} has the greatest effect on flyrock.
3.	2018	Hasanipanah et al. (2018)	Risk level of flyrock through rock engineering systems. In this study, 62 blast events (Ulu Tiram quarry) are analyzed; risk of flyrock for the mine = 32.95 (low to medium degree of vulnerability). Authors conclude that RES is better than MLR.
4.	2016	Armaghani et al. (2016)	Flyrock simulated using Monte Carlo method. 62 data points analyzed, mean simulated flyrock by MC is 236.3 m, while this value was achieved as 238.6 m for the measured one. Sensitivity analysis shows that powder factor is the most influential parameter on flyrock.
5.	2016	Nieble and Penteado (2016)	This article describes the technique employed in the blasting plans, which were carried out successfully
6.	2014	Faramarzi et al. (2014)	Modeling flyrock risk and distance prediction using rock engineering systems. 13 effective parameters used. 47 blasts analysis reported. RES, MVRM, dimensional analysis. RES-based model proved to be better.

(Continued)

232
14. Blast-induced flyrock: risk evaluation and management

TABLE 14.7 Flyrock risk analysis by various authors as found in published literature.—cont'd

Sl. No.	Year	Reference	Methods in short
7.	2013	Blanchier (2013)	Flyrock risks model claims to be a first useable quantitative model based on the characteristics of the blasting plan and of the generic laws utilized in casting and selective blasting operations and on a statistical approach of rockmass.
8.	2011	Ye et al. (2021)	Claim novel predicting and simulating methods for the flyrock. Two tree-based techniques, genetic programming (GP), and random forest (RF) were developed and applied to predict flyrock distance
9.	2011	Raina et al. (2011)	Flyrock risk model based on probability of flyrock and, consequences of flyrock defined in terms of a Threat Ratio. Classified the blasting operations from safe to unsafe.
10.	2010	Little and Blair (2010)	Mechanistic Monte Carlo models for flyrock, an axisymmetric and a three-dimensional trajectory impacting two-dimensional structures. Include prescribed distributions of launch speed, direction and size of fragments, and air drag and impact on pit walls. A one in a million chance that flyrock fragments will exceed 640 m with no pit walls. However, with pit walls, the distance is 560 m.
11.	2007	Little (2007)	Discussed consequence-based approach. "Risk matrix"—based method with controls and uncertainty ratings given. Stressed "Effective Blast Emission Management" to reduce and/or eliminate safety risks and to manage public perception of flyrock. A combination of quantitative and qualitative method is regarded as the best practice.
12.	2007	Schwengler et al. (2007)	Case study of an open cut near habitats. The general industry-accepted 600 m Blast Danger Zone (BDZ) might not favor production. Stressed optimization of BDZ with 1. Quality Team assures adherence to specified blast design, 2. Use of high-quality stemming material with little fines 3. Training and support blast crews, and 4. Support from management to ensure best blasting practice. Report reduction of BDZ after monitoring 176 blasts to 200–250 m from 600 m, ensuring risk management.
13.	2004	Baliktsis and Baliktsis (2004)	Flyrock risk prevention—provide theoritical basis for such evaluation and ideas to a perfectly apply such findings in a blasting project.
14.	1995	Davies (1995)	See text following Table 14.7 for details
15.	1994	Kleine et al. (1994)	Discuss variability in blast variables, their probabilities, and results with measurement methods and probability distribution of results thus estimating the uncertainty in achieving targets. The probability of a perfect blast and its cost consequences can be together used to calculate the risk. Parameters can be fixed to quantify the blast performance in terms of fragmentation, throw, and flyrock.

Risk due to flyrock is important as it can be used to fix the "Flyrock Danger Zone," generally referred to as "Blast Danger Zone—BDZ" or "Security Area" or "Blast Exclusion Zone." Mathematically, risk of any operation is a product of the probabilities of an event and the cost of the event. This can be defined in simple terms as follows:

$$Risk = \text{Probability of an event } P(E)$$
$$\times \text{ Consequence of an event } C(E)$$

Or

$$Risk = P(E) \times C(E)$$

Although the equation of risk appears to be a straightforward one, there are several components in risk related to Flyrock as defined below:

1. The general BDZ is regarded as a circular space around the blast area that can have different values as proposed by several regulatory bodies world over. This is termed as consequence-based BDZ. The basis of such BDZ is purely on the distance of maximum flyrock travel distance based on occurrence and has no scientific basis. Lundborg (1981) calculated that at a distance of 600 m, while blasting a 76 mm diameter blasthole, the probability of being struck by a flyrock was the same as that being struck by a lightning (1 in 10,000,000). However, this probability may not hold good for cases where blasting is carried out in close proximity of habitats. He, however, demonstrated that the probability of the flyrock distance assumes a Weibull distribution.
2. Some BDZ definitions state that flyrock should not be casted more than the half of the distance from blast to an "Object(s) of Concern" that may include residential, nonresidential areas, structures, and other habitats. This method ignores the presence of mine personnel and costly equipment within such BDZ.

3. The number of blasts that a mine or quarry or excavation conducts in a unit period of time is associated with the risk potential for flyrock.
4. The distances of the Objects of Concern vary from mine to mine and blast to blast and have a strong bearing on defining the BDZ.
5. The nature of the Objects of Concern also varies and has a definite relationship with the consequences of a flyrock.
6. The flyrock can travel in any direction though the concept has been a subject of debate and accordingly some modifications in the zone were proposed by Richards and Moore (2004) and Raina et al. (2013b). The probabilities of flyrock traveling in a particular direction with respect to blast face need to be identified and defined properly.
7. The probability that a flyrock will be generated has two aspects:
   a. Flyrock can occur due to the geological factors as defined earlier.
   b. Flyrock can occur due to the improper blast design or deployment of such a design in actual application with several reasons of objective or subjective nature.
8. Whether an Object of Concern is exposed to flyrock or not also contributes to the consequences.
9. The sum total of the consequences is very subjective in nature and extremely difficult to quantify. Based on such findings, Raina et al. (2011) devised an alternate method to define the consequences of flyrock and termed it as Threat Ratio. Another method of Kleine et al. (1994) that defines the cost in terms of safety can also be used as a converse of the consequences.

Davies (1995) gives an approach to set the "Blast Danger Zones" by considering the incidence of flyrock, which is calculated from the available data, and the probability that a predicted distance will be exceeded. While utilizing the formula established by (Roth, 1979) for calculation of flyrock distance and its exit velocity, he explained the target impact frequency for both the individual and societal or group risks of flyrock impact.

Normally short-distance flyrock usually travels less than 300 m. However, when there is some abnormality in a blast or rock formation, flyrock has been known to travel much further than the calculated distance known as "wild flyrock" (Fletcher and D'Andrea, 1990; Persson et al., 1994). The frequency of impact by "wild flyrock" at a constant distance, for single shot provided by Davies (1995), is given in the following equation:

$$f = N_b f_f P_{(R)} P_{(T)} P(T_e)$$

where, f is target impact frequency (impact/year), N_b is total number of blasts per year, f_f is frequency of flyrock per blast, $P_{(R)}$ is probability of wild flyrock traveling the target distance, $P_{(T)}$ is probability of wild flyrock traveling on an impact trajectory, and $P(T_e)$ is probability of target exposure. While providing some data on flyrock probabilities, Davis (1995) argues that such probabilities should be increased by a factor of 2—3 to account for underreporting of such events. The main disadvantage of this method is that it is difficult to obtain realistic values for each component (Little and Blair, 2010). While defining the mechanistic method of assessment of flyrock risk, Little (2007) defined the consequence and risk matrix for flyrock while simulating conditions as per the criterion of Richards and Moore (2004) and concluded that the main reason for flyrock generation is the mismatch between the energy available and the work to be done. In any case, the equation above simply provides the probabilities and not the risk involved. Hence, the cost of the event is necessary which can be taken into account by methods provided by Kliene et al. (1994) and Raina et al. (2011), as alternatives. The following method can be used to determine the risk:

1. Establish uncertainties in terms of probability or failure of a risk criterion or probabilities of flyrock exceeding a distance. This has to have a strong basis of actual data from a mine and all other variables should be known.
2. Establish or devise a method to quantify the consequences of flyrock exceeding certain distance. This has to be enumerated in clear terms. Alternatively, the threat ratio or the distance of permissible flyrock to the distance of Object of Concern can be used to quantify the consequences. This can be normalized for the purpose of standardization.
3. Establish a relationship between the probabilities and the consequences of a flyrock. Fig. 14.21 can be used as a guideline to establish the risk domain.

7.1 Risk matrix and blast danger zone

Blast Danger Zone (BDZ) or Blast Exclusion Zone is defined as "the area that is determined by the risk assessment process, to ensure that all the expected/foreseen dangers and effects of the blast are maintained within a controlled area. The Blast Exclusion Zone may be layered, with an inner zone parameter being defined as the minimum distance for equipment and an

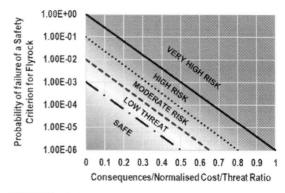

FIGURE 14.21 Relationship between flyrock probabilities and consequences. *Modified after Raina, A.K., Chakraborty, A.K., Choudhury, P.B., Sinha, A., 2011. Flyrock danger zone demarcation in opencast mines: a risk based approach. Bull. Eng. Geol. Environ. 70, 163—172. https://doi.org/10.1007/s10064-010-0298-7*

outer zone parameter being defined as a personnel zone" (AEISG, 2011). The method is scientific in nature in comparison to guidelines that are based on consequences, e.g (MoL&A GOI, 2019), India.

Any risk analysis should finally yield the risk matrix which can be utilized to have a control over any phenomenon. The approach as discussed above can be used in case of flyrock also. The three major things that have been discussed are as follows:

1. The probability of flyrock traveling a particular distance from blast face or probability of failure of a safety rule for flyrock.
2. The consequence of the flyrock in terms of cost or other relevant measure that can be devised by mine management through self-evaluation.
3. Identify the zone from flyrock risk matrix and decide on BDZ or methodology for blasting.

A schematic representation of the above methodology is provided in Fig. 14.22. This involves a comprehensive evaluation of the blast area, geology, and drilling and blasting conditions.

The objective of such an exercise is to provide a tool to the blasters and management to have a thorough check on the outcome of blasting in terms of its performance and control of the unwanted effects like flyrock. It should be noted that the process is a cyclic one yet continuous and the best method to control unwanted effects of blasting like flyrock should be addressed holistically and not in isolated manner. Local constraints and conditions that vary over a wide range should form part of such risk management regime.

8. Maturity model for flyrock risk assessment

Flyrock can result in serious bodily injury, injuries, fatalities, or damage to property or even disaster involving multiple fatalities and/or damage to the property in areas surrounding a blast. Socio-economic impact of flyrock due to blasting on the organization is thus eminent. There are legal implications for the persons involved in blasting operation including top management of the organization that is governed by the local regulations of respective country. It is vital to have an insight into the management practices that ultimately define the production vis-a-vis blasting practice and hence safety of personnel and objects around the mine. In order to assist the management in defining their maturity level with respect to blasting, Table 14.8 has been prepared considering different operational procedures and their compliance that can in turn help to evaluate the flyrock risk. The method has four basic tenets as defined below:

1. Restrain personnel entering blasting zone or keeping away all assets beyond scientifically defined Blast Danger Zone.
2. Automate process for blasting operation using information technology
3. Reengineer process of blast design to execution including security of the area
4. Redesign organization: Based on current practices, local regulation, nearby and worldwide blasting incidents, internal and external blasting audit, and organization can be strengthened through training of every person involved including top management

Maturity level of an organization is proposed into five stages as below:

Level 1 — mostly where blasting operation is started at a new place or construction site where few persons are aware of pros and cons of the hazards associated with blasting and flyrock. Regulatory governance may be poor.

Level 2 — organization where blasting operation is carried out complying with statutory requirements. However, certain deficiencies, such as security or any other provision, are absent.

FIGURE 14.22 Risk management for flyrock prevention.

IT tools are not used for execution of any blast design or risk assessment.

Level 3 — organization where all statutory provisions are made. Regulatory authority has a good governance system. Use of information technology is used into certain processes such as inspection of blasting face. "What if" analysis of critical incidents is carried out periodically and thus strengthening the safety processes.

Level 4 — organization where all processes are adequately matured and well defined. Security arrangements for restraining people entering blasting zone will modify the processes of blasting operation to execution, periodic training is provided on safety, and technical competence in blasting for all personnel associated with blasting operation. Drones are used for monitoring and logging of blasting operation.

TABLE 14.8 Maturity model for flyrock risk assessment.

Maturity level	Restrain personnel entering blasting zone	Automate process	Reengineer process	Redesign organization
5	Achieves zero accident status	All blast design to execution with electronic surveillance	Shows leadership in blast design to execution	Policy document and good governance
4	Retrain employees and local community, Good security arrangements	Use of drones for supervision of blast	Adoption of the best practices for drilling, blasting	Training of personnel based on past incidence in industry
3	Meeting statutory norms of warning at entry points, use of blasting shelter, etc.	Use of phones/WA for cautioning people. Digital photographs for inspection of blasting face	Modification done periodically based on any incident	Additional supervision, what if analysis for critical flyrock
2	Basic knowledge, Inadequate security	Use of sirens for warning	Basic document on blasting procedure	Statutory organization for blasting
1	Unorganized	IT infrastructure absent	Any blast procedure document absent	Organization structure absent

Level 5 − organization where leadership is shown in every aspect of issues related to flyrock and blasting operation using IT tools. Organization shows zero accident/incident rate related to blasting. Electronic surveillance from blast design to execution is adopted. Good policy document related to blasting operation and environmental issues exists. Excellent implementation mechanism of governance exists.

A hypothetical case is presented here for a group of mines in a particular region or country. Table 14.9 shows how different blasting sites of Mine A, Mine B, and Mine C may appear based on guidelines prepared in Table 14.8. Such types of charts or dashboards are useful tools for corporate office of mining company or statutory authorities to understand risk levels at different mine sites where blasting operation is carried out.

The reports based on the above method can be combined to devise a risk criterion and identify the associated risks. Accordingly, a Blast Danger Zone can be made by each mine in accordance with the probabilities with directional attributes as shown in Fig. 14.23. There are case studies that have used risk methodology to customize the Blast Danger Zone (Schwengler et al., 2007) and hence improve productivity.

The methods presented here are devised and proposed to give the BDZ a scientific basis in which independent variables of risk can be evaluated and measured to have a tangible analysis in quantitative terms. This will provide the mine management a tool to implement a BDZ that conforms to their maturity and standards and not just based on consequence.

8.1 Best practices

Predrilling inspection, blast design, and charging: Some of the best practices (Table 14.10) include that blaster should inspect the site where blasting block is identified. Blaster should inspect blast face and bench top, prior to marking of blast holes for drilling. Inspection

14. Blast-induced flyrock: risk evaluation and management

TABLE 14.9 Comparative maturity status of different mines for flyrock risk management—a qualitative approach.

Maturity level	Restrain personnel entering blasting zone	Automate process	Reengineer process	Redesign organization
5	△			
4		△		△
3	▭	▭	△	
2			▭	▭
1	●	●	●	●
Risk assessment		Mine A: △	Mine B: ▭	Mine C: ●

should include identification of any cavity, back break, overhang, softer strata, slip planes, faults, or other irregularities. Based on previous back break, excessive toe or undercut in the face suitable drilling should be planned. Essential parameters such as hole diameter, hole depth, subdrilling, burden, spacing, and angle of each hole should be clearly communicated through drilling plan prepared for each blast. Based on inspection by blaster, drillers logs, and laser profile of blasting face, geological anomaly can be identified.

Prior to explosive charging, blaster should inspect the blast face and blastholes from the top of the face. Helper should remain at the bottom of the blasting face floor. During explosive charging of holes, suitable alteration should be done to explosive charge. Nowadays, laser profile survey is available which further improves accuracy in blast design. Laser survey can identify excessive or inadequate burden (Bajpayee et al., 2004). If there is any geological discontinuity or crack present at blasthole, explosives gases are likely to escape resulting in flyrock.

Table 14.10 given below can be used to identify the potential flyrock zones in a blast face and accordingly, given control measures can be adopted to cap the flyrock.

9. Need for future research

Almost all research so far is based on flyrock measured taking all blastholes into account

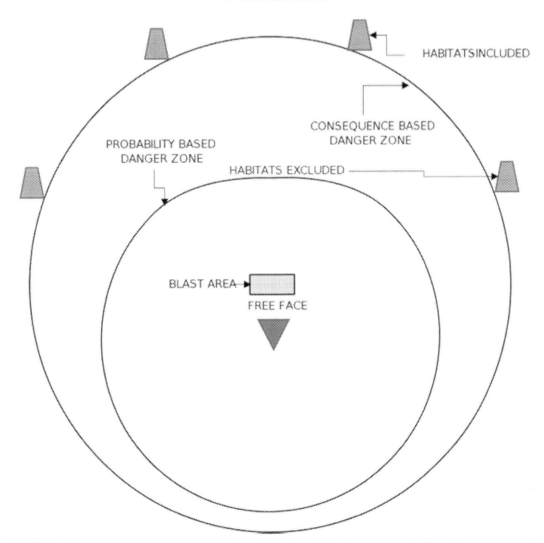

FIGURE 14.23 Definitions of consequence and probability based blast danger zones for flyrock impact prevention.

wherein critical controllable parameters such as burden, stemming length, bench height, specific charge, etc., are average values for a given blast. However, only single hole may contribute to maximum flyrock coupled with uncontrollable geological conditions or rock mass properties. Hence, it is essential to identify such conditions which may result in maximum flyrock for each blast separately. Such research will help for better blast design and control of flyrock.

TABLE 14.10 Causes and corresponding control measures for prevention of flyrock.

Geological factors	Possible causes of flyrock	Possible control measures
• Discontinuity	• Explosive gases escaping through faults, folds, joints	• Predrilling, precharging inspection of blasting site • Integration of drillers feedback, inspection for blast site into charging of holes • Recording actual flyrock for future reference • Safety precautions during blasting
• Voids	• Excessive concentrated charge in void area • Excessive explosives acting like bomb	• Integration of drillers feedback into blast design • Stemming in cavity portion to avoid excessive charge • Controlled charging of hole based on available information
• Fracture,	• Extent of fracture and density of fracture • Weak mass created around fracture	• Integration of drillers feedback into blast design • Reduce explosive charge in fractured zone
• Join spacing - Closely spaced (<0.1 m), intermediate spaced (0.1–1 m), and (>1 m) widely spaced	• Closely spaced joint makes rock weaker • Widely spaced joint makes rock stronger	• Reduce powder factor • Stronger rock can have higher powder factor
• Joint orientation Dip out of face (DOF) Dip into face (DIF) Strike normal to face (SNF) Horizontal joints (H)	• DOF - At toe portion, burden will be higher as face will be dipping outside the face. • DIF - At toe portion, burden will be lower as face will be dipping inside the face. • SNF/H – Joints will make face weaker	• DOF - Excess charge at toe and minimum at the top of the hole to avoid flyrock from top portion of the hole. • DOF - Minimum charge at toe to avoid flyrock • DOF/DIF - Inclined drilling with calculated charging • SNF/H–Density of joints to be observed • Flyrock to be recorded
• Joint aperture	• Small aperture <10 mm • Medium aperture 10–25 mm • Larger aperture >25 mm	• In front row of wider aperture faces, explosives charge needs to be controlled
• Mud seam or lowering of strength of rock	• Weaker rock between stronger rock	• Needs to identify weaker portion based on drillers feedback • Stemming in weaker portion or reducing explosives chare
• Intrusions	• Intrusions are stronger rock as compared to host rock • Uneven distribution of explosives energy as compared to strength of rock	• Needs to identify intrusion portion based on drillers feedback and face observation • Appropriate blast design
• Weathering	• In blasting site, rock is fresh, slightly weathered, moderately weathered, highly weathered, and completely weathered. • If explosives charge not controlled based on degree of weathering	• Rock mass may be classified based on profile of the site • Powder factor to be reduced from fresh to completely weathered rock • Completely weathered rock may not need blasting which may be ripped or excavated with excavator

10. Conclusions

Flyrock studies have reached a maturity level at this juncture. At the outset, a need exists to use standard terminology as presented in Rustan (2010) to avoid confusion and to improve traceability and accessibility of the literature in published domain. Accordingly, an attempt has been made to dispel the confusion in flyrock terminology that blasting may adopt in the interest of science. Also, definitions of flyrock and associated features are presented over here while stressing on new inclusions like rebound of flyrock and the total flyrock distance. A fresh review of existing flyrock distance prediction models along with a comprehensive compilation of intelligent methods used for such prediction has been presented based on the significant work that has been carried out on different aspects of flyrock, with a critical approach. Various inputs thereof have been summarized in this work while trying to bring out the essence of such studies. Although many models are claimed to be highly effective in their own domain, the basic issue of aliasing of the variables and the postrelease effects have been brought out in this chapter. The performance of intelligent models will however require thorough validation for universal applications.

A comprehensive account of risk analysis methods used in flyrock control has been included while detailing and stressing some very important and noteworthy methods as risk needs to be looked into its totality. A peep into different methods and strategies used for control measures with risk assessment and impact of geology on the flyrock is part of this work. A new concept of Risk Maturity Level of an organization has been introduced to help blast management community to introspect and fix their spheres.

References

Adhikari, G.R., 1999. Studies on flyrock at limestone quarries. Rock Mech. Rock Eng. 32. https://doi.org/10.1007/s006030050049.

Adikari, A.M.H.V., Hmtga, P., 2018. A study on environmental issues due to the development of Nuwara Eliya Badulla Road. In: ISERME. ISERME, p. 105.

AEISG, 2011. Code of Good Practice: Blast Guarding in an Open Cut Mining Environment.

Aghajani-Bazzazi, A., Osanloo, M., Azimi, Y., 2010. Flyrock prediction by multiple regression analysis in Esfordi phosphate mine of Iran. In: Proceedings of the 9th International Symposium on Rock Fragmentation by Blasting, FRAGBLAST 9.

Amini, H., Gholami, R., Monjezi, M., et al., 2012. Evaluation of flyrock phenomenon due to blasting operation by support vector machine. Neural Comput. Appl. 21. https://doi.org/10.1007/s00521-011-0631-5.

Armaghani, D.J., Hajihassani, M., Mohamad, E.T., et al., 2014. Blasting-induced flyrock and ground vibration prediction through an expert artificial neural network based on particle swarm optimization. Arabian J. Geosci. 7 (12), 5383–5396. https://doi.org/10.1007/s12517-013-1174-0.

Armaghani, D.J., Hajihassani, M., Monjezi, M., et al., 2015. Application of two intelligent systems in predicting environmental impacts of quarry blasting. Arabian J. Geosci. 8 (11), 9647–9665. https://doi.org/10.1007/s12517-015-1908-2.

Armaghani, D.J., Mahdiyar, A., Hasanipanah, M., et al., 2016. Risk assessment and prediction of flyrock distance by combined multiple regression analysis and Monte Carlo simulation of quarry blasting. Rock Mech. Rock Eng. 49 (9), 3631–3641. https://doi.org/10.1007/s00603-016-1015-z.

Ash, R.L., 1973. The Influence of Geological Discontinuities on Rock Blasting. University of Minnesota, USA, pp. 1–289. Ph.D. Thesis.

Asl, P.F., Monjezi, M., Hamidi, J.K., Armaghani, D.J., 2018. Optimization of flyrock and rock fragmentation in the Tajareh limestone mine using metaheuristics method of firefly algorithm. Eng. Comput. 34 (2), 241–251. https://doi.org/10.1007/s00366-017-0535-9.

Bajpayee, T.S., Rehak, T.R., Mowrey, G.L., Ingram, D.K., 2002. A summary of fatal accidents due to flyrock and lack of blast area security in surface mining, 1989 to 1999. In: Proceedings of the 28th Annual Conference on Explosives and Blasting Technique, Feb. 10-13, Las Vegas, 2. ISEE, USA, pp. 105–118.

Bajpayee, T.S., Rehak, T.R., Mowrey, G.L., Ingram, D.K., 2004. Blasting injuries in surface mining with emphasis on flyrock and blast area security. J. Saf. Res. 35 (1), 47–57. https://doi.org/10.1016/j.jsr.2003.07.003.

Bakhtavar, E., Nourizadeh, H., Sahebi, A.A., 2017. Toward predicting blast-induced flyrock: a hybrid dimensional analysis fuzzy inference system. Int. J. Environ. Sci. Technol. 14 (4), 717–728. https://doi.org/10.1007/s13762-016-1192-z.

Baliktsis, E., Baliktsis, A., 2004. Flyrock risk prevention—from theory and ideas to a perfectly applied blasting project. In: Proceedings of the 1st International Conference on Advances in Mineral Resources Management and Environmental Geotechnology. Hania, Crete, Greece, pp. 17–23. Hania, Greece.

Bandyopadhyay, P.K., Roy, S.C., Sen, S.N., 2003. Risk assessment in open cast mining - an application of Yager's methodology for ordinal multiobjective decisions based on fuzzy sets. Jpn. J. Ind. Appl. Math. 20, 311–319. https://doi.org/10.1007/bf03167425.

Berta, G., 1990. Explosives : An Engineering Tool.

Beyglou, A., Johansson, D., Srhunnesson, H., 2017. Target fragmentation for efficient loading and crushing -The Aitik case. J. South African Inst. Min. Metall. 117. https://doi.org/10.17159/2411-9717/2017/v117n11a10.

Bhandari, S., 1997. Engineering Rock Blasting Operations. Balkema, Rotterdam.

Bhatawdekar, R.M., Danial, J.A., Edy, T.M., 2018. A review of prediction of blast performance using computational techniques. In: Proceedings of the ISERME. ISERME, p. 37.

Bhatawdekar, R.M., Edy, M.T., Danial, J.A., 2019. Building information model for drilling and blasting for tropically weathered rock. J. Mines Met. Fuels 67.

Bhowmik, S., Raina, A.K., Chakraborty, A.K., et al., 2004. Flyrock prediction and control in opencast mines. Eng. Min. J. 5, 10–16.

Bird, F.E., Germain, G.L., 1996. Loss Control Management: Practical Loss Control Leadership. Det Norske Veritas Inc., USA, pp. 1–5.

Blanchier, A., 2013. Quantification of the levels of risk of flyrock. In: FRAGBLAST 10—Proceedings of the 10th International Symposium on Rock Fragmentation by Blasting, pp. 549–553.

Buffington, G.L., 2000. The art of blasting on construction and surface mining sites. In: ASSE Professional Development Conference and Exposition 2000.

Calnan, J.T., 2015. Determination of Explosive Energy Partition Values in Rock Blasting through Small-Scale Testing. ProQuest Diss Theses.

Chernigovskii, A.A., 1986. Application of Directional Blasting in Mining and Civil Engineering, Second revised and enlarged edition. Appl Dir blasting Min Civ Eng Second Revis Enlarg Ed.

Chiapetta, R.F., Bauer, A., Dailey, P.J., Burchell, S.L., 1983. The use of high-speed motion picture photography in blast evaluation and design. In: Proceeding's 9th International Symposium on Rock Fragmentation by Blasting. International Society of Explosive Engineers, Dallas, pp. 31–40. January 31 - February 4.

Davies, P.A., 1995. Risk-based approach to setting of flyrock "danger zones" for blast sites. Trans Inst Min Metall Sect A 104. https://doi.org/10.1016/0148-9062(95)99212-g.

Dehghani, H., Shafaghi, M., 2017. Prediction of blast-induced flyrock using differential evolution algorithm. Eng. Comput. 33. https://doi.org/10.1007/s00366-016-0461-2.

Dehghani, H., Pourzafar, M., Asadi zadeh, M., 2021. Prediction and minimization of blast-induced flyrock using gene expression programming and cuckoo optimization algorithm. Environ. Earth Sci. 80. https://doi.org/10.1007/s12665-020-09300-z.

Dick, R.A., Fletcher, L.R., D'Andrea, D.V., 1983. Explosives and Blasting Procedures Manual. United States Bureau of Mines. Information Circular.

Ebrahimi, E., Monjezi, M., Khalesi, M.R., Armaghani, D.J., 2016. Prediction and optimization of back-break and rock fragmentation using an artificial neural network and a bee colony algorithm. Bull. Eng. Geol. Environ. 75 (1), 27–36. https://doi.org/10.1007/s10064-015-0720-2.

Esposito, G., Mastrorocco, G., Salvini, R., et al., 2017. Application of UAV photogrammetry for the multi-temporal estimation of surface extent and volumetric excavation in the Sa Pigada Bianca open-pit mine, Sardinia, Italy. Environ. Earth Sci. 76. https://doi.org/10.1007/s12665-017-6409-z.

Faradonbeh, R.S., Armaghani, D.J., Monjezi, M., Mohamad, E.T., 2016a. Genetic programming and gene expression programming for flyrock assessment due to mine blasting. Int. J. Rock Mech. Min. Sci. 88. https://doi.org/10.1016/j.ijrmms.2016.07.028.

Faradonbeh, R.S., Jahed Armaghani, D., Monjezi, M., 2016b. Development of a new model for predicting flyrock distance in quarry blasting: a genetic programming technique. Bull. Eng. Geol. Environ. 75. https://doi.org/10.1007/s10064-016-0872-8.

Faradonbeh, R.S., Armaghani, D.J., Amnieh, H.B., Mohamad, E.T., 2018. Prediction and minimization of blast-induced flyrock using gene expression programming and firefly algorithm. Neural Comput. Appl. 29 (6), 269–281. https://doi.org/10.1007/s00521-016-2537-8.

References

Faramarzi, F., Mansouri, H., Farsangi, M.A.E., 2014. Development of rock engineering systems-based models for flyrock risk analysis and prediction of flyrock distance in surface blasting. Rock Mech. Rock Eng. 47. https://doi.org/10.1007/s00603-013-0460-1.

Fattahi, H., Hasanipanah, M., 2021. An integrated approach of ANFIS-grasshopper optimization algorithm to approximate flyrock distance in mine blasting. Eng. Comput. 1–13. https://doi.org/10.1007/s00366-020-01231-4.

Fletcher, L.R., D'Andrea, D.V., 1990. Control of flyrock in blasting. J. Explos. Eng. 7.

Gate, W., Ortiz, L., Florez, R., 2005. Analysis of rock fall and blasting back break problems. In: American Rock Mechanics Conference, vol. 4. Anchorage, Alaska, pp. 1290–1298.

St George, J.D., Gibson, M.F.L., 2001. Estimation of flyrock travel distances: a probabilistic approach. In: AusIMM EXPLO 2001 Conference. Australasian Institute of Mining and Metallurgy, Hunter Valley, pp. 409–415.

Ghasemi, E., 2018. Prediction of blasting-induced flyrock using M5P tree technique. J. Analytical Numer. Methods Min. Eng. 8. https://doi.org/10.29252/anm.8.16.45.

Ghasemi, E., Sari, M., Ataei, M., 2012. Development of an empirical model for predicting the effects of controllable blasting parameters on flyrock distance in surface mines. Int. J. Rock Mech. Min. Sci. 52, 163–170. https://doi.org/10.1016/j.ijrmms.2012.03.011.

Ghasemi, E., Amini, H., Ataei, M., Khalokakaei, R., 2014. Application of artificial intelligence techniques for predicting the flyrock distance caused by blasting operation. Arabian J. Geosci. 111 (9), 1531. https://doi.org/10.1007/s12517-012-0703-6.

Grundstrom, C., Kanchibotla, S.S., Jankovich, A., Thornton, D., 2001. Blast fragmentation for maximizing the sag mill throughput at Porgera gold mine. In: Proceedings of the Annual Conference on Explosives and Blasting Technique.

Guo, H., Nguyen, H., Bui, X.N., Armaghani, D.J., 2019a. A new technique to predict fly-rock in bench blasting based on an ensemble of support vector regression and GLMNET. Eng. Comput. Aug. 1–15.

Guo, H., Zhou, J., Koopialipoor, M., et al., 2019b. Deep neural network and whale optimization algorithm to assess flyrock induced by blasting. Engineering with Computers. Eng. Comput. 1–14.

Guo, H., Zhou, J., Koopialipoor, M., et al., 2021. Deep neural network and whale optimization algorithm to assess flyrock induced by blasting. Eng. Comput. 37. https://doi.org/10.1007/s00366-019-00816-y.

Gupta, R.N., 1990. Surface blasting and its impact on environment. In: Trivedy, R.K., Sinha, M.P. (Eds.), Impact of Mining on Environment. Ashish Publishing House, New Delhi, pp. 23–24.

Hajihassani, M., Jahed Armaghani, D., Sohaei, H., et al., 2014. Prediction of airblast-overpressure induced by blasting using a hybrid artificial neural network and particle swarm optimization. Appl. Acoust. 80, 57–67. https://doi.org/10.1016/j.apacoust.2014.01.005.

Han, H., Jahed Armaghani, D., Tarinejad, R., et al., 2020. Random forest and Bayesian network techniques for probabilistic prediction of flyrock induced by blasting in quarry sites. Nat. Resour. Res. 29. https://doi.org/10.1007/s11053-019-09611-4.

Hasanipanah, M., Bakhshandeh Amnieh, H., 2020. A fuzzy rule-based approach to address uncertainty in risk assessment and prediction of blast-induced flyrock in a quarry. Nat. Resour. Res. 29, 669–689. https://doi.org/10.1007/s11053-020-09616-4.

Hasanipanah, M., Faradonbeh, R.S., Armaghani, D.J., et al., 2017a. Development of a precise model for prediction of blast-induced flyrock using regression tree technique. Environ. Earth Sci. 76, 27. https://doi.org/10.1007/s12665-016-6335-5.

Hasanipanah, M., Jahed Armaghani, D., Bakhshandeh Amnieh, H., et al., 2017b. Application of PSO to develop a powerful equation for prediction of flyrock due to blasting. Neural Comput. Appl. 28, 1043–1050. https://doi.org/10.1007/s00521-016-2434-1.

Hasanipanah, M., Jahed Armaghani, D., Bakhshandeh Amnieh, H., et al., 2018. A risk-based technique to analyze flyrock results through rock engineering system. Geotech. Geol. Eng. 36, 2247–2260. https://doi.org/10.1007/s10706-018-0459-1.

Hendron, A.J., 1978. Engineering of rock blasting on civil projects. Int. J. Rock Mech. Min. Sci. Geomech. Abstr. 15. https://doi.org/10.1016/0148-9062(78)90198-5.

Hudaverdi, T., Akyildiz, O., 2019. A new classification approach for prediction of flyrock throw in surface mines. Bull. Eng. Geol. Environ. 78, 177–187. https://doi.org/10.1007/s10064-017-1100-x.

Hustrulid, W., 1999. Blasting Principles for Open Pit Mining.

IME, 1997. Glossary of Commercial Explosives Industry Terms. Institute of Makers of Explosives.

Jahed Armaghani, D., Tonnizam Mohamad, E., Hajihassani, M., et al., 2016. Evaluation and prediction of flyrock resulting from blasting operations using empirical and computational methods. Eng. Comput. 32, 109–121. https://doi.org/10.1007/s00366-015-0402-5.

Janković, A., Valery, W., 2002. Mine to mill optimisation for conventional grinding circuits: a scoping study. J. Min. Metall. Sect. A Min. 38.

Jhanwar, J.C., Jethwa, J.L., Reddy, A.H., 2000. Influence of air-deck blasting on fragmentation in jointed rocks in an open-pit manganese mine. Eng. Geol. 57 (1−2), 13−29. https://doi.org/10.1016/S0013-7952(99)00125-8.

Jimeno, C.L., Jimeno, E.L., Carcedo, F.J., De Ramiro, Y., 1997. Drilling and blasting of rocks. Environ. Eng. Geosci. 1−391. https://doi.org/10.2113/gseegeosci.iii.1.154.

Kalaivaani, P.T., Akila, T., Tahir, M.M., et al., 2020. A novel intelligent approach to simulate the blast-induced flyrock based on RFNN combined with PSO. Eng. Comput. 36. https://doi.org/10.1007/s00366-019-00707-2.

Kanchibotla, S.S., Valery, W., Morrell, S., 1999. Modelling fines in blast fragmentation and its impact on crushing and grinding. In: Explo '99−A Conference on Rock Breaking. Australasian Institute of Mining and Metallurgy, Kalgoorlie, Australia, pp. 137−144.

Kang, L.S., Paulson, B.C., 1998. Information management to integrate cost and schedule for civil engineering projects. J. Construct. Eng. Manag. 124. https://doi.org/10.1061/(asce)0733-9364(1998)124:5(381).

Karami, A.R., Mansouri, H., Farsangi, M.A.E., Nezamabadi, H., 2006. Backbreak prediction due to bench blasting: an artificial neural network approach. J. Mines Met. Fuels 54.

Kecojevic, V., Radomsky, M., 2005. Flyrock phenomena and area security in blasting-related accidents. Saf. Sci. 43 (9), 739−750. https://doi.org/10.1016/j.ssci.2005.07.006.

Khandelwal, M., Monjezi, M., 2013. Prediction of flyrock in open pit blasting operation using machine learning method. Int. J. Min. Sci. Technol. 23, 313−316. https://doi.org/10.1016/j.ijmst.2013.05.005.

Kleine, T., Roberts, B., Cameron, A., Forsyth, B., 1994. The use of probability and risk in blast design. In: 1st North American Rock Mechanics Symposium, NARMS 1994, pp. 269−276.

Konya, C.J., Walter, E., 1991. Rock Blasting and Overbreak Control. Security.

Koopialipoor, M., Fallah, A., Armaghani, D.J., et al., 2019. Three hybrid intelligent models in estimating flyrock distance resulting from blasting. Eng. Comput. 35 (1), 243−256. https://doi.org/10.1007/s00366-018-0596-4.

Kumar, N., Mishra, B., Bali, V., 2018. A novel approach for blast-induced fly rock prediction based on particle swarm optimization and artificial neural network. In: Proceedings of International Conference on Recent Advancement on Computer and Communication. Springer, Singapore, pp. 19−27.

Lam, P.T.I., 1999. A sectoral review of risks associated with major infrastructure projects. Int. J. Proj. Manag. 17. https://doi.org/10.1016/s0263-7863(98)00017-9.

Li, D., Koopialipoor, M., Armaghani, D.J., 2021. A combination of fuzzy delphi method and ANN-based models to investigate factors of flyrock induced by mine blasting. Nat. Resour. Res. 30 (2), 1905−1924. https://doi.org/10.1007/s11053-020-09794-1.

Little, T.N., 2007. Flyrock risk. In: Australasian Institute of Mining and Metallurgy Publication Series. Wollongong, NSW, 3 − 4, September, pp. 35−43.

Little, T.N., Blair, D.P., 2010. Mechanistic Monte Carlo models for analysis of flyrock risk. In: Proceedings of the 9th International Symposium on Rock Fragmentation by Blasting, FRAGBLAST 9, pp. 641−647.

Liu, L., Katsabanis, P.D., 1997. A numerical study of the effects of accurate timing on rock fragmentation. Int. J. Rock Mech. Min. Sci. 34. https://doi.org/10.1016/S1365-1609(96)00067-8.

Livingston, C.W., 1956. Fundamentals of rock failure. In: The 1st US Symposium on Rock Mechanics (USRMS). American Rock Mechanics Association.

Lu, X., Hasanipanah, M., Brindhadevi, K., et al., 2020. ORELM: a novel machine learning approach for prediction of flyrock in mine blasting. Nat. Resour. Res. 29. https://doi.org/10.1007/s11053-019-09532-2.

Lundborg, N., 1974. The Hazards of Flyrock in Rock Blasting.

Lundborg, N., 1981. Probability of Flyrock.

Lundborg, N., Persson, A., Ladegaard-Pedersen, A., Holmberg, R., 1975. Keeping the lid on flyrock in open-pit blasting. Eng. Min. J. 176, 95−100. https://doi.org/10.1016/0148-9062(75)91215-2.

Mackenzie, A., 1966. Cost of explosives—do you evaluate it properly? Min. Congr. J. 52, 32−41.

Marto, A., Hajihassani, M., Jahed Armaghani, D., et al., 2014. A novel approach for blast-induced flyrock prediction based on imperialist competitive algorithm and artificial neural network. Sci. World J. 2014, 643715. https://doi.org/10.1155/2014/643715.

Masir, R.N., Ataei, M., Motahedi, A., 2021. Risk assessment of flyrock in surface mines using a FFTA-MCDM combination. J Min Environ 12, 191−203.

McKenzie, C.K., 2009. Flyrock range & fragment size prediction. In: 35th Annual Conference on Explosives and Blasting Technique, International Society of Explosives Engineers. International Society of Explosive Engineers, Denver, CO, pp. 1−17.

Mishra, A.K., Mallick, D.K., 2013. Analysis of blasting related accidents with emphasis on flyrock and its mitigation in surface mines. In: FRAGBLAST 10−Proceedings of the 10th International Symposium on Rock Fragmentation by Blasting.

Mohamad, E.T., Armaghani, D.J., Hajihassani, M., et al., 2013a. A simulation approach to predict blasting-induced flyrock and size of thrown rocks. Electron. J. Geotech. Eng. 18, 5561−5572.

Mohamad, E.T., Armaghani, D.J., Motaghedi, H., 2013b. The effect of geological structure and powder factor in flyrock accident, Masai, Johor, Malaysia. Electron. J. Geotech. Eng. 18 X.

Mohamad, E.T., Murlidhar, B.R., Armaghani, D.J., et al., 2016. Effect of geological structure and blasting practice in fly rock accident at Johor, Malaysia. J Teknol 78. https://doi.org/10.11113/jt.v78.9634.

Mohamad, E.T., Yi, C.S., Murlidhar, B.R., Saad, R., 2018. Effect of geological structure on flyrock prediction in construction blasting. Geotech. Geol. Eng. 36. https://doi.org/10.1007/s10706-018-0457-3.

MoL&A GOI, 2019. Gazettee Notification Regarding Metalliferous Mines Regulations. Government of India, India.

Monjezi, M., Bahrami, A., Yazdian Varjani, A., 2010a. Simultaneous prediction of fragmentation and flyrock in blasting operation using artificial neural networks. Int. J. Rock Mech. Min. Sci. 47. https://doi.org/10.1016/j.ijrmms.2009.09.008.

Monjezi, M., Rezaei, M., Yazdian, A., 2010b. Prediction of backbreak in open-pit blasting using fuzzy set theory. Expert Syst. Appl. 37 (3), 2637–2643. https://doi.org/10.1016/j.eswa.2009.08.014.

Monjezi, M., Bahrami, A., Varjani, A.Y., Sayadi, A.R., 2011a. Prediction and controlling of flyrock in blasting operation using artificial neural network. Arabian J. Geosci. 4. https://doi.org/10.1007/s12517-009-0091-8.

Monjezi, M., Ghafurikalajahi, M., Bahrami, A., 2011b. Prediction of blast-induced ground vibration using artificial neural networks. Tunn. Undergr. Space Technol. 26 (1), 46–50. https://doi.org/10.1016/j.tust.2010.05.002.

Monjezi, M., Mehrdanesh, A., Malek, A., Khandelwal, M., 2013. Evaluation of effect of blast design parameters on flyrock using artificial neural networks. Neural Comput. Appl. 23. https://doi.org/10.1007/s00521-012-0917-2.

Monjezi, M., Hashemi Rizi, S.M., Majd, V.J., Khandelwal, M., 2014. Artificial neural network as a tool for backbreak prediction. Geotech. Geol. Eng. 32. https://doi.org/10.1007/s10706-013-9686-7.

Murlidhar, B.R., Armaghani, D.J., Mohamad, E.T., 2020a. Intelligence prediction of some selected environmental issues of blasting: a review. Open Construct. Build Technol. J. 14. https://doi.org/10.2174/1874836802014010298.

Murlidhar, B.R., Kumar, D., Jahed Armaghani, D., et al., 2020b. A novel intelligent ELM-BBO technique for predicting distance of mine blasting-induced flyrock. Nat. Resour. Res. 29. https://doi.org/10.1007/s11053-020-09676-6.

Nguyen, H., Bui, X.N., Choi, Y., et al., 2021. A novel combination of whale optimization algorithm and support vector machine with different Kernel functions for prediction of blasting-induced fly-rock in quarry mines. Nat. Resour. Res. 30. https://doi.org/10.1007/s11053-020-09710-7.

Nieble, C.M., Penteado, J.A., 2016. Risk management: blasting rock near concrete inside a subway station in a densely populated urban environment. In: ISRM International Symposium—EUROCK 2016, pp. 1269–1273.

Ouchterlony, F., 2005. The Swebrec© function: linking fragmentation by blasting and crushing. Inst. Min. Metall. Trans. Sect. A Min. Technol. 114. https://doi.org/10.1179/037178405X44539.

Pal Roy, P., 2005. Rock Blasting: Effects and Operations. CRC Press.

Persson, P.A., Holmberg, R., Lee, J., 1994. Rock Blasting and Explosives Engineering.

Prakash, A., Murthy, V.M.S.R., Singh, K.B., 2013. Rock excavation using surface miners: an overview of some design and operational aspects. Int. J. Min. Sci. Technol. 23. https://doi.org/10.1016/j.ijmst.2013.01.006.

Rad, H.N., Hasanipanah, M., Rezaei, M., Eghlim, A.L., 2018. Developing a least squares support vector machine for estimating the blast-induced flyrock. Eng. Comput. 34. https://doi.org/10.1007/s00366-017-0568-0.

Raina, A.K., 2014. Modelling the Flyrock in Opencast Blasting under Difficult Geomining Conditions. Indian Institute of Technology—ISM.

Raina, A.K., Murthy, V.M.S.R., 2016a. Prediction of flyrock distance in open pit blasting using surface response analysis. Geotech. Geol. Eng. 34, 15–28. https://doi.org/10.1007/s10706-015-9924-2.

Raina, A.K., Murthy, V.M.S.R., 2016b. Importance and sensitivity of variables defining throw and flyrock in surface blasting by artificial neural network method. Curr. Sci. 111. https://doi.org/10.18520/cs/v111/i9/1524-1531.

Raina, A.K., et al., 2006. Flyrock Prediction and Control in Opencast Metal Mines in India for Safe Deep-Hole Blasting Near Habitats - A Futuristic Approach. Nagpur, Dhanbad, India.

Raina, A.K., Chakraborty, A.K., Choudhury, P.B., Sinha, A., 2011. Flyrock danger zone demarcation in opencast mines: a risk based approach. Bull. Eng. Geol. Environ. 70, 163–172. https://doi.org/10.1007/s10064-010-0298-7.

Raina, A.K., Murthy, V.M.S.R., Soni, A.K., 2013a. Relevance of shape of fragments on, flyrock travel distance: an insight from concrete model experiments using ANN. Electron. J. Geotech. Eng. 18, E:899–907.

Raina, A.K., Soni, A.K., Murthy, V.M.S.R., 2013b. Spatial distribution of flyrock using EDA: an insight from concrete model tests. In: FRAGBLAST 10—Proceedings of the 10th International Symposium on Rock Fragmentation by Blasting, pp. 563–568.

Raina, A.K., Murthy, V.M.S.R., Soni, A.K., 2014. Flyrock in bench blasting: a comprehensive review. Bull. Eng. Geol. Environ. 73, 1199–1209. https://doi.org/10.1007/s10064-014-0588-6.

Raina, A.K., Murthy, V.M.S.R., Soni, A.K., 2015. Flyrock in surface mine blasting: understanding the basics to develop a predictive regime. Curr. Sci. 108, 660–665. https://doi.org/10.18520/CS/V108/I4/660-665.

Rezaei, M., Monjezi, M., Yazdian Varjani, A., 2011. Development of a fuzzy model to predict flyrock in surface mining. Saf. Sci. 49 (2), 298–305. https://doi.org/10.1016/j.ssci.2010.09.004.

Richards, A., Moore, A., 2004. Flyrock Control - by chance or design. In: Proceedings of the 30th Annual Conference on Explosives and Blasting Technique, V1. International Society of Explosive Engineers, New Orleans, pp. 335–348.

Ricketts, J.T., Loftin, M.K., Merritt, F.S., 2004. Standard Handbook for Civil Engineers.

Rosenthal, M.F., Morlock, G.L., 1987. Blasting Guidance Manual.

Roth, J.A., 1979. A Model for the Determination of Flyrock Range as a Function of Shot Conditions. NTIS, U.S. Dept. of Commerce, USA, pp. 1–61. PB81222358.

Roy, M.P., Paswan, R.K., Sarim, M.D., et al., 2016. Rock fragmentation by blasting: a review. J. Mines Met. Fuels 64 (9), 424–431.

Rustan, A. (Ed.), 2010. Mining and Rock Construction Technology Desk Reference.

Saghatforoush, A., Monjezi, M., Shirani Faradonbeh, R., Jahed Armaghani, D., 2016. Combination of neural network and ant colony optimization algorithms for prediction and optimization of flyrock and back-break induced by blasting. Eng. Comput. 32 (2), 255–266. https://doi.org/10.1007/s00366-015-0415-0.

Sanchidrián, J.A., Segarra, P., López, L.M., 2007. Energy components in rock blasting. Int. J. Rock Mech. Min. Sci. 44 (1), 130–147. https://doi.org/10.1016/j.ijrmms.2006.05.002.

Schwengler, B., Moncrieff, J., Bellairs, P., 2007. Reduction of the blast exclusion zone at the black star open cut mine. In: Australasian Institute of Mining and Metallurgy Publication Series, pp. 51–58.

Scott, A., Morrell, S., Clark, D., 2002. Tracking and quantifying value from "mine to mill" improvement. In: Australasian Institute of Mining and Metallurgy Publication Series.

Shea, C.W., Clark, D., 1998. Avoiding tragedy: lessons to be learned from a flyrock fatality. Coal Age 130, 51–54.

Singh, S.P., Xavier, P., 2005. Causes, Impact and Control of Overbreak in Underground Excavations. Tunnelling and Underground Space Technology. https://doi.org/10.1016/j.tust.2004.05.004.

Singh, P.K., Roy, M.P., Paswan, R.K., et al., 2016. Rock fragmentation control in opencast blasting. J. Rock Mech.

Geotech. Eng. 8, 225–237. https://doi.org/10.1016/j.jrmge.2015.10.005.

Siskind, D.E., Kopp, J.W., 1995. Blasting accidents in mines: a 16 year summary. In: Proceedings of the 21st Annual Conference on Explosives and Blasting Technique. International Society of Explosives Engineers, Cleveland, pp. 224–239.

Sobko, B., Lozhnikov, O., Levytskyi, V., Skyba, G., 2019. Conceptual development of the transition from drill and blast excavation to non-blasting methods for the preparation of mined rock in surface mining. Rud. Geol. Naft. Zb. 34. https://doi.org/10.17794/rgn.2019.3.3.

Stojadinović, S., 2013. Coupled Neural Networks and Numeric Models for Flyrock Safe Distance Definition. University of Belgrade.

Stojadinović, S., Lilić, N., Obradović, I., et al., 2016. Prediction of flyrock launch velocity using artificial neural networks. Neural Comput. Appl. 27. https://doi.org/10.1007/s00521-015-1872-5.

Tripathy, G.R., Shirke, R.R., Kudale, M.D., 2016. Safety of engineered structures against blast vibrations: a case study. J. Rock Mech. Geotech. Eng. 8 (2), 248–255. https://doi.org/10.1016/j.jrmge.2015.10.007.

Trivedi, R., 2015. Prediction and Control of Blast Induced Flyrock in Surface Mines Using Artificial Neural Network and Adaptive Neuro-Fuzzy Inference System. Indian Institute of Technology Bombay.

Trivedi, R., Singh, T.N.N., Raina, A.K.K., 2014. Prediction of blast-induced flyrock in Indian limestone mines using neural networks. J. Rock Mech. Geotech. Eng. 6, 447–454. https://doi.org/10.1016/j.jrmge.2014.07.003.

Trivedi, R., Singh, T.N., Gupta, N., 2015. Prediction of blast-induced flyrock in opencast mines using ANN and ANFIS. Geotech. Geol. Eng. 33. https://doi.org/10.1007/s10706-015-9869-5.

Trivedi, R., Singh, T.N., Raina, A.K., 2016. Simultaneous prediction of blast-induced flyrock and fragmentation in opencast limestone mines using back propagation neural network. Int. J. Min. Miner. Eng. 7, 237. https://doi.org/10.1504/ijmme.2016.078350.

Venkatesh, H.S., Bhatawdekar, R.M., Adhikari, G.R., Theresraj, A.I., 1999. Assessment and mitigation of ground vibrations and flyrock at a limestone quarry. In: Proceedings of the Annual Conference on Explosives and Blasting Techniquee. International Society of Explosives Engineers, pp. 145–152.

Verakis, H., 2011. Floyrock: a continuing blast safety threat. J. Explos. Eng. 28.

Verakis, H., Lobb, T., 2007. Flyrock revisited an ever present danger in mine blasting. In: 33rd Annual Conference on Explosives and Blasting Technique. International Society of Explosive Engineers.

Wallace, J., 2001. Back to school on construction blasting rules of thumb revisited. J. Explos. Eng. 18.

Yari, M., Bagherpour, R., Jamali, S., Shamsi, R., 2016. Development of a novel flyrock distance prediction model using BPNN for providing blasting operation safety. Neural Comput. Appl. 27. https://doi.org/10.1007/s00521-015-1889-9.

Yari, M., Bagherpour, R., Jamali, S., 2017. Development of an evaluation system for blasting patterns to provide efficient production. J. Intell. Manuf. 28. https://doi.org/10.1007/s10845-015-1036-6.

Ye, J., Koopialipoor, M., Zhou, J., et al., 2021. A novel combination of tree-based modeling and Monte Carlo simulation for assessing risk levels of flyrock induced by mine blasting. Nat. Resour. Res. 30, 225–243. https://doi.org/10.1007/s11053-020-09730-3.

Zhou, Z., Cheng, R., Cai, X., et al., 2019. Comparison of pre-split and smooth blasting methods for excavation of rock wells. Shock Vib. 2019, 3743028. https://doi.org/10.1155/2019/3743028.

Zhou, J., Aghili, N., Ghaleini, E.N., et al., 2020a. A Monte Carlo simulation approach for effective assessment of flyrock based on intelligent system of neural network. Eng. Comput. 36, 713–723. https://doi.org/10.1007/s00366-019-00726-z.

Zhou, J., Koopialipoor, M., Murlidhar, B.R., et al., 2020b. Use of intelligent methods to design effective pattern parameters of mine blasting to minimize flyrock distance. Nat. Resour. Res. 29, 625–639. https://doi.org/10.1007/s11053-019-09519-z.

CHAPTER 15

The importance of environmental sustainability in construction

Beste Cubukcuoglu

Institute of Building Materials Research, RWTH Aachen University, Aachen, Germany

1. Introduction

Sustainability aims to achieve the balance between the society, environment, and economy by adopting and implementing required policies, strategies, and technologies. The concept of sustainable development was first defined in 1987 at the Brundtland World Commission as "meets the needs of the present without compromising the ability of future generations to meet their own needs" (Brundtland, 1987).

In order to understand the concept of sustainability in civil engineering, the climate, and air quality problems, water supply, quality and treatment issues, and water resources management should be understood and possible solutions to each issue should be investigated. Sustainable development is related to construction; hence, unsustainable development is robbing our grandchildren assets which we waste (Ehnert, 2009).

The understanding of sustainable development issues including agricultural and forestry resources and the investigation of energy use and sustainable energy options would widen the perspective of engineers regarding the importance of environmental sustainability in civil engineering applications. Understanding the importance of using recyclable materials in construction and exploring environmentally sustainable development options for industry and creating a sustainable built environment would be the total benefit to the society, world, and the environment. Sustainability can be explained as continuous improvement of life quality that protects and balances the ecological, social, and economic environment (Seidel, 2007; Liue et al., 2020).

It is obvious that we cannot expect to have sustainable development without *environmental sustainability*. The environmental performance could be enhanced by sustainable planning, design, and construction with considerations on climate and air quality; water supply, quality, and treatment; energy use and sustainable energy options; recyclable construction materials (resource-efficient building materials for a sustainable built environment); and many others. Environmental sustainability aims to balance the three pillars of economic and social development with environmental protection (Ding, 2008). It is very well known that the society with environmental sustainability manners would satisfy the basic needs of its people for food, clean water, and shelter into the indefinite future without depleting Earth's natural resources to prevent current and future generations of humans and other species from meeting their basic needs (Seidel, 2007; Zahibi et al., 2012; Braham and Casillas, 2020).

It is clear that environmental sustainability in construction projects and most importantly in a built environment would be provided if only the engineering manners work in cooperation with the architectural concepts and applications. It is obvious that work in cooperation during the implementation of a project would surely bring benefits to the whole project at the end. Sustainable architectural methods, design options, and approaches should be integrated with sustainable engineering ideas to promote the environmental sustainability in construction industry to a further and better level. For this purpose, of course, it is required to understand and investigate effectively the environmental issues but most importantly their causes to find alternative approaches to overcome those problems, and hence, become closer to a sustainable environment by means of concepts, ideas, and surely by their implementation.

2. Environmental issues, their causes, and sustainability

The environmental issues can be listed mainly as land, water, and air pollution. Climate change including global warming is mainly caused by fossil fuels that are the main issues that human beings face due to the heavily polluted environment. Climate change due to global warming and acid rains, environmental degradation, and habitat destruction can be listed as the other environmental issues which need to be overcome to provide environmental sustainability worldwide. It is obvious that the major causes of environmental problems can be listed as

unplanned and unexpected demographic changes due to the growth in population, waste resource use, poverty, poor environmental accounting, and of course ecological ignorance (Ding, 2008).

Political action and controls, scientific advances which can provide controls and communications, the construction which has provided the infrastructure for civilized communities, and the participation of communities in the enhancement of their built environment could be considered as the required precautions to be followed to achieve sustainable development throughout the world.

2.1 Air quality and sustainable development

The majority of the air pollution comes from a single class of activities by burning fossil fuels. Hence, at first, the reduction of the adverse effect of the energy sector on the environment should be eliminated. This could be only possible by, increasing the use of renewable energy, improving energy efficiency and enhancing the sustainable development goals for the prevention of ozone depletion. In order to reduce the dependency on fossil fuels, sustainable alternative ways of power generation should be encouraged. Renewable technologies such as wind power, fuel cells, solar power, and few others can be used as an alternative source of energy which produce power with no direct air pollution at all, which may lead to a decline in coal-fired energy generation.

2.2 Energy efficiency and sustainable development

The environmental impacts of unsustainable energy generation can be listed as air pollution caused by the emission of gases from combustion; climate change (related to increased emissions of greenhouse gases which include CO_2, methane -CH_4, N_2O); radioactivity from rocks (mining for fuel ores may release radioactivity to the surface or water); landform changes; ash disposal; and impacts on water (can occur in deep mining if the poor quality or saline groundwater is pumped out).

The engineers should examine how construction can help to solve the problems both of meeting energy demand and reducing pollution and exhaustion of natural resources which energy production causes. An estimate of global energy demand is required which will have to be satisfied by the end of this century. A review of the sources of energy and systems of supply and their environmental consequences should be undertaken. Identification of the options for meeting future energy demands and of the difficulties in doing this sustainably is also a must. Civil engineering solutions for obtaining fuels include locating and construction conversion facilities and dealing with wastes for the various types of energy production that will remain important or become so, reducing adverse environmental impacts and sustaining the natural resources. There are solutions and opportunities for energy conservation such as more efficient energy use, structural design and provision of insulation and temperature control in buildings and public utility systems (natural cooling and passive heating), and conservation of scarce resources by not using natural gas for electricity generation. Renewable sources often require even more comprehensive construction to control the flow of water, capture the force of offshore or high wind, or channel geothermal or solar heat energy into mechanical, electrical, or chemical conversion units.

Sustainable Energy addresses the challenges of generating electricity from renewable energy resources rather than fossil fuels without compromising the needs of the future generation and surely offers significant public health benefits.

2.3 Water quality and sustainable development

The sources of water contamination can be listed as sediments, infectious agents, organic material, chemicals, minerals, radioactive substances, and heat (such as released in power station cooling). The most important factor which causes water contamination is related to human activity; sewage; and other oxygen-demanding wastes. Controlling water contamination and provide effective and efficient water quality is another requirement to achieve sustainable development. Quality of water can be enhanced by a range of solutions to meet different economic, hydrological, and cultural conditions:

- Systems, such as for cities as at present, in which all the water is treated to a high standard
- Dual systems providing treated and untreated water separately and at a different cost
- Untreated systems, such as for irrigation schemes, in which special treatment plants and storage tanks are located at each settlement

Land space should be provided for water facilities. Just as water has to be allocated to various uses, so areas of land must be allotted providing the necessary space for water source development, conveyance, and drainage. Wastewater treatment requires more space for the process requirements. Freshwater, particularly of potable or near-potable quality, is an increasingly scarce resource in several regions of the world. Water recycling and reuse will become important water resource tools in these regions if all demands are to be met. Water resource considerations will become more and more a regional consideration rather than that of a single country. Water scarcity will undoubtedly lead to regional tensions during this century. The development is inevitable to meet the needs of the human being but it should be planned and applied sustainably. For instance, dams are part of development, but it should be undertaken in a way that it does not damage the environment, land, and space of inhabitants. Man-made dams and reservoirs have become an integral part of community infrastructure, primarily for water storage. Dams are associated in certain conditions with navigation, more often with flood control and almost universally with generating electricity. Dams and reservoirs play an important role and are a part of sustainable development. Without the stable water resource and river control provided by reservoirs, the current population and development levels would be unsustainable. Further economic development and population growth will require increased efficiency in using current water resources. Because of the cost and energy requirements for the long-distance transfer of water or desalination, it is inevitable that further water storage in sustainable reservoirs must form a key part of future development planning. Globally, irrigation is one of the oldest and most consuming beneficiaries of water control. Floods provided water, nutrient, and sediment and this is retained sufficiently with any timely rainfall, to grow the annual crop. However, the provision of reservoirs and continuous regulated water supply has allowed more and more crop production per year. Many large dams are elements of multipurpose projects. While very few are primarily constructed for ecological enhancement, the way in which reservoirs are operated or dams provide facilities, such as for fish migration, can have a major influence on nature conservation and even commercial benefit.

There are different functions of reservoir operations, such as the reduction in power generation when a reservoir is lowered to leave space for flood storage or flush out sedimentation. The key function of reservoirs is to store water and to release water and to release it in sufficient quantity and quality when it is required. Dams have always performed a vital role in storing

and controlling water to support civilized communities. The demand for water resources is increasing in parallel to global population growth. Under these circumstances, dams must form part of an integrated water management strategy for a given area. Demands and the potential reuse of water must also be carefully assessed and incorporated in overall planning.

2.4 Built environment and climate change

Air pollution damages building materials such as organic coatings and polymers. It also has damages to stones, brick, concrete and mortar, metals, and glass which are widely used in the construction of buildings. Ordinary construction materials produce large amounts of CO_2 during construction stage and throughout a building's life.

The idea of reasonable structure consolidates and incorporates an assortment of procedures during the plan, development, and activity of the structure projects. The utilization of green structure materials and items addresses one significant technique in the plan of a structure. Green structure materials offer explicit advantages to the structure proprietor and building tenants as the decrease in support/substitution costs over the existence of the structure, energy protection, and improved inhabitant well-being and efficiency where it gives more prominent plan adaptability.

Building components includes walls, roofs, and floors which are all composed of different types of materials like wood, concrete, metal, rubber, soils and rocks, bricks, and blocks. It could be useful if houses are built by using decomposable materials such as wood, gypsum, concrete, and natural stones. Using such materials as construction materials would be beneficial for the health of the environment as they decompose in a natural and completely clean way, which only leaves CO, water, and some sort of natural biomass.

There are opportunities for more sustainable use of concrete in the future such as the use of less easily corroded forms of reinforcement (ceramics, fiberglass, and stainless steel) and the use of binder other than cement such as polymers to meet special strength requirements. Efficient long-life structures should be designed for best use of available materials and to minimize or eliminate the use of raw materials which are in danger or eventual exhaustion. The responsibility of engineers is to plan means of eliminating or reducing adverse impacts and to estimate the extra capital cost or loss in revenue or output that will result.

3. The role of engineers in sustainable development

The engineers play a vital role in achieving sustainable development with success. The engineers should design, construct, and contribute with such a manner that would allow them to reduce adverse environmental and social aspects of developments. The engineers should design and construct in a way that they improve the environmental performance of the buildings and contribute to a high quality of life. The engineers have the responsibility to contribute to a high quality of life, and should help society to move toward to a more sustainable lifestyle (Spence and Mullisan, 1995).

Civil engineers always take an action in dealing with the challenges that the society may face and try their best to overcome all obstacles that a society faced in the past, present, and surely in the future. Sustainable development is another issue that civil engineers must deal with the challenges it might come up with (Bilec et al., 2007).

Sustainable Civil Engineering is more than standard designs, use of ordinary materials, methods, and plans. Sustainable engineering is all about creating a design and construction practices which are cost-effective, healthier, and easy to maintain. This branch of engineering is committed to developing smarter and more innovative designs that can meet the needs of the people but most importantly protect the environment (Seidel, 2007; Braham and Casillas, 2020). Civil engineers play an important role in sustainable planning, design, and construction by choosing the right materials and methods, and they should have the ability to precise location and layout of buildings, pavements and drains, and provide the entire supporting infrastructure for the building environment as part of green engineering (Hill and Bowen, 1997; Bourdeau et al., 1997; Graham, 1997).

When engineers focus on the sustainable project design and buildings they should pay attention to integrate natural systems, urban systems, site characteristics, buildings, energy use, economic considerations, use the land wisely, maintain and build on existing structures and land, design sustainable infrastructure systems, and most importantly minimize the impact of buildings by reducing the building "footprint." The engineers should plan their project and design buildings for resource efficiency by recycling materials, using solar energy, minimizing or recycling waste during construction, and should plan in coordination with the local communities (Braham and Casillas, 2020).

Reasonable arranging, plan, and development are just conceivable using green materials, frameworks, and strategies. Excellent feasible offices can lessen squander age and contamination, upgrade inhabitant solace, well-being and profitability, diminish energy and water use, and neighborhood framework, affect and expand life span and proficiency. Buildings are important for health and quality of life for the occupants. Achieving healthy indoor climate, including the air quality, humidity, visual comfort, and acoustics within the building are pretty important factors which need to be considered as a part of sustainable design (Guy and Kibert, 1997; Bauer et al., 2007; Plank, 2008).

4. Conclusions

Sustainable construction basically is a definition that is given to a construction that is built incorporating with the design site elements such as water conservation, energy efficiency, local resources, material conservation, waste reduction, indoor environment quality, innovation in design, and sustainable design selection. These elements of sustainability are very crucial and dependent on the site selection as well. The main purpose of these elements is to improve the conditions of mankind, and that is by considering the sustainable design elements in construction and maintenance, through the lifetime of the structure. Quite a number of ideas can be used in reducing energy consumption, generating better lighting, water management, and improving the social aspect of a built environment.

Sustainability is essential, as it saves materials and energy resources, protects the environment and air quality, embellishes the view of the buildings, and ensures good life quality for the coming generations.

References

Bauer, M., Mösle, P., Schwars, M., 2007. Green Building: Guideline for Sustainable Architecture.

Bilec, M., Ries, R., Matthews, H.S., 2007. Sustainable development and green design — who is leading the green initiative. J. Prof. Iss. Eng. Ed. Pr. 133, 265–269.

Bourdeau, L., Halliday, S., Huovila, P., Richter, C., 1997. Sustainable development and the future of construction. In: Proceedings Second International Conference on Buildings and the Environment, vol. 2. CSTB and CIB, Paris, pp. 497–504. June.

Braham, A., Casillas, S., 2020. Fundamentals of Sustainability in Civil Engineering. CRC Press, ISBN 9780367420253, p. 272.

Brundtland, G.H., 1987. Report of the World Commission on Environment and Development: Our Common Future. Oslo.

Ding, G.K.C., 2008. Sustainable construction, the role of environmental assessment tools. J. Environ. Manag. 86, 451–464.

Ehnert, I., 2009. Chapter 2 Linking the Idea of Sustainability to Strategic HRM. Springer Science and Business Media LLC.

Graham, P., 1997. Assessing the sustainability of construction and development activity: an overview. In: Proceedings Second International Conference on Buildings and the Environment, vol. 2. CSTB and CIB, Paris, pp. 647–656. June.

Guy, G.B., Kibert, C.J., 1997. Developing indicators of sustainability: US Experience. In: Proceedings Second International Conference on Buildings and the Environment, vol. 2. CSTB and CIB, Paris, pp. 549–556. June.

Hill, R.C., Bowen, P.A., 1997. Sustainable construction: principles and a framework. Construct. Manag. Econ. 15, 223–239.

Liu, Z., Pyplacz, P., Ermakova, M., Konev, P., 2020. Sustainable construction as a competitive advantage. Sustainability 12, 5946.

Plank, R., 2008. The principles of sustainable construction. IES J. Part A Civ. Struct. Eng. 1 (4), 301–307.

Seidel, J., 2007. Sustainability in Civil Engineering. GRIN Publishing, ISBN 9783638804226.

Spence, R., Mulligan, H., 1995. Sustainable development and the construction industry. Habitat Int. 19 (3), 279–292.

Zahibi, H., Habib, F., Mirsaeedie, L., 2012. Sustainability in building and construction: revising definititions and concepts. Int. J. Emerg. Sci. Eng. 2 (4), 570–578.

CHAPTER

16

Rock mass classification for the assessment of blastability in tropically weathered igneous rocks

Ramesh Murlidhar Bhatawdekar[1,2], Edy Tonnizam Mohamad[1], Mohd Firdaus Md Dan[3], Trilok Nath Singh[4], Pranjal Pathak[2], Danial Jahed Armagahni[5]

[1]Geotropik-Centre of Tropical Geoengineering, Department of Civil Engineering, Universiti Teknologi Malaysia, Johor Bahru, Johor, Malaysia; [2]Department of Mining Engineeing, Indian Institute of Technology, Kharagpur, West Bengal, India; [3]Department of Infrastructure and Geomatic, Faculty of Civil and Environmental Engineering, Universiti Tun Hussein Onn Malaysia (UTHM), Johor, Darul, Takzim, Malaysia; [4]Earth Science Department, Indian Institute of Technology Bombay, Mumbai, Maharashtra, India; [5]Department of Urban Planning, Engineering Networks and Systems, Institute of Archi-tecture and Construction, South Ural State University, Chelyabinsk, Russia

1. Introduction

Weathering of rocks in tropical region can reach up to 100 m depth beneath the ground surface especially in lower ground area. Definition of weathering can be stated as a series natural action on rock structure involving physical decomposition and breakdown of chemicals in minerals of rock causing erosion of the rock. Organic fluids with water coupled with temperature act as denuding process. At higher elevation or at sea side wind also causes erosion in the rock. Fracture in open form and various kinds of discontinuities in rocks are observed

due to physical decomposition (Hall et al., 2012). Water can easily filter into the discontinuities present in the rock mass and accelerate the process of decomposing and disintegrating rock resulting into weathered rock (Ehlmann et al., 2008; Velde and Meunier, 2008). Regolith or zone of weathering zone is formed through continued process of dissolving rock minerals into water and decaying of rock (Dethier and Lazarus, 2006; Dethier and Bove, 2011). Chemical decomposition can be defined through series of process oxidizing, saturating with carbon dioxide, addition of water to chemical molecule, or forming homogeneous mixture in

Risk, Reliability and Sustainable Remediation in the Field of Civil and Environmental Engineering
https://doi.org/10.1016/B978-0-323-85698-0.00027-7

liquid form (Lednicka and Kalab, 2012; Freire-Lista et al., 2015). The change in properties of minerals such as feldspar and biotite in granite is an indicator of chemical decomposition (Dearman, 1974; Eggleton and Banfield, 1985).The quartz grain continues to remain intact in granite during weathering process. However, the feldspar is converted into clay and biotite is changed to chlorite (Borrelli et al., 2016). The aggressive reaction of weathering in tropical igneous rock leads to the formation of complex heterogeneous zones that are developed with various characteristics of engineering and geological properties, geometry of discontinuity pattern, and ground water condition. In the construction projects such as tunnels, tunnel boring machine (TBM), cutting of slopes on the highways, foundation, earth excavation, blasting is carried in most of igneous rocks consisting of granite. Construction quality is significantly affected and challenges continue to exist due to weathering nature of granite.

1.1 Statement of the problem

Rock mass classification has been developed by several previous researchers since 1946 to deal with various civil engineering works such as tunnel support, slope stability, excavatability, rippability, foundation structures, and blastability works.

2. Literature review

2.1 Review of rock mass classification system

Strata Variation There is a lot of variation in condition of strata from bench to bench based on rock type. Variation in geo-mechanical characteristics is maximum in overburden benches due to geology and degree of weathering.

Blast hole drilled in competent strata in lower portion and upper portion of the hole has weaker rock with different degree of weathering and having fractures. One boulder is embedded in weaker strata in the upper portion of the hole. Thus, charging of hole with explosives is challenging. Lower portion of the hole would be fragmented due to uniform competent strata. Strong massive boulder embedded in upper portion of the strata which cannot be easily broken. The pocket charge could break massive boulder in weaker strata. Thus, with above pattern of charging, lower portion of competent strata would be fragmented, upper portion would be displaced, and massive boulder would be broken with the pocket charge.

Another possibility is when a layer of weak fractured or soft rock lies between two layers of hard competent rock masses—one on top and one below the soft formation, e.g., a layer of thinly bedded shale or sand between two layers of hard compact sandstones. In such cases, best blast results are obtained when the explosives charges are located in boreholes at the location of hard layers only, separated by a deck of stemming material.

In such layered strata, if one were to charge the entire borehole simply with a single column of explosives, the explosives energy would go first into the weak strata compressing it, decaying in the process, and leaving very little for breaking the hard strata, resulting in poor fragmentation and possible fly rock. Variable strata of waste, friable ore, and massive strong ore need change in charging pattern.

3. Blastability index

3.1 Comparison of assigned parameters for BI by Lilly (1986) versus Ghose (1988)

Table 16.1 shows the specific gravity influence parameter for BI by Lilly (1986) and density t/m^3 by Ghose (1988). Specific gravity values are assumed for BI by Lilly (1986) and SGI

3. Blastability index

257

TABLE 16.1 Comparison of SGI and density parameters (Ghose, 1988; Lilly, 1986).

Parameters	Researcher	Ranges				
Specific gravity influence (SGI)	Lilly (1986)	1.4	1.8	2.2	2.4	2.6
		−15	−5	5	10	15
Density (t/m^3)	Ghose (1988)	<1.6	1.6−2.0	2.0−2.3	2.3−2.5	>2.5
		20	15	12	6	4

TABLE 16.2 Comparison of JPS and DSR parameters (Ghose, 1988; Lilly, 1986).

Parameters	Researcher	Ranges				
Joint plane spacing (JPS)	Lilly (1986)	0.09	0.3	0.5	1.3	2.1
		10	20	20	50	50
Discontinuity spacing ratio (DSR)	Ghose (1988)	<0.2	0.2−0.4	0.4−0.6	0.6−2.0	>2.0
		35	25	20	12	8

parameter is calculated. Lilly (1986) does not specify relationship between BI and powder factor and it is to be determined for each site. Density parameter for BI by Ghose (1988) decreases with the increase in density. There is reverse impact of SGI as compared to density parameter on BI.

Table 16.2 shows comparison of joint plane spacing (JPS) by Lilly (1986) and discontinuity spacing ratio (DSR) by Ghose (1988). JPS values have been assumed and rating has been calculated based on Table 16.3. JPS rating increases with the increase in spacing. There is a sudden change in rating value from 20 to 50. DSR rating

TABLE 16.3 Comparison of JPO parameters by Lilly (1986) and Ghose (1988).

Parameters	Researcher	Ranges				
Joint plane orientation (JPO)	Lilly (1986)	DIF		SNF	DOF	HOR
		40		30	20	10
Joint plane orientation (JPO)	Ghose (1988)	DIF	SAF	SNF	DOF	HOR
		20	15	12	10	8

is calculated based on Table 16.4. DSR rating decreases with the increase in spacing and DSR rating values gradually decreases with increase in spacing.

Table 16.3 shows the comparison of joint plane orientation (JPO) by Lilly (1986) and JPO by Ghose (1988). JPO rating values calculated by both researchers are comparable based on Table 16.3 and Table 16.4, respectively. BI index calculated by Lilly (1986) has multiplication factor of 0.5. BI calculated by Ghose (1988) shows lower BI has higher PF. Values for strike acute angle to face are not calculated by Lilly (1986).

With the advancement of computational processing power, instead of qualitative subjective assessment of various parameters for blastability, computational techniques are used by various researchers. Latham and Lu (1999) stated that the structure of discontinuity in rock mass and in situ rock characteristics have direct influence on the performance of blasting. The intrinsic resistance of the rock mass to blasting is represented by a blastability designation (BD) for developing predictability through model.

258   16. Rock mass classification

TABLE 16.4   Assessment of RMC for excavation.

Author	Assessment criteria for excavation			Common criteria to blastability and gaps with blastability
ISRM (1981)	Factors for RMC (i) Nature of rock, **(ii)** Number of joints, **(iii)** Spacing between different joints, **(iv)** Angle and direction of joints, **(v)** Status and condition of surface for each set of joints, **(vi)** Extent of weathering.			These parameters influence blasting. Degree of weathering is considered. However, there is exponential change in geomechanical properties with respect to degree of weathering and needs to be studied.
Abdullatif and Cruden (1983)	Use of RMR system for judging quality of rock mass quality. Estimation of level of easiness in production and rate of production.			Rock mass quality and RMR have direct relation for blastability of rock.
Scoble and Muftuoglu (1984)	**(i)** The extent of weathering, **(ii)** Strength of in situ rock mass, **(iii)** Spacing between joints **(iv)** Spacing of bedding planes for a layered rock mass.			(i) Every parameter stated has an impact on blastability, (ii) Orientation of joint spacing with respect to face also affect blastability.
Gribble and McLean (1985)	**(i)** Relationship between SHRN and UCS of intact rock, **(ii)** Relationship of above strength to rippability.			(i) Schmidt hammer hardness (rebound number) is useful to determine strength of rock in the field, (ii) Rock discontinuity is not considered.
Romana (1995)	**(i)** Dry density, **(ii)** Point load strength, **(iii)** Geological strength index.			These parameters influence blasting. GSI does not account for angle and orientation of joints with respect to the blasting face.
Pettifer and Fookes (1994)	**(i)** Method of working, **(ii)** Type of excavation equipment, **(iii)** Individual properties of rocks, **(iv)** Rock strength by point load index, **(v)** Discontinue characteristics, **(vi)** Individual rock block size, **(vii)** Detailed chart for excavatability as compared to Franklin et al. (1971) and correlated with rippability.			(i) Method of working and type of excavation equipment not related to blastability, (ii) Rock strength, block size, and characteristics of discontinuity affect blastability, (iii) Explosive strength and different delay patterns will have impact on fragmentation.
Hoek and Karzulovic (2000)	GSI	About 40	60	>60	GSI does not account for angle and orientation of joints with respect to the blasting face.
	Rock mass strength	1 Mpa	10 Mpa	>15 Mpa	
	Methodology for rock excavation	Digging	Ripping	Blasting	

TABLE 16.4 Assessment of RMC for excavation.—cont'd

Author	Assessment criteria for excavation	Common criteria to blastability and gaps with blastability
Tsiambaos and Saroglou (2010)	Assessment of excavatability based on GSI Chart. GSI charts are developed with PLI strength less than or more than 3 MPa. PLI of 3 MPa represents 70 MPa of UCS of intact rock.	Blast design of hard of medium hard rock can be done on the basis of these charts. However, joint orientation needs to be considered in jointed rock mass for blasting.
	GSI excavation chart with $I_{s50} > 3$ MPa or intact rock strength greater than 70 MPa.	**(i)** For hard or very rock, guidelines are suitable for blasting, **(ii)** Joint orientation is important in blasting which is not considered here.
Kirsten (1982, 1988) Excavatability of natural materials	**(i)** Strength of parent material, **(ii)** In situ density, **(iii)** Degree of weathering, **(iv)** Seismic velocity, **(v)** Block size, **(vi)** Shape of excavation relative to excavating equipment, **(vii)** Block shape, **(viii)** Block orientation, **(ix)** Joint roughness, **(x)** Joint gouge, **(xi)** Joint separation.	**(i)** Type of equipment and shape of excavation are related to excavatability and not blasting, **(ii)** Joint gouge and separation will affect boulder formation after blasting, **(iii)** Other parameters are related to blasting.

Azimi et al. (2010) developed prediction model for BD for rock mass by applying fuzzy sets. For blasting operation, blastability is an important sensitive characteristic of the rock mass. The main objectives of blast design are balancing between optimization of powder factor, good fragmentation, and minimization of environmental effect which depend upon blastability of rock mass. Azimi et al. (2010) suggested 12 input rock mass characteristics which are connected with blastability. New methodology of "effective rules" is developed by applying Mamdani fuzzy inference system to BD classification systems. The fuzzy set model showed consistent results as compared to conventional methods.

4. Comparative function based RMC for blastability

Based on various research findings, various RMCs for different functions such as exaction, rippability, and slope stability are discussed. Comparative study with respect to blastability is also discussed in subsequent sections.

4.1 Judgment of RMC for excavation

Table 16.4 is a review of RMC developed for excavation explaining common criteria to blasting and gaps with blastability.

The system developed by ISRM (1981) used the following parameters for rock mass classification: (i) Nature of rock, (ii) Number of joints, (iii) Spacing between different joints, (iv) Angle and direction of joints, (v) Status and condition of surface for each set of joints, and (vi) Extent of weathering. Although these parameters influence blasting, there is exponential change in geomechanical properties with respect to degree of weathering and this needs to be studied. The excavation system developed by Abdullatif and Cruden (1983) for level of difficulty for excavation vis-a-vis productivity based on RMR and quality of rock mass and blastability depends upon these two parameters. Scoble and Muftuoglu (1984) system used the following parameters (i) extent of weathering, (ii) strength of in situ rock mass, (iii) spacing between joints and (iv) spacing of bedding planes for a layered rock mass. The foregoing four parameters stated in the excavation system of Scoble and Muftuoglu (1984) affect blastability. Also, the orientation of joint spacing with respect to face also affects blastability. Gribble and McLean (1985) developed an excavation system by exploiting correlation between (SHRN) and (UCS) of in situ rock as well as correlation between UCS and rippability. SHRN is a useful tool to determine the strength of rock at the site of investigation. However, rock discontinuity was not considered during evaluation. The excavation system developed by Romana (1995) used the following parameters: (i) dry density, (ii) point load strength, and (iii) geological strength index (GSI). These parameters influence blasting although GSI does not consider orientation of joints and angle respect to the blasting face. Pettifer and Fookes (1994) developed an excavation system considering, (i) method of working, (ii) type of excavation equipment, (iii) individual properties of rocks, (iv) rock strength by point load index (PLI), (v) discontinue characteristics, (vi) individual rock block size, and (vii) detailed chart for excavatability as compared to Franklin et al. (1971) and correlated with rippability. There is a difference between assessment of excavation and blastability. Horse power of equipment and

technique of excavation are important for excavation and are not directly related to blastability. However, strength of rock mass, size of rock block, and phenomenon of discontinuity have an impact on blastability, while fragmentation depends upon type and strength of explosives, variation in delay patterns, and type of delays utilized. The system developed by Hoek and Karzulovic (2000) considered GSI, rock mass strength, and methodology for rock excavation. However, joint orientation is not the evaluation criteria, while estimation of GSI value. Tsiambaos and Saroglou (2010) used GSI charts for rock masses excavatability assessment. The charts used are GSI excavation chart based on PLI strength lower than 3 MPa and (PLI) strength greater than 3 MPa. As per many researchers (PLI) strength 3 MPa is equivalent to 70 MPa. This bench mark rock strength is suitable for design of blasting for medium and hard rocks. Blasting performance depends upon angle and orientation joints with respect to the blasting face. In estimation of GSI of blasting face, this aspect of joint orientation is not considered. Hence, along with GSI, other parameters need to be considered in jointed rock for blastability purpose. Kirsten (1982, 1988) developed an excavation system for natural materials considering (i) strength of parent material, (ii) intact rock density, (iii) extent of weathering, (iv) Pwave velocity, (v) Size of block, (vi) placement of excavator or loading equipment and particular shape and direction of excavation, (vii) optimum shape of block, (viii) orientation and direction of block with blasting face, (ix) whether joints are rough or smooth, (x) gouging of joints, and (xi) separation between joints. The type of equipment and shape of excavation used in their system are related to excavatability and not blasting. Other parameters are related to blasting and joint gouge and separation will affect boulder formation after blasting.

Table 16.5 shows various criteria used by researchers to evaluate rippability, its similarities, as well as differences with blastability. In

TABLE 16.5 Rippability rock mass classification system.

Author	Criteria for evaluation of rippability		Knowledge gap/Similarity with blastability
Caterpillar Handbook	P-wave velocity for different rock types and different models whether rock is rippable or not rippable P-wave velocity increases with size or HP of ripper dozer model from D6 to D11		Degree of weathering and other material properties are not considered. This provides broad guidelines whether particular site ripper dozer is suitable or not
Karpuz (1990) and Basarir and Karpuz (2004)	A rippability classification system for Coal Measures and marls: • The seismic P-wave velocity • Uniaxial compressive strength or the point load index • The Schmidt hammer hardness • The average discontinuity spacing		• Factors considered for rippability are also important for blasting • Discontinuity orientation with respect to face is important for blasting

Singh et al. (1987)	A rippability index for Coal Measures
	 • Rock type
	 • Rock fabric
	 • Intact rock strength
	 • Rock abrasiveness
	   * Degree of weathering
	   * Structural features of rock mass
	   * Seismic wave velocity
	   * Rock intrinsic and other properties
			Rippability is comparable with blasting in soft, weathered, jointed rock mass. Rock abrasiveness will have little impact on blasting. Hard and very hard rock mass

Church (1981), Caterpillar (2001)	• P-wave seismic velocity in a wide variety of rocks provides ripper performance charts		P-wave velocity also affects blasting. There are other parameters, joint orientation and filling material, that can affect blasting
Amin et al. (2009)	Material properties for classification of rippability		Material properties are important for blasting
	Property	Resistance against	UCS, PLI, surface hardness, and tensile strength are important properties for blasting. P-wave velocity is not essential. Due to discontinuities, P-wave velocity may vary in tropically weathered rock
	UCS	Fracturing and loading	
	PLI	Fracturing and loading	
	SDI	slaking, degree of cement bonding	
	Surface hardness	Impact and abrasion	
	Tensile strength	Fracturing, degree of cement bonding	
	P-wave velocity	Denseness, pulverization	

Continued

262   16. Rock mass classification

TABLE 16.5   Rippability rock mass classification system.—cont'd

Author	Criteria for evaluation of rippability	Knowledge gap/Similarity with blastability
Basarir (2007)	A rippability classification system for marl based on fuzzy logic	Rock mass classification for blasting in tropically weathered rock needs to be developed
	Inputs − P velocity, PLI, UCS, SHR, and average discontinuity spacing to develop rippability classification system for each parameter	PLI, UCS, SHR, and discontinuity spacing
Basarir et al. (2008)	Rippability classification system for marl/coal bearing rocks based on specific energy consumption	Rock mass classification for blasting in tropically weathered rock needs to be developed
	Each type is classified into P velocity, graphical, grading, direct, and indirect method for rippability to estimate final class for rock mass classification based on specific energy	Specific energy is important in blasting for breaking rock. Powder factor, charge/m
Rashidi et al. (2014)	Bulldozer's productivity using linear mixed models	These are technical parameters related to dozer and material properties or
	Inputs type and blade capacity, operating condition, ground grade%, dozing distance, and operating time	
Mohamad et al. (2017a)	Predicting ripper productivity with ANN-based GA algorithm	ANN-based GA algorithms are used for predicting blast performance fragmentation as output
	PLI, sonic velocity, joint spacing, and weathering zone are inputs for predicting productivity of ripper	PLI or UCS or TS/E, joint spacing or block size, blast parameters, and powder factor are utilized for predicting blast fragmentation. Degree of weathering has not been considered for blast fragmentation
Mohamad et al. (2017b)	Predicting ripper productivity with laboratory test results through regression	Regression and laboratory test results are used for prediction of blast fragmentation
	Laboratory tests consisted of PLI, BTS, SDI, and P-wave velocity as input	PLI and BTS are important parameters for blast fragmentation. SDI and P-wave velocity is not directly correlated with blast fragmentation. Degree of weathering not included as input.

Caterpillar's handbook, P-wave velocity was used to ascertain whether rock is rippable or not. The P-wave velocity increases with size or HP of ripper dozer model from D6 to D11. Although the criteria provide broad guidelines whether particular site ripper dozer is suitable or not, the degree of weathering and other material properties were not considered in it. Karpuz

(1990), Basarir and Karpuz (2004) provided a rippability classification system for coal measures and marls. Their system used properties like seismic P-wave velocity, uniaxial compressive strength or the PLI, Schmidt hammer hardness, and the average discontinuity spacing. The factors which were considered for rippability as well as discontinuity orientation with respect to face are important for blasting. Singh et al. (1987) provided a rippability index for coal. Their index used rock type, rock fabric, intact rock strength, rock abrasiveness, degree of weathering, structural features of rock mass, seismic wave velocity and rock intrinsic, and other properties to evaluate rippability. The rippability is comparable with blasting in soft, weathered, and jointed rock mass and rock abrasiveness has little impact on blasting of hard and very hard rock mass, Church (1981) and Caterpillar (2001) provided a ripper performance charts by measuring P-wave seismic velocity in a wide variety of rocks. The P-wave velocity affects blasting, while other parameters like joint orientation and filling material can affect blasting. Amin et al. (2009) used material properties like UCS, PLI, SDI, surface hardness, tensile strength, and P-wave velocity for classification of rippability. While UCS, PLI, surface hardness, and tensile strength are important properties for blasting, P-wave velocity is not essential. This is because P-wave velocity may vary in tropically weathered rock due to discontinuities. Basarir (2007) developed a rippability classification system for marl based on fuzzy logic. The system takes P velocity, PLI, UCS, SHR, and average discontinuity spacing as inputs to develop rippability classification system for each parameter. For blasting in tropically weathered rock, rock mass classification needs to be developed. Basarir et al. (2008) developed another rippability classification system for marl/coal bearing rocks based on specific energy consumption. In the system, each type is classified into P velocity, graphical, grading, direct, and indirect method for rippability to estimate final class for rock mass classification based on specific energy. In tropically weathered rock needs, rock mass classification for blasting needs to be developed, while specific energy is important in blasting for breaking rocks. Rashidi et al. (2014) developed a rippability classification system to ascertain bulldozer's productivity using linear mixed models. The system took P velocity, PLI, UCS, SHR, and average discontinuity spacing as inputs.

Mohamad et al. (2017a) used ANN-based GA algorithm which used PLI, sonic velocity, joint spacing, and weathering zone as inputs for predicting ripper productivity. While PLI or UCS or TS/E, joint spacing or block size, blast parameters, powder factor can be utilized for predicting blast fragmentation, degree of weathering has not been considered. Mohamad et al. (2017b) also used laboratory test results through regression for predicting ripper productivity. The laboratory tests conducted used PLI, BTS, SDI, and P-wave velocity as inputs. The regression and laboratory test results can also be used for prediction of blast fragmentation. Also, PLI and BTS are important parameters for blast fragmentation, while SDI and P-wave velocity are not directly correlated with blast fragmentation.

5. Assessment of slope stability with rock mass classification

Table 16.6 shows the slope stability assessment versus common criteria to blastability and gaps with blastability.

Romana (1985) used the "Slope Mass Rating" (SMR) which is calculated from RMR by summation of a factorial adjustment factors built upon the relative orientation of joints and slope. The second adjustment factor was built upon the excavation method.

$$SMR = RMR_B + (F1 \times F2 \times F3) + F4.$$

where

F1 builds upon parallelism between slope face strike and joints. Its value varies from 1.00 to 0.15.
F2 refers to joint dip angle in the planar mode of failure. Its value changes from 1.00 to 0.15.
F3 follows the relationship of joints dips into the slope.
F4 empirical adjustment factor for excavating method.

16. Rock mass classification

TABLE 16.6 Slope stability assessment versus common criteria to blastability and gaps with blastability.

Author	Criteria for slope assessment	Common criteria to blastability and gaps with blastability
Romana (1985)	The "Slope Mass Rating" (SMR) is calculated from RMR by summation of a factorial adjustment factors built upon the relative orientation of joints and slope. Second adjustment factor is built upon the excavation method. $SMR = RMR_B + (F1 \times F2 \times F3) + F4$.	Joints at blasting face are considered for blastability. Blasting depends upon RMR, while slope depends upon SMR. Excavation method is not considered for blastability or for blasting.
	F1 is built upon parallelism between slope face strike and joints. Its value varies from 1.00 to 0.15.	Angle of joints with respect to the blasting face is not considered for blastability.
	F2 refers to joint dip angle in the planar mode of failure. Its value changes from 1.00 to 0.15.	Breaking of rock in any weak plane is important in blasting. Planer mode of failure is not considered in blasting.
	F3 follows the relationship of joints dips into the slope.	Joint dips with respect to blasting are divided into four types: SNF, SAF DOF and HOR.
	F4 empirical adjustment factor for excavating method.	Blasting or blastability does not depend upon excavating method.
Romana (1993)	SMR classes are described into five categories based on stability of slope: completely unstable, unstable, partially stable, stable, and very stable.	Rock mass is classified based on hardness/strength of rock, density, degree of weathering, etc., which are important properties for blasting.
Romana (1995)	SMR values according to probability of failure. Failures are divided into 4 types: planer, wedge, toppling, and major. Possible, major, minor, none. Type of measures for slope stability are as follows: excavation, drainage, concrete, reinforcement, protection, and support not required.	Blast performance depends upon properties of rock mass, explosives, and blast design. Various methods exist for prediction of blast performance: fragmentation, flyrock, ground vibration, and air over pressure.
Pantelidis (2009)	Assessment of rock slope stability through rock mass classification systems.	Rock mass classification for tropically weathered rock needs to be developed.
	Each possible failure type (planar, wedge and toppling, differential erosion, nonstructurally controlled).	Failure of rock in blasting depends upon rock mass properties, discontinuities, and strength of rock mass.
	Each failure examined separately with failure factors: dip and orientation, condition of discontinuities for the stability of slope.	Any rock failure is positive impact on blast fragmentation. Flyrock and AOp may have adverse effect if rock fails too early as compared to blast design.
	Ground water condition, climatic condition examined which is triggering factor for slope stability.	Management of blasting operation can have impact due to climatic condition such as fog or rain. There are no triggering factors for blasting.

TABLE 16.6 Slope stability assessment versus common criteria to blastability and gaps with blastability.—cont'd

Author	Criteria for slope assessment	Common criteria to blastability and gaps with blastability
	Earthquake can have triggering effect on slope stability.	There is no impact due to earthquake due to blasting.
Riquelme et al. (2016)	Evaluation of rock stability on the sustainable basis through systematic approach.	Evaluation of blast performance (fragmentation, flyrock, PPV, and AOp) through systematic approach.
	Ascertain RMR of target slope	Classify tropical weathered RMC
	Find out SMR values by applying adjustment factor of Romana (1985) to RMR.	Find out BI, PF, and blast design parameters for target blast performance.
	Determine peak friction angle of discontinuity with tilt test.	Determine weathering index through site testing.
	Classify rock slope stability into various classes of stability.	Classify predicted blast performance results.
Hack (2011)	Slope stability probability classification (SSPC). Three-step classification based on properties of materials and discontinuities with specific parameters* (excavation method and degree of weathering)	For blasting, instead of probability classification, prediction methods exist. For fly rock, probabilistic approach is suggested by Raina et al. (2015)
	Exposure rock mass (ERM) − Exposure-specific parameters* are removed	Degree of weathered tropical rock mass determination is important
	Reference rock mass (RRM) − Site-specific parameters* are removed	Comparing properties of weathered rock with fresh rock mass will support blast evaluation.
	Slope rock mass (SRM) and effect of slope geometry (orientation and height) are determined to assess slope stability.	Face orientation and geometry has an impact on blastability
Basahel and Mitri (2017)	Assessment of slope stability through RMC.	Rock mass classification for tropically weathered rock for blasting needs to be developed.
	Rock slope conditions with known input parameters (slope failures and material properties) are compared with original SMR, Chinese SMR, continuous SMR, graphical SMR, and hazard index to get results of slope stability that are compared with each rock mass classification system.	Degree of weathering and their correlation with blast performance parameters need to be developed.

In this system, although joints at blasting face are considered for blastability, blasting depends upon RMR, while slope depends upon SMR. In addition to this, excavation method, angle of joints with respect to the blasting face, and planer mode of failure are not considered for blastability or blasting.

Romana (1993) gave another system for slope assessment where SMR classes were classified into following five categories based on stability of slope: completely unstable, unstable, partially stable, stable, and very stable. The properties like hardness/strength of rock, density, degree of weathering, etc., based on which the rock mass was classified are also important properties for blasting. The system used by Romana (1995) ascertained SMR values according to the probability of failure. The failures were divided into four types: planer, wedge, toppling, and major. The blast performance depends upon the properties of rock mass, explosives, and blast design. The slope stability assessment system given by Pantelidis (2009) assesses rock slope stability through rock mass classification systems. It includes possible failure type (planar, wedge and toppling, differential erosion, nonstructurally controlled) with each failure examined separately with failure factors—dip and orientation, and condition of discontinuities for the stability of slope. It was seen that ground water condition, climatic condition, and earthquake are triggering factors for slope stability. In case of blasting, failure of rock depends upon rock mass properties, discontinuities and strength of rock mass. Any rock failure is positive impact on blast fragmentation. Flyrock and AOp may have adverse effect if rock fails too early as compared to blast design. Any rock failure has positive impact on blast fragmentation. Flyrock and AOp may have adverse effect if rock fails too early as compared to blast design. Although there are no triggering factors for blasting, management of blasting operation can have an impact due to climatic condition such as fog or rain. The system given by Riquelme et al. (2016) evaluates rock stability on sustainable basis through systematic approach. In this system, RMR of target slope is first ascertained. The SMR values are then found by applying adjustment factor of Romana (1985) to RMR. The peak friction angle of discontinuity is then determined with tilt test. This is followed by classification of rock slope stability into various classes of stability. Similar to rock stability, blast performance (fragmentation, flyrock, PPV, and AOp) can also be evaluated through systematic approach. For this, tropical weathered RMC is first classified. Then, blast design parameters such as BI and PF are found for target blast performance. This is followed by determination of weathering index through site testing. The predicted blast performance results can then be classified. Hack (2011) gave slope stability probability classification (SSPC) system which is a three-step classification based on properties of materials and discontinuities with specific parameters (excavation method and degree of weathering). In the system, Exposure rock mass (ERM)—Exposure-specific parameters and reference rock mass (RRM)—Site-specific parameters are removed. Only, slope rock mass (SRM) and effect of slope geometry (orientation and height) are determined to assess slope stability. For blasting, instead of probability classification, prediction methods exist. For fly rock, probabilistic approach is suggested by Raina et al. (2015). Also, degree of weathered tropical rock mass determination is important and comparing properties of weathered rock with fresh rock mass will support blast evaluation. Basahel and Mitri (2017) assessed slope stability through RMC. In the system, rock slope conditions with known input parameters (slope failures and material properties) are compared with original SMR, Chinese SMR, continuous SMR, graphical SMR, and hazard index to get results of slope stability that are compared with each rock mass classification system. In blasting, degree of weathering and their correlation with blast performance parameters and the rock mass classification for tropically weathered rock need to be developed.

6. Development of weathering classification systems for tropically weathered igneous and andesite rocks

6.1 Weathering classification scheme

Weathered igneous rocks in tropical region are mostly formed in heterogeneous weathered profile with very complicated structures. It consists of various grades of weathering with formation of isolated boulders in various shapes and sizes in between fresh rock and residual soil zones. Since 1957, a few weathering classification systems have been developed by various researchers for civil engineering purposes. In most weathering classification systems that have been established, boulder was used as one of the main parameter in the weathering classification scheme for either soft granite rock (Martin and Henchert, 1986; Ruxton and Berry, 1957) or andesite (Patino et al., 2003). 'Small scale' refers to rock material. 'Classification of rock material' refers to alteration of individual minerals, loss of grain bonding, enlargement of micro and macrostructure. On the other hand, the 'large scale' refers to rock mass zones which normally consist of mixtures of various grades of weathering that possess distinctive characteristics.

Weathered igneous mass of tropical region is commonly formed in several weathering profiles with obvious difference in physicals and engineering characteristics. The weathered igneous mass has been qualitatively classified into 4—6 weathering zones due to the differences by most of the previous researchers (Moye, 1955; Ruxton and Berry, 1957; Little, 1969; Dearman, 1974, 1995; Komoo, 1989; Tuğrul and Gürpinar, 1997; Tsidzi, 1997; Fookes, 1997; Alavi et al., 2014a,b, 2016). Based on the study done by Patino et al. (2003), Arıkan et al. (2007), and Oguchi (2001), the physical characteristics of granite and andesite are quite similar with the presence of boulders in these zones (Zones 3—5) as reported by Ruxton and Berry (1957), Tsidzi (1997) and Md Dan et al. (2015). Both types of boulders surrounded by the formation of concentric weathered sheets are called as rindlets or onion skin.

Table 16.7 shows the weathering classification scheme of igneous rocks as proposed by Md Dan (2016) for engineering design purposes which consist of weathering zone characteristics, joint characteristics, and physical characteristics of boulder that are formed in moderately (Grade III) to completely (Grade V) weathered zones. According to Ceryan (1999), the first step to propose a classification is to determine the related parameters of the classification purposes. Therefore, several physical and engineering properties were used to establish this classification. Furthermore, weathering classification based on physical and engineering properties is more practical and easy to classify and apply in engineering design than those which are based on chemistry and petrography (Momeni et al., 2015).

6.2 Engineering properties of weathered igneous rocks

Weathering processes are very complex and mostly produce unpredictable changes in engineering properties. Level complexity that is formed from the weathering profile depends on some factors especially original geological characteristics such as lithology and joint pattern, topography, climatic influences, and fluctuations in groundwater (Martin & Henchert, 1986). Therefore, weathering classification is vital to group complex rock material of various rock types into similar engineering properties. Table 16.9 shows the engineering properties of granite and andesite from fresh (Grade I) to completely weathered (Grade V) that were classified based on the previous studies Table 16.10 shows porosities of granite and andesite in various weathering grades based on previous studies. On the other hand,

268   16. Rock mass classification

TABLE 16.7   Classification of igneous rock in tropical region for engineering design purposes (Md Dan, 2016).

Weathering profile	MWZ	MWG	Dp (m)	Classification of weathered igneous Mass	
				Weathering zone characteristics	
				Discoloration	Friability
	RS	VI	0.1—4.5	Completely soil. Medium to dark brown colors.	Like soil
	CW	V	4.0—11.5	Completely discoloration along relict joint walls. Light/dark brown to dark orange/reddish colors	Very friable. It is breakable with hand
	HW	IV	10.0—16.00	Highly to completely discoloration along relict joint walls yellowish pale/light orange to pinkish/light brown colors	Friable it is breakable with slight hammer blow
	MW	III	12.0—18.5	Moderate to highly discoloration along relict joints walls yellowish pale/light orange to pinkish/light brown colors	Intact but breakable when knocked with hammer
	SW	II	14.0—21.0	Slight discoloration along joint walls light gray to greyish colors	Intact give clinking sound when knocked with hammer; it is breakable along joints walls
	F	I	>20.00	No discoloration along joint walls. Medium gray to light gray colors	Very intact. Give clinking sound when knocked By hammer and hardly breakable

Table 16.11 shows uniaxial compressive strength of granite and andesite in various weathering grades based on previous studies.

7. Site study of tropically weathered igneous rocks

7.1 Andesite quarry, Java, Indonesia

Andesite quarry under this study is situated in Java, Indonesia which supplies construction aggregates to Jakarta and around for developing construction aggregates.

7.1.1 Geology

Based on the local geological mapping, it can be seen that in the andesite intrusion investigation is the result of the spread evenly across the body of the hill. Rocks are found around the body of the intrusion are alluvial and weathered andesite.

Alluvial comprises of highly and completely weathered andesite rock. Surface soil or top soil is found in layered form. Debris of weathered rock is found in different locations valley portion, on the banks and inside river bed, paddy fields. It is also found in the hilly area that has experienced very high weathering.

Classification of weathered igneous Mass					
Weathering zone characteristics	**Joint characteristics**				**Boulder characteristics**
Texture	Joint set	Joint spacing	Inclination	Shape	Feature
Completely destroyed	Not available	Not available	Not available	Not available	Not found
Original texture of parent rock is still preserved	Relict joints are filled completely with clay minerals			Rounded to well-rounded shapes	40% of relict boulders perhaps presented. 60% of boulders are surrounded by rindlets with whitish gray/yellowish pale to light/dark orange with whitish or yellowish spots
	Mode: 4 nos. Mean: 4.67 nos. Range: 3–6 sets	Mean: 1.38 ± 0.59 m. Range: 0.2–2.80m	V:78.5° D: 48.50° H: 14.0°		
Original texture of parent rock is preserved	Relict joints apertures are filled with soil debris or/and clay minerals			Subrounded to rounded shapes	10% of relict boulders perhaps presented almost 90% of boulders surrounded by rindlets with whitish gray to yellowish pale/light orange colors with black spots
	Mode: 3 nos. Mean: 4.25 nos. Range: 3–7 sets	Mean: 1.25 ± 0.57 m. Range: 0.15–2.80 m	V:77.0° D: 52.5° H: 17.0°		
Original texture of parent rock is preserved and intact	Joints aperture is filled with or without friable materials/soil debris			Angular to subangular shapes	100% of boulders are formed in block shapes surrounded by discontinue within some interlocking systems
	Mode: 4 nos. Mean: 4.75 nos. Range: 3–4 sets	Mean: 1.07 ± 0.56 m. Range: 0.10–2.85 m	V:74.0° D: 46.0° H: 15.0°		
Texture of parent rock is strongly preserved and very intact	Iron stains are developed along joint wall and aperture			Not available	Rock zone consists of slight to moderate discontinuities
	Mode: 3 nos. Mean: 3.5 nos. Range: 3–4 sets	Mean: 1.27 ± 0.54 m. Range: 0.10–2.95 m	V:77.5° D: 50° H: 18.5°		
Rock texture is strongly preserved and very intact	No visible sign of weathering along joints walls.			Not available	Rock zone consists of no or slightly discontinuities
	Mode: 4 nos. Mean: 4 nos. Range: 3–4 sets	Mean: 1.77 ± 0.55 m. Range: 0.30–3.10 m	V:77.5° D: 50° H: 18.5°		

D, dipping joint; D_p, depth (m); H, horizontal joint; MWG, Mass weathering grade; MWZ, Mass weathering zone; V, vertical joint.

Andesite in the investigation includes the igneous group, moderate to fine grained. The upper part of the body of the intrusion near the soil surface weathering experienced until that turned into soil nature. Weathering andesite is found in the southern part with a thickness of more than 10 m, and is in the northwestern part of the thickness of the weathering of andesite only about 1–2 m.

Andesite in the study area is highly jointly, so it is a rock block separated by a hefty field (joint). These blocks of rock are sized between 0.1 and 2 m and the zone of gravel to gravel-sized stocky.

TABLE 16.8 Basic properties of granite and andesite based on previous studies.

| | | Dry density of weathered granite (g/cm^3) | | | | | | | | | |
| | | Grade I | | Grade II | | Grade III | | Grade IV | | Grade V | |
Reference	Type of rock	Max	Min	Max	Min	Max	Min	Max	Min	Max	Min
Irfan and Dearman (1978a,b)	Granite	2.61	2.61	2.58	2.56	2.52	2.51	2.52	2.51	2.33	2.00
Irfan and Powell (1985)	Granite	2.85	2.70	2.76	2.65	2.75	2.60	2.68	2.20	2.20	—
Gupta, and Rao (2001)	Granite	2.75	—	2.69	—	2.54	—	—	—	1.97	—
Arel and Tugrul (2001)	Granite	2.67	2.56	2.61	2.55	2.57	2.43	2.47	2.32	2.42	2.05
Begonha and Braga (2002)	Granite	2.65	2.62	2.62	2.60	2.60	2.42	2.41	2.34	—	—
Thuro and Scholz (2004)	Granite	2.64	2.62	2.62	2.58	2.58	2.54	2.56	2.50	2.52	1.80
Aydin and Basu (2005)	Granite	2.69	2.57	2.59	2.36	2.52	2.37	2.47	2.13	—	—
Ceryan et al. (2008)	Granite	2.71	2.62	2.67	2.60	2.63	2.52	2.60	2.50	2.35	2.17
Basu et al. (2009)	Granite	2.67	2.65	2.65	2.63	2.58	2.46	2.11	—	—	—
Heidari et al. (2013)	Granite	2.66	2.63	2.73	2.54	2.51	2.31	2.53	2.17	2.31	2.08
Mert (2014)	Granite	2.65	2.51	2.60	2.49	2.54	2.35	2.36	2.18	2.28	2.04
Md Dan et al. (2016)	Granite	2.64	2.56	2.61	2.51	2.56	2.31	2.32	1.78	1.92	1.60
Arikan et al. (2007)	Andesites	2.72	2.39	2.66	2.27	2.55	2.10	2.46	2.01	—	—
Koca and Kıncal (2016)	Andesites	—	—	2.53	2.39	2.40	2.24	2.17	2.05	—	—

TABLE 16.9 Dry unit weight of granite and andesite in various weathering grades based on previous studies.

| | | Dry unit weight of weathered granite (k N/m^3) | | | | | | | | | |
| | | Grade I | | Grade II | | Grade III | | Grade IV | | Grade V | |
Reference	Type of rock	Max	Min	Max	Min	Max	Min	Max	Min	Max	Min
Irfan and Dearman (1978a,b)	Granite	25.60	25.60	25.31	25.11	24.72	24.62	24.72	24.62	22.86	19.62
Irfan and Powell (1985)	Granite	27.96	26.49	27.08	26.00	26.98	25.51	26.29	21.58	21.58	—
Gupta and Rao (2001)	Granite	26.98	—	26.39	—	24.92	—	—	—	19.33	—
Arel and Tugrul (2001)	Granite	26.19	25.11	25.60	25.02	25.21	23.84	24.23	22.76	23.74	20.11
Begonha and Braga (2002)	Granite	26.00	25.70	25.70	25.51	25.51	23.74	23.64	22.96	—	—
Thuro and Scholz (2004)	Granite	25.90	25.70	25.70	25.31	25.31	24.92	25.11	24.53	24.72	17.66
Aydin and Basu (2005)	Granite	26.39	25.21	25.41	23.15	24.72	23.25	24.23	20.90	—	—
Ceryan et al. (2008)	Granite	26.59	25.70	26.19	25.51	25.80	24.72	25.55	24.54	23.04	21.24
Basu et al. (2009)	Granite	26.19	25.00	26.00	25.80	25.31	24.13	20.70	—	—	—
Heidari et al. (2013)	Granite	26.09	25.80	26.78	24.92	24.62	22.66	24.82	21.29	22.66	20.40
Mert (2014)	Granite	26.00	24.62	25.51	24.43	24.92	23.05	23.15	21.39	22.37	20.01
Md Dan et al. (2016)	Granite	25.90	25.11	25.60	24.62	25.11	22.66	22.76	17.46	18.84	15.70
Arikan et al. (2007)	Andesites	26.68	23.45	26.09	22.27	25.02	20.60	24.13	19.72	—	—
Koca and Kıncal (2016)	Andesites		—	24.82	23.45	23.54	21.97	21.29	20.11	—	—

TABLE 16.10 Porosities of granite and andesite in various weathering grades based on previous studies.

		Porosity, n (%)									
		Grade I		Grade II		Grade III		Grade IV		Grade V	
Reference	Type of rock	Max	Min	Max	Min	Max	Min	Max	Min	Max	Min
Irfan and Dearman (1978a,b)	Granite	1.56	–	2.45	–	4.64	–	5.44	–	22.80	–
Gupta and Rao (2001)	Granite	–	0.61	2.09	–	7.89	–	–	–	24.41	–
Arel and Tugrul (2001)	Granite	5.78	5.36	7.36	5.60	11.56	6.20	14.68	6.30	16.57	12.14
Begonha and Braga (2002)	Granite	1.14	0.72	2.07	1.59	3.94	2.41	–	–	–	–
Thuro and Scholz (2004)	Granite	1.00	0.50	2.50	1.00	4.50	2.50	6.00	3.50	13.00	5.00
Aydin and Basu (2005)	Granite	5.78	5.36	7.36	5.60	11.56	6.20	14.68	6.30	16.57	12.14
Ceryan et al. (2008)	Granite	7.33	1.31	18.23	7.54	19.21	7.35	25.38	6.32	–	–
Basu et al. (2009)	Granite	2.46	1.38	2.91	1.73	5.73	3.13	6.43	3.23	19.02	11.62
Heidari et al. (2013)	Granite	1.82	0.91	2.76	2.49	5.05	4.41	11.14	5.48	18.61	14.89
Mert (2014)	Granite	0.77	0.73	1.08	1.02	5.80	3.49	15.42	12.30	27.69	14.80
Md Dan et al. (2016)	Granite	1.63	0.10	3.86	0.27	12.18	3.35	18.74	6.08	25.39	10.92
Arikan et al. (2007)	Andesites	12.30	5.10	14.20	8.90	18.70	9.30	25.00	13.20	–	–
Koca and Kıncal (2016)	Andesites	–	–	16.00	5.00	32.00	5.00	36.00	3.00	–	–

Exploration at andesite quarry was carried out. Interpretation of body geometry andesite intrusions in the investigation based on the results of geological mapping and exploration drilling is relatively straight-shaped dome. Based on the interpretation, then block model was made with dimensions intrusion body block to the $X = 10$, $Y = 10$, and $Z = 5$ m.

7.1.2 Structural geology of andesite quarry in Java, Indonesia

Field investigation showed that a crack in the form of fracture was found between the blocks of rock. As discontinuity is very close by and tight, crushed massive rock was observed. Besides, occasionally vertical long discontinuities were observed.

Various measurements of geological structures were done during geological mapping. The data were used for RMC and slope stability analysis. Geological structural data were useful for finding the major direction competent rock and discontinuous strata.

Detail geotechnical investigation was carried out from toe to crest in different benches of andesite quarry at an elevation of 100–120 m above mean sea level. Various rock mass conditions were classified on the basis of the following factors:

- Nature of rock and UCS
- RQD
- Joint orientation, joint spacing, and massive joints
- The ground water table during dry and wet season

Observations and measurements of the detailed structure as listed in the table field, and further data will be analyzed by the system of rock mass rating (RMR) to determine the

TABLE 16.11 Uniaxial compressive strength of granite and andesite in various weathering grades based on previous studies.

		Uniaxial compressive strength, σ_c (MP$_a$)									
		Grade I		Grade II		Grade III		Grade IV		Grade V	
Reference	Type of rock	Max	Min	Max	Min	Max	min	Max	Min	Max	Min
Irfan and Dearman (1978a,b)	Granite	288.00	283.00	251.00	219.00	197.00	187.00	123.00	58.00	11.00	—
Baynes and Dearman (1978)	Granite	262.00	232.00	163.00	105.00	46.00	—	26.00	—	5.00	—
Irfan and Powell (1985)	Granite	275.00	175.00	225.00	125.00	150.00	5.00	15.00	2.50	2.50	—
Gupta and Rao (2001)	Granite	—	132.80	—	102.70	—	53.01	—	—	2.54	—
Arel and Tugrul (2001)	Granite	151.30	101.20	129.50	66.60	79.40	14.60	21.30	11.10	3.50	1.20
Begonha and Braga (2002)	Granite	157.00	130.60	132.70	96.60	135.20	60.00	29.40	20.20	—	—
Thuro and Scholz (2004)	Granite	—	—	250.00	120.00	120.00	50.00	50.00	25.00	25.00	0.50
Tuğrul (2004)	Granite	151.00	102.00	130.00	67.00	79.00	15.00	21.00	11.00	3.50	1.20
Aydin and Basu (2005)	Granite	196.50	116.30	106.30	31.10	26.80	13.60	41.70	6.30	—	—
Ceryan et al. (2008)	Granite	194.30	126.30	170.90	86.90	99.80	33.80	69.30	9.30	3.20	1.20
Basu et al. (2009)	Granite	214.00	153.00	134.00	161.00	88.00	73.00	—	—	—	—
Heidari et al. (2013)	Granite	176.30	127.70	167.30	42.40	36.60	22.20	28.30	5.00	—	—
Mert (2014)	Granite	170.60	115.20	136.82	60.50	82.30	13.60	24.20	12.20	3.80	2.50
Momeni et al. (2015)	Granite	161.60	127.20	144.50	121.60	59.20	45.80	15.80	4.10	3.07	1.80
Md Dan et al. (2016)	Granite	—	—	—	—	90.62	42.51	38.67	6.84	5.31	1.04
Arikan et al. (2007)	Andesites	204.20	57.70	110.60	28.40	60.60	19.80	35.40	6.30	—	—
Koca and Kıncal (2016)	Andesites	—	—	76.51	45.41	42.37	27.57	16.52	12.37	—	—

classification of the rock mass and strength. Rock mass classification results are tabulated in Table 16.12 below:

RMR system provided rock mass strength results. The final slope of pit was designed by correcting on the basis of general orientation of joints in stout field. General direction of the field measurement results was based on stocky andesite rock mass.

7.1.3 Quality of andesite

Production quality results andesite processing must be able to meet the desired requirements of the market. According to the market survey, the quality results of processed andesite is shown in the following table (Table 16.13):

The results of uniaxial compressive strength test (UCS) for andesite are from 88.9 to 107.4 MPa (construction), resilience destroyed 13% (heavy construction), and absorption 2.18% (heavy construction). The test results of andesite are compared with quality standards, the andesite is included in class I–II (Heavy – Fair). Thus, it can be concluded that andesite can be used for a building foundation and the moderate grade Aggregate Concrete Mix.

Andesite quarry will be converted into a large storage of water reservoir at the end of life of the

7. Site study of tropically weathered igneous rocks

TABLE 16.12 Classification for rock mass Rating (RMR).

No	Particulars	Rating
1	Rock strength [UCS from 89 to 105 MPa]	12.0
2	Average RQD (90%)	20.0
3	Spacing between discontinuities from 0.6 to 2m	15.0
4	Condition of discontinuity condition (Rough surface slightly, separation less than 1 mm, weathered walls slightly)	25.0
5	Water condition (dipping)	4.0
6	Total rating	76.0
7	Rock mass classification (RMC)	Klas II (Good Rock)
8	Rock mass cohesion, C_m (kPa)	300 to 400
9	Friction angle of rock mass, Φ_m (degree)	35 to 45

TABLE 16.13 Average quality andesite rock.

No	Parameter	Unit	Value	Quality standard		
				Heavy	Fair	Light
1	Uniaxial Compressive Strength	MPa	100.25	150	100—150	80—100
2	Endurance	%	13.0	16	16—24	24—30
3	Absorption	%	2.18	3	3—5	5—8

quarry. The same will be useful for improving water table as well as supply of water to the local community.

7.2 Granite quarry, Thailand

Bangkok is well known for its world class infrastructure consisting of airports, North South highway, high rise buildings, metro rail, and fly-over bridges. Aggregate quarries with varying capacity from 2 to 5 million t per annum supply construction aggregates for manufacturing concrete. Construction aggregate consists of limestone or granite or basalt which is relatively hard rock. Most of the quarries are situated 100—200 km North of Bangkok or central portion of Thailand where large cement plants also exist. Very large quarries are captive in nature and supply limestone for manufacturing cement. The selected quarry under this study is granite quarry which is igneous rock and tropically weathered in Thailand.

7.2.1 Geology of Thailand granite deposit

Exploration program was planned to drill 16 bore holes in granite deposit with a grid pattern of 500 × 500 m grid after geological mapping was completed. Block model was developed based on exploration data, resistivity survey, and drill core log. The overburden (OB) is consisted with Sandy Clay (SC) and Weathering granite (W.gr) and respective thickness is 4.3 and 16.9 m as per drill core log model (Table 16.14).

7.2.2 Observations based on geological mapping, exploration data, and block model for planning mining operation at granite quarry

Selected quarry is granite quarry where exploration data show that upper layer has sandy clay at average thickness 10 m. Middle layer is weathered granite at average thickness 20 m. Lower

TABLE 16.14 Key highlights of fresh granite at Thailand (Geological report, 2013).

Physical properties			Chemical properties	
Particulars	Unit	Granite	Element	Analysis
Solid density	t/cu m	2.62−2.63	$SiO_2\%$	72.04
Bulk density	t/Cu m	1.22−1.26	$Al_2O_3\%$	15.43
Crushability	%	40−45	$Fe_2O_3\%$	3.98
Loss Angles Abrasive Index	%	19−23	$CaO\%$	2.04
			$MgO\%$	0.63
Abrasiveness	Grams/t	1280−1380	$K_2O\%$	4.43
Dynamic fragmentation	%	24−26	$Na_2O\%$	2.89
			$SO_3\%$	0.04
UCS	MPa	40−70	$LOI\%$	0.8

layer consists of fresh granite at average thickness >80 m (Aggregate grade). 90%−95% aggregates granite consists of medium to coarse grained texture, comprised mainly of feldspar, quartz, biotite, and muscovite. 5%−10% of waste consists of highly to completely weathered granite, sheared rock, and fracture. Local community is around mining lease area and controlled blasting is not practicable solution. Dry density of granite is 2.6. Silica in granite is 71%. Compressive strength varies from 40 to 70 MPa based on degree of weathering. Based on known information, construction aggregate can be produced from granite quarry.

7.3 Andesite quarry in Cambodia

Cambodia's industrial capacity will grow in future years as domestic manufacturing replaces imports and inward investment takes advantage of cheaper labor costs and advantageous tax incentives. As an emerging economy, the country is predominantly centered on manufacturing (textiles) and natural resources (precious metals, gems) and now proven oil reserves have been confirmed off the south coast. It is predicted that economic growth in Cambodia shall be more than economic growth of Thailand in last 20 years or Vietnam in 10 years. Phnom Penh is the capital city of Cambodia where infrastructure development is taking places. Other places where demand of aggregate exist are Sihanoukville, Kandal, Takev, and Siem Reap.

Igneous rocks such as basalt, andesite, and granetoids are the main source of aggregates in Cambodia. An exploration program at one of Andesite deposit is situated on a hill with height of 300 m and length of 1.2 km. Initial geological mapping was carried out in demarcated area. Five boreholes of 100 m depth were drilled. Exploration program included resistivity survey with traverse lines. 1−3 m of soil as overburden was observed.

Andesite of different degree of weathering was observed. Samples were collected and various physiomechanical tests were carried out. Andesite is generally found fresh, slightly weathered, moderately weathered, highly weathered, and completely weathered. RQD varies from 40 to 83%. Geological strength index varies from 35 to 70%. Rock density varies from 2.25 to 2.67 g/cc. PLI varies from 0.8 to 8 MPa. SHRN varies from 22 to 46. Following are some of the results of quarry face survey and test results of laboratory results tabulated in Table 16.15.

7.4 Malaysia

A world class infrastructure is developed in and around Kuala Lumpur, Johor Bahru, and

TABLE 16.15 Rock mass properties observed at Andesite in Cambodia.

Parameter	Unit	Value
RQD	%	40%−83%
GSI	%	35%−70%
Rock density	g/cc	2.25−2.67
PLI	MPa	0.8−8
SHRN	Number	22−46

7. Site study of tropically weathered igneous rocks

TABLE 16.16 State wise aggregate production in Malaysia.

State	Yearly production (Million T)	State	Yearly production (Million T)
Johor	6.7	Perlis	1.2
Kedah	3.3	Pualau Pinang	4.1
Kelantan	2.6	Sabah	4.8
Melaka	1.9	Sarawak	7.6
Negeri Sembilan	4.8	Selangor	28.5
Pahang	3.2	Terengganu	5.6
Perak	13.8	Grand Total	88.1

many other cities in Malaysia. This country has three major bands of granite eastern, western, and central band. Table 16.16 shows state wise production of aggregates. Malaysia produces 88.1 million t of aggregates.

Names of the quarries where studies have been carried out to collect geophysical properties of aggregates at laboratory and quarry faces are listed in Table 16.17. Due to tropical weathering, granite is found in different state of weathering from fresh, slightly weathered, moderately weathered, highly weathered, and completely weathered. Residual soil is found of varying thickness and is found in these deposits. Rock quality designation (RQD) varies from 25% to 90%. GSI varies from 17% to 75%. RMR varies from 22 to 82. Rock density of aggregate varies from 2.21 to 2.73 g/cc. UCS is determined for fresh to moderately weathered granite. Young's modulus of the said sample varies from 8.41 to 29.88 GPa. PLI and Schmidt hammer rebound number are determined for slightly weathered granite to completely weathered granite. Porosity and rate of water abortion in weathered granite and rate of decrease in strength due to moisture decrease from completely weathered granite to fresh granite (Mohamad et al., 2011, 2016a). Compressive strength and elasticity modulus of tropically weathered granite are predicted based on rock mass properties such as porosity, PLI, P-wave velocity, and SHRN (Armaghani et al., 2016).

TABLE 16.17 A list of quarries in Malaysia under this study (Armaghani et al., 2015a,b; Armaghani et al., 2018a,b; Faradonbeh et al., 2016a,b; Mohamad et al., 2012, 2013, 2016; Mahdiyar et al., 2020; Koopialipoor et al., 2019).

Quarry name	Latitude	Longitude
Bukit Indah	1°93′ 12″ N	103°35′ 08″ E
Hulu Langat	3°7′ 0″N	101°49′ 1″ E
Kota Tinggi	1°44′ 12″ N	103°54′ 08″E
Kulai	1°39′ 21″ N	103°36′ 11″ E
Masai	1°29′ 42″N	103°52′ 28″E
Pengerang	1°22′ 58″N	104°7′ 58″E
Putri Wangsa	1°35′ 32″ N	103°48′ 4″ E
Senai Jaya	1°36′ 00″ N	103°39′ 00″ E
Taman Bestari	1°60′ 41″ N	103°78′ 32″ E
Trans Crete	1°31′ 21″N	103°52′ 60″E
Ulu Choh	1°31′ 48″ N	103°32′ 41″ E
Ulu Tiram	1°36′ 41″ N	103°49′ 20″ E

Deep hole drilling and blasting are carried out in aggregate quarries. Bulk emulsion explosives are used for charging of holes. Nonelectric detonators are used for minimizing environmental effect due to the blasting.

Aggregates produced in Malaysia are exported to Singapore by road and waterways. Table 16.16 shows standard being followed for importing aggregates in Singapore.

7.4.1 Standard for imported aggregates in Singapore

Building Construction Authority (BCA) is government agency which has set standard for imported aggregates in Singapore. BCA has enforced a set of test procedures for imported aggregates (BCA).

7.4.1.1 Stage I—preimport test

Key features of preimport test are location map of aggregate quarry showing name of mine/quarry, geological report, and tests as prescribed in Table 16.18 from accredited laboratory, aggregate samples (150 g) in three containers.

7.4.1.2 Stage II—confirmation test

Confirmation test is similar to Stage I test. The Stage II test is essentially conducted by an Authorized Analyst designated by the BCA and the cost of the tests is borne by the BCA. Importer can import aggregates for 1 year after Stage II confirmation test meets the standard.

7.4.1.3 Stage III—random test

Random test is conducted for future of any of the consignments to check if standards are being met as per Table 16.8. Random test may be conducted at unloading point for chloride content.

7.4.1.4 Research study on Alkali—Silica Reaction carried out in Malaysia

During 1940, Alkali—Silica Reaction (ASR) was first noticed and investigated into the United States. Research on ASR continues across the world as field and laboratory conditions differ from place to place. Time required knowing the status of ASR also may vary from couple of days to couple of years. The three main components of ASR are alkalis in cement, silica from aggregates, and moisture. Malaysia being in the tropical region, moisture in any concrete structure is likely to be present as rainfall takes place during the whole year. Various research studies have been carried out in Malaysia and details are given in Table 16.19.

8. Comparison of tropically weathered igneous rocks in Indonesia, Thailand, Cambodia, and Malaysia

All igneous rock deposits under this study are selected from wet tropical region where climatic

TABLE 16.18 Test methods and test standards for imported coarse and fine aggregates.

Description	Test method	Relevant standards	Specification
Petrographic Examination	ASTM C295	ASTM C33	Aggregates should not contain any constituents that are deleterious to concrete.
Alkali—silica reactivity (Mortar Bar Method)	ASTM C1260	ASTM C33	Only aggregates with expansion value not greater than 0.20% shall be used.
Water-Soluble Chloride Content	EN1744—1:1998 Clause 7	SS EN 12620	Chloride ion content ≤0.01% by mass of chloride ion of combined aggregate.
Acid-Soluble Sulfates Content	EN1744—1:1998 Clause 12	SS EN 12620	Sulfate content of the aggregates and filler aggregates for concrete ≤0.8% by mass for aggregates

TABLE 16.19 Studies carried out on Alkali Silica Reaction (ASR) in Malaysia.

Researcher	Brief particulars of research findings
Sum and Sahat (1990)	Chalcedony, tridymite and strained quartz, and crypto to microcrystalline quartz are reactive minerals in tuff rocks
Beng (1992)	Reactive minerals—chalcedony, fibrous quartz, opal, SiO_2 rich volcanic glass, and tridymite contributing to ASR
Fatt and Beng (2007)	Damaging impact in concrete due to ASR can be minimized if reactive minerals (strained and microcrystalline quartz) are kept below 12% in deformed granite.
Yaacob et al. (1994)	The agglomeratic tuff is holocrystalline—mainly composed of quartz, alkali feldspar, plagioclase, and lithic-clasts of andesite, rhyolite, and sandstone. The rhyolite is holocrystalline, crypto- to microcrystalline quartz with feldspars and sericite. Quartz clasts are very fine to coarse in grain size. Sandstone may contain chert clasts
Fatt (2011)	Mylonitic, cataclastic granites in faulting zone contain strained quartz, fine secondary sericite, and microcrystalline quartz—highly reactive minerals
Fatt et al. (2013)	Kuala Lumpur fault Zone (Bukit Lagong area) has small quantum deformed granite and thus severe impact ASR would not be observed.
Mohamad et al. (2016b)	Studying geology and preparing mine plan helps in monitoring and moderating impact of ASR
Murlidhar et al. (2016)	Different rock types studied based on geology and investigation of potential rock type contributing to ASR
Murlidhar (2016)	Influence of local geological setting on ASR

conditions are similar. Rainfall takes place intermittently throughout the year. Due to humid and hot conditions, rate of weathering is higher as compared to other regions. Degree weathering has direct impact on rock mass properties of rock. In tropical region, weathered rocks are classified into five classes namely W_I, W_{II}, W_{III}, W_{IV}, and W_V. W_I = Fresh rock, W_{II} = Slightly weathered rock, W_{III} = Moderately weathered rock, W_{IV} = Highly weathered rock, W_V= Completely weathered rock W_I = Fresh rock, W_{II} = Slightly weathered rock, W_{III} = Moderately weathered rock, W_{IV} = Highly weathered rock, and W_V= Completely weathered rock. All five weathering grease rocks are found in Thailand, Cambodia, and Malaysia. In Andesite deposit from Indonesia, WI, W_{II}, and W_{III} grades of weathering were observed. There may be weathered rock outside license area of the andesite quarry. Andesite deposits under this study are from Indonesia and Cambodia.

Granite deposits under this study are from Thailand and Malaysia. Various studies have been carried out in Malaysia and rock mass properties have been compiled in this study based on the previous researchers (Table 16.20).

8.1 Rock mass assessment for blastability purpose

In the beginning of this chapter, rock mass classification systems are reviewed for excavation, rippability, and slope stability. All the rock mass classification systems have commonly related to rock mass strength. Excavation and rippability have to break rock with any equipment such as excavator, backhoe, or dozer where horsepower and capability of equipment play an important role. On the other hand, rock is broken in all directions with explosives during blasting. In slope stability, slope to remain stable, rock

TABLE 16.20 Comparison of tropically weathered igneous rocks in Indonesia, Thailand, Cambodia, and Malaysia.

| Particulars | | | | | |
Type of rock and country	Unit	Andesite rock in Indonesia	Granite rock in Thailand	Andesite rock in Cambodia	Granite rock in Malaysia
Whether in tropical region		Yes	Yes	Yes	Yes
Weathered rock types		W_I, W_{II}, W_{III}	W_I to W_V	W_I to W_V	W_I to W_V
Thickness of weathered rock type	m	$W_I = 80$ to 100 $W_{II} = 3$ to 20 $W_{III} = 1$ to 10	$W_I = 60$ to 80 $W_{II} = 3$ to 12 $W_{III} = 2$ to 5 $W_{IV} = 1$ to 2 $W_V = 1$ to 2	$W_I = 40$ to 100 $W_{II} = 5$ to 16 $W_{III} = 3$ to 8 $W_{IV} = 2$ to 6 $W_V = 1$ to 2	$W_I = 70$ to 100 $W_{II} = 10$ to 25 $W_{III} = 7$ to 12 $W_{IV} = 2$ to 8 $W_V = 1$ to 10
RQD	%	50 to 100	40 to 83	43 to 88	25 to 90
RMR		76	35 to 70	27 to 55	22 to 82
GSI	%	40 to 75	30 to 65	35 to 70	25 to 75
Rock density		2.41 to 2.74	2.28 to 2.64	2.47 to 2.61	2.21 to 2.67
UCS	MPa	88.7 to 107.4	46.6 to 62.2	43.2 to 59.8	41.3 to 68.7
E	GPa	12.3 to 37.8	15.3 to 32.5	9.3 to 31.7	8.41 to 29.88
PLI	MPa	2.1 to 10	0.2 to 7	0.8 to 8.2	0.1 to 9.1
SHRN	No.	28 to 51	16 to 40	22 to 46	18 to 42
Porosity	%	1.1 to 3.3	4.7 to 12.1	2.5 to 4.6	3.2 to 15.5

Remarks: W_I = Fresh rock, W_{II} = Slightly weathered rock, W_{III} = Moderately weathered rock, W_{IV} = Highly weathered rock, W_V = Completely weathered rock, RQD = Rock quality designation, RMR = Rock mass rating, UCS = Uniaxial Compressive Strength, E = Young's modulus of elasticity, PLI = Point load index, SHRN = Schmidt hammer rebound number.

mass strength is crucial. Water content or moisture, impact of ground vibration, or seismic activity can have severe impact on slope stability. On the other hand, in blasting, watery holes are not usually common. Only certain sites using Ammonium Nitrate Fuel Oil (ANFO) may be affected due to water in the blasting if adequate precautions are not taken. In tropically weathered rock, strength of rock is reduced by different extent in dry and wet condition based on degree of weathering. Thus, factors related to the assessment of blastability differ as compared to already developed rock mass classification systems for other applications.

In tropical region, rocks are classified into five grades of weathering. Physical and engineering properties of rock mass properties vary from place to place and rock type. This classification is suitable for preliminary assessment for blastability purpose. However, various studies have shown that strength of rock has direct impact on blastability. Hence, UCS or Young's modulus can be considered as one of the parameters for blastability purpose. However, it is not possible to obtain core sample for highly weathered rock. Hence, rock mass properties such as SHRN and PLI represent strength of rock for the purpose of tropically weathered rock. Such tests can be carried out easily in the field for every blasting face. Rock density has direct impact on blastability. RQD and GSI can be easily estimated in the field for every blasting face. Porosity or water absorption can be also easily determined. Thus, after evaluation of degree of weathering, rock density, RQD, GSI, SHRN, PLI, porosity, and BI become important parameters for the assessment of blastability purpose.

9. Conclusion

1. Igneous rocks were studied from Indonesia, Thailand, Cambodia, and Malaysia which are in the tropical region. Weathered rocks are classified into five classes namely W_I, W_{II}, W_{III}, W_{IV}, and W_V. W_I = Fresh rock, W_{II} = Slightly weathered rock, W_{III} = Moderately weathered rock, W_{IV} = Highly weathered rock, and W_V = Completely weathered rock
2. Weathered andesite from Indonesia and Cambodia and weathered granite from Thailand and Malaysia were studied. Andesite in Indonesia has W_I, W_{II}, and W_{III}. Other places all five grades of weathered rocks are found.
3. Andesite in Indonesia was found stronger based on RQD% and UCS as compared to other locations.
4. Weathered rocks classification is useful for preliminary assessment of blastability. However, further classification can be done based on actual rock mass properties.
5. UCS and Young's modulus of elasticity are important parameters which represent strength of rock and can be correlated to blastability. However, there is difficulty in obtaining cores for highly weathered and completely weathered rocks. Hence, indirect way of PLI and SHRN can provide strength parameters from the field
6. Rock density, RQD, GSI, SHRN, PLI, porosity, and BI parameters are useful for rock mass classification assessment of topically weathered rocks for blastability purpose.
7. There is a need for future research for strengthening assessment of blastability

purpose in tropically weathered rocks based on the following:
(a) Preparing weathering profile of each rock type in different locations
(b) Collecting more samples from each site and conducting laboratory tests for physio-mechanical properties
(c) Moisture content has direct impact on strength of weathered igneous rock and needs further investigation.

References

Abdullatif, O.M., Cruden, D.M., 1983. The relationship between rock mass quality and ease of excavation. Bull. Int. Assoc. Eng. Geol. - Bulletin de l'Association Internationale de Géologie de l'Ingénieur 28 (1), 183–187. https://doi.org/10.1007/BF02594813.

Alavi, N.K.A.S.V., Edy Tonnizam, M., Komoo, I., 2014. Dominant weathering profiles of granite in southern Peninsular Malaysia. Eng. Geol. 183, 208–215. https://doi.org/10.1016/j.enggeo.2014.10.019.

Alavi, N.K.A.S.V., Edy Tonnizam, M., Komoo, I., Kalatehjari, R., 2014. A typical weathering profile of granitic rock in Johor, Malaysia based on joint characterization. Arabian J. Geosci. https://doi.org/10.1007/s12517-014-1345-7.

Alavi, N.K.A.S.V., Tugrul, A., Gokceoglu, C., Armaghani, D.J., 2016. Characteristics of weathering zones of granitic rocks in Malaysia for geotechnical engineering design. Eng. Geol. 200, 94–103. https://doi.org/10.1016/j.enggeo.2015.12.006.

Amin, M.M., Huei, C.S., Hamid, Z.A., Ghani, M.K., 2009. Rippability assessment of rock based on specific energy and production rate. In: 2nd Construction Industry Research Achievement International Conference (CIRAIC 2009), vol. 9.

Arel, E., Tugrul, A., 2001. Weathering and its relation to geomechanical properties of Cavusbasi granitic rocks in northwestern Turkey. Bull. Eng. Geol. Environ. 60 (2), 123–133.

Arıkan, F., Ulusay, R., Aydın, N., 2007. Characterization of weathered acidic volcanic rocks and a weathering classification based on a rating system. Bull. Eng. Geol. Environ. 66 (4), 415–430. https://doi.org/10.1007/s10064-007-0087-0.

Armaghani, D.J., Momeni, E., Abad, S.V.A.N.K., Khandelwal, M., 2015. Feasibility of ANFIS model for prediction of ground vibrations resulting from quarry blasting. Environ. Earth Sci. 74 (4), 2845–2860.

Armaghani, D.J., Hajihassani, M., Sohaei, H., Mohamad, E.T., Marto, A., Motaghedi, H., Moghaddam, M.R., 2015. Neuro-fuzzy technique to predict air-overpressure induced by blasting. Arabian J. Geosci. 8 (12), 10937—10950.

Armaghani, D.J., Mohamad, E.T., Momeni, E., Monjezi, M., Narayanasamy, M.S., 2016. Prediction of the strength and elasticity modulus of granite through an expert artificial neural network. Arabian J. Geosci. 9 (1), 48.

Armaghani, D.J., Hasanipanah, M., Amnieh, H.B., Mohamad, E.T., 2018. Feasibility of ICA in approximating ground vibration resulting from mine blasting. Neural Comput. Appl. 29 (9), 457—465.

Armaghani, D.J., Hasanipanah, M., Mahdiyar, A., Abd Majid, M.Z., Amnieh, H.B., Tahir, M.M., 2018. Airblast prediction through a hybrid genetic algorithm-ANN model. Neural Comput. Appl. 29 (9), 619—629.

Aydin, A., Basu, A., 2005. The Schmidt hammer in rock material characterization. Eng. Geol. 81 (1), 1—14.

Azimi, Y., Osanloo, M., Aakbarpour-Shirazi, M., Bazzazi, A.A., 2010. Prediction of the blastability designation of rock masses using fuzzy sets. Int. J. Rock Mech. Min. Sci. 47 (7), 1126—1140. https://doi.org/10.1016/j.ijrmms.2010.06.016.

Basahel, H., Mitri, H., 2017. Application of rock mass classification systems to rock slope stability assessment: a case study. J. Rock Mech. Geotech. Eng. 9 (6), 993—1009. https://doi.org/10.1016/j.jrmge.2017.07.007.

https://www.bca.gov.sg/ImportersLicensing/others/test_requirements.pdf.

Basarir, H., 2007. Numerical analysis of support performance and rock-support interaction considering different rock conditions.

Basarir, H., Karpuz, C.E.L.A.L., 2004. A rippability classification system for marls in lignite mines. Eng. Geol. 74 (3—4), 303—318.

Basarir, H., Karpuz, C., Tutluoglu, L., 2008. Specific energy based rippability classification system for coal measure rock. J. Terramechanics 45 (1—2), 51—62.

Basu, A., Celestino, T.B., Bortolucci, A.A., 2009. Evaluation of rock mechanical behaviors under uniaxial compression with reference to assessed weathering grades. Rock Mech. Rock Eng. 42 (1), 73—93.

Baynes, F.J., Dearman, W.R., 1978. The relationship between the microfabric and the engineering properties of weathered granite. Bull. Int. Assoc. Eng. Geol. 18, 191—197.

Begonha, A., Braga, M.A.S., 2002. Weathering of the oporto granite: geotechnical and physical properties. Catena 49, 57—76.

Beng, Y.E., 1992. The mineralogical and petrological factors in alkali silica reactions in concrete. Bull. Geol. Soc. Malaysia 31, 1—20.

Borrelli, L., Coniglio, S., Critelli, S., La Barbera, A., Gullà, G., 2016. Weathering grade in granitoid rocks: the san Giovanni in Fiore area (Calabria, Italy). J. Maps 12 (2), 260—275. https://doi.org/10.1080/17445647.2015.1010742.

Caterpillar, 2001. Caterpillar Performance Handbook of Ripping 2001, 32nd ed. Caterpillar Tractor Company Preoria Illinios, p. 32.

Ceryan, S., 1999. Weathering and Classification of Harsit Granitoid and the Effect of Weathering on Engineering Properties. Ph.D. Thesis. Karadeniz Technical University, Trabzon, Turkey, p. 300 (in Turkish).

Ceryan, S., Zorlu, K., Gokceoglu, C., Temel, A.B.İ.D.İ.N., 2008. The use of cation packing index for characterizing the weathering degree of granitic rocks. Eng. Geol. 98 (1—2), 60—74.

Church, H.K., 1981. Excavation Handbook. McGraw-Hill, New York, USA.

Dearman, W.R., 1974. Weathering classification in the characterisation of rock for engineering purposes in British practice. Eng. Geol. 33—42.

Dearman, W.R., 1995. Description and classification of weathered rocks for engineering purposes: the background to the BS5930:1981 proposals. Q. J. Eng. Geol. 28, 207—242. https://doi.org/10.1144/GSL.QJEGH.1995.028.P3.05.

Dethier, D.P., Bove, D.J., 2011. Mineralogic and geochemical changes from alteration of granitic rocks, boulder creek catchment, Colorado. Vadose Zone J. 10 (3), 858. https://doi.org/10.2136/vzj2010.0106.

Dethier, D.P., Lazarus, E.D., 2006. Geomorphic inferences from regolith thickness, chemical denudation and CRN erosion rates near the glacial limit, boulder creek catchment and vicinity, Colorado. Geomorphology 75 (3—4), 384—399. https://doi.org/10.1016/j.geomorph.2005.07.029.

Eggleton, A., Banfield, F., 1985. The alteration of granitic biotite to chlorite. Am. Mineral. 70, 902—910. Retrieved from. http://www.minsocam.org/msa/collectors_corner/amtoc/toc1985.htm.

Ehlmann, B.L., Heather, A.V., Bourke, M.C., 2008. Quantitative morphologic analysis of boulder shape and surface texture to infer environmental history: a case study of rock breakdown at the Ephrata Fan, Channeled Scabland, Washington. J. Geophys. Res.: Earth Surf. 113, 1—20. https://doi.org/10.1029/2007JF000872.

Faradonbeh, R.S., Armaghani, D.J., Abd Majid, M.Z., Tahir, M.M., Murlidhar, B.R., Monjezi, M., Wong, H.M., 2016a. Prediction of ground vibration due to quarry blasting based on gene expression programming: a new model for peak particle velocity prediction. Int. J. Environ. Sci. Technol. 13 (6), 1453—1464.

Faradonbeh, R.S., Armaghani, D.J., Monjezi, M., 2016b. Development of a new model for predicting flyrock distance in quarry blasting: a genetic programming technique. Bull. Eng. Geol. Environ. 75 (3), 993—1006.

Fatt, N.T., Beng, Y.E., 2007. Potential alkali-silica reaction in aggregate of deformed granite. Bull. Geol. Soc. Malaysia 53, 81—88.

Fatt, N.T., 2011. Effects of fault deformation on the quality of granite aggregates. In: National Geoscience Conference.

Fatt, N.T., Raj, J.K., Ghani, A.A., 2013. Potential alkali-reactivity of granite aggregates in the Bukit Lagong Area, Selangor, Peninsular Malaysia. Sains Malays 42 (6), 773–781.

Fookes, P.G., 1997. Tropical residual soils geological society engineering group working party report. Q. J. Eng. Geol. Hydrogeol. Geol. Soc. London 23, 4–101. https://doi.org/10.1144/GSL.QJEG.1990.023.001.01.

Franklin, J.A., Broch, E., Walton, G., 1971. Logging the mechanical character of rock. Trans. Inst. Min. Metall. 80 (770), 1–9.

Freire-Lista, D.M., Gomez-Villalba, L.S., Fort, R., 2015. Microcracking of granite feldspar during thermal artificial processes. Period. Mineral. 84 (3A), 519–537. https://doi.org/10.2451/2015PM0029.

Geological Report of Granite Quarry in Thailand, 2013.

Ghose, A.K., 1988. Design of Drilling and Blasting Subsystems - A Rock Mass Classification Approach. Mine Planning and Equipment Selection. Rotterdam: AA Balkema.

Gribble, C., McLean, A., 1985. Geology for Civil Engineers, second ed. (2nd, revised ed.). George Allen and unwin.

Gupta, A.S., Rao, S.K., 2001. Weathering indices and their applicability for crystalline rocks. Bull. Eng. Geol. Environ. 60 (3), 201–221.

Hack, R., 2011. RQD, Slope Stability Classification, Weathering. Engineering Geology Course. Duluth: University Twente, The Netherlands.

Hall, K., Thorn, C., Sumner, P., 2012. On the persistence of "weathering.". Geomorphology 149 (150), 1–10. https://doi.org/10.1016/j.geomorph.2011.12.024.

Heidari, M., Momeni, A.A., Naseri, F., 2013. New weathering classifications for granitic rocks based on geomechanical parameters. Eng. Geol. 166, 65–73.

Hoek, E., Karzulovic, A., 2000. Rock mass properties for surface mines. In: Hustrulid, W., McCarter, M., Zyl, D. (Eds.), Slope Stability in Surface Mining. Society for Mining, Metallurgy, and Exploration, Littleton, pp. 59–70.

Irfan, T.Y., Dearman, W.R., 1978a. The engineering petrography of a weathered granite in Cornwall, England. Q. J. Eng. Geol. Hydrogeol. 11 (3), 233–244.

Irfan, T.Y., Dearman, W.R., 1978b. Engineering classification and index properties of a weathered granite. Bull. Int. Assoc. Eng. Geol. - Bulletin de l'Association Internationale de Géologie de l'Ingénieur. 17 (1), 79–90.

Irfan, T.Y., Powell, G.E., 1985. Engineering geological investigations for pile foundations on a deeply weathered granitic rock in Hong Kong. Bull. Int. Assoc. Eng. Geol. - Bulletin de l'Association Internationale de Géologie de l'Ingénieur. 32 (1), 67–80.

ISRM, 1981. Rock characterization testing and monitoring. In: Brown, E. (Ed.), ISRM Suggested Methods. Pergamon Press Ltd, Oxford, pp. 113–116.

Karpuz, C., 1990. A classification system for excavation of surface coal measures. Min. Sci. Technol. 11 (2), 157–163.

Kirsten, H.A.D., 1982. A classification system for excavating in natural materials. Civil Eng. = Siviele Ingenieur 24 (7), 293–308.

Kirsten, H., 1988. Case histories of groundmass characterization for excavatability. In: Rock Classification Systems for Engineering Purposes. ASTM International.

Koca, M.Y., Kıncal, C., 2016. The relationships between the rock material properties and weathering grades of andesitic rocks around İzmir, Turkey. Bull. Eng. Geol. Environ. 75 (2), 709–734.

Komoo, I., 1989. Engineering properties of the igneous rocks in Peninsular Malaysia. In: Proc. 6th Regional Conference on Geology. Mineral and Hydrocarbon Resources of Southeast Asia, Jakarta, Indonesia, pp. 445–458.

Koopialipoor, M., Fallah, A., Armaghani, D.J., Azizi, A., Mohamad, E.T., 2019. Three hybrid intelligent models in estimating flyrock distance resulting from blasting. Eng. Comput. 35 (1), 243–256.

Latham, J.P., Lu, P., 1999. Development of an assessment system for the blastability of rock masses. Int. J. Rock Mech. Min. Sci. 36 (1), 41–55. https://doi.org/10.1016/S0148-9062(98)00175-2.

Lednicka, M., Kalab, Z., 2012. Evaluation of granite weathering in the Jeronym mine using non-destructive methods. Acta Geodyn. Geomater. 9 (2), 211–220.

Lilly, P.A., 1986. An Empirical Method of Assessing Rock Mass Blastability. The Aus.

Little, A.L., 1969. The engineering classification of residual tropical soils. In: Proceedings of the 7th International Conference on Soil Mechanics and Foundation Engineering Specialty Session on Engineering Properties of Lateritic Soils, Mexico, vol. 1, pp. 1–10 (1).

Mahdiyar, A., Jahed Armaghani, D., Koopialipoor, M., Hedayat, A., Abdullah, A., Yahya, K., 2020. Practical risk assessment of ground vibrations resulting from blasting, using gene expression programming and Monte Carlo simulation techniques. Appl. Sci. 10 (2), 472.

Martin, R.P., Henchert, S.R., 1986. Principles for description and classification of weathered rock for engineering purposes. Geol. Soc. Eng. Geol. Special Publication 2 (January 1986), 299–308. https://doi.org/10.1144/GSL.1986.002.01.53.

Md Dan, M.F., 2016. Physical Classifications and Engineering Characteristics of in Situ Boulders in Tropically Weathered Granite. PhD Thesis. Universiti Teknologi, Malaysia (October).

Md Dan, M.F., Edy Tonnizam, M., Akip Tan, S.N.M., Komoo, I., 2015. Physical field characterization of boulders in tropical weathering profile - a case study in Ulu Tiram, Johor Malaysia. Jurnal Teknologi 76 (2), 59—65.

Mert, E., 2014. An artificial neural network approach to assess the weathering properties of sancaktepe granite. Geotech. Geol. Eng. 32 (4), 1109—1121.

Mohamad, E.T., Fauzi, M., Isa, M., Komoo, I., Gofar, N., Saad, R., 2011. Effect of Moisture Content on the Strength of Various Weathering Grades of Granite.

Mohamad, E.T., Hajihassani, M., Armaghani, D.J., Marto, A., 2012. Simulation of blasting-induced air overpressure by means of artificial neural networks. Int. Rev. Model. Simul. 5, 2501—2506.

Mohamad, E.T., Armaghani, D.J., Hajihassani, M., Faizi, K., Marto, A., 2013. A simulation approach to predict blasting-induced flyrock and size of thrown rocks. Electron. J. Geotech. Eng. 18 (B), 365—374.

Mohamad, E.T., Armaghani, D.J., Hasanipanah, M., Murlidhar, B.R., Alel, M.N.A., 2016. Estimation of air-overpressure produced by blasting operation through a neuro-genetic technique. Environ. Earth Sci. 75 (2), 174.

Mohamad, E.T., Latifi, N., Arefnia, A., Isa, M.F., 2016. Effects of moisture content on the strength of tropically weathered granite from Malaysia. Bull. Eng. Geol. Environ. 75 (1), 369—390.

Mohamad, E.T., Faradonbeh, R.S., Armaghani, D.J., Monjezi, M., Majid, M.Z.A., 2017a. An optimized ANN model based on genetic algorithm for predicting ripping production. Neural. Comput. Appl. 28 (1), 393—406.

Mohamad, E.T., Armaghani, D.J., Ghoroqi, M., 2017b. Ripping production prediction in different weathering zones according to field data. Geotech. Geol. Eng. 35 (5), 2381—2399.

Momeni, A.A., Khanlari, G.R., Heidari, M., Sepahi, A.A., Bazvand, E., 2015. New engineering geological weathering classifications for granitoid rocks. Eng. Geol. 185, 43—51. https://doi.org/10.1016/j.enggeo.2014.11.012.

Moye, D., 1955. Engineering geology for the snowy mountains scheme. J. Inst. Eng. 27 (10—11), 287—298. https://doi.org/10.1007/s13398-014-0173-7.2.

Murlidhar, B.R., 2016. In: Rock Mass Classification for Karst Limestone of Cambodia for Blasting 3rd International Conference by China Aggregate Association Held at Guanzhou from 13—15 Dec'16.

Murlidhar, B.R., Tonnizam Mohamad, E., Armaghani, D.J., 2016. Potential alkali silica reactivity of various rock types in an aggregate granite quarry. Measurement 81, 221—231.

Oguchi, C.T., 2001. Formation of weathering rinds on andesite. Earth Surf. Process. Landforms 26 (8), 847—858. https://doi.org/10.1002/esp.230.

Pantelidis, L., 2009. Rock slope stability assessment through rock mass classification systems. Int. J. Rock Mech. Min. Sci. 46 (2), 315—325. https://doi.org/10.1016/j.ijrmms.2008.06.003.

Patino, L.C., Velbel, M.A., Price, J.R., Wade, J.A., 2003. Trace element mobility during spheroidal weathering of basalts and andesites in Hawaii and Guatemala. Chem. Geol. 202 (3—4), 343—364. https://doi.org/10.1016/j.chemgeo.2003.01.002.

Pettifer, G.S., Fookes, P.G., 1994. A revision of the graphical method for assessing the excavatability of rock. Q. J. Eng. Geol. Hydrogeol. 27 (2), 145—164. https://doi.org/10.1144/GSL.QJEGH.1994.027.P2.05.

Raina, A.K., Murthy, V.M.S.R., Soni, A.K., 2015. Flyrock in surface mine blasting: understanding the basics to develop a predictive regime. Curr. Sci. 108 (4), 660—665.

Rashidi, A., Nejad, H.R., Maghiar, M., 2014. Productivity estimation of bulldozers using generalized linear mixed models. KSCE J. Civil Eng. 18 (6), 1580—1589.

Riquelme, A., Tomás, R., Cano, M., Abellán, A., 2016. Using open-source software for extracting geomechanical parameters of a rock mass from 3D point clouds: discontinuity set extractor and SMRTool. In: ISRM International Symposium-EUROCK 2016. Ürgüp. International Society for Rock Mechanics and Rock Engineering.

Romana, M., 1985. New adjustment ratings for application of Bieniawski classification to slopes. In: Proceedings of the International Symposium on Role of Rock Mechanics. Zacatecas, pp. 49—53.

Romana, M.R., 1993. A geomechanical classification for slopes: slope mass rating. In: Hudson, J. (Ed.), Rock Testing and Site Characterization, pp. 575—600. https://doi.org/10.1016/b978-0-08-042066-0.50029-x.

Romana, M., 1995. The geomechanical classification SMR for slope correction. In: 8th ISRM Congress. International Society for Rock Mechanics and Rock Engineering, Tokyo.

Ruxton, B.P., Berry, L., 1957. Weathering of granite and associated erosional features in Hong Kong. Bull. Geol. Soc. Am. 68 (10), 1263—1292. https://doi.org/10.1130/0016-7606(1957)68.

Singh, R.N., Denby, B., Egretli, I., 1987. Development of a new rippability index for coal measures excavations. In: The 28th US Symposium on Rock Mechanics (USRMS). OnePetro.

Sum, C.W., Sahat, A.M., 1990. Potential alkali-silica reactivity of Tuffaceous rocks in the Pengerang Area. Johor. https://doi.org/10.7186/BGSM26199009.

Scoble, M.J., Muftuoglu, Y.V., 1984. Derivation of a diggability index for surface mine equipment selection. Min. Sci. Technol. 1 (4), 305—322. https://doi.org/10.1016/S0167-9031(84)90349-9.

Thuro, K., Scholz, M., 2004. Deep weathering and alteration in granites-a product of coupled processes. In: Elsevier Geo-Engineering Book Series, vol. 2. Elsevier, pp. 785—790.

Tsiambaos, G., Saroglou, H., 2010. Excavatability assessment of rock masses using the Geological Strength Index (GSI). Bull. Eng. Geol. Environ. 69 (1), 13—27. https://doi.org/10.1007/s10064-009-0235-9.

Tsidzi, K.E.N., 1997. An engineering geological approach to road cutting slope design in Ghana. Geotech. Geol. Eng. 15, 31—45.

Tuğrul, A., 2004. The effect of weathering on pore geometry and compressive strength of selected rock types from Turkey. Eng. Geol. 75 (3—4), 215—227.

Tuğrul, A., Gürpinar, O., 1997. A proposed weathering classification for basalts and their engineering properties (Turkey). Bull. Int. Assoc. Eng. Geol. - Bulletin de l'Association Internationale de Géologie de l'Ingénieur 55 (1), 141—149.

Yaacob, S., Beng, Y.E., Razak, H.A., 1994. Potential Alkali-Silica Reaction in Some Malaysian Rock Aggregates and Their Test Results. https://doi.org/10.7186/BGSM35199402.

Velde, B., Meunier, A., 2008. The Origin of Clay Minerals in Soils and Weathered Rocks. Verlag Berlin Heidelberg, Springer. Springer Berlin Heidelberg.

Further reading

Bhandari, S., 1997. Engineering Rock Blasting Operations. A.A. Balkema, Rotterdam.

Chatziangelou, M., Christaras, B., 2013. Blastability index on poor quality rock mass. Int. J. Civ. Eng. 2 (5), 9—16.

Chatziangelou, M., Christaras, B., 2016. A geological classification of rock mass quality and blast ability for widely spaced formations. J. Geol. Resour. Eng. 4, 160—174. https://doi.org/10.17265/2328-2193/2016.04.002.

Christaras, B., Chatziangelou, M., 2014. Blastability Quality System (BQS) for using it, in bedrock excavation. Struct. Eng. Mech. 51 (5), 823—845. https://doi.org/10.12989/sem.2014.51.5.823.

Geological Report of Andesite Quarry at Java, Indoensia, 2008.

Geological Report of Andesite Quarry in Cambodia, 2017.

Gupta, R.N., 1990. A Method to Assess Charge Factor Based on Rock Mass Blastability in Surface Mines. MINTECH Publications.

Hudson, J., 1992. Rock Engineering Systems: Theory and Practice. High Plains Press (JAH), Chichester.

Md Dan, M.F., Edy Tonnizam, M., Komoo, I., Alel, M.N.A., 2014. Physical characteristics of boulders formed in the tropically weathered granite. Jurnal Teknologi 72 (3), 75—82.

Rakishev, B.R., 1981. A new characteristic of the blastability of rock in quarries. Soviet Mining 17 (3), 248—251. https://doi.org/10.1007/BF02497198.

Scott, A., Cocker, A., Djordjevic, N., Higgins, M., La Rosa, D., Sarma, K.S., Wedmaier, R., 1996. Open Pit Blasting Design Analysis and Optimization. Julius Kruttschnitt Mineral Research Centre, Queensland.

Tangchawal, S.A.N.G.A., 2006. Planning and Evaluation for Quarries: Case Histories in Thailand. IAEG2006, Nottingham, United Kingdom.

Terzaghi, K., 1946. Rock Defects and Loads on Tunnel Supports, vol. 25. Harvard University, Massachusetts.

CHAPTER 17

Best river sand mining practices vis-a-vis alternative sand making methods for sustainability

Ramesh Murlidhar Bhatawdekar[1,2], Trilok Nath Singh[3], Edy Tonnizam Mohamad[1], Rajesh Jha[4], Danial Jahed Armagahni[5], Dayang Zulaika Abang Hasbollah[1]

[1]Geotropik-Centre of Tropical Geoengineering, Department of Civil Engineering, Universiti Teknologi Malaysia, Johor Bahru, Johor, Malaysia; [2]Department of Mining Engineering, Indian Institute of Technology, Kharagpur, West Bengal, India; [3]Earth Science Department, Indian Institute of Technology Bombay, Mumbai, Maharashtra, India; [4]Aggregate Innovations Pvt Ltd, New Delhi, India; [5]Department of Civil Engineering, Faculty of Engineering, University of Malaya, Kuala Lumpur, Selangor, Malaysia

1. Introduction

On a weight-to-weight basis, of all materials extracted from the Earth, except fossil fuels and biomass, sand and gravel constitute the largest quantity. Sand mining accounts for 85% of all mining projects on Earth simply because sand is used for everything from buildings to roads to watches, transmitters, jewelry, glass (Cortes et al., 2008; Ahmad et al., 2014), hydraulic fracking, etc., besides sand extraction for rare earth minerals or extracting quartz for making solar panels which have gained prominence in last decades. Maximum amount of sand is however used in construction—almost 79% of the construction materials consist of sand. Other than construction activity, significant quantities of sand are being extracted for land reclamation, shale gas extraction, and nourishment of beaches (Yong et al., 1991). No doubt sand and gravel as raw material has highest consumption volume second only to water and air.

Sand, which is used in the construction business, is formed by weathering and erosive processes over thousands of years (Bradley, 1970; Hudson, 1992). Generally, a river rises in a hill/mountain/highland. In its upper course, a river flows very fast because of sloping terrain.

A fast moving water erodes the river course (its bed and banks), thereby creating sediments (sand being a type sediment).The action of erosion takes place either by Abrasion, Attrition, Corrosion, or the Hydraulic action. In a river system, the largest and most angular rocks are farthest upstream nearest to source material and the smallest and roundest are at the river mouth and in underwater channels.

Dredging the beds of rivers and extracting the sand from the bottom of it to obtain sand is one method besides desilting which any way is inevitable. River sand is perfect fit compared to other types of sand as it falls in the right index of having good roundness, fineness, density, water absorption, etc. Sometimes as seen in Western Australia, there are a lot of big "yellow sand" mines. Yellow sand is normal (white) sand with a little iron in it. It is universally used in all construction projects here in truly vast quantities. There is no processing of the sand; it is just dug up and put in trucks to be stockpiled or sent straight to site.

Since nature has endowed us with huge quantities of desert sand and marine sand besides river sand, it is therefore counterintuitive to think that there is a possibility of having a global sand shortage—however, this is exactly the case today.

It is a very real issue—unbelievable as it may sound, in spite of being the most abundant resource on earth, almost 90% of the sand available on earth cannot be used as the desert sand grains are too fine and smooth to offer interlocking and multidirectional chemical linkages. If such smooth fine-grained sand is used, the slurry would slip and concrete would become weaker. If bridges are constructed with such smooth and fine-grained sand, and when it becomes wet, the structure would soften and in the event of overloading, the bridges could collapse. Marine sand from the seas also tends to be fine and rounded. Besides, when structures are built over sea, the chlorine present in the sea water will corrode the iron and steel

and thereby reduce the carrying capacity of the structures. Furthermore, sea sand does not have sufficient compressive strength or tensile strength, so it is not advisable to use sea sand in any construction activity. Also, the salt in sea sand tends to absorb moisture from atmosphere, bringing dampness.

The availability of sand in real sense therefore boils down to its appropriateness for construction activities and its ability to produce concrete that meets the design and engineering specifications required. At the end of the day it is a chemical compound but all sand is not "just sand."

The ever-increasing size of cities and taller and taller high rises crave gargantuan amounts sand. Many places are now forced to import sand from other countries in order to meet the demands. Dubai, for example, is surrounded by plenty of desert sand but still for aforementioned reasons it has to import sand for construction.

One significant example of reclamation of watery areas into useable land is that of Singapore, where since 1960, the total land area was increased by 20%—25%. In order to achieve this, they brought in sand from Indonesia and Malaysia, their neighbors. Likewise, in Chek Lap Kok, Hong Kong, in order to reclaim land and develop an international airport, they used huge quantities of sand (Jefferson et al., 2009; Jiang and Lin, 2010). In India, by bringing in sand from other areas, rocky shorelines have been converted to sandy beaches. All of these relied on river sand.

Now since the number of rivers is limited and the location of availability of good river sand is also limited, we are naturally facing a deficit in the river sand volume. As the extraction is so high, we are not allowing enough time for the formation of river sand. Reports suggest that we are mining twice as much sand as can be replaced in river beds by natural processes. Apart from the fact that sand mining is terrible for the environment, if current trends continue, we may be in danger of running out of useable sand. Adding to the scarcity cry is the incompatibility of most sands with required specifications along. The disparity between where appropriate sand is available and where it is needed also creates the shortage, a catch-22 situation where economy and ecology both take the hit.

So, sand scarcity induced by high consumption and unscientific extraction and its limited availability, which has been constantly challenged by our unsatiated greed, has surely caused irreparable and visible damages to environment at large. We humans are abetting an ecocide, as Vince Beiser in his book, *The World in a Grain*, has beautifully chronicled innovation, greed, and heedless waste of human civilization.

2. Global sand scenario and environment accountability

However, there is a demand for sand from 10 billion people of the world, and to meet the demand, countries must execute effective policy, regulation, and management of sand mining. Until now, policy and decision makers in public and private sectors have not addressed this serious issue. A constructive dialogue is called for between nations to address the need for generating and conserving sand resources and finding solutions. Some initiative has been taken by the United Nations in its Environmental Program—an expert round-table conference was held at Geneva, Switzerland, on October 11, 2011, on this subject. The need to find potential solutions to mitigate the negative impact of mining sands and aggregates and generating support for responsible consumption was emphasized in this conference. Views from experts and key messages were compiled and programs on further research and consultation were chalked out to tackle this issue in 2019 and beyond. It was agreed that new materials and novel construction techniques should be developed in the context of Environmental

Accountability. Establishing an effective way of monitoring of sand excavation and reducing consumption of sand by finding substitutes for sand in the construction industry were decided upon. Manufactured sand, for instance, is fast gaining recognition and acknowledged as partial to total replacement of sand.

3. Environmental impacts of sand mining

Excessive mining of gravel and sand from riverbeds causes river degradation. It reduces the river bottom levels which in turn leads to floodplain erosion (river bank erosion). At a macrolevel, steady depletion of riverbed sand and seacoast sand results in deepening of rivers, widening of river mouths and estuaries, and inlets along the coast (Lawler, 1993; Hubble, 2004). A summary of case studies in India is given below:

Rivers in peninsular India are perennial and face severe environmental and ecological damage.

3.1 Kerala (India)

Bharathapuzha River, one of the 44 rivers in the state, was a flourishing river but the same has dried up. The fate of the Pamba river, third largest river in Kerala, its water capacity has decreased. Palar and its tributaries, Rivers namely Cheyyar, Araniyar, Kosathalaiyer andthe tributaries of Cauvery, are also facing the water resources depletion.

3.2 Karnataka (India)

Papagni riverbed and catchment is illegally and excessively mined and the river is almost dried up due to depletion of water levels and environmental degradation of villages of river banks.

3.3 Maharashtra (India)

Godavari is a major river of peninsular India, and around six crore people depend on its water. It has been heavily mined in the Marathwada region of Maharashtra, so it is almost dried up most of the year. As per the rules, only 1 meter can be dug up for mining, but it has been dug up 7–9 m leading to drought-prone area. Due to sand mining, other sources like wells and ponds too dried up.

3.4 Tamilnadu (India)

Sand mining in the red hills and Cholavaram lakes hindered the water flow thereby posing a threat to Chennai water supply. It was also noticed that river bridges and railway tracks are damaged by illegal sand mining. Chennai is the most affected city by developmental activities and sand mining. Lakes are mined and developed for domestic purposes. Chennai is the city of lakes; earlier it had the capacity to sustain floods and cyclones due to rich environment and water bodies filling out excess water in the event of flood. In 2016, Chennai faced devastating floods and in 2019 the city went dry with no source of water. Mining activities not only destroy ecosystems but also affect humans on large scale. Chennai is an example of man-made damage to environment.

3.5 Goa (India)

Goa is the smallest state in India has. It has 3 rivers and 44 tributaries. The rivers are Galgibag, Mandovi, Saleri, Canacona, Zuari, Sal, Talpona Tiracol, and Colvale. These rivers and their tributaries cover the entire state of Goa. Most of these rivers originate in Maharashtra and Karnataka. Majority of sand is extracted from Tiracol, Chapora, Guleli, etc. Prior to 1993, sand was mined traditionally with an iron bucket and boats; after economic reforms, Goa's tourism-based industry developed multifold resulting in many infrastructural developments and sand mining with dredging vessels in sea and use of heavy industrial machines to sand mining; this lead to huge damage to coastal aquifers and ecology.

A glimpse of the impact on water storage potential of sand is described below:

- If a 1 m thick layer of sand is removed from a 100 m wide riverbed over a stretch of 1 km, it has the potential of storing 1500 m^3 of water; then the specific yield of sand is said to be 15% or 0.15
- Thus, excavation of this stretch of sand implies that a natural in-channel storage of water of 1.5 million liters is lost
- With every truck load of sand (20 tonnes equal to 12 m^3 sand), a storage potential of 1.8 m^3 of ground water is permanently lost
- Sand mining from river beds harms the environment in two ways—loss of storage capacity of ground water at the location of mining and the means for releasing water downstream to the same extent
- When extrapolated to the entire region, it means there is a loss of ~17 million cubic meter of water storage capacity lost in sandy aquifers

In addition to above losses, there are impacts on sea and river such as the loss of habitat, loss of corals and consequent natural disasters like the flooding of shores, an observable shift in climate patterns, river course change, islands submersion, and ecological damage to mangroves. Further losses consisting of irreversible damage to the biodiversity of coastal areas, riverbanks, delta regions, aquatic life of the sea need to be emphasized.

Rivers in peninsular India are perennial and face severe environmental and ecological damage. The sand mining in Narmada River continues even during the monsoon ban period, riding over the ban on mechanized mining by the National Green tribunal. Illegal mining is rampant and mafia is aggressive in their endeavor. Fig. 17.1 shows river sand mining in Narmada River.

3.6 Anguillan beach (Caribbean Islands)

On the Northeast coast of the Caribbean Islands, between the windward point and Savannah Bay, naturally formed sand dunes across the beach used protect the coastline and its vegetation. However, heavy trucks and earth-moving equipment have been working in major sand removal activity, with no regulation whatsoever. The sand dunes have been reduced to hardly three foot mounds, and those too are being eroded by sea water erosion and of course, sand mining. The cliff on the coastline is now unprotected and unstable.

3.7 Ghana coast

Construction industry near Ghana's coastal regions is dependent on mining of sand and pebbles from the coastal areas. Large-scale construction activity for making bridges, roads, and other buildings is causing environmental damage of the coastal areas at an alarming rate.

3.8 Kwale coast (Kenya)

In Kenya, on the coastal areas adjacent to the Indian Ocean, sand mining resulted in serious health issues and titanium poisoning. Mangroves and coral reefs were damaged and several endangered species like the Stable Antelopes, Tiger Sharks, Colobus Monkey, and Marlins lost their natural habitats.

3.9 Waikato beaches (New Zealand)

Vegetation clearance for sand mining around Waikato beaches damaged sand dunes. Many species of fishes are extinct in the Waikato coastal area. Countries are exploiting other countries' natural environment through sand trade.

3.10 Singapore

Singapore spends 70 bn dollars a year on sand imports, making it the largest importer of sand in the world, to expand its land area through reclamation. In the last 15 years, Singapore has

FIGURE 17.1 River sand mining in Narmada river, India.

increased its land area by 22%. Most of the sand is imported from the Cambodia, Vietnam, Myanmar, and Indonesia. In Cambodia and Vietnam, riverine ecosystems are severely destroyed by sand mining.

3.11 Middle Eastern countries

Despite being covered in sand, the Middle East countries like Saudi Arabia, United Arab Emirates, Kuwait, and Qatar are the major importers of sand for construction and developing the new islands. Importing huge quantities of sand is transforming the landscapes and lifescape of developing countries. Most of the sand is imported from Australia.

4. Best river sand mining practices

The following section describes best river sand mining practices in different countries.

4.1 India

"Sand" is one of the minor minerals in India which is governed by State. Telangana state is one of the leading states in terms of best river sand mining practices and serves as a role model for other states. Fig. 17.2 shows the location of Telangana state and districts in the State.

Salient features of Telangana State Government Policies are as follows:

- River bed sand mining
- Desiltation of irrigation projects
- Decasting sand from private land near rivers to avoid flooding
- Promotion of manufactured sand

Details of sand mining in Telangana State:

- Sand mining for large rivers, thickness of sand bed can be reduced from 8 m to 2 m, and for small streams thickness of sand bed up to 1 m.
- Vehicles not to ply over flood banks of river.
- River sand mining permitted up to ground water table in monsoon.

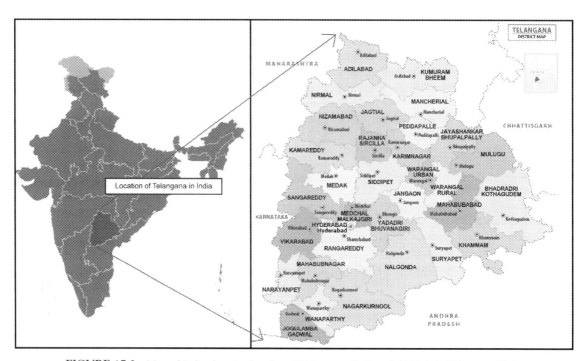

FIGURE 17.2 Map of India showing location of Telangana State and districts in Telangana State.

- All government departments connected with river sand mining jointly inspect the area and provide recommendations—Ground water department, department of mines and geology, or any other department.

Sand mining or transportation of sand is not permitted under following conditions:

- In the areas where ground water is affected or over exploited ground water.
- Near bridges, dams, or drainage structures like weirs up to 500 m from the structures
- Near river banks, up to 15 m from the bank or 1/5th the width of the streambed
- Within rivers or streams where the sand layer has become less than 2.00 m in thickness
- May be permitted in large perennial rivers (Godavari, Krishna, Pennar) without affecting irrigation, water for drinking, or industry.

Mechanism evolved for curbing Illegal Sand Mining in Telangana State:

- Establishment of check posts at vulnerable locations.
- Arresting the unauthorized transportation; registration of Transport Associations and vehicles transporting sand with Telangana State Mining Development Corporation (TSMDC) is mandatory to discriminate from bringing sand from unauthorized sources.
- The State has constituted mobile squads for each district, in which officers from Police, Revenue, Transport and Mines and Geology Departments to prevent illegal sand mining, and their activities are directly supervised by District Collectors.
- Other than M/s TMDC Ltd., no other organization is allowed to extract and transport sand.
- Fine is made 3 times for second offense. Any vehicle other than TSMDC registered for transporting sand is seized.

- Mechanism developed to dispose off seized sand.

4.2 Malaysia

The following case study is described from the case study of CUCIGALI PASIR SUNGA.

Initial preparation of the operating site

Entrance Exit: The contractor is required to provide access to the operating site. The intersection between the dirt road and the tar road should be paved with gravel for about 50 m. This is to prevent dirt from the soil and dust from contaminating paved roads.

Water and Electricity Supply: Operators must ensure that the operating site has access to water and electricity supplies. For areas that do not have TNB electricity supply facilities, operators must provide alternative electricity supply such as generator sets. Clean water supply should be provided to workers at the operating site as most operating sites do not have access to clean water supply sources.

Cabin Site and Weight Bridge: Appropriate location should be identified before equipment is brought into the operating site to facilitate installation. Cabins and Weighing Bridges are usually placed near the entrance to the operating site. This makes it easier to control traffic and avoid confusion when the docket is issued to the sand customer. Operators also need to provide fuel (diesel) and genset storage sites to avoid environmental pollution.

Sand Stack Storage Site: Sand operators must provide an open area with a distance of at least 20 m from the river bank to be used as a storage area for sand piles before selling to customers.

Sand Washing Site: Sand extracted from rivers is mostly mixed with impurities such as rubbish, wood of chips, and other domestic waste and is known by the term dirty sand. Dirty sand is usually poor quality and can only be sold at a minimal price. To improve the quality and selling price of dirty sand, sand operators have an alternative to wash the sand. Sand washing can be done using a trough and water pump. The sand washing site should be at least 20 m from the river bank or any water source. Sand operators need at least 2.5 acres of land to build sand washing sites. **Excavator Site (RB):** To facilitate the movement of the excavator (RB) during the river excavation operation, the path on the river bank should be cleaned at the beginning of the operation.

Method of operation

There are several methods used to remove sand from the river bed. One among the commonly used methods is to use an excavator or better known as RB. For rivers with a width of more than 40 m, the use of gravel pumps placed on a float (Pontoon) is more suitable because sand dredging should be in the middle of the river ("Middle third").

Excavation method (Ruston Bucycrus)

For rivers with a width of less than 40 m, the use of RB Excavators is recommended by DID because the impact on the slope of the river bank during dredging is less than that of hydraulic excavators.

With RB, the slope of the river bank will be preserved as well as the work of extracting sand from the riverbed is done. The use of RB requires a large area of cliffs to place sandstones extracted from the riverbed. The length of the "boom" can reach up to 15 m. The pile of dirty sand will be taken to the sand washing site by truck. The sand brought to the washing site will be dumped on the "ramp" for washing using a rapid jet of water; the sand will be separated from the mixture of foreign matter and collected at the edge of the trough to be separated and dried. Fig. 17.3 shows recommended procedure of removal of sand from river bed with dragline excavator.

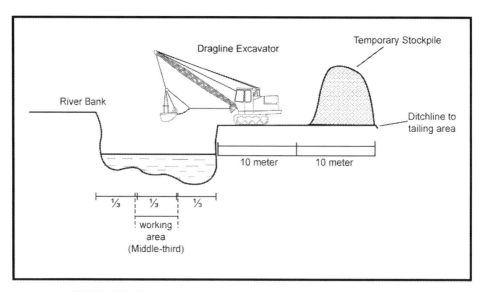

FIGURE 17.3 Recommended dredging procedure for removal of sand.

Fig. 17.4 shows RB excavator on the bank of river. Dirty sand having clay and mud is separately stacked as shown in Fig. 17.5. On the other hand, sand washing bars are reshown in Fig. 17.6. Sand washing process is shown in Fig. 17.7.

Gravel pump method

Gravel pumps are used to extract sand from riverbanks having a width of the more than 40 m. RB is not suitable for use because it cannot reach the middle part of river. For the use of gravel pumps, the required area of sand processing area is at least 1 ha. This is because the construction of a larger waste reservoir pond is needed to prevent pollution to the quality of river water. The horsepower of the pump used depends on the width and depth of the river. The suction strength of the pump depends on the distance between the gravel pump and the sand processing site. Sand extraction using pebble pump is shown in Fig. 17.8. Sand wash bar is shown in Fig. 17.9. Sand processing bar (box) is shown in Fig. 17.10. The capacity of the pond must be sufficient to accommodate the quantity of silt that comes out after processing the sand through the built trough.

The size of the constructed trough is determined by the amount of dirty sand to be washed. There are various types of trenches used for sand washing purposes. The constructed troughs have floors made of iron rods ("grizzly bars") arranged at a distance of 8–10 mm from each other. There is also a floor built using iron mesh ("wire mesh") to get a finer quality of sand. Fig. 17.11 shows schematic diagram showing sand extraction pump.

Table 17.1 shows summary of common sand mining methods from river.

FIGURE 17.4 RB excavator.

FIGURE 17.5 Dirty sand stack after digging.

FIGURE 17.6 Sand washing bar.

FIGURE 17.7 Sand washing process.

FIGURE 17.8 Sand extraction using pebble pump.

FIGURE 17.9 Sand wash bar (Trommel).

FIGURE 17.10 Sand processing bar (Box).

It is clear that river extraction remains an important source of aggregates in most countries, particularly where access to hard rock resources is geologically unavailable or unreasonably restricted.

River extraction can be carried out in full harmony with nature and can provide flood control and can foster biodiversity. Sand requirement of each country varies as well as local conditions of availability of sand are different.

4. Best river sand mining practices

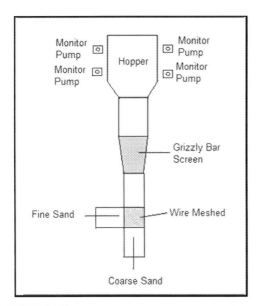

FIGURE 17.11 Schematic diagram showing sand extraction using pebble pump.

TABLE 17.1 Summary of sand mining and washing methods.

Name of method	Purpose	Type of equipment used
Mechanized sand mining	Sand extraction	Back hoe excavator, tipping trucks
RB excavation method	Sand extraction	Ruston Bucycrus excavator
Sand extraction with gravel pump	Sand extraction	Pebble pump
Sand extraction with pebble pump	Sand extraction, washing	Pebble pump, grizzly bar screen, wire mesh screen
Sand wash bar (Trommel)	Sand washing	Sand wash bar Trommel make
Sand wash bar (box)	Sand washing	Grizzly bar

Responsible river extraction requires very detailed prior study of the river hydrogeology with particular focus on annual/seasonal replacement of sedimentation to ensure that the extraction is sustainable.

It is also necessary to ensure that river flooding patterns are not negatively impacted. In fact, judicious river extraction can help alleviate flooding by creating buffer water storage.

It is also important to study the impacts of rainy and dry seasons. In certain countries, extraction is prohibited during the monsoon season.

Besides, the baseline biodiversity study is also important, not just within the river, but also in its surroundings. The case studies show that biodiversity can be enhanced by well-planned river extraction.

4.3 Canada (Alberta)

In Alberta, sand and gravel predominates, with 40% coming from rivers. River extraction now requires extensive hydraulic and hydrogeological studies. A detailed Code of Practice is now being developed with the industry. Extraction is not allowed in the main river channel, debate on "nonactive" areas. Policy guidelines for aggregate operations within surface water bodies are shown in Fig. 17.12.

4.4 Colombia

Forty percent of all aggregates come from river mining, up to 90 percent in some regions. This is fully legal, though there is a large illegal sector. Very detailed code is being developed. Sedimentation is closely analyzed. Fig. 17.13 shows plant type of structure of river. In the origin, streams join together to form a bigger stream. River can be explained as youth, old age, and maturity stage in delta region. Profile of river shows gradient in different places. Legal framework of Colombia is shown in Fig. 17.14. The code requires analysis of river geology, geomorphology, geotechnics, vegetation cover, hydraulics, hydrogeology, and location with

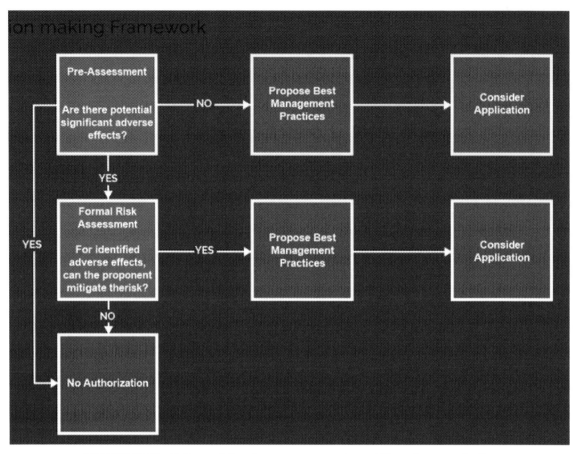

FIGURE 17.12 Policy guidelines for aggregate operations within surface water bodies.

respect to local communities. Characterization of exploitation of areas is shown in Fig. 17.15. Determination is made with respect to mining and environmental legislation. Operational compliance is closely monitored by regional authorities and has to account for wet (El Niño) as well as dry (La Niña) periods.

4.5 China

In the past, most of the extraction was from rivers. This has been phased back, now with 70% sourced as crushed stone. The remaining 30% still comes from rivers. This activity is now very strictly controlled in very detailed Technical Code. The focus is on river integrity, flood prevention, and safety.

Technical code of safety production of river sand mining in Hebei Province

In recent years, with the increasing number of river sand mining operations, the scale of sand mining is expanding and safety problems are becoming more prominent. It is important to ensure community safety from river flooding promotes industry development and maintains social license through sustainable mining of river sand. In October 2016, the Provincial Water Conservation Development organized the relevant

4. Best river sand mining practices

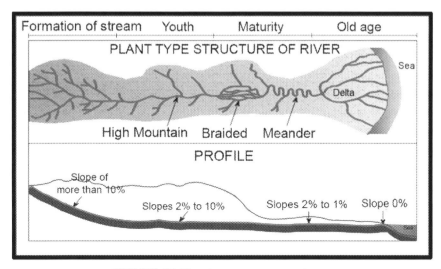

FIGURE 17.13 Plant type structure of river.

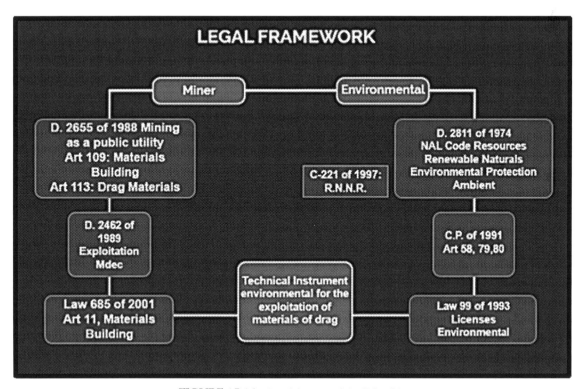

FIGURE 17.14 Legal framework in Colombia.

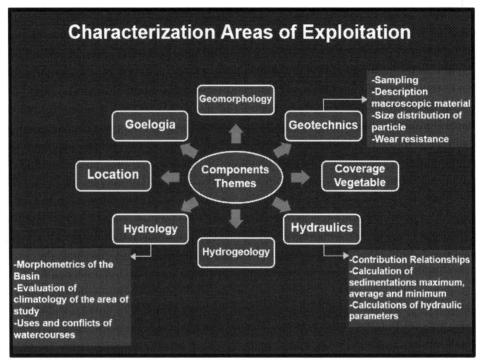

FIGURE 17.15 Characterization of areas of exploitation in Colombia.

technical personnel to investigate the current situation of river sand mining in province. They conducted extensive data collection from China and abroad, looked at actual management of river sand mining in the province, conducted many investigations and consultation, to complete the "Technical specification for safe mining of the river." On September 6 2017, the Provincial Department of Quality and Technical Supervision jointly issued the Technical Specification for the Safe Technical Supervision and jointly issued the Technical Specification for the Safe Production of River Sand Mining (DB13/T 2549-2017), which was officially implemented on September 16 2017.

The implementation of the Technical Specification for the Safe Production of River Sand Mining (DB13/T 2549-2017) provides a strong technical basis for the management of river sand mining in the province. It will effectively regulate and guide the safe production of river sand mining and has important and far reaching significance for protecting the community from river flooding, ensuring the safety of buildings and villages and protecting the ecological environment.

4.6 New Zealand

Main area of river extraction in New Zealand is around Canterbury (South Island). Aggregate supplies meet 40% of local market. Original concerns were river flow and flooding. Now, the focus is on biodiversity preservation with significant successes achieved in fish spawning and bird nesting. Detailed Code of Practice has been developed. The Quarry Code of Practice is supplemented by aggregate industry.

Guiding principles

The objective of this standard code is achieving good environmental practice by:

(i) Recognizing and protecting the natural character of braided river beds

(ii) Considering habitat and morphological diversity

(iii) Minimizing discharges of sediment or contaminants

(iv) Mitigating/avoiding effects of activities on native bird breeding, chick-rearing, and associated habitat

(v) Recognizing the sensitivities of archaeological or historic sites

(vi) Being aware of and responding to cultural values

(vii) Recognizing and considering emergencies contingencies

(viii) Mitigating against the transfer of aquatic pests

(ix) Managing flood and erosion risks.

4.7 United Kingdom

In the United Kingdom, river dredging is allowed only for flood control purposes, a significant issue. Of more relevance is the sea-dredging of aggregates, which supplies 14 million tonnes/year or 5% of the UK market. The impacts of sea-dredging have been extensively studied in terms of potential physical biological, social, and economic impacts. Based on these studies, the BMAPA Good Practice Guidance (2017) has been developed and is rigorously applied and transparently reported. In parallel, a PR communication document has been developed to ensure public understanding of the benefits of sea-dredging. Figs. 17.11 and 17.12 show dredging at the United Kingdom.

5. Sustainability

Regardless of this danger of running out of sand, our civilization is dependent on this natural resource. So sand mining will not stop; we should better adopt ways to recycle sand as much as possible (Gallagher and Peduzzi, 2019). The legal framework is not enough until illegal practices are curbed.

Sustainability has been the buzz word and the words like sustainability, drone or satellite imagery, and CCTVs have relegated to becoming just fancy words. On ground, nothing concrete is done toward monitoring, persecuting illegal, sand mining activities. The only truthful way of sustainability therefore is when the waste of one process becomes the input for other. It is good that we live in a world of substitution. We can use manufactured sand in the place of river sand and still achieve the same sand properties. We need to think of a circular economic system, even in regards to sand mining. Only then we can ensure that we do not run out of sand. The legislative ambiguity and dilutions are the major challenges in any country vis-a-vis sand mining. In India, for example, Sand is a minor mineral and the sand mining projects are B2 projects which do not require scoping and public consultation. There is overdelegation of powers through latest amendments and the laws in vogue lack implementation on ground.

6. The alternatives

Solutions for a sustainable future:

Sand is an important economic good in the present world. Millions of people in many developing countries work in sand mining activities. However, sand mining is sustainable only if it is extracted in a scientific way up to prescribed quantities. Beyond that, sand mining poses a threat of irreversible damage to ecosystems by decreasing natural resilience capacity. In the present, markets' demand around the world for sand is very high. Many technologies are in development for the alternatives to river sand as its extraction is environmentally unviable at present extraction rates for the future. Regulations by the governments in developing countries are encouraging the exploitation of sand resources and damaging the ecosystems in the process and illegal mining of sand emerged as an alternative black market for sand in Asia, Africa, and Latin American countries (Gavriletea, 2017; Koehnken and Rintoul, 2018).

Realistic alternatives to very high extraction of sand are grouped broadly as follows:

- Adapting circular economy, instead of linear economy to reduce or avoid using valuable resources
- Use of Alternative materials
- Best legal and administrative frameworks and its implementation to minimize extraction impacts.

Some specific alternatives could be enumerated as listed below:

1. In place of in-stream mining or in-channel mining of sand, one could use sand from desilting dams, reservoirs, or resort to off-channel dry mining or mining deposits from flood plains
2. Sand in concrete or mortar could be partially replaced by reusing or recycling waste from construction or demolition sites as manufactured sand. In this chapter, we have taken up a case study and presented the use of waste host rock of some minerals in India which are lying in dump yard for decades now.
3. Usage of fly-ash concrete or fly-ash blocks or gypsum blocks could be termed as "Green Concrete" and encouraged to be used instead of conventional concrete blocks, burnt bricks, or geopolymer concrete.
4. Waste products from various industries like quarry dust from mining, foundry sand, sheet glass powder, slag from copper smelting, slag from stainless steel, and blast furnace slag (granulated) could be converted to manufactured sand

5. Another unique method of bringing down the usage of sand is to relook at construction materials differently:
   - Energy efficient, biomass-based materials technology
   - Infrastructure sectors where it can work: water harvesting, buildings, retaining walls, road, etc.
   - It has a potential of bringing down the use of sand by 30%
6. Engineered bamboo to construct walls, slabs, roofs in residential and public building
7. Waste plastics for partial replacement of sand in concrete/mortar
8. Aggregates made from End-of-life Tires (UNEP, 2019)
9. Geocell Reinforcement - Another form of high strength synthetic plastic reinforcement for roads.
10. In the developed nations Geogrid and Geocell Reinforced Roads are fast replacing concrete roads and pavements. Fig. 17.16 shows geogrid reinforcement in roads

To elaborate on the possibility of some important research studies on the alternative sources of sand, we discuss the following:

6.1 Recycle

Whenever an old building is destroyed, we can reuse the construction and demolition material to erect the new building in its place. There has been much advancement in using C&D material in recent years. Similarly, demolished material from disasters can be used to construct new buildings, rather than using virgin material. **Development indicators like GDP should be replaced by** economies of the world who can adopt different indicators such as Genuine Progress Indicator (like Maryland and Vermont in the United States) that incentivize reuse of materials (by marking it as a positive development) and dis-incentivize use of virgin material.

FIGURE 17.16 Geogrid reinforcement in roads.

This can greatly reduce the demand for sand in construction. After all, environmental accountability has to be developed. Research on developing artificial sand from recycled bottles is another area that the authors are exploring and need support.

6.2 Use of sand dredged out of reservoirs

Sand deposits in reservoirs are a big headache for dam operators. Leaving the sand there would mean less reservoir capacity. Removing it incurs a cost. Dam operators and sand miners can both benefit by mining this sand rather than extracting virgin sand from floodplains.

In India, for example, illegal and unauthorized extraction of sand without valid permits or challans granted to approved leases is rampant (Sharma, 1999; Martinez-Alier et al., 2015; Rege, 2016). It has been painfully noticed that when there is rainfall, river or stream flows but dries immediately after that. Earlier, despite getting rainfall for only 40–50 days over 4 months in a year, our major rivers still flowed round the year because they received groundwater. Now indiscriminate river sand mining has removed the sand, and the river cannot store water. Logically, it can be concluded that river forms aquifer system. Aquifers act as storage house for water and water flows downstream. Rivers normally get groundwater and not through recharging mechanism which is general understanding. It is simple actually—a river is the lowest point in an area and therefore groundwater flows from the surrounding areas toward the rivers. Look at all rivers today, even though they are dry on top, there are wells in the river bed which supply water. If the sand is gone, this whole system will be destroyed and rivers will act only as pipes.

6.3 Start with partial replacement of sand

In India, any materials whose characteristics confirms to *IS:383–1970* can be replaceable (IS 383, 2016). On research, only M Sand has been recognized as the equivalent replacement for sand. There are certain slags which can be considered for partial replacement of sand. Besides, the unavailability of slags for mass consumption is also to be noted.

6.3.1 Copper slag

India produces 6—6.5 million tonnes of copper slag out of 33 million worldwide production of copper slag. Sand can be replaced up to 50% in concrete, while maintaining criteria of good strength and durability is the finding of Peduzzi S. Al-Jabri et al. (2011). On the other hand, Central Road Research Institute (CRRI) recommends consuming copper slag up to 40% while maintaining cohesiveness in pavement grade concrete. Advantage of concrete with copper slag is that as compressive strength of the said concrete is 20% higher as compared to conventional concrete.

6.3.2 Granulated blast furnace slag

India currently produces 10 million tonnes of blast furnace slag from iron and steel industry. There is an increase in compressive strength of cement mortar with increase in GBFS by replacing cement (Nataraja et al., 2013). Various researchers' study show that GBFS can be used up to 75%.

6.4 Bottom ash

India produces more than 100 million tonnes of coal ash. 80%—85% of total coal ash is fly ash and balance is bottom ash. There is good demand for fly-ash in cement industry. However, there is very little demand for bottom ash. As per the study conducted in Malaysia by Mohd Syahrul Hisyam, special concrete can be produced by replacing natural sand up to 30% with sand produced from bottom ash.

6.5 Foundry sand

As per the 42nd census survey of World Casting Production conducted during 2007, India is the fourth largest producer of foundry sand (7.8 million tonnes per annum). Even though there is high silica in foundry sand, metal industry discards most of its foundry sand. One of the research studies shows that up to 30% foundry sand can be used for production of concrete in sustainable manner (Bhimani et al., 2012).

6.6 Construction and demolition waste

With the aging of most of the metro cities in India, construction demolition waste will continue to increase year to year basis. Most of the cities face problem of disposal of construction and demolition waste. Delhi already has a recycling unit in place and plans to open more to handle its disposal problem. Construction demolition waste (crushed and sieved) can replace sand to the amount of 25%. Thus, it is absolutely necessary to shift from linear economic model to that of circular economy where waste of one industry is used by other thus reducing the carbon footprint. In case of C&D waste, however, availability at a particular location is to be ensured before considering them as replacement to sand in a mix design.

6.7 Quarry dust

In India, various opportunities of replacing natural sand with importing sand, quarry dust, and filtered sand have been explored (Radhikesh Nanda et al., 2010). Most of the quarry dust is not utilized and dumped as waste as there are finer particles more than 150 micron. However, solutions are available by using classifiers and limiting microfines up to 25% to maintain good strength of the concrete.

6.8 M sand from natural stone

Many metro cities in India have manufactured sand (m sand) units. 100% of natural sand can be replaced with M Sand.

7. Comparison of river sand and manufactured sand

Table 17.2 shows at a glance comparison of properties of river sand vis-a-vis M sand.

7.1 Revisiting aggregate and sand standards

In Singapore, ready mixed concrete is produced using partial replacement with sand produced from copper slag. Hence, it has become essential to replace existing aggregate standard set by Building Construction Authority (BCA) for aggregates as well as sand.

In India, various parts of the country face actuate shortage of sand due to booming construction activity in metro cities as well as mega infrastructure development. Locally, various alternative waste materials are available and can be converted into aggregates and sand with proper Research and Development studies in the nearest laboratories. India has to revisit standards so that various alternative materials can be accommodated and converted into aggregates and m sand. Even though new building material may not be economical,

TABLE 17.2 Properties of river sand vis a vis M sand.

Properties	River sand	M sand
Definition	A type of natural sand which is available naturally and is extracted from river banks or river beds	M sand or manufactured sand is manufactured fine or crushed aggregate. It is a type of sand that is artificially made in industries by crushing natural stone
Source	Naturally available	Quarry site
Manufacturing	Extracted by digging	By crushing
Shape and size	Rounded in shape and has a smooth surface	Cubical and angular in shape and has a rough texture
Wastage	Sieving wastage − More wastages	Already sieved − Wastage in form of ultrafines
Moisture content	Moisture is present between sand particles	Less or no moisture content
Strength	Compressive and flexural strength of river sand for same mix ratio as compared to m sand is less	High in case of well-graded M sand
Setting time	The workability of concrete is more	Less
Permeability	When used in concrete permeability is poor as compared to M sand	Permeability of concrete with M sand is very poor
Water absorption	The water absorption of river sand is 1.5%−3%	The water absorption is between 1% and 4%
Presence of silt and clay	Silt and clay present	No silt or clay content
Presence of oversized material	Contains oversized material, hence washing and sieving necessary	No oversize as it is manufactured artificially

(Continued)

17. Best river sand mining practices vis-a-vis alternative sand making methods for sustainability

TABLE 17.2 Properties of river sand vis a vis M sand.—cont'd

Properties	River sand	M sand
Presence of marine products	May contain sea shells, tree barks, grass, algae, etc., which needs to be removed	M sand does not contain marine products in it, thus saving removal time and cost
Presence of micro fine particles	As it is formed naturally, they contain less microfine particles	Contains microfine particles which need to be separated using wet or dry classifiers
Slump	More slump	Less slump
Bulk density	1.44 gm/cm^3	1.75 gm/cm^3
Specific gravity	2.3–2.7, depends on rock in catchment area	2.5–2.9, depends on parent rock
Bulkage	Bulkage correction is required in design mix	No need of bulkage correction when M sand is used
Ability to hold surface water	The ability to hold surface water by river sand is up to 7%	The ability to hold surface water is up to 10%
Adulteration	Owing to shortage, chances of adulteration with saline sand in coastal area are common	Adulteration of m sand is less. Quality vis-a-vis presence of microfines must be strictly checked from time to time for quality assurance
Quality	Quality differs from one to other	The quality of M sand is better as it is manufactured in a controlled environment
Environmental factors	More ecological impacts	Less ecological impacts
Availability	Becoming scarce. Constraints of location.	Can be manufactured at closer to construction sites so more assured supply
Cost	More expensive	Cheaper
Usage	RCC, plastering, Brickwork	RCC, brickwork, blockwork

consuming such waste material will have long-term benefit of ecological balance (Gandhimathi and Nidheesh, 2013).

8. A case study on sand from waste rocks

8.1 Manganese and soap stone mines in India

In almost every mineral bearing region, OB mining and land degradation are inseparably connected. Across India and elsewhere, minerals like magnesite, manganese limestone, and soapstone have a cover of hard and compacted waste rock which is treated as Overburden (OB) and dumped out. A study was undertaken in two of these mines to study the compatibility of these rocks as alternative sources of sand and even aggregate. Manganese mines of MOIL near Nagpur and Soapstone mines of ASDC (Golcha group) near Udaipur, Rajasthan, in India were extensively studied and findings are promising as these can not only prove to be alternative source of material for sand (both M sand and/ or stowing sand) but also produce suitable aggregates. This would not only promote mineral conservation to a large extent by putting less pressure on mining of natural stone specially for sand production but also free up land parcels blocked under these huge OB dumps lying unutilized for decades. The replacement can go up to 50%.

8.2 Manganese mines of MOIL

M/s MOIL Limited (MOIL) operates seven underground mines (Kandri, Munsar, Beldongri, Gumgaon, Chikla, Balaghat, and Ukwa mines) and four opencast mines (Dongri Buzurg, Sitapatore/Sukli, Persoda, and Tirodi).

The associated rocks or the host rocks of Dongri Buzurg mines, the largest open cast manganese mines in India, formed the core of this study. The Dongri Buzurg mine is situated at Village Balapur Hamesha, taluka/mandal Tumsar, in Bhandara district of the state of Maharashtra, approximately 130 Km from Nagpur, which is the most prominent among nearby cities. These mines are operated by fully mechanized opencast mining method and spread over 234 Hectares. Dongri Buzurg mines produce approximately 300,000 MT of manganese ore per year. Gondite is regional metamorphic host rock for manganese ore identified as manganiferous and noncalcareous rock. Other salient feature of Gondite rock includes bedded characteristics having spessartite (a manganese almandine garnet) and quartz with or without manganese silicates. These mines have a stripping ratio ranging from 12 to 15 and mining operations are in existence for last over 100 years. Dongri Buzurg has extracted large quantity of waste during the course of mining of manganese. This waste is dumped into the designated dump yards both inside and outside the mining lease area (Shome et al., 2020). All the four dead dumps and the two active dumps are found to have a mix of hard rocks, weathered rocks, deleterious material, and soil, segregation of which for any purpose can currently be dearer, particularly when plenty of rocks are yet to be freshly mined (Fig. 17.17).

MOIL generates huge stocks of unused rock during the course of its manganese mining and dumps them onto nonmineralized zones as waste dumps. Now it had problems associated with this waste rock disposal system which was primarily that of (a) land degradation and

FIGURE 17.17 Gneiss rock sample at Soapstone mine.

search for new dump yards and (b) good rock was getting wasted and, in a way, added to woes of mineral conservation. Prominent exposures of hard suitable Gneiss are seen between CH 25 and CH 42 (Fig. 17.18).

The gneisses in between these chainages are crudely foliated granitoid type. Certain rocks show strong foliation, streaked with banded Gneiss varying thickness of dark bands enriched in biotite and light bands consisting of quartz and Felspathic material. Darker rocks containing biotite were clearly avoided and all grayish

FIGURE 17.18 Weathered gneiss rock sample at soapstone mine.

white or pinkish gray rocks were collected for aggregate and M sand crushing.

These host rocks were tested and found to have low water absorption and high strength characteristics and therefore present a strong case for qualifying as aggregate in pavement construction in accordance with MORTH and BIS standards as far as physico-mechanical properties is concerned. Further, it is being innocuous to alkali and it also makes strong case for concrete grade aggregate.

8.3 Soapstone mines of ASDC

M/s Associated Soapstone Distributing Company Private Limited (ASDC) is the leading producer of soapstone in India. Their mining activities for extraction of soap stone are concentrated in few districts of the southern part of the Indian state of Rajasthan. The mines under purview of this study are located at Devpura Village, Sarada Tehsil of Udaipur and Devla, and Bhungabhat villages of Dhariawad Tehsil of Udaipur. ASDC produces approximately 300,000 tons of soapstone per year from these mines. These mines have a stripping ratio ranging from 25 to 30, averaging to 27 and mining operations being in existence for last 10 years. ASDC has extracted an approximate volume of 75 million cum of waste during the course of mining of soapstone. This waste is dumped into the designated dump yards inside mining lease area. In addition, an estimated volume of 100 Million cum of such waste rock-mass still remains to be mined out and disposed from in situ depositions to win planned quantity of soapstone over the balance life of mining lease (Fig. 17.19).

Devpura area is located at 240 18′ 01.69″ N Latitude and 730 46′ 18.29″ E longitude. Devpura falls under Sarada Tehsil of Udaipur District of Rajasthan. It houses five different mining zones with a cumulative lease hold area of 150 ha approximately. Quartzite is present here overlying mineralized zone bearing soapstone. Majority of soapstone production at

FIGURE 17.19 Soapstone mine.

Devpura area is achieved from Ganesh and Mahadev zones which together have a leasehold expanse of 80 ha approximately (Fig. 17.20).

This area not only makes available five to seven million tons of rock mass in waste dump yard, but also promises progressive availability of more than 25 million tons of in situ quartzite deposits overlying mineralized zones which have to be removed in due course of operation to exploit proven reserves of soapstone (Fig. 17.21).

The in situ volume of waste rock mass in these regions exceeds 25 million tons. The physical appearance of the rock mass is sound and compact with a reddish color imparted to its exterior presumably due to oxidation and chemical weathering of iron contained in the rock. Natural deposits in these regions contain approximately 10%—15% of soil and other disintegrated strata which is not useful for manufacturing of aggregates. The top layers with 1—3-meter thickness look friable and noncompact and need to be removed and stocked separately to avoid contamination of good and crushable rock mass. The western most part of the deposit has the most massive and compact sheet deposits which require no segregation and can directly be used for crushing purpose. The test results are tabulated in Table 17.3.

9. Need of future research

This study was limited to finding different ways of converting waste rocks and waste products of one industry to use in other industry and myriad of opportunities and challenges thrown up by circular economy, adopting which is critical for sustenance of ecological balances. A need is felt to fill knowledge and data gaps leading to informed public discourse and acceptability. Extraction of sand with chemical method does not exist. However, the possibility of extraction sand from tailings after beneficiation may be an opportunity for further research. Various studies have been carried out on impact

FIGURE 17.20 Host rock at hanging wall of MOIL mine.

FIGURE 17.21   Waste rock dump yard at MOIL mine.

TABLE 17.3   Comparison of test results of rock samples from Golcha and MOIL mines.

S.No.	Name of test	Test results		Limits (as per IS:383-2016/MORT&H 5th revision)	
		MOIL	GOLCHA		
1	Water absorption	0.21	0.66	Max 2.0	
2	Specific gravity	2.63	2.43	—	
3	Crushing value	21.0	24.0	<30 for wearing course	<45 for nonwearing surface
4	Impact value	19.0	22.0	-do-	-do-
5	Los Angeles abrasion	22.0	19.0	-do-	-do-
6	Alkali reactivity	Innocuous	Innocuous	—	

of manufactured sand in the performance of concrete (Elavenil and Vijaya, 2013; Jadhav and Kulkarni, 2013; Vijayaraghavan and Wayal, 2013; Nanthagopalan and Santhanam, 2011). Further research is required for various alternative sand vis-a-vis performance of the concrete.

It is important to carry out further research in the area of converting sand from other than river sources and relook at the policy and legal/institutional frameworks as to how can we make them more ecosystem sensitive and democratic.

10. Conclusions

As per moderate estimate, in the present era, extraction of sand has grown rapidly three times as compared to the replenishment of sand by nature. There is huge pressure on natural sand. Regardless of this fact, the fact remains that as long as demand of sand exists, there will be supply, be it legal or illegal. We are running out of sand, but realization has not touched the global conscience and regulatory governance is not adequate. It is necessary to rethink how the demand for construction industry can be met. Innovation in construction technology and alternative sources is the only key, which could reduce sand consumption and ease the pressure on natural sand. It includes a radical rethink of our infrastructure and construction projects, redesigning buildings or constructing only what is needed, focusing on alternative sources and sustainability. A task force to assess these alternative sources must be made in each country and further research in this field has become absolutely necessary.

In conclusion we can say that

1. Engineering solutions alone would not be enough to overcome the impending river sand resource crunch.
2. The domination of popular and convenient paradigm that there are infinite sand resources would have to be challenged through constructive policy and legal dialogue and implementing of alternative solutions.
3. The alternative solutions as suggested in this chapter and many more that are being researched globally have the potential, but the solutions would only optimize the consumption of sand by the construction industry.
4. Implementation on priority of the "Strict though Just" policy, institutional, and legal measures with respect to river sand extraction and usage is the need of the day.
5. Concurrently standardization of engineering designs and construction practices and user acceptance to the eco-friendly alternatives

would play an important role in saving our river ecosystem.

Acknowledgments

Authors are thankful to Prof. Dr. Edy Tonnizam Director Geotropik, Universiti Teknologi Malaysia, for encouraging in preparing this book chapter. Authors are thankful to Aggregate India for providing information on case studies based on waste rock. Authors are thankful to Global Aggregate Information Network (GAIN) and their members viz ASOGRAVAS Colombia, China Aggregate Association (CAA), AQA, New Zeland, and UPEG, Europe, for their valuable contribution. Authors are thankful to KSSB Consult Sdn. Bhd for providing details of river sand practices in Malaysia.

References

Ahmad, T., Ahmad, R., Kamran, M., Wahjoedi, B., Shakoor, I., Hussain, F., Riaz, F., Jamil, Z., Isaac, S., Ashraf, Q., 2014. Effect of thal silica sand nanoparticles and glass fiber reinforcements on epoxy-based hybrid composite. Iran. Polym. J. (Engl. Ed.) 24 (1), 21—27. https://doi.org/10.1007/s13726-014-0296-x.

Al-Jabri, K.S., Al-Saidy, A., Taha, R., February 2011. Effect of copper slag as a fine aggregate on the properties of cement mortars and concrete. Construct. Build. Mater. 25 (2), 933—938. https://doi.org/10.1016/j.conbuildmat.2010.06.090.

Bhimani, D.R., Pitroda, J., Bhavsar, J.J., June 2012. A Study on Foundry Sand: Opportunities for Sustainable and Economical Concrete. https://doi.org/10.15373/22778160/January2013/64.

Bradley, W.C., 1970. Effect of weathering on abrasion of granitic gravel, Colorado river (Texas). Geol. Soc. Am. Bull. 81 (1), 61—80.

Cortes, D.D., Kim, H.K., Palomino, A.M., Santamarina, J.C., 2008. Rheological and mechanical properties of mortars prepared with natural and manufactured sands. Cement Concr. Res. 38 (10), 1142—1147. https://doi.org/10.1016/j.cemconres.2008.03.020.

Elavenil, D.S., Vijaya, B., 2013. Manufactured sand, a solution and an alternative to river sand in concrete manufacturing. Int. J. Civil Eng. Res. Dev. 3 (1).

Gallagher, L., Peduzzi, P., 2019. Sand and Sustainability: Finding New Solutions for Environmental Governance of Global Sand Resources. Geneva.

Gandhimathi, R., Nidheesh, P.V., January 2013. Use of furnace slag and welding slag as replacement for sand in concrete. Int. J. Energy Environ. Eng. 4 (1). https://doi.org/10.1186/2251-6832-4-3.

Gavriletea, M., 2017. Environmental impacts of sand exploitation. Analysis of sand market. Sustainability 9 (7), 1118. https://doi.org/10.3390/su9071118.

Hubble, T.C.T., 2004. Slope stability analysis of potential bank failure as A result of toe erosion on weir-impounded lakes: an example from the Nepean river, new South Wales, Australia. Mar. Freshw. Res. 55 (1), 57—65. https://doi.org/10.1071/MF03003.

Hudson, J., 1992. Rock Engineering Systems: Theory and Practice. High Plains Press (JAH), Chichester.

IS 383:2016, 2016. I.: Indian Standard: Coarse and Fine Aggregate for Concrete Specification (Third Revision).

Jadhav, P.A., Kulkarni, D.K., 2013. Effect of replacement of natural sand by manufactured sand on the properties of cement mortar. Int. J. Civ. Struct. Eng. 3 (3), 621.

Jefferson, T.A., Hung, S.K., Würsig, B., 2009. Protecting small cetaceans from coastal development: impact assessment and mitigation experience in Hong Kong. Mar. Pol. 33 (2), 305—311. https://doi.org/10.1016/j.marpol.2008.07.011.

Jiang, L., Lin, H., 2010. Integrated analysis of SAR interferometric and geological data for investigating long-term reclamation settlement of Chek Lap Kok airport, Hong Kong. Eng. Geol. 110 (3—4), 77—92. https://doi.org/10.1016/j.enggeo.2009.11.005.

Koehnken, L., Rintoul, M., 2018. Impacts of sand mining on ecosystem structure, process and biodiversity in rivers. WWF Rev. 159.

Lawler, D.M., 1993. The measurement of river bank erosion and lateral channel change: a review. Earth Surf. Process. Landforms 18 (9), 777—821. https://doi.org/10.1002/esp.3290180905.

Martinez-Alier, J., Temper, L., Demaria, F., 2015. Social metabolism and environmental conflicts in India. In: Nature, Economy and Society. Springer, India, pp. 19—49.

Nanthagopalan, P., Santhanam, M., 2011. Fresh and hardened properties of self-compacting concrete produced with manufactured sand. Cement Concr. Compos. 33 (3), 353—358.

Nataraja, M.C., Dileep Kumar, P.G., Manu, A.S., Sanjay, M.C., May 2013. Use of granulated blast furnace slag as fine aggregate in cement mortar. IJSCER 2 (No. 2).

Radhikesh Nanda, P., Amiya Das, K., Maharana, N.C., 2010. Stone crusher dust as a fine aggregate in concrete for paving blocks. Int. J. Civ. Struct. Eng. 1 (3), 613—620.

Rege, A., 2016. Not biting the dust: using A tripartite model of organized crime to examine India's sand mafia. Int. J. Comp. Appl. Crim. Justice 40 (2), 101—121. https://doi.org/10.1080/01924036.2015.1082486.

Sharma, M.L., 1999. Organised crime in India: problems & perspectives. Resour. Mater. Ser. 54, 82—129.

Shome, S.D., Manekar, D.G., Tiwari, M.S., 2020. Use of technology for sustainable development—a case study. Int. J. Adv. Sci. Eng. Tech. 11 (6).

Vijayaraghavan, N., Wayal, A.S., 2013. Effects of manufactured sand on compressive strength and workability of concrete. Int. J. Struct. Civil Eng. Res. 2 (4), 228—232.

Yong, K.Y., Lee, S.L., Karunaratne, G.P., 1991. Coastal reclamation in Singapore: a review. In: Urban Coastal Area Management: The Experience of Singapore. ICLARM Conference Proceedings, pp. 59—67.

Further reading

Arun, P.R., Sreeja, R., Sreebha, S., Maya, K., Padmalal, D., 2006. River sand mining and its impact on physical and biological environments of Kerala rivers, Southwest coast of India. Eco-Chronicle 1 (1), 1—6.

Ashraf, M.A., Maah, M.J., Yusoff, I., Wajid, A., Mahmood, K., 2011. Sand mining effects, causes and concerns: a case study from Bestari Jaya, Selangor, peninsular Malaysia. Sci. Res. Essays 6 (6), 1216—1231.

Bhatawdekar, R.M., Singh, T.N., Mohamad, E.T., Armaghani, D.J., Hasbollah, D.Z.B.A., 2020. November). River sand mining vis a vis manufactured sand for sustainability. In: Proceedings of the International Conference on Innovations for Sustainable and Responsible Mining. Springer, Cham, pp. 143—169.

Department of Irrigation and Drainage (DID), 2009. River Sand Mining Management Guideline, Malaysia.

Gari, S.R., Newton, A., Icely, J.D., 2015. A review of the application and evolution of the DPSIR framework with an emphasis on coastal social-ecological systems. Ocean Coast Manag. 103, 63—77.

Meenakshi Sudarvizhi, S., Ilangovan, R., 2011. Performance of copper slag and ferrous slag as partial replacement of sand in concrete. Int. J. Civ. Struct. Eng. 1 (No 4).

Myers, N., Mittermeler, R.A., Mittermeler, C.G., Da Fonseca, G.A.B., Kent, J., 2000. Biodiversity hotspots for conservation priorities. Nature 403 (6772), 853—858. https://doi.org/10.1038/35002501.

Patten, D.T., 1998. Riparian ecosystems of semi-arid North America: diversity and human impacts. Wetlands 18 (4), 498—512. https://doi.org/10.1007/BF03161668.

Pedersen, B., Hotvedt, O., 2009. Production and Utilisation of Manufactured Sand: State-Of-The-Art Report. Blindern.

Pettingell, H., 1975. The Kemco RC7. Quarry Manag. 36 (12), 11—14.

Pilegis, M., Gardner, D., Lark, R., 2016. An investigation into the use of manufactured sand as A 100% replacement for fine aggregate in concrete. Materials 9 (6), 440. https://doi.org/10.3390/ma9060440.

Global Aggregate Information Network, www.GAIN. Rajesh Jha, Aggregate India, New Delhi, 2018. Feasibility study report for production of aggregate and M sand from waste dump at the Dongri Buzurg mine, MOIL.

Saviour, M.N., 2012. Environmental impact of soil and sand mining: a review. Int. J. Sci. Environ. 1 (3), 125—134.

Sreebha, S., Padmalal, D., 2011. Environmental impact assessment of sand mining from the small catchment rivers in the Southwestern coast of India: a case study. Environ. Manag. 47 (1), 130—140. https://doi.org/10.1007/s00267-010-9571-6.

Steinberger, J.K., Krausmann, F., Eisenmenger, N., 2010. Global patterns of materials use: a socioeconomic and geophysical analysis. Ecol. Econ. 69, 1148—1158.

Wigum, B.J., 2015. Summary Report. Production of High Quality Manufactured Aggregate for Concrete. Blindern.

CHAPTER 18

Learning lessons from river sand mining practices in India and Malaysia for sustainability

B.R.V. Susheel Kumar[1], Muhammad Faiz Bin Zainuddin[2], Dato Chengong Hock Soon[3], Ramesh Murlidhar Bhatawdekar[4,5]

[1]Mining Engineers Association of India, Hyderabad, Telangana, India; [2]KSSB Consult Sdn Bhd, Tingkat LPH, Shah Alam, Selangor Darul Ehsan, Malaysia; [3]MQA and Training Development Committee, Sectorial Training Committee, Malaysia Quarry Association, Kuala Lumpur, Selangor, Malaysia; [4]Geotropik-Centre of Tropical Geoengineering, Department of Civil Engineering, Universiti Teknologi Malaysia, Johor Bahru, Johor, Malaysia; [5]Department of Mining Engineering, Indian Institute of Technology, Kharagpur, West Bengal, India

1. Introduction

Sand is defined as a "minor mineral" and widely used for building and construction (Ministry of Mines Government of India, 2018a). There are several sources of sand availability as follows:

- **i.** River bed sand extraction;
- **ii.** Removing sand from patta lands close to rivers while making the land feasible for agriculture;
- **iii.** Desiltation of irrigation projects;
- **iv.** Promotion of manufactured sand from rocks;
- **v.** Sand from overburden of coal mines;
- **vi.** Importing of sand from nearby countries.

India needs approximately 700 million tonnes of sand per year. The demand of sand increases 6%–7% yearly; however, the production is seasonally uneven and causes scarcity in several states. Fig. 18.1 shows process of sand mining. The most crucial element in the process is standards and regulation. It must be noted that a few states specifically differentiate their policies for sand or M-Sand.

2. Objectives

- **i.** Need-supply evaluation
  Methodologies used to calculate sand demand vary among states. Besides, some of

---

Risk, Reliability and Sustainable Remediation in the Field of Civil and Environmental Engineering
https://doi.org/10.1016/B978-0-323-85698-0.00002-2

© 2022 Elsevier Inc. All rights reserved.

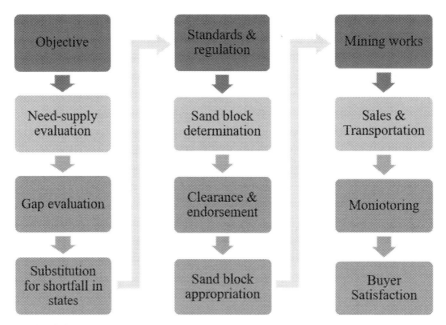

FIGURE 18.1 Sand mining process (Ministry of Mines Government of India, 2018b).

the methods implemented are ineffective; thus, the estimated amount varies significantly compared to demand estimated by scientific methods (Ministry of Mines Government of India, 2018a).

ii. Need approximation methodologies
Scientific need-supply evaluation and resultant gap aid in selection of business models and development of allocation policy for sand as well as alternatives of sand (Ministry of Mines Government of India, 2018a). There are two techniques for approximate sand demand as follows:
- Depends on cement usage: The sand demand is approximated by multiplying cement usage with conversion factor which is the cement to sand ratio. In general, the value of conversion factor is 2.5. If a company uses 10 million tonnes of cement, then it needs 25 million tonnes of sand (Ministry of Mines Government of India, 2018a).
- Substitutions for natural sand: Several substitutions for natural sand shall be considered depending on the shortfall. The substitutions should be promoted with several considerations such as availability during monsoon or peak period, minimize reliance on natural sand, and have ability to reduce price of sand alternatives while protecting natural resources (Ministry of Mines Government of India, 2018a).

3. Gap evaluation depending on need-supply evaluation based on district survey report

Government prepares District Survey Report based on Ministry of Environment, Forest and Climate Change's Sustainable Sand Mining Management Guidelines 2016 (Ministry of Mines Government of India, 2018a). It requires a replenishment study which provides information on yearly sand deposition rate in streams, deposition reach of rivers, and sum of obtainable resources in each state. Although the necessity of replenishment study is acknowledged, it is not carried out extensively at state level. It shows that proper training is needed for relevant authorities who carry out the study. Experts in fields of geology, environment, and hydrology shall be requested to train the relevant staff and conduct first few series of the study in the near future.

3.1 River sand

Sand mainly comprises varying quantities of weathered rock materials which are transported to coastline through streams or wind and/or shells and other hard fragments precipitated by marine organisms. Hence, it accounts processes of numerous periods. Strahler's system of classification designates a segment with no tributaries as a first-order stream. Where two first-order stream segments join, they form a second-order stream segment and so on (Strahler, 1964).

3.1.1 Determination of river sand areas/ blocks

The viable sand reaches shall be determined based on Water, Land, and Trees Act 2002 and 23(1) and (2) of Water, Land, and Trees Rules 2004. The mining and transportation of sand shall be informed and barred in regions which overexploit groundwater and/or the process distress groundwater system. Otherwise the following conditions shall be applied for utilization (Government of Andhra Pradesh, 2010):

i. Mining and transportation using mechanical means shall be prohibited in notified regions except for local use of village or town near rivers. Besides, mining shall be prohibited to specified regions for fourth-order rivers. In addition, it shall be allowed for streams of fifth-order onwards such as Godvari, Pennar, Krishna, etc., as long as it does not affect existing sources for irrigation, industry, and domestic (Government of Andhra Pradesh, 2010).
ii. Sand leaseholders shall ensure a safe distance of 500 m from mining location to any existing structures such as bridges and dams (Government of Andhra Pradesh, 2010).

iii. Trucks transporting sand should avoid flood banks unless at crossing junctions, bridges, or on metal pavements (Government of Andhra Pradesh, 2010).
iv. Mining shall be prohibited within 500 m of groundwater abstraction structures for irrigation and domestic (Government of Andhra Pradesh, 2010).
v. The removal depth shall be extended to 2 m in rivers with >8 m of sand. Extension to 2 m depth shall not be done if rivers with 8 or less than 8 m of sand depth. The removal depth shall be limited to 1 m in minor rivers with sand deposition of 3–8 m. Meanwhile, mining shall be banned in rivers with <2 m of sand deposition (Government of Andhra Pradesh, 2010).
vi. Quarrying shall be prohibited within 15 m or 1/5 of width of river bed from shore whichever is further (Government of Andhra Pradesh, 2010).
vii. Quarrying shall be limited to depths above water table measured during monsoon season, otherwise it disrupts the water table (Government of Andhra Pradesh, 2010).
viii. Observation stations shall be established to observe yearly sand deposition quantity along every stream.

3.1.1.1 Process of identification of specified sand-bearing areas

The Ground Water Department shall periodically conduct detail hydrological and hydrogeological studies in the states and prepares ground water overexploited villages where sand mining will be prohibited till replenishment. Under District Level Sand Committee (DLSC), there is inspection team in each district comprising officials from Revenue, Ground Water, Irrigation, and Mining Departments which take up joint inspection of sand-bearing areas and submit feasibility reports with quantity and depth of the sand mining to be conducted. Based on feasibility, the District Level Sand Committee (DLSC) issues notice for submission of statutory clearances. After submission of the clearances, the District Administration accords permission for commencing the quarry with due adherence to the clearances. Fig. 18.2 shows the process of determination of specified areas with sand.

3.1.2 Alternatives of natural sand

i. M-Sand indicates superiority over natural sand based on tests of durability, permeability, and shrinkage. Besides, Rapid Chloride Permeability Test indicates better robustness of M-Sand mixes (Ministry of Mines Government of India, 2018b).

ii. Sand shall be extracted from overburden of coal mines with minimal expense. One of the states comprises 164 M-Sand manufacturing plants which yield 20 million tonnes yearly. It is resulted from extensive promotion of M-Sand. Besides, other states are also promoting M-Sand and offering multiple incentives to launch M-Sand production units (Ministry of Mines Government of India, 2018b).

iii. Importation of sand from nearby countries is one of the alternative measures. Several states such as Karnataka, Tamil Nadu, and Kerala have already started sand import.

Karnataka has established regulations for import of sand as well (Ministry of Mines Government of India, 2018a).

3.1.3 Geological report

A comprehensive geological report including information of a region such as infrastructure and environment, Differential Global Positioning System (DGPS) survey, geology and geomorphology, river system, laboratory results of sand samples, etc., shall be prepared by geologists prior to tenders. Potential zones of quarry lease shall be determined and delimited based on DGPS, topographic, and geological maps developed through Total Station. Also, the boundaries shall be physically delimited using boundary pillars (Ministry of Mines Government of India, 2018b).

3.1.4 Business model for allocation

The business model is mainly governed by objectives and established conditions of mining. The common models applied are as follows:

i. Notified or controlled pricing model
This model is to assign mining corporations or SHGs or Panchayats based on appointment. Local government acquires the clearances and approvals, whereas the mining is carried out either by the government itself or through hired contractors. Market price is set by the government and revenue is collected accordingly.

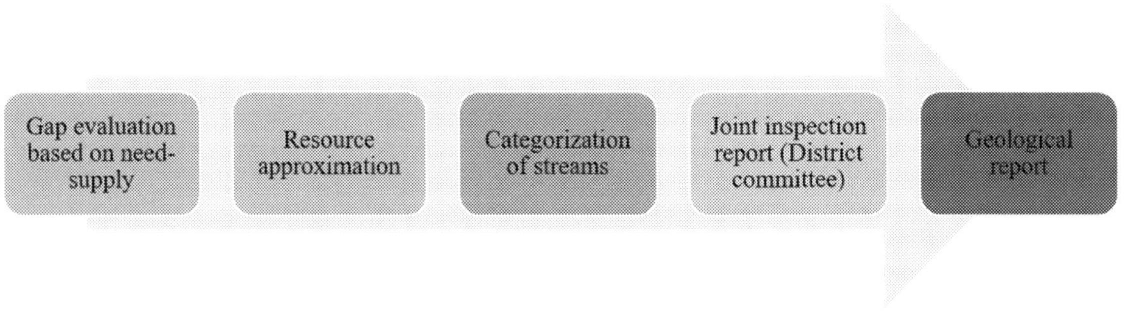

FIGURE 18.2   Process of identification of specified sand-bearing areas (Ministry of Mines Government of India, 2018a).

ii. Market model

The allocation is estimated using tender, auction, or tender cum auction mode in this model.

3.1.5 Clearances and endorsements, recommendations for MoEFCC

Clearances and endorsements are acquired by the state government. Yet, it is left to project advocates in many states. The accountability of pursuing clearances and endorsements is recommended to be given to the contactors, while the department acts as facilitator. A certain time limit shall be set for the clearances process. Also, the authority in charge is responsible to complete it within the set time. In addition, the applications should be made online to ease the process. Existing procedure shall be continued in several states with state departments or Public Sector Units (Ministry of Mines Government of India, 2018a).

3.1.6 Operations

Mining processes must be only carried out as per approved mining plan. Also, it must fulfill with all the regulations specified in Environmental and Other Statutory Clearance (Ministry of Environment, 2020). Mine owner is responsible to ensure compliance of operational method including trade, dispatch, depository, reserve reconciliation, and transportation based on government's monitoring standard procedure.

Operational control of sand reaches counts on the selected model for allocation. Contractor, the successful bidder, is in charge of operations in competitive bidding model, whereas the operations are controlled by selected authorities who carry out mining by themselves or through contractors in nomination model (Ministry of Mines Government of India, 2018a).

Regardless of authorities in control or allocation modal, the mining process shall be carried out by following rules and regulations of environment clearance, conditions of lease deed, and mining procedures. Mechanical mining shall be carried out in rivers which are more than fifth-order upon endorsement of mining operation and notification from the Ministry of Environment, Forest and Climate Change. Withdrawal processes shall be carried out following guidelines of Sustainable Mining Management Guidelines (2016) and allotted circulars set by the Ministry of Environment, Forest and Climate Change although these guidelines are needed to be reviewed (Ministry of Mines Government of India, 2018a).

3.1.7 Sales and transportations

The mechanisms of online sale of sand are subjected to availability of free market in state and price regulation by government (Ministry of Mines Government of India, 2018a). The licensed brokers in a state shall register on internet portal or mobile app by providing information such as stockyard, location, and weekly demand based on endorsed mining plan. Upon registration, the online system would show details of stockyards and price including delivery charges. Thereafter, the dealer shall update availability and price frequently (Ministry of Mines Government of India, 2018a).

The consumers shall purchase sand through mobile app, online website, telephone customer service, and licensed traders. They shall register on portals using Aadhar card (identity card) only (Ministry of Mines Government of India, 2018a). They can select the best available option for them in portals as entire list of stockyards with details such as location of stockyard, available sand quantity, and price will be shown. Besides, purchasing can be done through telephone customer service. Therefore, stockyards shall be developed close to primary consumption hubs in the state depending on approximated need (Ministry of Mines Government of India, 2018a).

3.1.8 Online sale

Establishment of online sale for sand shall be prepared in the context of municipal cities or towns based on terms suggested by Urban Ministry or Census. Every state shall adopt the online sale within a year and advances more subjected to its infrastructure competencies (Ministry of Mines Government of India, 2018a).

3.1.9 Offline sale provisions

Offline sale shall be provided to low demand hubs and villages which is determined by states depending on populace or connectivity. The low demand hubs shall be supplied by local licensed dealer who might do online sale as well (Ministry of Mines Government of India, 2018a).

3.1.10 Synthesis of Ministry of Environment, Forest and Climate Change's Sustainable Sand Mining Management Guidelines, 2016

Securing availability of sand and gravel which are major construction components is vital for nation's infrastructural growth. Streams are exceptional sources although there are various sources of these components. With increasing demand for these materials, it is crucial to protect physical character of the steams, so they can play their natural roles.

3.1.11 Ministry of Environment, Forest and Climate Change notification dated on October 7, 2014

Disputes of sand mining such as destruction of rivers' health, hiking prices, and shortfall continue even after various actions taken by both Central and State Government. Courts such as Hon'ble Supreme Court, respective High Courts of the States, and National Green Tribunal (NGT) have banned mining in certain occasions (Ministry of Mines Government of India, 2018a). Latest occasions where court interfered are as follows:

i. Haryana: A set of new regulations were outlined stating the observation made by Hon'ble Apex Court which were approved on February 27, 2012, following Deepak Kumar's case (Ministry of Mines Government of India, 2018a).

ii. Rajasthan: Supreme Court ordered a blanket ban on mining of sand and bajri on November 2017 because the mines did not carry out environmental clearance (Ministry of Mines Government of India, 2018a).

iii. Tamil Nadu: Chennai High Court instructed to halt sand mining activity in the state within a half year on November 2017 (Ministry of Mines Government of India, 2018a).

iv. Uttar Pradesh: NGT banned illegal mechanized sand mining in Gonda and Faizabad on May 2016. Also, it ordered an investigation on unlicensed activities. In addition, it also instructed the state government to safeguard Yamuna riverbeds in Kanpur from mechanized sand mining on June 2017 (Ministry of Mines Government of India, 2018a).

v. Uttarakhand: High court banned mining in the state on March 28, 2017, for 4 months during when no new lease or license can be issued (Ministry of Mines Government of India, 2018a).

vi. Maharashtra: NGT order on May 30, 2017, is related to illegal sand mining from middle river Bhima, Sholapur (Ministry of Mines Government of India, 2018a).

4. Availability of sand and regulatory mechanism to meet local requirements

Streams of I, II, and notified overexploited III order shall cater the local requirement exclusively in the surrounding villages for domestic needs free of cost and exemption from payment of seigniorage fee. The regulatory mechanism shall be evolved by the District Administration having movement of sand as a unit. In any unit when no Ist–IIIrd order rivers are available to meet local necessities, the administration shall identify up to 5.00 Ha area in IVth–VIth order streams. The sale price of sand for local use shall be either equivalent to the actual cost or the cost as decided by the district administration. The district administration in consultation with the State Public Sector Undertaking (PSU) or state-owned "**Sand Mining Corporation**" shall fix the sale price for sand extraction from IVth–VIth order streams and sand extraction through desiltation to cater the exclusively local needs. Every district administration based on the conditions evolved mechanism for making availability of sand for local use. Establishing a "**Sand Mining Corporation**" shall be mandatory as the river sand has to be monitored by the respective governments in order to supply at affordable prices to the people and maintain ecosystem for the availability of sand to the coming generations.

4.1 Sand monitoring committee

The government shall constitute the DLSC to determine and monitor sand-bearing areas in streams of more than fourth order by appointing district administration as chairman. The Government shall constitute State Level Sand Committee as well (Ministry of Mines Government of India, 2018a).

4.1.1 Implementation of information technology

Role of technology is inevitable when aiming for a transparent system as it aids in enhancement, monitoring, and control the system. It can be incorporated in all the stages of sand mining process to enhance system's transparency. Therefore, this section aims to determine the potential incorporation of IT to enhance the system through technical examination of all stages of the process. Current IT systems in the primary sectors such as allocation of sand reaches, orders, extraction monitoring, and delivery process were examined comprehensively.

Various technological equipment is utilized to monitor and improvise the entire operation from allocation of sand-bearing sites, monitoring process, and stockyards to sales and transportation as well as collecting buyer feedback. Upon incorporation of auction system of allotment, utilization of technology became a norm in allocation of sand reaches. Yet, its usage is still narrow in other stages. Incorporation of IT in various stages is discussed subsequently.

4.1.2 Practice of IT in allocation

The sand reaches allotment shall be carried out through online auction or competitive bidding. However, in certain instances, public or SHGs notify allocation of sand reaches for mining without royalty fee. Yet, these allotments are given to Panchayats, mining companies, or Public Works Department on appointment basis instead of contractors.

4.1.3 Use of IT in ordering

Several states developed online portals or app for sand purchasing as usage of IT in ordering become widespread now. Besides, state collects customer feedback through People First Grievance Redressal center. Upon the consumer registering and logging in the website, they can select active sand reaches which are listed based on their keyed in information. Then, they can make online payment for their purchase and the delivery will be done. Upon a successful transaction, a digital receipt is produced. In addition, a digital waybill is generated at the stockyard once receipt is submitted. This system guarantees no artificial inflation in sand price. Also, it makes sure the government obtains royalty and taxes from the sale. It also provides insight of sand usage pattern and current demand.

5. Replenishment of sand

There are several topics to be considered during replenishment study as follows:

i. Study shall cover at least a distance of 1 km in both upstream and downstream direction from the possible mining location.
ii. The sediment sampling must comprise bed materials and change in load of bed material from before to after the mining process.
iii. Utilizing the studied cross-section to development of sediment rating curve at the upstream end of the potential reach.
iv. Identification of viable high flow period which replenish mined quantity using historical or gauged flow rating curve.
v. Computation of extraction quantity using sediment rating curve and high flow period upon identifying permissible mining depth.

6. Curbing illegal sand mining

Various governments have taken multiple steps to curb illegal mining as follows (Ministry of Mines Government of India, 2018a):

i. Prohibited extraction from undeclared areas
ii. Illegal abstraction of more than allowable limits
iii. Illegal transportation of sand without permit

6.1 Monitoring

Monitoring ensures legality and environmental sustainability. Therefore, state government is required to develop and launch an effective system to monitor the process including mined quantity and transportation. In addition, transportation must be supervised to supply the sand within a realistic charge (Ministry of Mines Government of India, 2018a). Stockyards should be managed by authorities in charge such as leaseholders or contactors. They must be located within 500 m from motor able roads to ease the process. Also, stockyards must be capable of storing 3 months of stock to ensure continues supply during monsoon season. Geofencing is needed to alert entry of unauthorized vehicles (Ministry of Mines Government of India, 2018a). Sand carrying vehicles must possess a valid transport permit produced at stockyard after confirmation of payment with a scan code. It ensures that the permit is not duplicated or used multiple times. On the other hand, Project Proponent should ensure the availability of IT equipment (computer, CCTV, wifi modem, etc.), generator, access control to mine lease areas, and arrangement for weight depending on trailers' capacity used at mines (Ministry of Mines Government of India, 2018a).

Mobile application and/or bar code scanners must be given to authorities in charge of monitoring to check transport permit anywhere on road (Ministry of Environment, 2020). Upon scanning the QR code or key in vehicle number in the application or send text message to a predefined number, the particulars of permit including plot information, vehicles details, valid duration, etc., shall be acquired. It shows any forgery of transport permit can be traced instantly. The system could generate multiple reports on daily lifting and user performance. Hence, the loaded vehicles shall be traced from source to destination (Ministry of Environment, 2020). Fig. 18.3 shows levels of monitoring process.

6.2 Stringent mechanism evolved in curb illegal extraction

In case of more than two times, the vehicle/machinery shall be seized/confiscated by authorized officers. The seized sand shall be disposed by the concerned officer as nominated by the District Administration and Chairman of the DLSC as per the sale price and following approved guidelines. Adjudicating mechanism

6. Curbing illegal sand mining

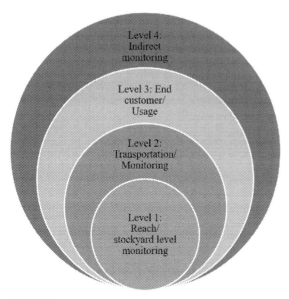

FIGURE 18.3 Levels of monitoring process (Ministry of Mines Government of India, 2018a).

on complaints regarding illegal extraction and sand transportation, the government constituted district and state level committees under the Collector and the Principal Secretary for redressal of the matter through enquiry by a team comprising of Regional Development Officer, Deputy Superintendent of Police, and Assistant Director of Mines and Geology. Any vehicle(s)/machinery found involved in unauthorized stocking/storing/trading of sand by other than Telangana State Mineral Development Corporation and Irrigation Department shall be seized/confiscated (Ministry of Mines Government of India, 2018a).

Decasting:

i. The decasting sand in any patta land abutting river bed, to make the land suitable for cultivation by State PSUs.
ii. The DLSC will hand over the patta land to State PSUs for decasting of sand.
iii. State PSUs enter into agreement with the pattadar and comply with such other conditions as deemed fit as per the procedure involved to conduct decasting sand from patta land.

Desilting:

Desiltation to source sand by Irrigation Department for internal use or consumption in irrigation projects (Ministry of Mines Government of India, 2018a):

i. The Irrigation Department shall take up desiltation of Dams, Irrigation Projects to source sand for utilization in irrigation projects exclusively.
ii. Department shall define and demarcate the area to be desilted with DGPS coordinates for the purpose.
iii. There shall be joint inspection of the demarcated area by concerned Executive Engineer and Irrigation Department, and to ensure that the demarcated area to be desilted by Irrigation Department shall not overlap with any of the area(s) already under desiltation or likely to be desilted
iv. The Irrigation Department shall quantify the sand likely to be sourced by desilting process.
v. The demarcated area to be desilted by the Irrigation Department for exclusive consumption in irrigation projects shall be placed in the DLSC for approval.
vi. After approval of desiltation by the DLSC, the Irrigation Department shall carry out desiltation with due adherence to the Engineering Department Rules, Public Works Code, and Regulations and after necessary administrative and technical sanction from competent authority in the Irrigation Department.
vii. The Irrigation Department shall put in place a suitable administrative mechanism, as per the rules and codal provisions, at the field level to efficiently supervise the desiltation process and also to prevent any misuse of sand sourced from desiltation.
viii. The Irrigation Department shall pay seigniorage fee and contribution toward

District Mineral Foundation in advance and obtain the waybills for transportation of desilted sand to stockyard(s) concerned.

ix. Sand sourced from desiltation by Irrigation Department shall be exclusively utilized for internal consumption by the Irrigation Department and not be sold outside. If there is any abuse or misuse of the provision, the Irrigation Department shall be squarely responsible.

x. The Irrigation Department shall obtain Mineral Dealer License in the name of the project for the stockyards from the competent authority.

xi. The Irrigation Department shall dispatch sand from the stockyard(s) to the respective irrigation project site with transit pass concerned as per Mineral Dealer Rules.

xii. The Irrigation Department shall furnish a desiltation area-wise monthly statement on the quantity of sand de-silted and transported to stockyard as well as sand dispatched from the stockyard to the respective irrigation project site.

7. Manufactured sand (crushed stone sand)

Natural resources such as rivers, forests, and minerals are nation's wealth and must be conserved for future generations. Rivers are needed for persistence of mankind. Therefore, it should not be treated as only source of sand. Manufactured Sand (M-Sand) or crushed stone sand (CSS) is a substitute for natural sand which is produced by blasting rocks and crushing larger aggregates into sand size particles (Ministry of Mines Government of India, 2018a). Upon sieving produced sand, it is washed to eliminate fine particles. Then, a series of quality test is carried out before approving as useable aggregate for construction. M-Sand should follow IS Code 383:2016 to be used in construction. In addition, the code states that fine particles

of <150 microns should not be present in excess because they are harmful to be used in construction and not qualified for incentive claim (Ministry of Mines Government of India, 2018a).

It must be noted that use of M-Sand has increased due to shortfall of river sand and environmental concern. Illegal mining and depletion of natural sand caused execution of new environmental or land use legislations. Subsequently, obtaining river sand became challenging and costly. Also, occurrence of silt and clay in river sand which damages structures further amplified the demand for M-Sand because it has low risk of impurities. Meanwhile, M-Sand has higher unit weight, higher flexural strength, lower permeability, and better abrasion resistance. Besides, as the low cost by-products of rock crushing can be used to produce M-Sand, the wastage decreases, while product value increases. All these make M-Sand to be used widely in Southern states of India for construction. However, the usage is still limited in Northern India (Ministry of Mines Government of India, 2018a).

7.1 Terminology

The definition of coarse and fine aggregates is listed below based on IS Code 383:2016 (Ministry of Mines Government of India, 2018a):

i. Fine aggregate: Aggregates which pass through 4.75 mm IS Sieve.

ii. Natural sand: Fine aggregates disintegrated naturally from rocks and transported by wind or rivers.

iii. Crushed sand: There are two types of crushed sands such as Crushed Stone Sand (CSS) and Crushed Gravel Sand (CGS) which are produced from crushing hard stones and natural gravel, respectively, into fine products.

iv. Mixed sand: Produced by mixing natural sand with CSS/CGS in appropriate ration.

v. M-Sand: Fine aggregate crushing larger aggregates into sand size particles, followed by sieving, washing, and classifying. It can be created from Recycled Concrete Aggregates as well.

vi. Coarse aggregate: Particles retained on 4.75 mm IS Sieve with limited fine materials which can be obtained from the following:

- Naturally disintegrated uncrushed gravel or stone;
- Crushed gravel or hard stone;
- Mixing of naturally disintegrated uncrushed and crushed gravel or stone;
- Factory-made from other natural or processed sources through a series of process, e.g., Recycled Concrete Aggregate (RAC) or Recycled Aggregate (RA).

CSS is referred as M-Sand in IS 383:2016 because it is the well-known market name. The sand particles must have greater crushing strength and smooth surface with grounded edges. Also, the ratio of fines below 600 microns in coarse sand (zone I) should be between 15% and 34% with no organic impurities. Besides, silt content must be less than 2% for crushed sand. The allowable limit for fines (<75 microns) in M-Sand is less than 15%.

7.2 Technical specifications of aggregates for M-Sand

There are several specifications for M-Sand as the following (Ministry of Mines Government of India, 2018a):

i. Size: Fine aggregates shall be less than 4.75 mm when sieved. The percentage of sand passing through 600 μm sieve would determine the purpose of sand:
- Zone I: Coarse sand for concreting;
- Zone II–IV: Fine sand for plastering.

ii. Quality: Aggregates prior to crushing must be strong and durable. It must be free of veins, adherent coating, alkali, and vegetative. In addition, deleterious materials such as pyrites, mica, coal, shale, sea shells, and organic contaminants should not present in the aggregates because it would diminish the strength or durability of the concrete. Although the standard limit of deleterious materials is suggested in IS 383: 2016, the engineer may carry out additional testing to check acceptable performance to approve the aggregate quality. Also, it is suggested to avoid flaky, scoriaceous, and elongated aggregates as well. Several quality tests are required to be carried out for crushed sand such as

- Sieve analysis: To determine percentage of clay and silt;
- Optical microscopic study: To study the particle size and shape;
- Slump test: To determine the workability;
- Cube compressive test: To determine the compressive strength.

iii. Aggregate Crushing Value: It should be less than 30% for coarse aggregates for concrete on varying surfaces such as runways and pavements. For other application, it should be at least 30% based on IS 2386 (Part IV): 1963. Other parameters that need to be tested are aggregates impact value, abrasion value, and soundness (Ministry of Mines Government of India, 2018a).

7.3 Quarrying

The raw material of boulders is located in quarry similar to all mining industry (Ministry of Mines Government of India, 2018a). The materials will then be sent to plants for further processing. Suitable raw materials for M-Sand are deposits of granite, sandstone, quartzite, pegmatites, basalt, etc. State can determine and hoard the resources wholly for manufacture of M-Sand. The quarrying activity is discussed comprehensively in this section.

7.3.1 Drilling and blasting

The main task in a quarry is drilling in which a drill machine with powered bits of 105–115 mm diameter and compressor of 6–12 bars of pressure is commonly utilized. Holes are bored upright in a defined geometric configuration which differs depending on rock type, condition, and feed size of the crusher. Upon drilling, blasting will be carried out generally. The drilled bores are filled with explosives of adequate energy. It breaks the rock mass and yields favored rock fragmentation upon explosion.

7.3.2 Loading and transportation

The blasted rocks are loaded onto excavators with appropriate bucket capacity which commonly ranges between 0.9 cum to 1.1 cum. The quantity of excavators used for loading is influenced by capacity of plant to process and transportation. Tippers of 6–10 tyres are loaded by excavators with rock fragments of preferred size to be transported from quarry to plant.

7.3.3 Crushing

Tippers unload rock fragments in the crushing unit. The crushing process is classified into three stages as follows:

i. Primary Crusher: The rock fragments transported by tippers are unloaded into jaw crusher with feed size of 400–550 mm in common. The jaw crusher is considered as receiving bin of raw material from quarry.
ii. Secondary Crusher: It is fed by output of jaw crusher through belt conveyors and screens. This crusher is called cone crusher with feed size of 150 mm. It comprises two truncated cones of different diameters which are named Concave and Mantle. They are manufactured using manganese alloy. The rocks are crushed into 40 mm size at this crusher between these cones.

iii. Tertiary Crusher: The product of secondary crusher is transferred to Vertical Shaft Impactor (VSI) through belt conveyors and screens. The end product of VSI is called as M-Sand. River sand formation processes of hitting of rocks and attrition are replicated in VSI chamber. Thus, the end product is in well-graded particle shape. In addition, the attrition method remarkably eliminates surface roughness of fine particles. The cubicle structure of particles is set forth in the sizing chamber of VSI which makes the output great for construction. The size of end product ranges between 2 and 20 mm. M-Sand plants make sure optimize grading of fine particles for improved particle size distribution. Several plants control microfines (<75 micron) quantity to be less than 3% through Air Filter System and/or washing facility. The washing facility retains wetness of M-Sand to reduce absorption rate during concrete mixing which directly increases the workability. M-Sand manufactured by VSI crusher and washing system shows lowered water absorption compared with Crusher Dust.

7.3.4 Economics of M-Sand

The landed cost of M-Sand can be compared with landed cost of river sand under regular circumstances. The section shows production details of M-Sand plant situated in Chikkballapura district of Karnataka. It is capable of manufacturing 360 tonnes/h. Estimating the plant functions for 20 h/day and 300 days/year, the production is as follows:

$$\text{Overall plant capacity} = 360 \text{ tph} \times 20 \text{ h/day}$$
$$= 7200 \text{ tpd}$$

$$\text{Operational capacity} = 3500 \text{ tpd}$$
$$\times 300 \text{ days/year}$$
$$= 10,50,000 \text{ or } \sim 1 \text{ mtpa}$$

7.3.5 *Promoting M-Sand*

Primary purposes for encouraging M-Sand usage are as follows:

i. To preserve ecosystem by downsizing usage of river sand in a conservative way;
ii. To commercialize M-Sand as a substitute of river sand to cater the shortage due to increasing demand;
iii. To encourage Micro-Small and Medium Enterprises (MSME) in launching M-Sand plants which create employment opportunities and increase effective consumption of resources.

One of the important steps is to award industry status to the M-Sand manufacturing units. Also, suitable provisions in Minor Mineral Concession Rules of states shall be provided. It would assist in determining obtainability of raw materials and quarries' preference in allocation to the interested M-Sand manufacturers. Upon obtaining industrial status, units can be aided by incentives and concessions given in accordance to each State's Industrial Policy or MSME Policy.

7.4 Sand from overburden of coal mines

Overburden materials in coal mines must be isolated from coal seam prior to mining and usually dumped outside the site. However, 80% of the spoil is needed to backfill the mined area during closure of mine for land reclamation, while the remaining 20% shall be used for sand production. The processing of overburden produces 60%–65% of sand, 30%–35% of clay, and 5% of pebbles based on research studies carried out by the Central Institute of Mine and Fuel Research. Laboratory investigations shall be carried out to compute theoretical quid pro quo between sand recovery and its quality. Western Coalfields Limited (WCL) has started to segregate sand from the overburden. Besides, WCL has planned to launch sand segregation plant with daily capacity of 200 m^3 in Nagpur. In addition, it trades sand at economical price, only one fourth of market price.

This alternative can be executed in states with coal mines such as Gujarat, Chhattisgarh, Madhya Pradesh, Bihar, Telangana, Maharashtra, Jharkhand, etc. Also, this overburden substitute is extensible similar to accomplishment of WCL which launched plant with effective capacity. The need for sand in Maharashtra and Gujarat is 100 million tonnes and WCL itself can cater 45% of overall yearly demand which is 45.36 million tonnes/year. Government has ordered seven subsidiaries of Coal India Limited to deal with sand segregation from the overburden wastage which is approximately 283 million tonnes of sand. It is 35% of the total current sand usage in the nation. In addition to cater the sand demand, it also guarantees an effective technique of using of overburden wastage.

7.5 Import of sand for coastal cities

Importing sand is one of the alternatives to cater the need for sand. Several Asian nations such as Malaysia and Indonesia have abundant sand and needed to be removed to prevent floods. The removed sand shall be imported to cater the shortfall. Nonetheless, it is vital to ensure the sand complies with standard quality based on IS 383 and be free from phytosanitary issues. The quality shall be tested at both imported nation and arrival harbor.

i. Karnataka: The state has already begun sand imports and establishes regulations for the trade.
ii. Tamil Nadu: The state is getting ready to import sand.
iii. Kerala: The state has started sand trade from Malaysia which is retailed in loose at harbor price. However, the traded sand is expensive. Thus, it is appropriate for locations with great shortage only.

8. Sand mining in Malaysia

The primary source of sand is in-stream mining in Malaysia (Ministry of Natural Resources, 2009). This type mining is well known as the extraction points are located close to market and/or along the transportation pathway. Hence, it decreases the transportation fee. The impacts of mining are follows (Ministry of Natural Resources, 2009):

i. Excess mining of bed material more than the replenishment through transport destroys the bed near extraction zone.
ii. Bed degradation weakens the structures such as bridge supports and pipelines.
iii. Degradation leads to alteration of river morphology which affects aquatic environment.
iv. Exposed materials beneath gravel due to thinning of gravelly bed, if detrimental, may distress the quality of aquatic environment.
v. Groundwater levels decline when floodplain aquifer flows into river due to bed degradation. This could lead to destruction of riparian vegetation.
vi. Bed degradation decreases flooding by reducing flood heights. Thus, it reduces risk for population occupancy in floodplains. However, it can damage the nearby structures.
vii. Supply of overbank sediments to floodplains decreases with reduction in flood level.
viii. Accelerated bed degradation prompts collapse and erosion of bank as the elevation of bank increases.
ix. Gravel extraction decelerates sediment deposition in streams and retains channel's capability to carry flood waters.
x. Erosion in banks is higher with decrease in bar size and it depends on extraction quantity, distribution of removal, and geometry of bend.
xi. Downstream bards erode when obtain fewer bed materials than being transported by fluvial process during removal of gravel from bars.

There are several guidelines must be reflected before approving sand and gravel mining permits such as follows (Ministry of Natural Resources, 2009):

i. Guarantee protection of stream and surrounding environment's equilibrium;
ii. Avoid deposition at downstream close to hydraulic structures such as jetties and pipelines;
iii. Make sure the streams are in stable profile and secured from bank and bed erosion;
iv. Avoid hindering stream maintenance works carried out by the Department of Irrigation and Drainage or related organizations;
v. Freedom of river flow and water transport;
vi. Prevent river contamination as it degrades water quality.

Suggestions for in-stream mining depend on the following conceptions (Ministry of Natural Resources, 2009):

i. Allow mining quantity depending on calculated yearly replenishment;
ii. Set a certain elevation below which no withdrawal can take place;
iii. Restrict in-stream mining approaches to bar skimming;
iv. Mine sand and gravel from downstream of bar;
v. Fixate in-stream mining processes to reduce disrupted zones;
vi. Evaluate snowballing effects of sand and gravel withdrawal;
vii. Sustain stream's flood discharge capacity;
viii. Launch an abiding monitoring program;
ix. Limit activities which discharge fine sediment to streams;
x. Maintain riparian buffer at water edge and river bank;

xi. Restrict in-stream extraction to dry season only such as between May and September;

xii. Department of Irrigation and Drainage shall produce yearly status and trends report.

Floodplain mining suggestions depends on the following conceptions (Ministry of Natural Resources, 2009):

i. Floodplain mining must be avoided from the major stream;

ii. The utmost depth of floodplain withdrawal must be kept above the stream;

iii. Embankments of floodplain mining shall range between 3:1 and 10:1;

iv. Place accumulated topsoil above the 25-year return period or average reoccurrence interval level;

v. Floodplain pits shall be reinstated as wetland habitat or reclaimed for agriculture;

vi. A proposal which emphasizes long-term liability needed to be submitted;

vii. Launch an abiding monitoring program;

viii. Department of Irrigation and Drainage shall produce yearly status and trends report.

The loose bounds of alluvial passage deform due to flowing water. Besides, the bed changes its roughness when interacting with the flow. A dynamic equilibrium condition of boundary can be anticipated during steady and uniform flow. Sediment transport is the movement of bed materials which exceeded critical bed shear stress in the flow direction (Ministry of Natural Resources, 2009).

8.1 River morphology

There are plenty of engineering methods to diminish environmental issues related to sand and gravel mining in rivers, floodplains, and terraces (Ministry of Natural Resources, 2009). The method used must consider the characteristics of specific hydrologic system. The primary root of environmental issues on in-stream excavation is the withdrawal of sediment more than the replenishment rate. Coarse materials are bed loads which are transported along river through a series of actions such as rolling, sliding, or bouncing. Several researchers state that limiting withdrawal based on calculated yearly bed load shall protect the environment (Teo et al., 2017), whereas exact estimated quantity of annual replenishment shall be withdrawn in the first year of implementation of management plan. The yearly replenishment depends on intermittent nature of sediment transport. For instance, monitoring data would indicate quick replenishment of bed materials during monsoon seasons with high flow rate and sedimentation rate. Meanwhile, the replenishment rate is slower during drought season with low flow rate and sedimentation rate (Ministry of Environment, 2016).

"Redline" or the uttermost elevation below which extraction is prohibited shall be investigated at every mining site to prevent damages of structures such as bridges and pipelines as well as diminish vegetation impacts. A withdrawal spot with deposition level of 1 m above stream thalweg elevation based on site investigation shall be permitted by the Department of Irrigation and Drainage. Withdrawal from the downstream portion of the bar retaining the upstream one to two thirds of the bar and riparian vegetation while mining from the downstream third of the bar is recognized as a technique to protect waterway stability and narrow width of low flow stream essential for fish. Flood capacity in the streams must be retained in flood hazard locations near existing infrastructure (Ministry of Natural Resources, 2009). Reclamation plans for in-stream include the following (Ministry of Natural Resources, 2009):

i. A baseline investigation shall be carried out on current condition of cross-section between two monument endpoints set back from embankment top with elevations ascribed from the Department of Survey and Mapping Malaysia's bench mark.

ii. The suggested excavation cross-section information shall be delineated over the baseline information to highlight the vertical range of the planned mining.

iii. If the cross-section of the replenished bar is similar to baseline information, then it indicates that elevation of bar before withdrawal and after replenishment is equal.

iv. A proposal metric map presenting the aerial range of the mining and riparian buffers shall be developed.

v. A planting proposal shall be developed by a plant ecologist who is well aware of flora of the stream required to be restored.

8.1.1 Sand replenishment, geomorphology, and hydrology

Mining activity needs physical monitoring which includes geomorphic maps, waterway cross-sections survey, bed material measurements, longitudinal profiles and discharge, as well as sediment transport measurements. The physical monitoring provides information on replenishment rate and alteration in stream morphology, embankment erosion, as well as material size. The monitoring spots shall be situated close to permanent monumental areas in upstream, downstream, and within mining location. Besides, the longitudinal profile should cover mining area from upstream to downstream. Profile developed from thalweg survey with elevation referenced to Department of Survey and Mapping Malaysia's bench mark must be exhaustive enough to illustrate the channel morphology. Discharge and bed material measurements of suspended and bed load are carried out by the Department of Irrigation and Drainage to develop a statistically substantial database. In addition, knowledge of stream shall be further enhanced through long-term measurements with vast range of flows. Also, monitoring wells shall be developed in off-channel floodplain mining to measure monthly fluctuations in groundwater levels.

8.2 River modeling using HEC-RAS

HEC-RAS is a hydraulic analysis program in which a user interacts with the system using graphical user interface (GUI) (Burner et al., 1996). It can be used to do calculations such as steady and unsteady flow of surface water as well as sediment transport. In HEC-RAS, a project is termed as database of a specific river in which numerous types of analysis can be performed. In general, the information included in a project is plan data, geometric data, sediment transport, hydraulic design, and steady as well as unsteady flow data. Besides, it has been used to develop mathematical model of scour and deposition in Sungai Muda. Hence, it shows that HEC-RAS is an effective tool for modeling (Nor et al., 2014).

9. Conclusion

Sand is one of the crucial minerals which enhances economics and industry of a nation. As the demand increases significantly, mining is carried out in large scale and also illegally. However, exhaustive extraction of natural sand will give negative impact to the environment. Therefore, government must enforce rules and regulations further to guarantee proper scientific investigations that have been carried out which would provide sufficient knowledge on safe mining. Besides, alternatives of natural sand such as M-Sand, sand trade, and sand from overburden of coal mines shall be promoted widely. Besides preserving natural sand, it could cater the shortfall as well.

References

Brunner, G.W., June 22-28, 1996. Hec-ras (river analysis system). In: Proceedings of the North American Water and Environment Congress & Destructive Water. California, American Society of Civil Engineers, New York, United States, pp. 3782–3787.

Government of Andhra Pradesh, 2010. Impact of Sand Mining on Groundwater Regime. https://apsgwd.gov.in/SandMining.aspx. (Accessed 28 June 2021).

Ministry of Environment, 2016. Forest and Climate Change, Sustainable Sand Mining Management Guidelines 2016. http://perfactgroup.com/www/wp-content/uploads/2016/12/1940760443Final-Sustainable-Sand-Mining-Management-Guidelines-2016_new.pdf. (Accessed 28 June 2021).

Ministry of Environment, 2020. Forest and Climate Change, Enforcement & Monitoring Guidelines for Sand Mining. http://environmentclearance.nic.in/writereaddata/SandMiningManagementGuidelines2020.pdf. (Accessed 28 June 2021).

Ministry of Mines, Government of India, 2018a. Sand Mining Framework. https://www.mines.gov.in/writereaddata/UploadFile/sandminingframework260318.pdf. (Accessed 28 June 2021).

Ministry of Mines, Government of India, 2018b. Draft Sand Mining Recommendations. https://mines.gov.in/writereaddata/UploadFile/sandmining16022018.pdf?cv=1. (Accessed 28 June 2021).

Ministry of Natural Resources, 2009. and Environment Department of Irrigation and Drainage Malaysia, River Sand Mining Management Guideline. https://www.engr.colostate.edu/~pierre/ce_old/classes/ce717/Sand%20mining.pdf. (Accessed 28 June 2021).

Nor, D., Adnan, N., Ainul, Z., Zulkarnain, H., Mokhtar, E., August 8–9, 2014. Geospatial flood inundation modelling and estimation of Sungai Muda Kedah floodplain, Malaysia. In: Proceedings of the Arte-Polis 5 International Conference - Reflections on Creativity: Public Engagement and the Making of Place. Bandung, Indonesia.

Strahler, A., 1964. Quantitative geomorphology of drainage basins and channel networks. In: Chow, V. (Ed.), Handbook of Applied Hydrology. McGraw Hill, New York, pp. 439–476.

Teo, F.Y., Noh, M.M., Ghani, A.A., Zakaria, N.A., Chang, C.K., 2017. River sand mining capacity in Malaysia. In: Proceedings of the 37th IAHR World Congress, 13-18 August 2017, Kuala Lumpur, IAHR & Usains Holding Sdn Bhd, Kuala Lumpur, Malaysia, vol. 6865, pp. 538–546.

CHAPTER
19

Probabilistic response of strip footing on reinforced soil slope

Koushik Halder[1], Debarghya Chakraborty[2]

[1]Department of Civil Engineering, University of Nottingham, Nottingham, United Kingdom; [2]Department of Civil Engineering, Indian Institute of Technology Kharagpur, Kharagpur, West Bengal, India

1. Introduction

Shallow foundation system on a level ground is designed based upon two conditions, i.e., (i) load distributed by the foundation system should not exceed the ultimate bearing capacity of the underneath soil and (ii) foundation settlement should be within a permissible limit. However, on many occasions, owing to rapid urbanization and scarcity of land, structures such as low-rise buildings, transmission towers, bridge abutments, hanging cable cars etc., are constructed on the hilly terrain and sloping ground. Risk and vulnerability of such structures increase more as foundation induces slope instability in addition to reducing bearing capacity and increasing settlement. Performance of such structures also depends on the position of foundation from the slope edge, slope geometry, strength of soil, etc (Davis and Booker, 1973; Bathurst et al., 2003; Georgiadis, 2010; Chakraborty and Kumar, 2013; Leshchinsky, 2015; Halder et al., 2019; Beygi et al., 2020). Because of the above-mentioned complexities, overall design of foundation for structures resting on the slope becomes difficult to the design engineers.

Over the years, various researchers have carried out investigations on stabilization of loaded soil slope and suggested several methods to improve the performance of overall system. Typical examples include chemical grouting, modification of slope geometry, weaker soil replacement, installation of retaining walls or micropiles, and reinforcing slope filled with high tensile strength material. The reinforced soil technique has proved to increase both the load-bearing capacity of footing and slope stability (Selvadurai and Gnanendran, 1989; Lee and Manjunath, 2000; Yoo, 2001; El-Sawwaf, 2007; Mehrjardi et al., 2016; Halder and Chakraborty, 2018a,b, 2019a,b, 2020a). It is also advantageous in terms of overall economy and ease of construction with respect to other methods. Metal strips, metal bars, jute fibers, tire chips, and polymeric geosynthetics are used to reinforce the soil.

Risk, Reliability and Sustainable Remediation in the Field of Civil and Environmental Engineering
https://doi.org/10.1016/B978-0-323-85698-0.00021-6

© 2022 Elsevier Inc. All rights reserved.

However, these days, uses of metal strips and metal bars as reinforcement below the foundation are becoming obsolete. Among the polymeric geosynthetics, geotextile, geogrid, and geocell are widely used. In case of planar geosynthetics like geotextile and geogrid, frictional force develops at soil—reinforcement interface. In addition to tensile resistance, geogrid mobilizes passive resistance through bearing on its ribs. Owing to the frictional and passive resistance, planar geosynthetics restrains the lateral movement of soil in both sides of the footing and helps to propagate load in a wider and deeper area below the reinforcement leading to large performance improvement. Vertical settlement of the foundation also reduces. On the other hand, each individual geocell unit is connected to build a three-dimensional cage like structure, known as geocell mattress. By virtue of its honeycomb structure, geocell mattress contains a large amount of soil with respect to geogrid layer. Large confinement of soil in the geocell cage induces significant amount of frictional and passive resistance along the soil—geocell interfaces. Due to the development of large frictional resistance along the soil—geocell interfaces, geocell mattress acts like high stiffness material, which in turn spreads the footing load in a wider and deeper area.

Traditionally, deterministic numerical analysis of shallow foundation on the soil slope is carried out by considering uniform soil strength parameters throughout the domain. Based on their experience, design engineers use factor of safety to ensure safe design of overall system. The deterministic approach has the limitation in considering the same factor of safety for two identical slope geometry having different failure probability. On the other hand, different formational process, mineralogical components, and various types of loading induce randomness and heterogeneity in natural soil deposit. A significant variation in the soil strength parameters is observed with a small change in distance, horizontally, and vertically.

Therefore, consideration of uniform values of soil strength parameters all over the domain causes inaccuracy in the obtained results (Griffiths et al., 2002, 2009; Fenton and Griffiths, 2003; Haldar and Babu, 2008; Pramanik et al., 2019; Halder and Chakraborty, 2019c, 2020b,c; Krishnan et al., 2021).

This chapter presents a comprehensive discussion about the significant role of uncertainties related to soil shear strengths on the (i) bearing capacity of a strip footing located on the reinforced soil slope edge and (ii) stability of the reinforced slope. Probabilistic ultimate bearing capacity of a strip footing placed on the reinforced slope edge is determined by combining the lower bound finite element limit analysis and the Monte Carlo simulation (MCS). Besides that, finite difference software FLAC (Fast Lagrangian Analysis of Continua) (ITASCA, 2011) in combination with the MCS technique is used to predict probabilistic load carrying capacity at a specified settlement level of foundation placed on the geocell reinforced soil slope. For both the problems, random field method is used to incorporate spatial variability of soil strength parameters. Probabilistic stability analysis of the reinforced slope is performed by combining FLAC and the MCS technique.

2. Methodology

Spatial variations of soil strength properties are considered by modeling soil as a random field. The influence of soil randomness on the response of strip footing is incorporated through the MCS technique. Both random field modeling and the MCS technique are combined with the lower bound finite element limit analysis method and FLAC, respectively, to obtain the probabilistic ultimate collapse load and probabilistic load-carrying capacity at a particular settlement level of the strip footing. A brief discussion about the above-mentioned methods used for both deterministic and probabilistic analyses is presented in the following subsections.

2.1 Lower bound finite element limit analysis

The lower bound value of collapse load is obtained from statically admissible stress field. Formation of statically admissible stress field is very difficult for a practical stability problem with complex geometry, loading, and soil heterogeneity. For that reason, many researchers have used finite element method to discretize the stress field. Lower bound limit analysis always predicts safe collapse load. In addition to that, there are no requirements of (i) considering full stress—strain response of the soil material and (ii) assumption of the geometry of failure surface. Only shear strength parameters are required as inputs in the analysis. For that reason, in this chapter, the lower bound finite element limit analysis technique is used to compute the collapse load of the strip footing.

As this chapter predicts the lower bound bearing capacity of a strip footing placed on the soil slope, the plane strain lower bound finite element limit analysis formulation of Sloan (1988), Chakraborty and Kumar (2014) is employed. Three noded triangular element as shown in Fig. 19.1A discretizes the problem domain in two dimensions (x-y). All nodal stresses, i.e., the normal stresses in the x and y directions (σ_x and σ_y) and the shear stress (τ_{xy}) are basic unknowns. In order to formulate the statically admissible stress field, it is required to maintain (i) the static equilibrium of overall domain, employment of (ii) different stress boundary conditions over the boundary edges (Fig. 19.1B), and implementation of (iii) the yield criteria throughout the nodes. Apart from these, normal and shear stresses continuity are preserved along the shared edges of two neighboring elements. Reinforcement effect is included by restraining shear stresses and allowing normal stresses over the edges indicating reinforcement positions (Chakraborty and Kumar, 2014). Based on the satisfaction of above-mentioned conditions, various equality and inequality constraints are generated. An optimization problem is formulated which is solved by using second-order conic optimization technique proposed by Makrodimopoulos and Martin (2006).

2.1.1 Element equilibrium conditions

Static equilibrium conditions as expressed in Eq. (19.1) are employed on all elements and satisfaction of Eq. (19.1) generates two equality constraints, which are presented by Eq. (19.2).

$$\frac{\partial \sigma_x}{\partial x} + \frac{\partial \tau_{xy}}{\partial y} = 0 \qquad (19.1a)$$

$$\frac{\partial \sigma_y}{\partial y} + \frac{\partial \tau_{xy}}{\partial y} = \gamma \qquad (19.1b)$$

$$[A^{EL}]_{2 \times 9} \{\sigma^{EL}\}_{9 \times 1} = \{b^{EL}\}_{2 \times 1} \qquad (19.2)$$

In the above equations, γ is the unit weight of soil, $[A^{EL}]$ and $\{b^{EL}\}$ are known quantities, whereas $\{\sigma^{EL}\}$ is unknown stress vector.

2.1.2 Stress discontinuity conditions

In case of lower bound finite element limit analysis, nodes are distinct for all elements, which develops an edge of stress discontinuity between two adjacent triangles (Fig. 19.1C). However, in order to satisfy normal and shear stresses continuity over the edges, discontinuity criterion as expressed in Eq. (19.3) needs to be employed. Because of that, four numbers of equality constraints are generated.

$$[A^{DS}]_{4 \times 12} \{\sigma^{DS}\}_{12 \times 1} = \{b^{DS}\}_{4 \times 1} \qquad (19.3)$$

$$\text{where } [A^{DS}]_{4 \times 12} = \begin{bmatrix} S_d & -S_d & 0 & 0 \\ 0 & 0 & S_d & -S_d \end{bmatrix}_{4 \times 12};$$

$$[S_d]_{2 \times 3} = \begin{bmatrix} \sin^2 \lambda & \cos^2 \lambda & -\sin 2\lambda \\ -0.5 \sin 2\lambda & 0.5 \sin 2\lambda & \cos 2\lambda \end{bmatrix}$$

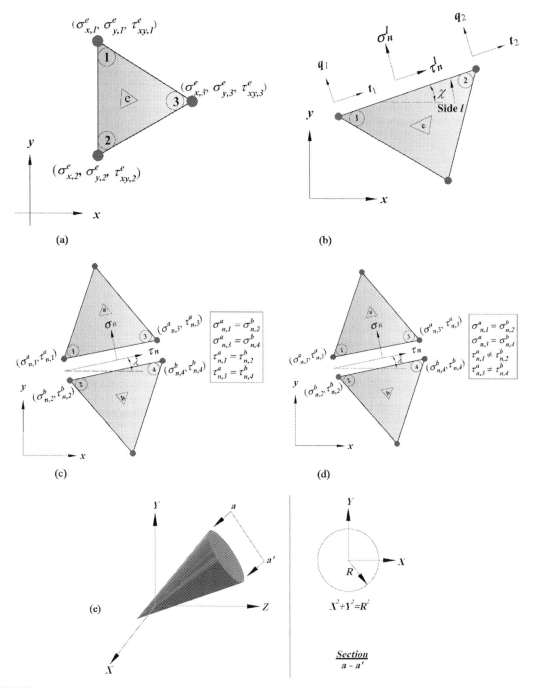

FIGURE 19.1 (A) Triangular element used in the numerical limit analysis; (B) Stress boundary conditions; (C) Stress discontinuity conditions; (D) Modified stress discontinuity conditions; and (E) Mohr–Coulomb yield criteria as second-order Cone.

$$\left\{\sigma^{DS}\right\}_{1 \times 12}^{T} = \left\{ \sigma_{x,1}^{a} \quad \sigma_{y,1}^{a} \quad \tau_{xy,1}^{a} \quad \sigma_{x,2}^{b} \quad \sigma_{y,2}^{b} \quad \tau_{xy,2}^{b} \quad \sigma_{x,3}^{a} \quad \sigma_{y,3}^{a} \quad \tau_{xy,3}^{a} \quad \sigma_{x,4}^{a} \quad \sigma_{y,4}^{b} \quad \tau_{xy,4}^{b} \right\}$$

$$\left\{b^{DS}\right\}_{4 \times 1} = \{0 \quad 0 \quad 0 \quad 0\}$$

Here, $[A^{DS}]$ and $\{b^{DS}\}$ are known and $\{\sigma^{DS}\}$ is unknown stress vector. Angle between the stress discontinuity edge and horizontal axis in anticlockwise direction is λ.

2.1.3 Modified stress discontinuity conditions to incorporate reinforcement

Reinforcement is not modeled by using explicit element. Following Chakraborty and Kumar (2014), effect of reinforcement is incorporated by continuing normal stresses and relaxing shear stress over the discontinuous edges formed by the elements lying above and below the reinforcement layer (Fig. 19.1D). Final form of the stress discontinuity conditions for reinforced cases is provided in Eq. (19.4). Interface between soil and reinforcement is assumed as fully bonded.

$$\left[A^{DS}\right]_{2 \times 12}\left\{\sigma^{DS}\right\}_{12 \times 1} = \left\{b^{DS}\right\}_{2 \times 1} \quad (19.4)$$

where
$$\left[A^{DS}\right]_{2 \times 12} = \begin{bmatrix} S_{dm} & -S_{dm} & 0 & 0 \\ 0 & 0 & S_{dm} & -S_{dm} \end{bmatrix}_{2 \times 12} ;$$
$$\left[S_{dm}\right]_{1 \times 3} = \begin{bmatrix} \sin^2 \lambda & \cos^2 \lambda & -\sin 2\lambda \end{bmatrix}$$

2.1.4 Stress boundary conditions

With the successful implementation of the stress boundary conditions over the edges generate four equality constraints on six nodal stresses as expressed in the following equations:

$$\left[A^{SB}\right]_{4 \times 6}\left\{\sigma^{SB}\right\}_{6 \times 1} = \left\{b^{SB}\right\}_{4 \times 1} \quad (19.5)$$

where

$$\left[A^{SB}\right] = \begin{bmatrix} M_s & 0 \\ 0 & M_s \end{bmatrix}_{4 \times 6} ; \; [M_s]_{2 \times 3}$$

$$= \begin{bmatrix} \sin^2 \chi & \cos^2 \chi & -\sin 2\chi \\ -0.5 \sin 2\chi & 0.5 \sin 2\chi & \cos 2\chi \end{bmatrix}$$

$$\left\{\sigma^{SB}\right\}^{T} = \left\{ \sigma_{x,1}^{SB} \quad \sigma_{y,1}^{SB} \quad \tau_{xy,1}^{SB} \quad \sigma_{x,2}^{SB} \quad \sigma_{y,2}^{SB} \quad \tau_{xy,2}^{SB} \right\}_{1 \times 6}$$

$$\left\{b^{SB}\right\}^{T} = \left\{ q_1 \quad t_1 \quad q_2 \quad t_2 \right\}_{1 \times 4}$$

In Eq. (19.5), χ is measured angle between the boundary edge and horizontal axis in the anticlockwise direction, q_1 and q_2 are the normal stresses acting on the boundary edge, and t_1 and t_2 are the shear stresses acting along the boundary edge. Here, the known quantities are $[A^{SB}]$ and $\{b^{SB}\}$, whereas the unknown is stress vector $\{\sigma^{SB}\}$.

2.1.5 Yield condition

The Mohr–Coulomb constitutive model is chosen to be applicable throughout problem domain in this chapter. Plane strain Mohr–Coulomb yield criterion is expressed in Eq. (19.6).

$$F_{MC} = \left(\sigma_x - \sigma_y\right)^2 + \left(2\tau_{xy}\right)^2$$
$$- \left[2c \cos \varphi - \left(\sigma_x + \sigma_y\right)\sin \varphi\right]^2 \leq 0 \quad (19.6)$$

After implementation of the Mohr–Coulomb yield criterion in terms of second order cones at each node, the Mohr–Coulomb yield criterion (refer: Fig. 19.1E) can be written as follows:

$$(\sigma_x + \sigma_y)\sin\varphi + \xi_1 = 2c\cos\varphi \qquad (19.7a)$$

$$-\sigma_x + \sigma_y + \xi_2 = 0 \qquad (19.7b)$$

$$-2\tau_{xy} + \xi_3 = 0 \qquad (19.7c)$$

In the above equations, ξ_1, ξ_2, and ξ_3 are the second-order cones. The canonical form of Eq. (19.7) is expressed below in Eq. (19.8).

$$A_i^{SCP}\sigma_i^{SCP} + \xi_i = b_i^{SCP} \quad i = 1, 2, \ldots, N_{nod}$$
$$(19.8)$$

where $\sigma_i^{SCP} = \left\{ \sigma_{x,i}^{SCP} \quad \sigma_{y,i}^{SCP} \quad \tau_{xy,i}^{SCP} \right\}^T$; $\xi_i = \left\{ \xi_{1,i} \quad \xi_{2,i} \quad \xi_{3,i} \right\}^T$; $b_i^{SCP} = \{ 2c\cos\varphi, \quad 0, \quad 0 \}^T$

$$A_i^{SCP} = \begin{bmatrix} \sin\varphi & \sin\varphi & 0 \\ -1 & 1 & 0 \\ 0 & 0 & -2 \end{bmatrix}, \quad N_{nod} = \text{total}$$

number of nodes.

2.1.6 Objective function

The objective function (collapse load) is evaluated with the integration of the normal stresses associated with the nodes representing footing position.

$$Q_{uV} = \int_{L_s} \sigma^{OBJN} dl \qquad (19.9)$$

In Eq. (19.9), Q_{uV} is the magnitude of collapse load acting per unit width of the footing along the footing–soil interface of length L_s. Here, σ^{OBJN} and dl denote average normal stress associated with the ith element of footing–soil interface and the length of the ith element, respectively. Assuming linear variation of stresses, matrix form of Eq. (19.9) is provided in Eq. (19.10).

$$Q_{uV} = \left\{ g^{OBJ} \right\}_{1 \times 6}^T \left\{ \sigma^{OBJN} \right\}_{6 \times 1} \qquad (19.10)$$

where, $\left\{ g^{OBJ} \right\}^T = \frac{L_s^i}{2} \{ \sin^2\omega \quad \cos^2\omega \quad -\sin 2\omega \quad \sin^2\omega\cos^2\omega - \sin 2\omega \}_{1 \times 6}$

$$\left\{ \sigma^{OBJN} \right\}^T = \left\{ \sigma_{x,1}^{OBJN} \quad \sigma_{y,1}^{OBJN} \quad \tau_{xy,1}^{OBJN} \quad \sigma_{x,2}^{OBJN} \quad \sigma_{y,2}^{OBJN} \quad \tau_{xy,2}^{OBJN} \right\}_{1 \times 6}$$

2.1.7 Final form of conic optimization

All equality constraints generated from the satisfaction of element equilibrium conditions, stress discontinuities, and stress boundary conditions are assembled. The global vector of objective function coefficients is obtained. The Mohr–Coulomb yield criterion is formulated as an equality constraint in the conic optimization. Second-order cones are used to present the nonlinear yield criterion as an equality constraint. The canonical expression of conic optimization problem is mentioned below.

$$\text{Maximization of } \{g\}^T \{X_{TOTAL}\} \qquad (19.11)$$

$$\text{Subjected to } [A_{TOTAL}]_{N_{Total} \times 2M_2} \{X_{TOTAL}\}_{2M_2 \times 1}$$
$$= \{B_{TOTAL}\}_{N_{Total} \times 1}$$

where
$$[A_{TOTAL}] = \begin{bmatrix} A^{EL} & 0 \\ A^{DS} & 0 \\ A^{SB} & 0 \\ A^{SCP} & I \end{bmatrix};$$

$$\{X_{TOTAL}\} = \left\{ \begin{array}{c} \sigma \\ \xi \end{array} \right\}; \text{ and } \{B_{TOTAL}\} = \left\{ \begin{array}{c} b^{EL} \\ b^{DS} \\ b^{SB} \\ b^{SCP} \end{array} \right\}$$

In Eq. (19.11), $[A_{TOTAL}]$ is the global equality matrix that consists of $[A^{EL}]$, $[A^{DS}]$, $[A^{SB}]$, $[A^{SCP}]$, and identity matrix $[I]$. The global vector of unknown quantities is symbolized by

$\{X_{\text{TOTAL}}\}$, which is composed of unknown stress vector $\{\sigma\}$ and vector of conic constraints $\{\xi\}$. $\{B_{\text{TOTAL}}\}$ is the global equality vector generated due to assembly of the right-hand sides of different equality constraints. In Eq. (19.11), $N_{\text{Total}} = N_1 + M_2$, where N_1 constitutes total number of unknowns evolved from the satisfaction of element equilibrium, stress boundary, stress discontinuity conditions, yield function, and $M_2 = 9 \times E$. An in-house developed MATLAB (2019) code is used to carry out numerical analysis. Conic optimization is performed by using MOSEK software (MOSEK, 2020).

2.2 Fast Lagrangian Analysis of Continua

FLAC utilizes Lagrangian explicit time marching scheme to convert the differential equation of the problem into a set of algebraic equations associated with the nodes of the discretized domain. Each term in the algebraic equations is unknown field variable. Algebraic equations are then solved in the framework of matrix algebra for specified initial or boundary conditions. FLAC divides the problem into small time steps and calculates nodal displacements and velocities from equations of motion. Strain rates are then computed from velocities and finally new stresses are calculated from strain rates. FLAC continues to compute the nodal parameters for several time steps until it reaches to a state of static equilibrium or plastic flow occurs. Finite difference mesh is generated by discretizing the problem domain with quadrilateral elements.

FLAC utilizes strength reduction method in the stability calculation of soil slope and provides factor of safety. In strength reduction method, a series of simulations are run for trail factor of safety value (F_{trial}). The initial value of soil shear strength (φ) is progressively reduced in each simulation by that F_{trail} value until the slope reaches to the failure state. FLAC uses "bracket and bisecting technique" where lower and upper brackets are set up for a F_{trail} value. Lower bracket corresponds to that F_{trail} value for which solution converges, and upper bracket corresponds to that F_{trail} value for which solution does not converge. The average of these two bracket values is selected in the next trail and simulation is run again. If the solution converges, then lower bracket value is substituted with the current F_{trail} value and if the solution does not converge then upper bracket value is substituted with the current F_{trial} value. Several simulations are carried out until the difference between upper and lower bracket values reach to a specified tolerance value.

2.3 Random field

The influence of soil spatial variability is incorporated in this chapter by modeling soil strength parameters as a zero-mean stationary random field model as proposed by Vanmarcke (1977). According to Vanmarcke (1977), three parameters, i.e., (i) mean (μ), (ii) standard deviation (σ_{sd}), and (iii) autocovariance function (ρ_{ac}) are required to describe a stationary random field. Mean of any data set is simply the average value of the data set, whereas standard deviation denotes the deviation in the data from the mean value. The autocovariance function is obtained by fitting empirical autocovariance data. The stationary criteria of the random field are characterized by an autocorrelation function, which is dependent on the interval between two spatial points rather than the absolute coordinates of those points. The autocorrelation function (ρ_{sp}) is estimated by normalizing autocovariance function with variance. Different researchers (Vanmarcke, 1977; DeGroot and Baecher, 1993; Lacasse and Nadim, 1996) proposed various autocorrelation functions in the past. This chapter utilizes the Markovian exponential autocorrelation function (ρ_{sp}) as expressed in Eq. (19.12).

Various authors (Halder and Chakraborty, 2019b, 2020a; Lacasse and Nadim, 1996; Griffiths and Fenton, 1997) used the same autocorrelation function.

$$\rho_{sp}\left[(x_k, x_l)(y_k, y_l)\right]$$
$$= \exp\left(-\frac{|x_k - x_l|}{L_x} - \frac{|y_k - y_l|}{L_y}\right) \qquad (19.12)$$

In Eq. (19.12), spatial coordinates are x_k, y_k; x_l, y_l, whereas spatial horizontal and vertical correlation distances are L_x and L_y. Generally, soil mass at two adjacent locations is highly correlated, whereas correlation reduces with increasing distance. Higher order of randomness in soil properties is denoted by smaller values of correlation lengths. Isotropic random field is generated by assuming $L_x = L_y$ and anisotropic random field is generated by assuming $L_x \neq L_y$. The range of variation of the magnitude of spatial correlation lengths for different types of soil parameters is provided in Phoon and Kulhawy (1999a,b), Duncan (2000).

2.4 Monte Carlo simulation

There are several methods such as first-order reliability method (FORM), second-order reliability method (SORM), point estimate method (PEM), etc., which are used for the calculation of the probability of failure (p_F). However, above-mentioned analytical methods are complex in nature and good knowledge of probability and statistics is required, whereas MCS is a probabilistic method that is comparatively easier to compute the probability of failure and a little background knowledge in probability and statistics is required. The advancement of computer makes MCS more computationally efficient. The basic steps of MCS method are provided in Haldar and Mahadevan (1999). Soil parameters are considered as random variables, which follow a random distribution such as normal, lognormal, beta, etc. In each simulation, random numbers corresponding to nth number of

random variables are generated. Based on the generated values of random numbers, output function (g_o) is calculated in each simulation. The same procedure is carried out for total N_T number of simulations. In order to calculate the failure probability (p_F) of the output function (g_o), a failure criterion ($g_o < 0$) is defined first. With the accomplishment of N_T simulations, total number of times (N_F) the value of g_o becomes less than zero is counted. Now, the failure probability of the output function is obtained through Eq. (19.13).

$$p_F = \frac{N_F}{N_T} \qquad (19.13)$$

The total number of simulations required to find the constant value of p_F is calculated as provided in Haldar and Mahadevan (1999). However, sometimes the total number of MCS required to obtain a constant value of p_F can also be chosen from the variation of μ and CoV of the output function with number of simulations (Haldar and Babu, 2008). The simulation after which mean and coefficient of variation of the output function changes insignificantly is chosen as the optimum number of MCS.

2.5 Coupling of lower bound limit analysis, random field, and Monte Carlo simulation

A concise discussion related to the random field generation of a single variable X_{r1} in the framework of lower-bound finite element limit analysis is presented below. Using the same methodology, random field for any number of uncorrelated random variables can be generated. The random variable X_{r1} is assumed to follow lognormal distribution. The mean $\left(\mu_{\ln X_{r1} X_{r1}}\right)$ and standard deviation $\left(\sigma_{\ln X_{r1} X_{r1}}\right)$ of X_{r1} are expressed by Eqs. (19.14) and (19.15), respectively.

$$\sigma^2_{\ln X_{r1}} = \ln\left(1 + CoV^2_{X_{r1}}\right) \qquad (19.14)$$

$$\mu_{\ln X_{r1}} = \ln\left(\mu_{X_{r1}}\right) - 0.5\sigma_{\ln X_{r1}}^2 \qquad (19.15)$$

The spatially distributed value of X_{r1j} for jth element is obtained with the implementation of following equation:

$$X_{r1j} = \exp\left(\mu_{\ln X_{r1}} + \sigma_{\ln X_{r1}} G_{X_{r1j}}\right) \qquad (19.16)$$

In Eq. (19.16), standard normal random field of X_{r1} is indicated by $G_{X_{r1}}$, which is obtained by the following procedures. In the first step, the autocorrelation function (ρ_{sp}) is calculated at the centroidal points of each element by using Eq. (19.12) for the discretized soil domain. As the centroidal point is considered, the values of x_k and y_k are calculated by Eq. (19.17). Similarly, the values of x_l and y_l are obtained for other location.

$$x_k = \frac{(x_1 + x_2 + x_3)}{3}; \; y_k = \frac{(y_1 + y_2 + y_3)}{3}$$
$$(19.17)$$

In Eq. (19.17), (x_1, y_1), (x_2, y_2), and (x_3, y_3) are coordinates of three nodes of the triangular element. With the computed values of ρ_{sp}, the correlation matrix (C_r) is constituted as expressed in Eq. (19.18).

$$[C_r] = \begin{bmatrix} 1 & \rho_{12} & \cdots & \rho_{1E} \\ \rho_{21} & 1 & \cdots & \rho_{2E} \\ \cdots & \cdots & \cdots & \cdots \\ \rho_{E1} & \rho_{E2} & \cdots & 1 \end{bmatrix} \qquad (19.18)$$

By using the Cholesky decomposition technique, $[C_r]$ is then decomposed into upper and lower triangular matrices which are expressed in Eq. (19.19).

$$KK^T = [C_r] \qquad (19.19)$$

The lower triangular matrix, K^T, is replaced in the following expression (Eq. 19.20) to obtain the standard normal random field, $G_{X_{r1}}$:

$$[G_{X_{r1}}] = [K^T]\{S_{r1}\} \qquad (19.20)$$

In the above expression, $\{S_{r1}\}$ is the vector that consists of independent standard random numbers $\{S_{r1}\} = \{S_{r11}, S_{r12}, \ldots, S_{r1E}\}^T$. In the final step, by substituting the value of $G_{X_{r1}}$ along with the known values of $\mu_{\ln X_{r1}}$ and $\sigma_{\ln X_{r1}}$ in Eq. (19.16), the spatial values of X_{r1} are obtained for each MCS. For each MCS, after obtaining the spatial dataset of X_{r1} corresponding to all the elements of the domain, limit analysis code is run, and the output function is estimated. Subsequently, the above-mentioned procedure is repeated for N_T number of times. After the completion of the MCS, statistical information of the output function is obtained in terms of mean, standard deviation, and failure probability.

2.6 Coupling of FLAC, random field, and Monte Carlo simulation

In addition to the estimation of probabilistic ultimate load-bearing capacity at a settlement of the strip footing using FLAC, this chapter also investigates the influence of cross-correlated random fields. A generalized approach for the creation of the cross-correlated random field of two variables, X_{cr1} and X_{cr2} in the framework of finite difference software FLAC is presented below. Using the same methodology, random field for any number of correlated random variables can be generated. Two random variables X_{cr1} and X_{cr2} are assumed to be correlated with each other and follow the lognormal distribution. The lognormally distributed correlated random fields of X_{cr1} and X_{cr2} are obtained by following expressions:

$$X_{cr1} = \exp\left(\mu_{\ln X_{cr1}} + \sigma_{\ln X_{cr1}} G_{crX_{cr1}}\right) \qquad (19.21)$$

$$X_{cr2} = \exp\left(\mu_{\ln X_{cr2}} + \sigma_{\ln X_{cr2}} G_{crX_{cr2}}\right) \qquad (19.22)$$

In the above Eqs. (19.21) and (19.22), $G_{crX_{cr1}}$ and $G_{crX_{cr2}}$ are the correlated standard normal random fields; $\mu_{\ln Xcr1}$ and $\mu_{\ln Xcr2}$ are mean of two random variables X_{cr1} and X_{cr2}, respectively;

and $\sigma_{\ln X_{cr1}}$ and $\sigma_{\ln X_{cr2}}$ are standard deviation of two random variables X_{cr1} and X_{cr2}, respectively. Like the random field generation in lower bound limit analysis framework, here in this case too, the autocorrelation function (ρ_{sp}) is first calculated by using Eq. (19.16) for the discretized soil domain. However, instead of calculating the value of ρ_{sp} at the centroidal position of each element, ρ_{sp} value is obtained at each node of the discretized domain. After the formation of correlation matrix (C_r) by following Eq. (19.18), it is then decomposed into upper and lower triangular matrices as per Eq. (19.19). The lower triangular matrix, K^T, is replaced in the following expressions to obtain two independent standard normal random fields, $G_{X_{cr1}}$ and $G_{X_{cr2}}$:

$$\left[G_{X_{cr1}}\right] = \left[K^T\right]\{S_{cr1}\} \qquad (19.23)$$

$$\left[G_{X_{cr2}}\right] = \left[K^T\right]\{S_{cr2}\} \qquad (19.24)$$

Here, $\{S_{cr1}\}$ and $\{S_{cr2}\}$ are the vectors that consist of independent standard random distribution for two random variables. Now, as the two random variables X_{cr1} and X_{cr2} are correlated, the correlated standard normal random fields, $G_{crX_{cr1}}$ and $G_{crX_{cr2}}$, are obtained by Eqs. (19.25) and (19.26).

$$L_c L_c^T = \begin{bmatrix} 1 & \rho_{xr} \\ \rho_{xr} & 1 \end{bmatrix} \qquad (19.25)$$

$$\begin{Bmatrix} G_{crX_{cr1}} \\ G_{crX_{cr2}} \end{Bmatrix} = \begin{bmatrix} L_{c11} & 0 \\ L_{c21} & L_{c22} \end{bmatrix} \begin{Bmatrix} G_{X_{cr1}} \\ G_{X_{cr2}} \end{Bmatrix} \qquad (19.26)$$

In the above Eq. (19.25), ρ_{xr} is the cross-correlation coefficient. Finally, cross-correlated spatial values of X_{cr1} and X_{cr2} are obtained for each MCS by substituting $G_{crX_{cr1}}$ and $G_{crX_{cr2}}$ in Eqs. (19.21) and (19.22). For each MCS, after obtaining the spatial dataset of X_{cr1} and X_{cr2} corresponding to all the elements of the domain, FLAC code is run, and the output function value is calculated.

Methodologies described in the previous sections are provided below in terms of flowchart. Fig. 19.2 represents the methodologies considered for determining probabilistic ultimate load-bearing capacity of footing and the probabilistic load-carrying capacity of footing under a settlement level.

3. Problem statement: Probabilistic bearing capacity of strip footing on reinforced soil slope

3.1 Objective

The objective is to find out the probabilistic bearing capacity of a rough, rigid, and surface strip footing placed on the reinforced cohesionless soil slope. A strip footing of width of B is placed on the edge of a slope with a slope angle of β ($= 20°$) as shown in Fig. 19.3. Probabilistic bearing capacity of a footing on slope is obtained in terms of dimensionless factor N_γ, which is related to soil unit weight (γ). It is to be mentioned here that Halder and Chakraborty (2018a, 2019a) presented optimum depths of reinforcement layers for various combinations of slope geometry and soil parameters. Maximum reinforcing efficiency is obtained with the placement of reinforced layers below the footing within the soil mass. In this chapter study, two layers of reinforcement, one at a critical depth of d_{1cr}/B from the footing base and another with optimum spacing of d_{2cr}/B from top reinforcement layer, are laid throughout slope. The Mohr−Coulomb constitutive model governs the failure criteria of soil. Dilatancy behavior of cohesionless soil is incorporated by nonassociated flow rule. Reinforcement is modeled as per the formulation of Chakraborty and Kumar (2014). Soil−reinforcement interface is considered fully rough. Reinforcement tearing is avoided by considering large tensile strength. Failure occurs only due to shearing between reinforcement and soil mass.

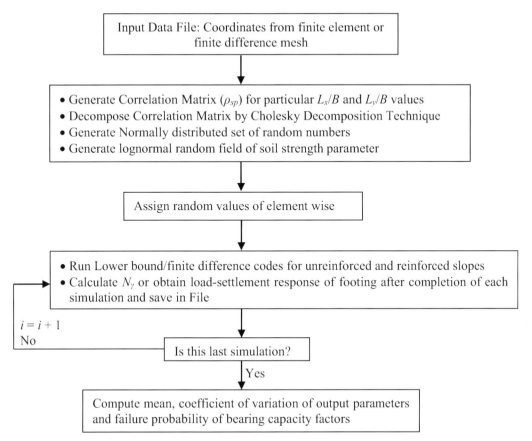

FIGURE 19.2 Algorithm used in the present study for probabilistic modeling.

3.2 Boundary conditions and FE mesh

Different problem domains are chosen depending upon spatial correlation distances. Fig. 19.3 shows a typical problem domain chosen for large values of correlation lengths. Problem domain is selected large to ensure plastic zone cannot extend to the boundary faces. Collapse load should also remain independent with the selected domain. For a probabilistic analysis, horizontal and vertical extents of problem domain should be larger than corresponding correlation lengths (L_x/B and L_y/B). Zero normal and shear stresses along sloping edge IJ and top horizontal edge GH are imposed since no surcharge pressure is considered. Zero normal and shear stresses are also maintained along the vertical edge JE. Footing–soil interface (HI) is rough and included by a condition $|\tau_{xy}| \leq (-\sigma_N \tan \varphi)$.

Three noded triangular elements are used for discretization of problem domain. Relatively finer meshing is done near the possible stressed zone surrounding the footing and coarser meshing is generated near the boundary edges. Fig. 19.4 shows discretized finite element mesh used for a cohesionless soil slope, having 20° slope angle and 30° friction angle of soil. In Fig. 19.4, total number of elements, nodes, footing–soil interface nodes, and discontinuities are expressed by E_l, N_n, N_{ni}, and D_{cs}, respectively.

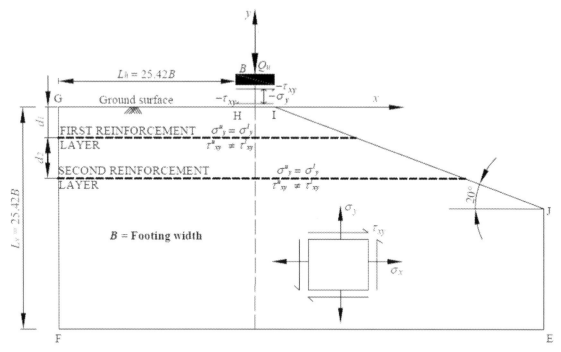

FIGURE 19.3 Problem domain and boundary conditions.

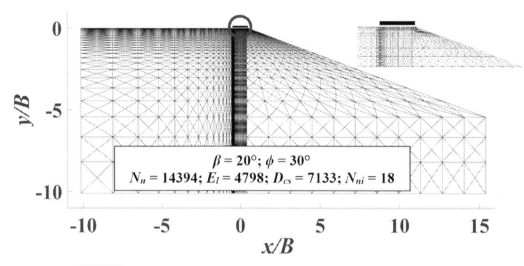

FIGURE 19.4 Finite element mesh used in the study along with zoomed view.

3.3 Results

3.3.1 Spatial distribution of soil strength parameters

The effect of uncertainty related to friction angle of soil on the ultimate collapse load of footing situated at the cohesionless soil slope is investigated. Random distribution of soil friction angle is considered to follow lognormal distribution. Two CoV_φ values are taken as 5% and 10%. Anisotropic random fields are generated by considering different values of L_x/B (=1.25, 5, 10, and 20) and L_y/B (=1.25, 5, 10, and 20). These wide ranges of parameters are in accordance with Phoon and Kulhawy (1999a,b). Slope angle (β) is kept fixed as $20°$. Mean deterministic soil friction angle (φ) of cohesionless soil is assumed $30°$ and $45°$. Since dilation angle of soil is one sixth times randomly generated friction angle of soil, it becomes a random variable. Following, Drescher and Detournay (1993), modified friction angle (φ^*) is obtained after considering soil dilatancy. It is to be mentioned here that for slope combination of $\beta = 20°$, $\varphi = 30°$, and $0.39B$ and $0.41B$, critical optimum depths of first and second layers of reinforcement are $0.23B$ and $0.24B$, respectively. For all the probabilistic analyses, reinforcements are at optimum positions. Spatially distributed modified friction angle (φ^*) is obtained for different probabilistic parameters. Fig. 19.5A shows sparse distribution of φ^* for lower correlation lengths. Lower values of correlation length denote high correlation between soil parameters over a small distance. As correlation length increases, modified friction angle becomes uniform as considered in deterministic analysis (refer Fig. 19.5B). Higher values of correlation length denote less correlation between soil parameters.

3.3.2 Probabilistic bearing capacity factor N_γ for footing on cohesionless soil

Fig. 19.6A–D shows different values of mean bearing capacity factor N_γ ($\mu_{N\gamma}$) obtained for different correlation lengths for a footing resting on the cohesionless soil slope with $\beta = 20°$ and $\varphi = 30°$. Mean bearing capacity factor is also obtained for similar slope inclination having $\varphi = 45°$ with different correlation lengths and illustrated in Fig. 19.6E and F. Mean bearing capacity factor associated with reinforced slope value is always greater than that of unreinforced slope. As an example, for double reinforced soil slope combination of $\beta = 20°$, $\varphi = 45°$, $L_x/B = 1.25$, $L_y/B = 1.25$, and $CoV_\varphi = 10\%$, mean bearing capacity factor is found to be 41.79 with respect to a value of 15.98 obtained for unreinforced slope. It is found out that for smaller correlation lengths, mean value of N_γ is lesser than that obtained from the deterministic analysis of both unreinforced and reinforced slopes. It signifies the importance of considering soil spatial variability in the bearing capacity estimation of strip footing, placed at cohesionless soil slope edge. As an example, mean N_γ value corresponding to $\beta = 20°$, $\varphi = 30°$, $L_x/B = 1.25$, $L_y/B = 1.25$, and $CoV_\varphi = 10\%$ is 5.12, whereas deterministic N_γ value is 5.41. However, with escalating correlation lengths, this difference reduces and become almost same for higher values of correlation lengths. For the identical reinforced slope, mean bearing capacity increases from 5.12 to 5.41 with enhancing L_x/B value from 1.25 to 20 and L_y/B value from 1.25 to 5. Spatial variation of modified friction angle in Fig. 19.5B also supports the above-mentioned phenomenon from the perspective of randomness. Smaller correlation length denotes sparsely scattered values of φ^* and higher correlation lengths indicate uniform distribution of φ^* like that considered in deterministic analysis. As CoV_φ increases, the mean bearing capacity factor decreases for both reinforced and unreinforced slopes. It is found in Fig. 19.6A corresponding to a reinforced slope with $\beta = 20°$, $\varphi = 30°$, $L_x = 1.25B$, and $L_y = 1.25B$ that magnitude of $\mu_{N\gamma}$ reduces by an amount of 3.42% with an increase in CoV_φ value from 5% to 10%. For both

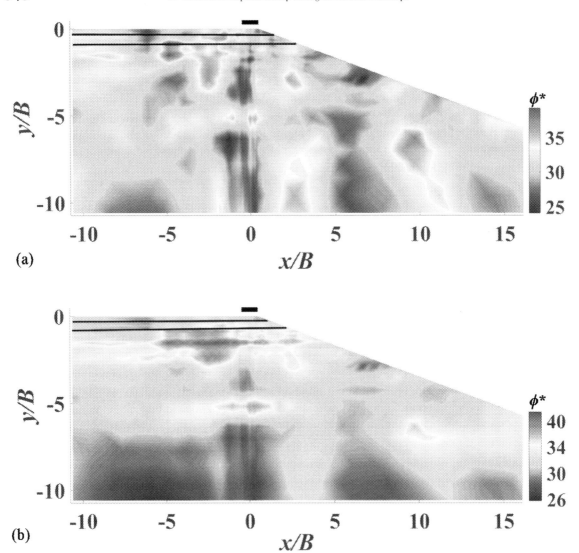

FIGURE 19.5 Spatial distribution plot of modified friction angle (φ) for reinforced slope of $\beta = 20°$ with (A) $L_x = 1.25B$, $L_y = 1.25B$, $CoV_\varphi = 10\%$; and (B) $L_x = 10B$, $L_y = 1.25B$, $CoV_\varphi = 10\%$.

unreinforced and reinforced slopes, bearing capacity increases with increasing friction angle of soil (Fig. 19.6E and F).

It is to be mentioned here that in the present section authors discuss the effect of uncertainties related to soil friction and dilation angle on the ultimate load-bearing capacity of strip footing placed on reinforced cohesionless soil slope. By using the similar methodologies, anyone can obtain probabilistic ultimate load-bearing capacity of strip footing on reinforced cohesive soil slope. One can follow Halder and Chakraborty (2020c) for more details (Fig. 19.7).

3. Problem statement: Probabilistic bearing capacity of strip footing on reinforced soil slope 347

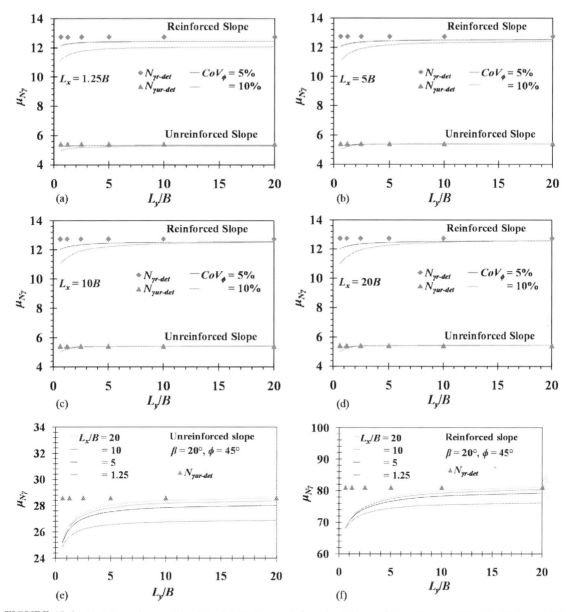

FIGURE 19.6 Variation of $\mu_{N\gamma}$ with L_y/B and CoV_φ for unreinforced and slope of $\beta = 20°$, $\varphi = 30°$ with (A) $L_x = 1.25B$; (B) $L_y = 5B$; (C) $L_x = 10B$; (D) $L_x = 20B$; variation of $\mu_{N\gamma}$ with L_x/B and L_y/B for (E) unreinforced slope of $\beta = 20°$, $\varphi = 45°$ (F) reinforced slope of $\beta = 20°$, $\varphi = 45°$. *Modified from Halder, K., Chakraborty, D., 2019c. Probabilistic bearing capacity of strip footing on reinforced soil slope. Comput. Geotech. 116, 103213-1-11.*

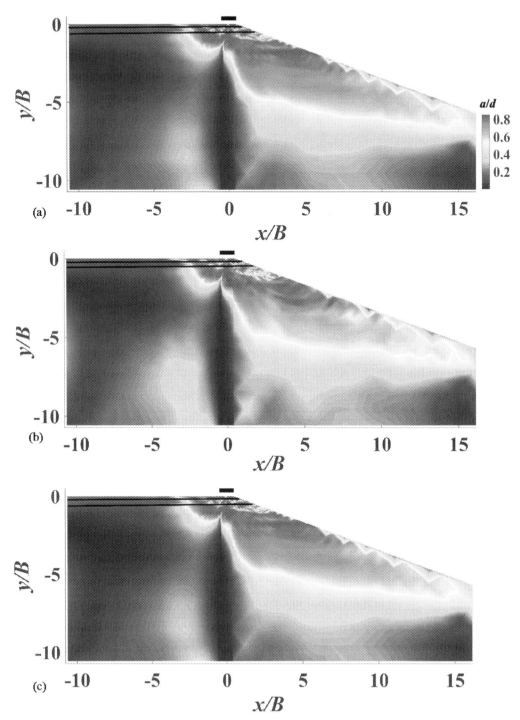

FIGURE 19.7 Failure pattern obtained from (A) deterministic analysis of reinforced slope with $\beta = 20°$, $\varphi = 30°$ probabilistic analysis with (B) $\beta = 20°$, $\varphi = 30°$ $CoV_\varphi = 10\%$, $L_x/B = 1.25$, $L_y/B = 1.25$; and (C) $\beta = 20°$, $\varphi = 30°$, $CoV_\varphi = 10\%$, $L_x/B = 10$, $L_y/B = 1.25$.

4. Problem statement: Probabilistic load carrying capacity of strip footing on geocell reinforced soil slope

4.1 Objective

This chapter finds probabilistic load–settlement response of a rigid, rough, and surface strip footing, resting on the soil slope. Geometry of slope used is illustrated in Fig. 19.8. Slope height is 10 m and angle of inclination (β) with horizontal axis is 30°. Footing width is 1 m ($B = 1$). A geocell mattress is embedded at a depth of 0.25B. Geocell mattress is expanded up to a length of 11.25B to achieve enough anchorage. Friction angle of soil is a random variable following log-normal distribution. Interdependency between c and φ is also investigated in this chapter. Mean load carrying capacity (μ_Q) of strip footing is calculated.

4.2 Boundary conditions and FD mesh

Depending upon the types of spatial correlation distances, dimensions of the problem domain are varied. Fig. 19.8 illustrates a sufficiently expanded problem domain in both horizontal and vertical directions. It is ensured that failure stresses do not reach to the boundary edges of the domain. Fig. 19.9 shows implemented displacement-based boundary conditions at the edges. Horizontal and vertical

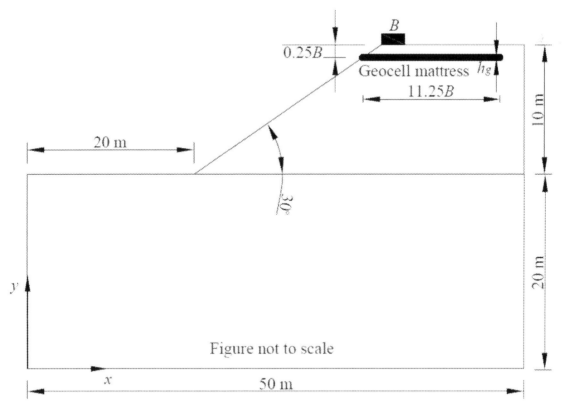

FIGURE 19.8 Schematic diagram of the geocell-reinforced slope. *Modified Halder, K., Chakraborty, D., 2020b. Influence of soil spatial variability on the response of strip footing on geocell-reinforced slope. Comput. Geotech. 122, 103533-1-13.*

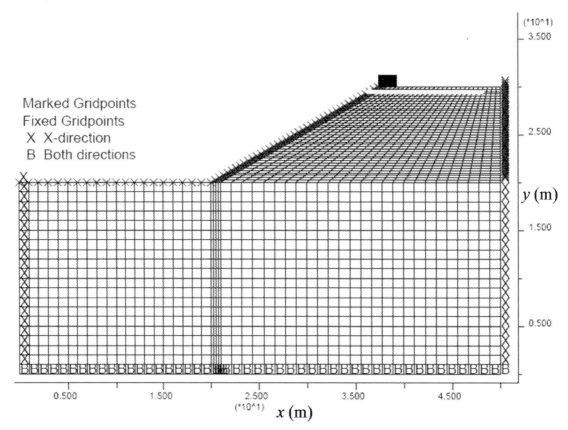

FIGURE 19.9 Finite difference mesh used in the numerical analyses. *Modified from Halder, K., Chakraborty, D., 2020b. Influence of soil spatial variability on the response of strip footing on geocell-reinforced slope. Comput. Geotech. 122, 103533-1-13.*

movements are restricted along bottom boundary edge. On the other hand, fixity in horizontal movement is maintained at both sides of domain. Problem domain is discretized into 1500 nonuniform zones by four noded quadrilateral elements. Relative finer meshing is done near the footing, whereas coarser mesh is chosen for the region far from the influence zone. However, length of finer mesh should be less than that of smaller correlation lengths and coarse mesh should have large length in comparison to the larger correlation lengths. The Mohr—Coulomb constitutive model defines stress—strain relationship of cohesive-frictional soil fill. Table 19.1 presents various soil parameters used in the numerical analysis. Properties of these parameters are taken from Bowles (2007), Mehdipour et al. (2013). Beam element accounts frictional and bending resistance of geocell mattress. One single run of three-dimensional modeling of the geocell mattress requires large time, which makes it computationally inefficient. Total running time of one single simulation is an important constraint in case of probabilistic analysis, as it requires to run several simulations. In contrast, computational time reduces significantly for the two-dimensional modeling of geocell with a little compromise in the achieved solution. Several research studies were carried out in the past with the consideration of geocell

4.3 Results

TABLE 19.1 Properties of different soil parameters.

Parameter	Value
Friction angle, φ (°)	30
Cohesion, c (kPa)	2
Dry unit weight, γ_{dry} (kN/m³)	20
Young's modulus, E (MPa)	25
Poisson's ratio, v	0.25

Modified from Halder, K., Chakraborty, D., 2020b. Influence of soil spatial variability on the response of strip footing on geocell-reinforced slope. Comput. Geotech. 122, 103533-1-13.

mattress in two-dimensional framework. Two approaches, namely, (i) equivalent stiffness approach (Rajagopal et al., 2001; Latha et al., 2006; Latha and Rajagopal, 2007) and (ii) flexible slab approach (Mehdipour et al., 2013) were used to model geocell as a beam element in FLAC. Attachment of a beam element with FLAC grid can be done in two ways: (i) direct connection of grids with nodes or (ii) use of interface element between beam and nodes representing soil. This chapter follows the former approach. Following Mehdipour et al. (2013), tensile yield strength and secant modulus of geocell mattress are considered 60 kN/m and 150 kN/m, respectively. Geocell mattress height is 0.2 m.

Two-dimensional numerical model is run for several cycles before it attains a steady state of static equilibrium. After that, displacements at all nodes are neutralized to zero. An insignificant vertical downward velocity is applied to replicate strip loading. After carrying out a few trials, optimized vertical downward velocity is chosen as 1.5×10^{-6} m/time step. Besides requirement of lesser time, optimized vertical downward velocity should not have influence on the load—settlement response of the footing. Numerical analysis is carried out for several cycles before attaining a plastic steady state. A state of plastic equilibrium is indicated by lower values of unbalanced forces or by the constant load of footing with increasing settlement.

4.3.1 *Spatial distribution of soil friction angle*

Spatial distribution of soil friction angle in the x and y directions as shown in Fig. 19.10A and B is obtained for smaller ($L_x/B = L_y/B = 1$) and larger magnitudes ($L_x/B = L_y/B = 10$) of correlation lengths and coefficient of variation of friction angle (CoV_φ) = 5%. Fig. 19.10A shows sparsely generated values of φ for smaller correlation lengths. Spatial distribution of φ becomes uniform with an increase in correlation lengths as depicted in Fig. 19.10B.

4.3.2 *Probabilistic load carrying capacity at a particular settlement level of footing*

It is to be mentioned here that before carrying out probabilistic analysis, deterministic load—settlement response is obtained with constant soil and slope parameters as: $\beta = 30°$, $c = 2$ kPa, and $\varphi = 30°$. Footing settlement (δ) corresponding to the ultimate load carrying capacity is first estimated from deterministic load—settlement plot. It is found out that for the abovementioned slope geometry and soil property, corresponding settlement is 5%. One can refer Halder and Chakraborty (2020b) for details. Then for the probabilistic analysis, for each MCS, load carrying capacity is evaluated based on footing settlement = 5%. Three correlation lengths in x direction (1, 10, and 50) are chosen are considered, whereas L_y/B values are varied as 1, 2.5, 5, 10, and 20. Values of CoV_φ are taken as 10% and 20%. For different values of L_x/B, CoV_φ, variation between μ_Q and L_y/B is depicted in Fig. 19.11A. It is found out from Fig. 19.11A that for a constant L_x/B value and varying L_y/B values, mean load carrying capacity first decreases, followed by the attainment of lowest value corresponding to a particular L_y/B and then it again starts increasing. By keeping L_y/B as constant, similar trend is visible for the variation of L_x/B. As CoV_φ value increases, mean load carrying capacity decreases significantly. For an

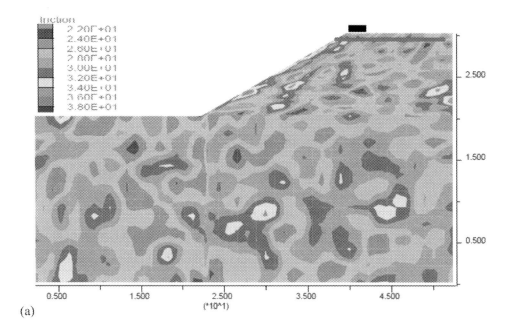

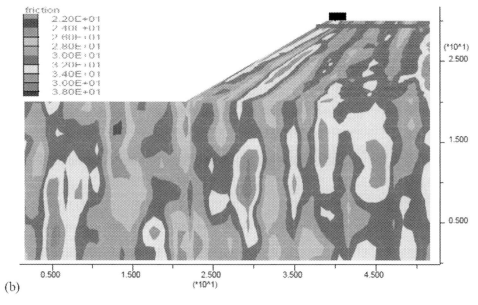

FIGURE 19.10 Spatial distribution of φ in a geocell reinforced slope of $\beta = 30°$ with $L_x/B = 1$, $L_y/B = 1$, $CoV_\varphi = 10\%$ and (B)$L_x/B = 1$, $L_y/B = 10$, $CoV_\varphi = 10\%$. *Modified from Halder, K., Chakraborty, D., 2020b. Influence of soil spatial variability on the response of strip footing on geocell-reinforced slope. Comput. Geotech. 122, 103533-1-13.*

4. Problem statement: Probabilistic load carrying capacity of strip footing on geocell reinforced soil slope

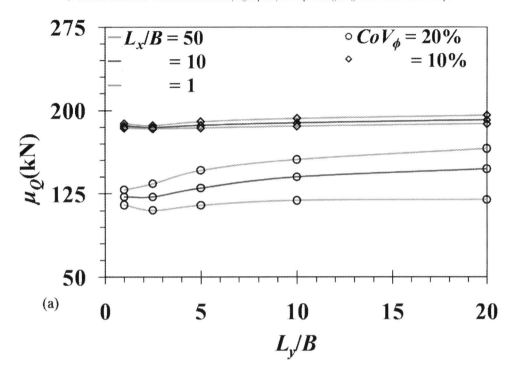

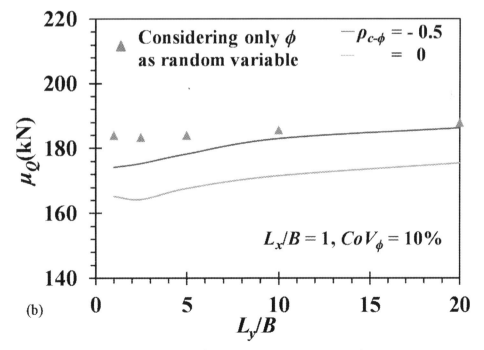

FIGURE 19.11 (A) Variation between μ_Q and L_y/B for different values of CoV_φ, L_x/B and (ii) variation of μ_Q with L_y/B, $\rho_{c-\varphi}$ for $CoV_\varphi = 10\%$, $L_x/B = 1$. Modified from Halder, K., Chakraborty, D., 2020b. Influence of soil spatial variability on the response of strip footing on geocell-reinforced slope. Comput. Geotech. 122, 103533-1-13.

instance, μ_Q value decreases from 184.62 to 122.26 kN with an increment in CoV_φ value from 10% to 20% by keeping $L_x/B = 10$ and $L_y/B = 2.5$.

Besides the spatial variability of shear strength parameters of soil (c and φ), there can be interdependency between each parameter. Various researchers (Griffiths et al., 2009; Cherubini, 2000) have considered the effects of cross-correlation between cohesion and friction angle of soil and found it significant. Negative cross-correlation between cohesion and friction angle is considered in all the previous studies. In this chapter, cross-correlation coefficient ($\rho_{c-\varphi}$) is assumed as -0.5 and 0. Fig. 19.11B shows variation between μ_Q and $\rho_{c-\varphi}$ for various magnitudes of L_y/B. Values of L_x/B ($= 1$) and CoV_φ ($= 10\%$) are kept fixed. Highest value of μ_Q is obtained for $\rho_{c-\varphi} = -0.5$. As an instance, the magnitudes of μ_Q are found to be 178.34 and 167.63 kN, respectively, for $\rho_{c-\varphi} = -0.5$ and 0 with $L_x/B = 1$, $L_y/B = 5$. Negative value of $\rho_{c-\varphi}$ indicates decrease in c value and increase in φ value. For that reason, mean load carrying capacity is higher for negative cross-correlation coefficient and as cross-correlation coefficient increases mean load carrying capacity reduces. However, irrespective of cross-correlation coefficient, mean load carrying capacity is always lesser with respect to that achieved by assuming φ as a single random variable.

4.4 Failure mechanism

Failure mechanism of soil mass under strip loading is discussed with the help of accumulated shear strain contours up to the ultimate loading stage of footing. Shear strain contours are plotted as shown in Fig. 19.12A and B for soil slope having spatially distributed soil friction angle. Shear strain value is maximum just below the footing. It then extends to the slope face. Fig. 19.12A and B illustrates erratic shear strain profile when spatial variability of soil

friction angle is considered. For a combination of $L_x/B = L_y/B = 1$, $CoV_\varphi = 10\%$, maximum accumulated shear strain is 4.2×10^{-2}. However, for a combination of $L_x/B = 1$, $L_y/B = 10$ and $CoV_\varphi = 10\%$ as shown in Fig. 19.12B, with increasing correlation length, accumulated shear strain reduces.

5. Problem statement: Probabilistic stability analysis of reinforced soil slope subjected to strip loading

5.1 Objective

Probabilistic stability analysis of both unreinforced and reinforced soil slopes under strip loading is carried out by considering uncertainties related to soil friction angle and soil unit weight. Slope is inclined with an angle of 30° with the horizontal axis and a surcharge loading over width B is applied to simulate strip loading. Fig. 19.13 shows that a single layer of geotextile reinforcement is laid within the slope at optimum depth. Soil is assumed to follow the Mohr–Coulomb constitutive model. Geotextile is modeled as cable element by assuming no bending resistance, only having axial resistance. Properties used to model soil and geotextile are as per Yang et al. (2016). In order to carry out probabilistic analyses, soil friction angle and unit weight are assumed to have lognormal distribution.

5.2 Boundary conditions and FD mesh

A problem domain is first selected after carrying out several numbers of trials, in such a way so that it would not be affected by the plastic zones, developed as a result of strip loading on the top of the slope. Stability of the slope should also not change too much with a change in the area of the problem domain. An explicit finite difference software FLAC is used to discretize the problem domain and because of that,

5. Problem statement: Probabilistic stability analysis of reinforced soil slope subjected to strip loading

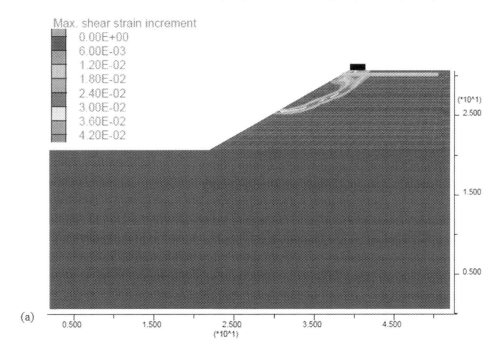

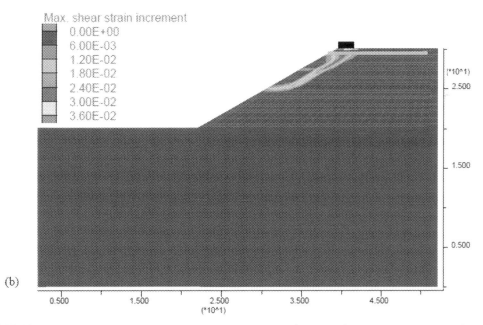

FIGURE 19.12 Accumulated shear strain plots for slope with (A) $L_x/B = 1, L_y/B = 1, CoV_\varphi = 10\%$; and (B) $L_x/B = 1, L_y/B = 10, CoV_\varphi = 10\%$. *Modified from Halder, K., Chakraborty, D., 2020b. Influence of soil spatial variability on the response of strip footing on geocell-reinforced slope. Comput. Geotech. 122, 103533-1-13.*

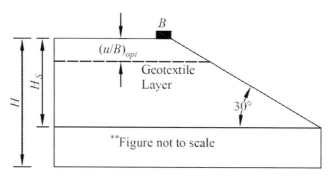

FIGURE 19.13 Problem domain considered for stability analysis of reinforced slope subjected to strip loading.

1865 number of quadrilateral zones is generated within the finite difference mesh. Both the movements along vertical and horizontal directions are constrained at the bottom of the problem domain, whereas movements along the horizontal directions are restricted on left- and right-side boundaries of the mesh.

5.3 Results

5.3.1 Probabilistic stability analysis

It is to be noted that all probabilistic analyses of reinforced slopes are performed by keeping reinforcement layers at an optimum depth. One can refer to Halder and Chakraborty (2018a, 2019a, 2020a) to find the details of optimum depth of reinforcement in a cohesionless soil slope subjected to strip footing. It is found that the magnitude of p_F increases with the increase in the value of CoV of φ. For an unreinforced slope, it is found out that when CoV of φ changes from 5% to 20%, p_F changes from 0.49 to 0.53. A drastic reduction in the magnitude of failure probability is observed with the utilization of reinforcement layer. As an example, p_F of unreinforced slope is found to be 0.52 for a slope with combination of $CoV_\varphi = 15\%$, and $CoV_\gamma = 5\%$. In contrast, for the similar slope combination but with the inclusion of geotextile reinforcement, failure probability reduces to 0.10.

Effects of uncertainties related to soil unit weight on the stability of reinforced and unreinforced slopes are explored by varying CoV_γ and keeping $CoV_\varphi = 10\%$. Failure probability increases with increasing value of CoV_γ. However, the effect is not as significant as observed in case of variation of CoV_φ.

6. Conclusions

This chapter first discusses the methodology of incorporating spatial variability of soil strength properties in any probabilistic analysis and computes the probabilistic ultimate collapse load of a strip footing resting on the reinforced soil slope. Ultimate collapse load is obtained by utilizing the lower bound finite element limit analysis. Soil properties are modeled as lognormally distributed random field model and probabilistic outcomes are obtained by running the MCS. The finite difference method, random field modeling, and MCS technique are coupled to examine influences of strength properties of soil on the probabilistic load-carrying capacity of the footing at a particular settlement. Randomness in soil properties is included in the probabilistic stability analyses of geotextile reinforced soil slopes subjected to strip loading and failure probability of the slopes is obtained by carrying out a number of MCS. Salient features of the present chapters are as followed:

(i) For smaller correlation lengths, mean probabilistic bearing capacity factor N_γ is found to be lesser than the deterministic bearing capacity factor. Deviation in the magnitude of mean bearing capacity factor escalates with the increment in the value of coefficient of variation of friction angle. Therefore, it is important to consider spatial variability of soil friction angle. As an example, of a reinforced slope with $\beta = 20°$, $\varphi = 30°$, $L_x = L_y = 1.25B$ deviation in the deterministic and probabilistic values of N_γ increases from 2.22% to 5.73% with increasing CoV_φ value from 5% to 10%. However, for higher correlation lengths, mean bearing capacity factor approaches to the deterministic value of N_γ.

(ii) Spatial variability and randomness of soil friction angle and soil cohesion have a significant role on the probabilistic load—settlement response of footing. Probabilistic mean load-carrying capacity (μ_Q) of the strip footing corresponding to a settlement value is found less than that obtained for soil slope with uniform friction angle. As an example, for a slope angle of 30°, mean load-bearing capacity varies between 184.05 and 187.98 kN by considering L_y/B values between 1 and 20 and keeping $L_x/B = 1$, $CoV_\varphi = 10\%$. The value of probabilistic mean load-carrying capacity decreases with escalating correlation coefficient between soil cohesion and friction angle. Irrespective of the values of φ, the magnitude of μ_Q first decreases, attains lowest value at a particular L_y/B value, and finally it increases again with increasing L_y/B value.

(iii) Possibility of slope failure increases with increasing randomness related to soil friction angle and unit weight for any position of footing on slope. This becomes more pronounced for an unreinforced slope. However, failure probability of the slope reduces significantly with the incorporation of geotextile layer. It is found out that CoV_φ has more impact on the calculated values of failure probability than that of CoV_γ.

References

Bathurst, R.J., Blatz, J.A., Burger, M.H., 2003. Performance of instrumented large-scale unreinforced and reinforced embankments loaded by a strip footing to failure. Can. Geotech. J. 40 (6), 1067−1083.

Beygi, M., Keshavarz, A., Abbaspour, M., Vali, R., Saberian, M., Li, J., 2020. Finite element limit analysis of the seismic bearing capacity of strip footing adjacent to excavation in c-φ soil. Geomechanics Geoengin. 1−14.

Bowles, L.E., 2007. Foundation Analysis and Design. McGraw-hill.

Chakraborty, D., Kumar, J., 2013. Bearing capacity of foundations on slopes. Geomechanics Geoengin. 8 (4), 274−285.

Chakraborty, D., Kumar, J., 2014. Bearing capacity of strip foundations in reinforced soils. Int. J. GeoMech. 14 (1), 45−58.

Cherubini, C., 2000. Reliability evaluation of shallow foundation bearing capacity on $c' − \varphi'$ soils. Can. Geotech. J. 37 (1), 264−269.

Davis, E.H., Booker, J.R., 1973. Some adaptations of classical plasticity theory for soil stability problems. In: Proceedings of the Symposium on the Role of Plasticity in Soil Mechanics. England, Cambridge.

DeGroot, D.J., Baecher, G.B., 1993. Estimating autocovariance of in-situ soil properties. J. Geotech. Eng. 119 (1), 147−166.

Drescher, A., Detournay, E., 1993. Limit load in translational failure mechanisms for associative and non-associative materials. Geotechnique 43 (3), 443−456.

Duncan, J.M., 2000. Factors of safety and reliability in geotechnical engineering. J. Geotech. Geoenviron. Eng. 126 (4), 307−316.

El-Sawwaf, M., 2007. Behaviour of strip footing on geogrid-reinforced sand over a soft clay slope. Geotext. Geomembranes 25 (1), 50−60.

Fenton, G.A., Griffiths, D.V., 2003. Bearing-capacity prediction of spatially random c-φ soils. Can. Geotech. J. 40 (1), 54−65.

Georgiadis, K., 2010. Undrained bearing capacity of strip footings on slopes. J. Geotech. Geoenviron. Eng. 136 (5), 677−685.

Griffiths, D.V., Fenton, G.A., 1997. Three-dimensional seepage through spatially random soil. J. Geotech. Geoenviron. Eng. 123 (2), 153−160.

Griffiths, D.V., Fenton, G.A., Manoharan, N., 2002. Bearing capacity of rough rigid strip footing on cohesive soil: probabilistic study. J. Geotech. Geoenviron. Eng. 128 (9), 743−755.

Griffiths, D.V., Huang, J., Fenton, G.A., 2009. Influence of spatial variability on slope reliability using 2-D random fields. J. Geotech. Geoenviron. Eng. 135 (10), 1367–1378.

Haldar, S., Babu, G.L.S., 2008. Effect of soil spatial variability on the response of laterally loaded pile in undrained clay. Comput. Geotech. 35 (4), 537–547.

Haldar, A., Mahadevan, S., 1999. Probability, Reliability and Statistical Methods in Engineering Design. John Wiley & Sons, New York.

Haldar, K., Chakraborty, D., 2018a. Bearing capacity of strip footing placed on the reinforced soil slope. Int. J. Geo-Mech. 18 (11), 06018025-1-15.

Haldar, K., Chakraborty, D., 2018b. Probabilistic stability analyses of reinforced slope subjected to strip loading. Geotech. Eng. J. SEAGS & AGSSEA 49 (4), 92–99.

Haldar, K., Chakraborty, D., 2019a. Effect of interface friction angle between soil and reinforcement on bearing capacity of strip footing placed on reinforced slope. Int. J. Geo-Mech. 19 (5), 06019008.

Haldar, K., Chakraborty, D., 2019b. Seismic bearing capacity of strip footing placed on a reinforced slope. Geosynth. Int. 26 (5), 474–484.

Haldar, K., Chakraborty, D., 2019c. Probabilistic bearing capacity of strip footing on reinforced soil slope. Comput. Geotech. 116, 103213-1-11.

Haldar, K., Chakraborty, D., 2020a. Effect of inclined and eccentric loading on the bearing capacity of strip footing placed on the reinforced slope. Soils Found. 60 (4), 791–799.

Haldar, K., Chakraborty, D., 2020b. Influence of soil spatial variability on the response of strip footing on geocell-reinforced slope. Comput. Geotech. 122, 103533-1-13.

Haldar, K., Chakraborty, D., 2020c. Probabilistic bearing capacity of strip footing on reinforced anisotropic soil slope. Geomech. & Eng. 23 (1), 15–30.

Haldar, K., Chakraborty, D., Dash, S.K., 2019. Bearing capacity of a strip footing situated on soil slope using a non-associated flow rule in lower bound limit analysis. Int. J. Geotech. Eng. 13 (2), 103–111.

ITASCA, 2011. FLAC 2D Version 7.0.411 Fast Lagrangian Analysis of Continua in 2 Dimensions. ITASCA Consulting Group Inc.

Krishnan, K., Halder, K., Chakraborty, D., 2021. Probabilistic shakedown analysis of cohesive soil under moving surface loads considering wheel-soil interface friction. Road Mater. Pavement Des. https://doi.org/10.1080/14680629.2021.1888777.

Lacasse, S., Nadim, F., 1996. Uncertainties in characterising soil properties. In: Proceedings of the Uncertainty in the Geologic Environment: From Theory to Practice, Madison, Wisconsin, vol. 1, pp. 49–75.

Latha, G.M., Rajagopal, K., 2007. Parametric finite element analyses of geocell-supported embankments. Can. Geotech. J. 44 (8), 917–927.

Latha, G.M., Rajagopal, K., Krishnaswamy, N.R., 2006. Experimental and theoretical investigations on geocell supported embankments. Int. J. GeoMech. 6 (1), 30–35.

Lee, K.M., Manjunath, V.R., 2000. Experimental and numerical studies of geosynthetic-reinforced sand slopes loaded with a footing. Can. Geotech. J. 37 (4), 828–842.

Leshchinsky, B., 2015. Bearing capacity of footings placed adjacent to $c'-\varphi'$ slopes. J. Geotech. Geoenviron. Eng. 141 (6), 04015022-1-13.

Makrodimopoulos, A., Martin, C.M., 2006. Lower bound limit analysis of cohesive-frictional materials using second-order cone programming. Int. J. Numer. Methods Eng. 66 (4), 604–634.

MATLAB, 2019. MATLAB Version 2019a, The MathWorks, Inc., Natick, MA, United States.

Mehdipour, I., Ghazavi, M., Moayed, R.Z., 2013. Numerical study on stability analysis of geocell reinforced slopes by considering the bending effect. Geotext. Geomembranes 37 (4), 23–34.

Mehrjardi, G.T., Ghanbari, A., Mehdizadeh, H., 2016. Experimental study on the behaviour of geogrid-reinforced slopes with respect to aggregate size. Geotext. Geomembranes 44 (6), 862–871.

MOSEK ApS, 2020. MOSEK optimization toolbox for MATLAB, Version 9.2.16. Copenhagen, Denmark.

Phoon, K.K., Kulhawy, F.H., 1999a. Characterization of geotechnical variability. Can. Geotech. J. 36 (4), 612–624.

Phoon, K.K., Kulhawy, F.H., 1999b. Evaluation of geotechnical property variability. Can. Geotech. J. 36 (4), 625–639.

Pramanik, R., Baidya, D.K., Dhang, N., 2019. Implementation of fuzzy reliability analysis for elastic settlement of strip footing on sand considering spatial variability. Int. J. Geo-Mech. 19 (12), 04019126.

Rajagopal, K., Krishnaswamy, N.R., Latha, G.M., 2001. Finite element analysis of embankments supported on geocell layer using composite model. In: Proceedings of 10th International Conference on Computer Methods and Advances in Geomechanics Tuscon, Arizona, pp. 1251–1254.

Selvadurai, A., Gnanendran, C., 1989. An experimental study of a footing located on a sloped fill: influence of a soil reinforcement layer. Can. Geotech. J. 26 (3), 467–473.

Sloan, S.W., 1988. Lower bound limit analysis using finite elements and linear programming. Int. J. Numer. Anal. Methods GeoMech. 12 (1), 61–77.

Vanmarcke, E.H., 1977. Probabilistic modeling of soil profiles. J. Geotech. Eng. Div. 103 (11), 1227–1246.

Vanmarcke, E.H., n.d. Random Fields: Analysis and Synthesis, MIT Press, Cambridge, Mass.

Yang, B.H., Lai, J., Lin, J.H., Tsai, P.H., 2016. Simulating the loading behavior of reinforced strip footings with a double-yield soil model. Int. J. Geomech. 16 (1), B6015001.

Yoo, C., 2001. Laboratory investigation of bearing capacity behaviour of strip footing on geogrid-reinforced sand slope. Geotext. Geomembranes 19 (5), 279–298.

CHAPTER
20

Multivariate methods to monitor the risk of critical episodes of environmental contamination using an asymmetric distribution with data of Santiago, Chile

Carolina Marchant[1,2], Víctor Leiva[3], Helton Saulo[4], Roberto Vila[4]

[1]Faculty of Basic Sciences, Universidad Católica del Maule, Talca, Chile; [2]ANID-Millennium Science Initiative Program-Millennium Nucleus Center for the Discovery of Structures in Complex Data, Santiago, Chile; [3]School of Industrial Engineering, Pontificia Universidad Católica de Valparaíso, Valparaíso, Chile; [4]Department of Statistics, Universidade de Brasília, Brasília, Brazil

1. Symbology, introduction, and bibliographical review

In this section, abbreviations, acronyms, notations, and symbols used in our work are defined in Table 20.1. In addition, here, we provide the introduction, bibliographical review on the topic, objectives, and outline of the present study.

1.1 Abbreviations, acronyms, notations, and symbols

Next, in Table 20.1, the symbology considered in this work is provided to facilitate its reading.

1.2 Introduction and bibliographical review

Since the birth of the statistical quality control in industrial processes (Shewhart, 1931), diverse tools have been widely used to monitor process behavior, discover problems in internal systems, and find solutions to production problems. Among these tools, the control charts are largely employed for this monitoring (Leiva et al., 2015). These charts alert when a process is out of control and then an action must be taken to bring it back to an in-control state.

The use of such charts has been expanded to other areas beyond industrial processes,

---

Risk, Reliability and Sustainable Remediation in the Field of Civil and Environmental Engineering
https://doi.org/10.1016/B978-0-323-85698-0.00024-1

© 2022 Elsevier Inc. All rights reserved.

20. Multivariate methods to monitor the risk of critical episodes

TABLE 20.1 Abbreviations, acronyms, notations, and symbols employed in the present document.

	Abbreviations/acronyms		Notations/symbols
ARL	Average run length	Φ	Standard normal CDF
ARL1	Out-of-control ARL	$\varnothing$	Standard normal PDF
ARL0	In-control ARL	Φ_m	m-variate standard normal CDF
CDF	Cumulative distribution function	φ_m	m-variate standard normal PDF
CL	Central line	H	Adapted Hotelling statistic
DF	Degrees of freedom	$\chi^2(\nu)$	Chi-square distribution with v DF
FAR	False alarm rate	$0_{m \times 1}$	$m \times 1$ null vector
FL	Fatigue-life	FL_m	m-variate FL distribution
GOF	Goodness-of-fit	$Bin\,(n,p)$	Binomial distribution
KS	Kolmogorov-Smirnov		of n, p parameters
log-FL	Logarithmic fatigue-life	A	FL shape parameter
log-FL$_m$	m-variate log-FL distribution	B	FL scale parameter
LCL	Lower control limit	$p = P(T > t)$	Binomial p parameter
LSL	Lower specification limit	N	Sample size
MD	Mahalanobis distance	p_0	Nonconforming fraction
N_m	m-variate normal distribution	p_1	Conforming fraction
PCI	Process capability index	$\mu, \boldsymbol{\mu}$	Uni-multivariate process mean
PDF	Probability density function	$\mu_0, \boldsymbol{\mu}_0$	Target means
PM	Particulate matter	μ_1	Shifted mean
PM2.5	PM levels with diameter $< 2.5\,\mu g/Nm^3$	$\mu g/Nm^3$	Micrograms per cubic meter
PM10	PM levels with diameter $< 10\,\mu g/Nm^3$	$R+$	Set of positive real numbers
PP	Probability versus probability	t_0	Dangerous level
UCL	Upper control limit	K	Control coefficient
USL	Upper specification limit	A	Proportional constant
LSL	Lower specification limit	$\lfloor x \rfloor$	Integer part of x
USL	Upper specification limit	C_p	PCI
SD	Standard deviation	C_{pl}	PCI with lower specification limit
CV	Coefficient of variation	C_{pu}	PCI with upper specification limit
CS	Coefficient of skewness	C_{pk}	Minimum between C_{pl} and C_{pu}
CK	Coefficient of kurtosis	Σ	Standard deviation of the process
		$x(q)$	q-th normal quantile

1. Symbology, introduction, and bibliographical review

TABLE 20.1 Abbreviations, acronyms, notations, and symbols employed in the present document.—cont'd

Abbreviations/acronyms		Notations/symbols
	$z(q)$	q-th standard normal quantile
	$t(q)$	q-th nonnormal quantile
	N(0,1)	Standard normal distribution
	$F(m.n-m)$	Fisher distribution with m and $(n-m)$ DFs
	$N = n \times k$	Combined sample
	P	False alarm rate
	CUSUM	Cumulative sum
	EWMA	Exponentially weighted moving average

including environmental and health risk assessment (Lund and Seymour, 1999; Manly and Mackenzie, 2000; Grigg and Farewell, 2004; Ferreira-Baptista and De Miguel, 2005; Woodall, 2006; Morrison, 2008; Lio and Park, 2008; Saulo et al., 2015; Marchant et al., 2018, 2019; Aykroyd et al., 2019; Cavieres et al., 2020; Puentes et al., 2021). In addition, at the present, the control charts are being utilized to monitor service quality processes associated with accounting, banking, business, call center management, education, finance, government, human resource management, marketing, and supply chain, among many others (Faltin, 2007; Akber, 2012; Aykroyd et al., 2019). Thus, these charts are powerful techniques for nonmanufacturing processes as well (Jemayyle and Ruhhal, 2009; Aykroyd et al., 2019). Evidently, products and services have several differences because the products are tangible and their characteristics may be measured. However, the services are intangible, with a considerable difficulty of measurement and quantification. Among all this diversity of applications, the use of control charts in environmental monitoring has shown to be

important and had a good agreement with decisions made by official authorities. This is one of the principal motivations for the application of the present work.

The power or effectiveness of a control chart is usually measured by the average run length (ARL), which is the average number of inspected samples required to alert an out-of-control condition after it has occurred. On the one hand, practitioners of control charts desire to indicate an out-of-control condition as fast as possible, when a process is out of control, that is, the out-of-control ARL (ARL1) must be as small as possible. On the other hand, when the process is in control, these practitioners want to have a fewest possible number of false alarms, that is, to have a large in-control ARL (ARL0). Control charts are classified by variables or attributes, depending on whether the data used to monitor the quality characteristic are generated by measurements (continuous data) or by attributes (qualitative data) as presence (nonconforming) or absence (conforming) of an event. In addition, process capability indexes (PCIs) are other statistical control tools widely used by industries to determine the quality of their products and the performance of their manufacturing processes (Leiva et al., 2015).

The fatigue-life (FL) distribution is a family of probabilistic models originating from the law of cumulative damage related to fatigue and strength of materials (Leiva and Saunders, 2015). This distribution describes the time spent until the extension of a crack exceeds a threshold leading to a failure. The FL distribution has two parameters of shape and scale, is unimodal, positively skewed, and useful for modeling data that take values greater than zero (Birnbaum and Saunders, 1969b). This distribution has been extensively studied because of its good properties and relation to the normal distribution. This relation consists of a random variable following the FL distribution which can be considered as a transformation of a random variable with standard normal distribution (Johnson et al., 1995 and Leiva, 2016). The FL distribution has been successfully applied to several areas of knowledge other than engineering, including air quality, due to that it has shown to have theoretical arguments to model environmental data as proven in Leiva et al. (2015).

1.3 Objectives and outline

This chapter aims (i) to analyze methodologies on control charts for attributes and variables using fatigue-life distributions; (ii) to introduce multivariate control charts based on these distributions; (iii) to present a methodology based on CPIs for FL processes; and (iv) to apply the proposed methodologies to monitor real data on air quality, which can be obtained from https://sinca.mma.gob.cl. The Mahalanobis distance and goodness-of-fit (GOF) tools are employed to assess the adequacy of the distributional assumptions. In addition, this distance is utilized to detect multivariate outliers and a wide discussion about this topic is carried in the section of application.

The remainder of this chapter is organized as follows. Section 2 presents background on univariate and multivariate FL distributions, including some new useful properties. Section 3 discusses methodologies about control charts for attribute and variables using FL distributions. Moreover, multivariate quality control charts based on such distributions and the Hotelling statistic are studied. Furthermore, this section presents, discusses, and applies a methodology based on PCIs for FL processes. Section 4 provides illustrations of the FL tools for statistical processes control with real air quality data from the city of Santiago, Chile. Finally, Section 5 gives conclusions for the main results of this work and recommendations for future investigation.

2. Uni and multivariate fatigue-life distributions

In this section, background on univariate and multivariate FL distributions and their logarithmic versions are provided.

2.1 Univariate fatigue-life distribution

Let T be a random variable or quality characteristic following a univariate FL distribution with shape parameter $\alpha \in \mathbb{R}_+$ and scale parameter $\beta \in \mathbb{R}_+$, denoted as $T \sim \text{FL}(\alpha, \beta)$. Note that β is also the distribution median. The cumulative distribution function (CDF) of T is stated as

$$F_T(t; \alpha, \beta) = \Phi(A(t; \alpha, \beta)), \quad t \in \mathbb{R}_+,$$

where Φ is a standard normal CDF and

$$A(t; \alpha, \beta) = \frac{1}{\alpha}\left[\left(\frac{t}{\beta}\right)^{\frac{1}{2}} - \left(\frac{\beta}{t}\right)^{\frac{1}{2}}\right].$$

The random variable T is a transformation of a standard normal distributed random variable. Namely, $T \sim \mathrm{FL}(\alpha, \beta)$ can be written as

$$T = T(Z; \alpha, \beta) = \beta\left\{\frac{\alpha Z}{2} + \left[\left(\frac{\alpha Z}{2}\right)^2 + 1\right]^{\frac{1}{2}}\right\}^2,$$
$$(20.1)$$

where Z is a random variable with standard normal distribution. Then, we have

$$Z = \frac{1}{\alpha}\left(\sqrt{\frac{T}{\beta}} - \sqrt{\frac{\beta}{T}}\right) \sim \mathrm{N}(0, 1).$$

Thus, the probability density function (PDF) of $T \sim \mathrm{FL}(\alpha, \beta)$ is expressed as

$$f_T(t; \alpha, \beta) = \varphi(A(t; \alpha, \beta))a(t; \alpha, \beta), \quad t \in \mathbb{R}_+,$$
$$(20.2)$$

where φ is the standard normal PDF and $a(t; \alpha, \beta)$, the derivative of $A(t; \alpha, \beta)$ with respect to t, is established by

$$a(t; \alpha, \beta) = \frac{1}{2\alpha\beta}\left[\left(\frac{\beta}{t}\right)^{\frac{1}{2}} + \left(\frac{\beta}{t}\right)^{\frac{3}{2}}\right].$$

Let $T \sim \mathrm{FL}(\alpha, \beta)$. Then, some properties of T are as follows:

(P1) $k T \sim \mathrm{FL}(\alpha, k\beta)$, with $k > 0$;
(P2) $1/T \sim \mathrm{FL}(\alpha, 1/\beta)$;
(P3) $V^2 = (T/\beta + \beta/T - 2)/\alpha^2 \sim \chi^2(1)$, that is, V^2 follows a chi-square distribution with one degree of freedom (DF);
(P4) $\mathrm{E}[T] = \beta(2 + \alpha^2)/2$ and $\mathrm{Var}[T] = (\beta\alpha)^2(4 + 5\alpha^2)/4$.

Maximum likelihood estimators for the univariate FL distribution parameters are unique and can be easily obtained, solving numerically the maximum likelihood equations (Birnbaum and Saunders, 1969a; Leiva and Saunders, 2015). The univariate FL distribution is implemented in the R software by the bs and gbs packages that can be downloaded from https://cran.r-project.org/src/contrib/Archive/bs and https://cran.r-project.org/src/contrib/Archive/gbs, respectively, or in http://www.victorleiva.cl and http://www.r-project.org (Core Team, 2019). In these R packages, different aspects related to probabilistic and statistical features, including estimation methods, of the fatigue-life distribution are computationally implemented and available to be utilized.

2.2 Log-fatigue-life distribution

If $T \sim \mathrm{FL}(\alpha, \beta)$, then its logarithm, $Y = \log(T)$, follows a log-fatigue-life (log-FL) distribution, that is, $Y = \log(T) \sim \log\text{-}\mathrm{FL}(\alpha, \mu)$, where $\mu = \log(\beta) \in \mathbb{R}$ and $\alpha \in \mathbb{R}_+$. Thus, the corresponding PDF of Y is expressed as

$$f_Y(y; \alpha, \mu) = \varphi(B(y; \alpha, \mu))\, b(y; \alpha, \mu), \quad y \in \mathbb{R},$$
$$(20.3)$$

with shape parameter α and mean μ, where $b(y; \alpha, \mu)$ is given by

$$b(y; \alpha, \mu) = \frac{1}{\alpha}\cosh\left(\frac{y - \mu}{2}\right),$$

which is the derivative with respect to y of

$$B(y; \alpha, \mu) = \frac{2}{\alpha}\sinh\left(\frac{y - \mu}{2}\right).$$

Let $Y \sim \log\text{-}\mathrm{FL}(\alpha, \mu)$. Then, some properties of Y are as follows:

(P5)
$Y = \mu + 2\operatorname{arcsinh}(\alpha W/2) \sim \log\text{-FL}(\alpha, \mu)$, with $W \sim N(0,1)$, that is, a log-FL distributed random variable may be obtained directly from a random variable with standard normal distribution or from a FL distributed random variable;

(P6)
$W = B(Y; \alpha, \mu) = (2/\alpha)\sinh((Y - \mu)/2) \sim N(0,1)$;

(P7) $W^2 = B^2(Y; \alpha, \mu) \sim \chi^2(1)$.

2.3 Multivariate fatigue-life distribution

The random vector $\boldsymbol{T} = (T_1, \ldots, T_m)^\top \in \mathbb{R}_+^m$ follows an m-variate FL distribution with parameters $\boldsymbol{\alpha} = (\alpha_1, \ldots, \alpha_m)^\top \in \mathbb{R}_+^m$, $\boldsymbol{\beta} = (\beta_1, \ldots, \beta_m)^\top \in \mathbb{R}_+^m$ and scale matrix $\boldsymbol{\Sigma} = (\sigma_{kl}) \in \mathbb{R}^{m \times m}$, if $T_i = T(V_i; \alpha_i, \beta_i)$, for $i \in \{1, \ldots, m\}$. Note that T is defined in Eq. (20.1) and $\boldsymbol{Z} = (Z_1, \ldots, Z_m)^\top \in \mathbb{R}^m \sim N_m(0_{m \times 1}, \boldsymbol{\Psi})$, with $\boldsymbol{\Psi} = (\rho_{kl}) \in \mathbb{R}^{m \times m}$ being a correlation matrix. Furthermore, the FL case implies the diagonal elements of $\boldsymbol{\Sigma}$ are equal to one, that is $\sigma_{kk} = 1$, for all $k = 1, \ldots, m$, and then $\boldsymbol{\Sigma} = \boldsymbol{\Psi}$. The m-variate FL distribution is denoted by $\boldsymbol{T} \sim \text{FL}_m(\boldsymbol{\alpha}, \boldsymbol{\beta}, \boldsymbol{\Psi})$. The CDF and PDF of $\boldsymbol{T}$ are, respectively, written as

$$F_{\boldsymbol{T}}(\boldsymbol{t}; \boldsymbol{\alpha}, \boldsymbol{\beta}, \boldsymbol{\Psi}) = \Phi_m(\boldsymbol{A}; \boldsymbol{\Psi}),$$

$$f_{\boldsymbol{T}}(\boldsymbol{t}; \boldsymbol{\alpha}, \boldsymbol{\beta}, \boldsymbol{\Psi}) = \varphi_m(\boldsymbol{A}; \boldsymbol{\Psi})\, a(\boldsymbol{t}; \boldsymbol{\alpha}, \boldsymbol{\beta}),$$

$$\boldsymbol{t} = (t_1, \ldots, t_m)^\top \in \mathbb{R}_+^m,$$

where Φ_m and φ_m are the m-variate standard normal CDF and PDF (Kumar et al., 2015), respectively, $\boldsymbol{A} = A(\boldsymbol{t}; \boldsymbol{\alpha}, \boldsymbol{\beta}) = (A_1, \ldots, A_m)^\top$, with $A_j = A(t_j; \alpha_j, \beta_j)$,

$$a(\boldsymbol{t}; \boldsymbol{\alpha}, \boldsymbol{\beta}) = \prod_{j=1}^{m} a\left(t_j; \alpha_j, \beta_j\right),$$

and both $A(t_j; \alpha_j, \beta_j)$ and $a(t_j; \alpha_j, \beta_j)$ are as given in Eq. (20.2).

Let $\boldsymbol{T} \sim \text{FL}_m(\boldsymbol{\alpha}, \boldsymbol{\beta}, \boldsymbol{\Psi})$ and $j, k \in \{1, \ldots, m\}$. Then, some properties of $\boldsymbol{T}$ are as follows:

(P8) Let $\boldsymbol{T}_2 = (T_j, T_k) \sim \text{FL}_2(\boldsymbol{\alpha}^{(j,k)}, \boldsymbol{\beta}^{(j,k)}, \boldsymbol{\Psi}^{(j,k)})$, where $\boldsymbol{\alpha}^{(j,k)} = (\alpha_j, \alpha_k)$, $\boldsymbol{\beta}^{(j,k)} = (\beta_j, \beta_k)$, and $\boldsymbol{\Psi}^{(j,k)} \in \mathbb{R}^{2 \times 2}$ is a matrix with ones in its diagonal and its other elements equal to the element (j,k) of the matrix $\boldsymbol{\Psi}$;

(P9) The elements of the vector of means and covariance matrix are as follows:

(i) $\mathrm{E}[T_j] = \beta_j(1 + \alpha_j \mathrm{E}^2/2)$;

(ii) $\mathrm{Cov}[T_j, T_k] = \alpha_j \alpha_k \beta_j \beta_k \left[\alpha_j \alpha_k \rho_{jk}^2 - + 4I\left(\alpha_j, \alpha_k, \rho_{jk}\right)\right]/4$, where $I\left(\alpha_j, \alpha_k, \rho_{jk}\right) = \mathrm{E}\left[Z_j Z_k \left((\alpha_j Z_j/2)^2 + 1\right)^{1/2} \left((\alpha_k Z_k/2)^2 + 1\right)^{1/2}\right]$, and (Z_j, Z_k) is bivariate normal distributed with correlation matrix $\boldsymbol{\Psi}^{(j,k)}$;

(iii) The variance-covariance (dispersion) matrix of $\boldsymbol{T}$ is stated as

$$D[\boldsymbol{T}] = \frac{1}{4}\Omega \odot (\boldsymbol{\Psi} \odot \boldsymbol{\Psi} \odot \boldsymbol{\Xi} + 4\varUpsilon),$$

where $\Omega = (\omega_{jk})$, $\boldsymbol{\Xi} = (\xi_{jk})$, and $\varUpsilon = (v_{jk})$ have elements $\omega_{jk} = \alpha_j \alpha_k \beta_j \beta_k$, $\xi_{jk} = \alpha_j \alpha_k$, and $v_{jk} = I\left(\alpha_j, \alpha_k, \rho_{jk}\right)$, respectively, with $\odot$ being the Hadamard product. If $T_1, \ldots, T_m$ are independent random variables, then $D[\boldsymbol{T}] = \mathrm{diag}(\in_{11}, \ldots, \in_{mm})$, with elements $\in_{jj} = \alpha_j^2 \beta_j^2 / 4\left(\alpha_j^2 + 4I(\alpha_j, \alpha_j, 1)\right)$ (Sánchez et al., 2020);

(P10) $k\,\boldsymbol{T} \sim \text{FL}_m(\boldsymbol{\alpha}, k\,\boldsymbol{\beta}, \boldsymbol{\Psi})$, with $k > 0$;

(P11) $\boldsymbol{T}^* = (1/T_1, \ldots, 1/T_m)^\top \sim \text{FL}_m(\boldsymbol{\alpha}, \boldsymbol{\beta}^*, \boldsymbol{\Psi})$, with $\boldsymbol{\beta}^* = (1/\beta_1, \ldots, 1/\beta_m)^\top$;

(P12) $\boldsymbol{A}^\top(\boldsymbol{T}; \alpha, \beta)\boldsymbol{\Psi}^{-1}A(\boldsymbol{T}; \alpha, \beta) \sim \chi^2(m)$.

Unlike parameter estimation for the univariate FL distribution, where uniqueness is guaranteed (Birnbaum and Saunders, 1969a; Aykroyd et al., 2018), in the m-variate case there is no certainty that the system of maximum likelihood equations has a unique solution. Therefore, care must be taken to ensure that numerical procedures yield a global maximum.

2.4 Multivariate log-fatigue-life distribution

Let $T = (T_1, \ldots, T_m)^{\mathrm{T}} \sim \mathrm{FL}_m(\alpha, \beta, \Psi)$. Then, $Y = (Y_1, \ldots, Y_m)^{\mathrm{T}} = (\log(T_1), \ldots, \log(T_m))^{\mathrm{T}}$ follows an m-variate log-fatigue-life distribution with shape parameters $\alpha = (\alpha_1, \ldots, \alpha_m)^{\mathrm{T}}$, mean vector $\mu_Y = \mathrm{E}[Y] = (\mathrm{E}[Y_1], \ldots, \mathrm{E}[Y_m])^{\mathrm{T}} = (\log(\beta_1), \ldots, \log(\beta_m))^{\mathrm{T}} \in \mathbb{R}^m$, and correlation matrix $\Psi \in \mathbb{R}^{m \times m}$. This is denoted by $Y \sim \log\text{-}\mathrm{FL}_m(\alpha, \mu_Y, \Psi)$. The CDF of Y is defined as

$$F_Y(y; \alpha, \mu_Y, \Psi) = \Phi_m(B; \Psi), \quad y = (y_1, \ldots, y_m)^{\mathrm{T}} \in \mathbb{R}^m,$$

where $B = B(y; \alpha, \mu_Y) = (B_1, \ldots, B_m)^{\mathrm{T}}$, with $B_j = B(y_j; \alpha_j, \mu_j)$, for $j \in \{1, \ldots, m\}$, being given as in Eq. (20.3). The PDF of Y is expressed as

$$f_Y(y; \alpha, \mu_Y, \Psi) = \varphi_m(B; \Psi) b(y; \alpha, \mu_Y), \quad y \in \mathbb{R}^m,$$

where $b(y; \alpha, \mu_Y) = \prod_{j=1}^m b(y_i; \alpha_j, \mu_j)$, with $b(y_j; \alpha_j, \mu_j)$ being given as in Eq. (20.3), for $j \in \{1, \ldots, m\}$. If $Y \sim \log\text{-}\mathrm{FL}_m(\alpha, \mu_Y, \Psi)$, then from **(P6)** the following two properties hold:

(P13) $DB(Y; \alpha, \mu_Y) \sim \mathrm{N}_m(0_{m \times 1}, D\Psi D)$, where $D = \mathrm{diag}(\alpha_1, \ldots, \alpha_m)$.

(P14) $B(Y; \alpha, \mu_Y)^{\mathrm{T}} \Psi^{-1} B(Y; \alpha, \mu_Y) \sim \chi^2(m)$.

The Mahalanobis distance is used to identify multivariate outliers and to assess the GOF in m-variate log-FL distributions. In this case, the Mahalanobis distance for observation i, using property **(P12)**, is expressed as

$$\mathrm{MD}_i(\theta) = \mathrm{B}(Y_i; \alpha, \mu_Y)^{\mathrm{T}} \Psi^{-1} B(Y_i; \alpha, \mu_Y), i \in \{1, \ldots, n\},$$

with $\theta = (\alpha^{\mathrm{T}}, \mu_Y^{\mathrm{T}}, \mathrm{svec}(\Psi)^{\mathrm{T}})^{\mathrm{T}}$, where "svec" denotes vectorization of a symmetric matrix. Random vectors from m-variate log-FL distributions can be generated using Algorithm 20.1 (Leiva et al., 2008).

3. Fatigue-life statistical process control

In this section, methodologies about control charts for attribute and variables based on the

---

ALGORITHM 20.1

Generator of m-variate log-FL random vectors

1. Make a Cholesky decomposition of Ψ as $\Psi = LL^{\mathrm{T}}$, where L is a lower triangular matrix with real and positive diagonal entries.
2. Generate m independent standard normal distributed random numbers $Z = (Z_1, \ldots, Z_m)^{\mathrm{T}}$.
3. Compute $W = (W_1, \ldots, W_m)^{\mathrm{T}} = LZ$.
4. Obtain the vector Y with components $Y_j = \mu_j + 2 \arcsin(\alpha_j W_j / 2)$, for $j \in \{1, \ldots, m\}$.
5. Repeat Steps 1 to 4 until the required vector of data is generated.

FL distributions are discussed. In addition, we study multivariate quality control charts based on such distributions and the Hotelling statistic. Furthermore, this section presents, discusses, and applies a methodology based on CPIs for FL processes.

3.1 Fatigue-life np charts

An np-chart is an adaptation of the control chart for nonconforming fraction when samples of equal size (n) are taken from the process. The np-chart is based on the binomial distribution as detailed below. In quality monitoring processes, one could be concerned about the random variable corresponding to the number (U) of times that the quality variable (T) exceeds a fixed value (t) established for the process, given an exceedance probability (p). Here, p can be computed by means of a continuous distribution of the quality variable T as $p = P(T > t) = 1 - F_T(t)$, where F_T is the CDF of T. Thus, $U \sim \text{Bin}(n, p)$ and then

$$P(U = u)$$
$$= \binom{n}{u} p^u (1-p)^{n-u}, \quad u \in \{0, 1, \dots, n\}$$

$$(20.4)$$

Based on Eq. (20.4), an np-chart is proposed with lower control limit (LCL), central line (CL), and upper control limit (UCL) given by

$$\text{LCL} = \max\left\{0, np_0 - k\sqrt{np_0(1-p_0)}\,\right\},$$
$$\text{CL} = np_0,$$
$$\text{UCL} = np_0 + k\sqrt{np_0(1-p_0)},$$

$$(20.5)$$

where k is a control coefficient (for example, $k = 2$ indicates a warning level and $k = 3$ a dangerous level), p_0 is the nonconforming fraction corresponding to a target mean μ_0 of the quality variable T, when the process is in control, and n is the size of the subgroup. Note that the nonconforming fraction is the probability that the random variable exceeds a dangerous level (t_0) and, therefore, this probability is $P(T > t_0) = 1 - F_T(t_0)$.

The FL distribution can be parametrized from (α, β) to (α, μ), switching from the median β in the original parametrization to its mean given by $\mu = \beta(2 + \alpha^2)/2$. Consider t_0 as proportional to μ_0, that is, $t_0 = a\mu_0$, relating them to establish the monitoring criterion, where $a > 0$ is a proportionality constant. Note that the target mean μ_0 and dangerous level t_0 may be taken from process specifications. Then, the fatigue-life CDF can be parametrized in terms of its mean and expressed in function of t_0 and a as

$$F_T(t_0; \alpha, a, \mu) = \Phi\left(\frac{1}{\alpha}\xi\left(\frac{a(1 + \alpha^2/2)}{\mu/\mu_0}\right)\right), \quad (20.6)$$

where $\xi(x) = \sqrt{x} - 1/\sqrt{x} = 2\sinh(\log(\sqrt{x}))$. Thus, when a monitoring process is in control ($\mu = \mu_0$) for a FL distributed quality random variable, from Eq. (20.6), the nonconforming fraction is stated as

$$p_0 = 1 - F_T(t_0; \alpha, a) = \Phi\left(-\frac{1}{\alpha}\xi\left(a(1 + \alpha^2/2)\right)\right).$$

$$(20.7)$$

Note that the specification of the point t_0 is equivalent to establishing the inspection point $a > 0$, because $t_0 = a\mu_0$. Algorithm 20.2 provides a criterion for monitoring processes using an np-chart for a quality variable $T \sim \text{FL}(\alpha, \mu)$.

Consider a shift in the process mean and assume that the new (shifted) process mean becomes $\mu_1 = a\mu_0$, for a shift constant $a > 0$. Assume that the value of the shift constant a is greater than one, because of the interest in the case where the mean of the quality variable to be monitored becomes greater than the target

---

ALGORITHM 20.2

np control chart based on the FL distribution

1. Take N subgroups of size n.
2. Collect n data $t_1, \ldots, t_n$ of the random variable of interest T for each subgroup.
3. Fix the target mean μ_0, the inspection constant a, and the control coefficient k.
4. Count in each subgroup of n data the number u of times that t_i exceeds $t_0 = a\mu_0$, for $i \in \{1, \ldots, n\}$.
5. Compute LCL $= \max\{0, n\widehat{p}_0 - k\sqrt{n\widehat{p}_0(1 - \widehat{p}_0)}\}$ and

   UCL $= n\widehat{p}_0 + k\sqrt{n\widehat{p}_0(1 - \widehat{p}_0)}$, which are

obtained from Eq. (20.5), where $\widehat{p}_0 = \Phi(-(1/\widehat{\alpha})\xi(a(1+\widehat{\alpha}^2/2)))$ is given as in Eq. (20.7), with $\widehat{\alpha} = \sqrt{2(\sqrt{s/r} - 1)}$ being the modified moment estimate of α (Leiva, 2016), for $s = (1/n)\sum_{i=1}^{n}t_i$ and $r = \left[(1/n)\sum_{i=1}^{n}(1/t_i)\right]^{-1}$.

6. Declare the process as out of control if $u \geq$ UCL or $u \leq$ LCL, or as in control if LCL $\leq u \leq$ UCL.

---

mean. Note that μ_0 and μ_1 are different means corresponding to in-control and out-of-control processes, respectively, but both of them are means of the FL distribution. Hence, the nonconforming fraction corresponding to the new mean of the random variable T is obtained from Eq. (20.6) as

$$p_1 = 1 - F_T(t_0; \alpha, a)$$
$$= \Phi\left(-\frac{1}{\alpha}\xi\left(\frac{a}{\alpha}\left[1 + \alpha^2/2\right]\right)\right). \quad (20.8)$$

As mentioned, in general, a process is said in-control if LCL $\leq U \leq$ UCL. Thus, when the process is actually in-control, the probability to be in control is given by

$$p_0^{(in)} = P(\text{LCL} \leq U \leq \text{UCL}|p_0)$$
$$= \sum_{u=\lfloor \text{LCL} \rfloor + 1}^{\lfloor \text{UCL} \rfloor} \binom{n}{u} p_0^u (1 - p_0)^{n-u}, \quad (20.9)$$

whereas if the process mean has shifted to the new mean μ_1, the probability given in Eq. (20.9) is expressed as

$$p_1^{(in)} = P(\text{LCL} \leq U \leq \text{UCL}|p_1)$$
$$= \sum_{u=\lfloor \text{LCL} \rfloor + 1}^{\lfloor \text{UCL} \rfloor} \binom{n}{u} p_1^u (1 - p_1)^{n-u}, \quad (20.10)$$

where $\lfloor x \rfloor$ denotes the integer part of the real number x.

Efficiency of the proposed criterion can be evaluated by using ARL0 and ARL1. As mentioned, ARL0 is defined as the expected number of observations taken from an in-control state until the chart falsely signals an out-of-control case. ARL0 is regarded as acceptable if it is large enough to keep the level of false alarms at a reasonable value. ARL1 is stated as the expected number of observations taken from an out-of-control state until the chart correctly signals an out of control. The ARL1 value should be as small as possible. In-control and out-of-control ARLs are respectively given by

$$ARL0 = \frac{1}{1 - p_0^{(\text{in})}}, \quad ARL1 = \frac{1}{1 - p_1^{(\text{in})}},$$

where $p_0^{(\text{in})}$ and $p_1^{(\text{in})}$ are defined in Eqs. (20.9) and (20.10), respectively. The control coefficient k is selected in a form such that ARL0 must be close to a specified ARL. Then, with the selected value of k, it is possible to obtain the value of ARL1 for the shift constant a given in Eq. (20.8).

3.2 Fatigue-life process capability indexes

A PCI represents the ratio between specification and variability ranges. This index is useful for analyzing the process variability related to product requirements or specifications. In general, standard versions of PCIs are employed for processes whose quality characteristics have a normal distribution. However, nonnormality is usually present in productive processes. Therefore, misinterpretation of the process capability is likely if it is ignored, possibly leading to inaccurate business decisions. Then, when an random variable X has a normal distribution, denoted by $X \sim N(\mu, \sigma^2)$, the PCIs C_p (capability of process), C_{pl} (capability of process based on the lower specification limit), C_{pu} (capability of process based on the upper specification limit), and C_{pk} (the minimum between C_{pl} and C_{pu}) are defined, respectively, as

$$C_p = \frac{\text{USL} - \text{LSL}}{6\sigma},$$

$$C_{pl} = \frac{\mu - \text{LSL}}{3\sigma},$$

$$C_{pu} = \frac{\text{USL} - \mu}{3\sigma}, \quad (20.11)$$

$$C_{pk} = \min\{C_{pl}, C_{pu}\},$$

where LSL is the lower specification limit and USL is the upper specification limit. Note that 6σ is a range such that the output percentage of the process falling outside the $\mu \pm 3\sigma$ limits is 0.27%. Moreover, C_{pl}, C_{pu} are one-sided PCIs and C_{pk} is the corresponding one-sided PCI for the specification limit closer to the process mean.

Under nonnormality, the PCIs established here in Section 3.2 are not adequate (Kane, 1986). There have been attempts to modify the PCIs in Section 3.2 by substituting 6σ by a range $x(0.99865) - x(0.00135)$, where $x(0.00135)$ and $x(0.99865)$ are the 0.135-th and 99.865-th quantiles of the corresponding nonnormal distribution, respectively. Note that the range $x(0.99865) - x(0.00135)$ covers a 99.73% of the distribution of the monitored process data. Then, a general method to calculate nonnormal PCIs is defined as

$$C_p' = \frac{\text{USL} - \text{LSL}}{t(q_2) - t(q_1)},$$

$$C_{pl}' = \frac{2(t(0.5) - \text{LSL})}{t(q_2) - t(q_1)},$$

$$C_{pu}' = \frac{2(\text{USL} - t(0.5))}{t(q_2) - t(q_1)},$$

$$C_{pk}' = \min\{C_{pl}', C_{pu}'\},$$

where $t(q_i)$ is the $q_i \times 100$-th quantile of the nonnormal distribution, which is assumed for the quality characteristic of the process to be monitored, with $i \in \{1, 2\}$.

Based on Section 3.2 and the FL distribution discussed in Section 2, FL PCIs were proposed in Leiva et al. (2014) and given by

$$C_{\mathrm{p}}^{\mathrm{FL}} = \frac{\mathrm{USL} - \mathrm{LSL}}{t(q_2) - t(q_1)} = \frac{2(\mathrm{USL} - \mathrm{LSL})/\beta}{\left[\alpha^2 z_2^2 + \alpha\, z_2 \sqrt{(\alpha\, z_2)^2 + 4}\right] - \left[\alpha^2 z_1^2 + \alpha\, z_1 \sqrt{(\alpha\, z_1)^2 + 4}\right]},$$

$$C_{\mathrm{pl}}^{\mathrm{FL}} = \frac{2(\beta - \mathrm{LSL})}{t(q_2) - t(q_1)} = \frac{2(\beta - \mathrm{LSL})}{\beta\alpha\left\{z_2\left[\frac{\alpha\, z_2}{2} + \sqrt{\left(\frac{\alpha\, z_2}{2}\right)^2 + 1}\right] - z_1\left[\frac{\alpha\, z_1}{2} + \sqrt{\left(\frac{\alpha\, z_1}{2}\right)^2 + 1}\right]\right\}},$$

$$C_{\mathrm{pu}}^{\mathrm{FL}} = \frac{2(\mathrm{USL} - \beta)}{t(q_2) - t(q_1)} = \frac{2(\mathrm{USL} - \beta)}{\beta\alpha\left\{z_2\left[\frac{\alpha\, z_2}{2} + \sqrt{\left(\frac{\alpha\, z_2}{2}\right)^2 + 1}\right] - z_1\left[\frac{\alpha\, z_1}{2} + \sqrt{\left(\frac{\alpha\, z_1}{2}\right)^2 + 1}\right]\right\}},$$

and

$$C_{\mathrm{pk}}^{\mathrm{FL}} = \min\left\{C_{\mathrm{pl}}^{\mathrm{FL}}, C_{\mathrm{pu}}^{\mathrm{FL}}\right\},$$

where $z_1 = z(q_1)$ and $z_2 = z(q_2)$, with $z(q_i)$ being the $q_i \times 100$-th quantile of $Z \sim \mathrm{N}(0,1)$, for $i \in \{1,2\}$. The maximum likelihood method is used to estimate the model parameters, and the normal and quantile bootstrap methods to obtain confidence intervals (CI) for the PCIs. Values of q_1 and q_2 are obtained by minimizing the variance of the corresponding FL PCI estimator.

3.3 Multivariate fatigue-life control charts

Multivariate monitoring is carried out by considering correlated quality characteristics and simultaneously determining whether these characteristics are in control or out of control. By using multivariate quality control charts for subgroups based on fatigue-life distributions and an adapted Hotelling statistic, a methodology is introduced here and applied in the next section. Once again, the corresponding parameters are estimated with the maximum likelihood method and the parametric bootstrapping is utilized to obtain the distribution of the adapted Hotelling statistics. Furthermore, the Mahalanobis distance is considered to detect multivariate outliers and used to assess the adequacy of the distributional assumption (Marchant et al., 2018, 2019).

Suppose that one is interested in modeling a dynamic process with m quality characteristics and that for each of them a sample of n observations from the evolving process is collected. Let $\boldsymbol{Y}_i = (Y_{i1}, \ldots, Y_{im})^{\mathrm{T}} \in \mathbb{R}^m$ denote a random vector associated with measured values corresponding to subset i, for $i \in \{1, \ldots, n\}$. Assume that $\boldsymbol{Y}_i$ follows an m-variate log-FL distribution, that is, $\boldsymbol{Y}_i \sim \log\text{-FL}_m(\boldsymbol{\alpha}, \boldsymbol{\mu}, \boldsymbol{\Psi})$, with the vectors $\boldsymbol{Y}_i$ being independent over time and $\boldsymbol{\mu}_0$ being the mean vector of the in-control process. To confirm that the process is in control it is necessary to test the hypothesis:

$$\mathrm{H}_0\colon \boldsymbol{\mu} = \boldsymbol{\mu}_0 = \left(\mu_{1_0}, \ldots, \mu_{m_0}\right)^{\mathrm{T}} \text{ versus } \mathrm{H}_1\colon \boldsymbol{\mu} \neq \boldsymbol{\mu}_0.$$

These hypotheses can be contrasted by using an adapted Hotelling H statistic constructed as follows. Using Property **(P10)** and considering that

$$\boldsymbol{b}_i = \left(2\sinh\left(\frac{Y_{i1} - \mu_{1_0}}{2}\right), \ldots, 2\sinh\left(\frac{Y_{ip} - \mu_{m_0}}{2}\right)\right)^{\mathrm{T}}$$
$$\sim \mathrm{N}_m(0_{m \times 1}, \boldsymbol{D\Psi D}),$$

for $i \in \{1, \ldots, n\}$, a Hotelling H statistic adapted for m-variate log-FL distributions can be obtained as

$$H = n(n-1)\overline{\boldsymbol{b}}^{\mathrm{T}} \boldsymbol{C}^{-1} \overline{\boldsymbol{b}}, \qquad (20.12)$$

with $\overline{\boldsymbol{b}} = \sum_{i=1}^{n} \boldsymbol{b}_i / n$ and $\boldsymbol{C} = \sum_{i=1}^{n} \boldsymbol{b}_i \boldsymbol{b}_i^{\mathrm{T}}$.

ALGORITHM 20.3

Computation of m-variate FL control limits

1. Collect r samples $\left(y_{1h}, \ldots, y_{nh}\right)^{\mathrm{T}}$ of size n for an in-control process, with $h \in \{1, \ldots, r\}$, assuming that the m-variate vector with the logarithms of the data follows a log-$FL_p(\alpha, \mu, \Psi)$ distribution.

2. Compute the maximum likelihood estimates of α, μ_Y, and Ψ using the data of the pooled sample of size $N = r \times n$ collected in Step 1, and check the distributional assumption using GOF tools.

3. Generate a parametric bootstrap sample $\left(y_1^*, \ldots, y_n^*\right)^{\mathrm{T}}$ of size n from an m-variate log-FL distribution using the maximum likelihood

   estimates obtained in Step 2 as the distribution parameters.

4. Compute H defined in Eq. (20.12) with $\left(y_1^*, \ldots, y_n^*\right)^{\mathrm{T}}$, which is denoted by H^*, assuming a target mean μ_0.

5. Repeat Steps 3 and 4 an enough large number of times, for example, $B = 10,000$ and obtain B bootstrap statistics of H, denoted by $H_1^*, \ldots, H_B^*$.

6. Fix ϱ as the false alarm rate (FAR) of the chart.

7. Use the B bootstrap statistics obtained in Step 5 to find the $100(\varrho/2)$-th and $100(1 - \varrho/2)$-th quantiles of the distribution of H, which are the LCL and UCL for the chart of FAR ϱ, respectively.

Note that, if $Y \sim \log\text{-}FL_m(\alpha, \mu, \Psi)$, then H has a Fisher distribution with m and $(n - m)$ DFs, that is, $H \sim F(m, n - m)$ (Kundu, 2015). Algorithm 20.1 can be used to construct the bootstrap distribution of the H statistic using random vectors generated from the m-variate log-FL distribution, and then, Algorithm 20.3 may be employed to state the corresponding control limits. Once these limits are stated, the m-variate FL control chart is utilized to identify whether the evolving

process remains in control or not. Consider a new vector of values of the quality characteristics, and let H_{new} be the corresponding Hotelling statistic calculated using Eq. (20.12). As the process evolves, a sequence of values H_{new} is produced. Algorithm 20.3 details how to construct m-variate control charts based on fatigue-life distributions for process monitoring (Algorithm 20.4).

ALGORITHM 20.4

Process monitoring using the m-variate FL chart

1. Collect a sample of size n, $y_1, \ldots, y_n$, from the process.

2. Calculate the H_{new} statistic from the sample obtained in Step 1.

3. Declare the process as in control if H_{new} falls between LCL and UCL obtained in Algorithm

   20.3; otherwise, the chart indicates an out-of-control state.

4. Repeat Steps 1–3 for each sample collected at regular time intervals.

4. Illustrations

In this section, real pollution and meteorological data sets are used to illustrate the methodologies presented in Section 3.

4.1 Data description

Santiago, the capital of Chile, is one of the cities with higher air contamination levels around the world. Its location and weather, when combined with high anthropological emissions, provoke critical air contamination conditions (Marchant et al., 2019; Cavieres et al., 2020). Here, we use data sets collected by the Chilean Metropolitan Environmental Health Authority corresponding to the random variables: particulate matter (PM) with a diameter smaller than 2.5 μm —PM2.5— (X1) levels in $\mu g/Nm^3$; PM with a diameter smaller than 10 μm —PM10— (X2) levels in $\mu g/Nm^3$; ambient temperature (X3) in degrees Celsius; wind speed (X4) in meters per second; and relative air humidity (X5) in percent. As mentioned, these data are available at https://sinca.mma.gob.cl and were collected in 2015 as 1 h (hourly) average values. We consider these random variables due to several studies indicated elevated PM levels during high pollution events in Santiago of Chile. These variables are often a function of weather conditions in central Chile and in Argentina, which at the local level generate a depression at the base of the inversion layer, an increase in the vertical thermal stability, lower humidity, and low-wind conditions. Thus, the pollutant dispersion decreases leading to poor ventilation of contaminated air (Cavieres et al., 2020). These random variables were observed in the Pudahuel monitoring station in Chile. For our data analytics, we selected this station because it had high levels of pollution in the period of critical events of air quality (01 April 2015 to 31 August 2015) in Santiago (Marchant et al., 2013). We employ the Chilean guideline values

TABLE 20.2 Primary quality guidelines for PM2.5 and PM10 levels in 24 h (MMA, 2011; CONAMA, 1998).

PM2.5 level	PM10 level	Indication
]0, 50[	]0, 150[	Good
[50, 80[	[150, 195[	Regular
[80, 110[	[195, 240[	Alert
[110, 170[	[240, 330[	Preemergency
≩ 170	≩ 330	Emergency

as targets in this case study. Table 20.2 provides the primary quality guidelines for PM2.5 and PM10 levels for 24 h.

4.2 Exploratory data analysis

First, we carry out a correlation analysis to detect if X1, X2, X3, X4, and X5 data are statistically associated. Fig. 20.1 shows the scatter plots for X1, X2, X3, X4, and X5. From this figure, we detect that in Pudahuel station data there are:

(i) High positive association between X1 and X2 (Pearson coefficient of correlation equal to 0.85);

(ii) Moderate to high positive association between X3 and X4 (Pearson coefficient of correlation equal to 0.683);

(iii) Moderate to high negative association between X3 and X5 (Pearson coefficient of correlation equal to -0.761);

(iv) Medium negative association between X1 and X3 (Pearson coefficient of correlation equal to -0.525), X4 and X5 (Pearson coefficient of correlation equal to -0.523); and

(v) Low negative or positive association between X1 and X4 (Pearson coefficient of correlation equal to -0.434), X2 and X4 (Pearson coefficient of correlation equal to -0.300), X2 and X3 (Pearson coefficient of correlation equal to -0.287), X1 and X5

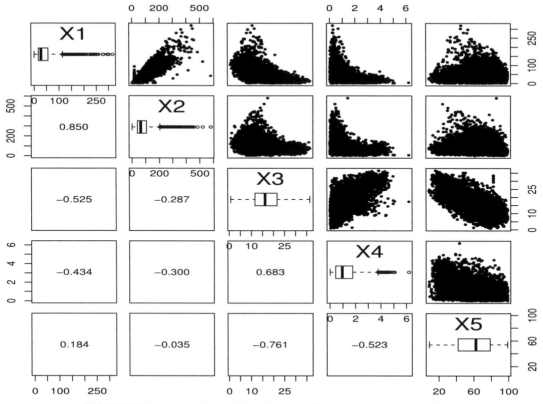

FIGURE 20.1 Scatter plot with boxplots for pollution and meteorological data.

(Pearson coefficient of correlation equal to 0.184), and X2 and X5 (Pearson coefficient of correlation equal to 0.035),

which confirm the need to use an m-variate control chart for jointly monitor these data sets. In addition, Fig. 20.1 shows boxplots for each of the variables under study. Note that we employ these 2D scatterplots to support the linear relationships between variables provided by correlation coefficients. However, we do not consider them to detect multivariate outliers due to limitations discussed in Section 4.5.

Table 20.3 provides descriptive statistics for X1, X2, X3, X4, and X5 in Phases I and II, including central tendency statistics, standard deviation (SD) and coefficients of variation (CV), skewness (CS), and kurtosis (CK).

Fig. 20.1 and Table 20.3 show that distributions with positive skewness, different degrees of kurtosis, and some univariate atypical data are appropriate, suggesting the use of FL distributions. Also, this table and figure evidence a considerable number of levels that exceed the Chilean guidelines for PM2.5 and PM10, that is, 50 μg/Nm^3 and 150 μg/Nm^3, respectively. Thus, these results indicate that the air pollution level of Santiago of Chile is worrying and dangerous from a toxicological perspective for the inhabitants of this city.

4.3 Fatigue-life np charts

We illustrate the use of the np control chart, described in Section 3.1, by utilizing the variable

4. Illustrations

TABLE 20.3 Summary statistics for pollution and meteorological data.

		N	Minimum	Maximum	Range	Median	Mean	SD	CV	CS	CK
Phase I	X1	1416	3	76	73	18	19.08	8.68	45.49	1.78	5.55
	X2	1416	10	214	204	51.50	54.75	24.13	44.06	1.30	3.61
	X3	1416	11.23	35.67	24.43	22.05	22.66	5.56	24.55	0.21	−1.12
	X4	1416	0.02	4.96	4.94	1.61	1.90	1.23	65	0.43	−1.02
	X5	1416	12.17	97	84.83	48.17	49.58	20.59	41.54	0.22	−1.04
Phase II	X1	720	5	318	313	70	80.14	53.13	66.30	1.22	1.87
	X2	720	10.50	522	511.50	121.25	139.11	89.92	64.64	1.11	1.22
	X3	720	0.77	24.67	23.90	11.45	11.72	5.27	44.91	0.21	−0.73
	X4	720	0.04	3.20	3.16	0.61	0.81	0.63	77.94	1.28	1.25
	X5	720	8.83	96.67	87.83	58.58	59.88	20.62	35.19	−0.32	−0.75

X2, that is, PM10 levels from the Pudahuel monitoring station. Initially, the months of January and February are utilized to generate the phase I control limits. We employ these months since their air pollution levels are stable (that is, assumed as an in-control process), because the meteorological and topographical conditions favor no saturation of PM levels. Then, those limits are adopted for phase II monitoring of the month of June. Specifically, this chart is used to monitor the number of nonconforming length measurements in $N = 30$ subgroups of size (that is, number of observations) $n = 24$, namely, 30 days with 24 observations each day corresponding to the hours of a day. The maximum permitted PM10 level according to Chilean guidelines is 150 μg/Nm^3. Then, values above this threshold are classified as nonconforming. Fig. 20.2 shows the np-chart for the data under analysis. From this figure, we note several points above the UCL so that the process is not in control for the month of June 2015.

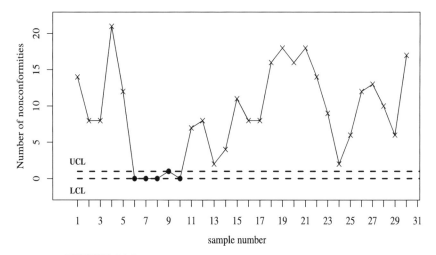

FIGURE 20.2 np chart for June 2015 PM10 levels with $t_0 = 150$.

4.4 Fatigue-life process capability indexes

We illustrate the use of PCIs with the same PM10 data utilized in the previous subsection. The data are for the month of June. A similar application employing PCI tools for PM10 data was carried out in Kaya and Kahraman (2009). Recall that the acceptable level of PM10, according to the government of Chile, is a maximum of $150 \ \mu g/Nm^3$, that is, the USL is 150.

The maximum likelihood estimates of the fatigue-life parameters α and β are $\widehat{\alpha} = 0.753$ and $\widehat{\beta} = 107.928$, respectively. Table 20.4 reports the results of FL PCIs based on a process with only the USL. In this table, the estimate of C_{pu}^{FL} and its corresponding bootstrap PCIs, along with sample median and mean, are reported. Note that the point and interval estimates indicate a poor quality condition according to Table 20.5. Thus, based on PCIs, we can conclude that the air quality of Santiago in June 2015 is not adequate for a healthy life.

4.5 Multivariate fatigue-life control charts

To calculate the control limits in Phase I, we utilize data for the months of January and February 2015 with $r = 59$, $n = 24$, $N = 1416$, $B = 10,000$ (bootstrap replications), and FAR $\varrho = 0.0027$.

The assumption that this data set follows a multivariate FL distribution is supported by

TABLE 20.4   USL, sample median and mean, and point and confidence interval (CI) estimates for PCIs C_{pu}^{FL} with the PM10 data.

USL	$\bar{x}$	$t(0.5)$	$\widehat{C}_{pu}^{FL}$	Normal bootstrap CI	Quantile bootstrap CI
150 μg/ Nm^3	139.110	121.250	0.273	(0.180; 0.312)	(0.245; 0.386)

TABLE 20.5   Quality conditions and C_p values.

Quality condition	
Super excellent	$C_p \geq 2.0$
Excellent	$1.67 \leq C_p < 2.00$
Satisfactory	$1.33 \leq C_p < 1.67$
Capable	$1.00 \leq C_p < 1.33$
Inadequate	$0.67 \leq C_p < 1.00$
Poor	$C_p < 0.67$

See Table 2 from Tsai, C.C., Chen, C.C., 2006. Making decision to evaluate process capability index cp with fuzzy numbers. Int. J. Adv. Manuf. Technol. 30, 334−339.

the empirical probability versus theoretical probability (PP) plot with Kolmogorov−Smirnov (KS) acceptance regions at 5% for the Mahalanobis distance shown in Fig. 20.3A. The KS test, although not particularly sensitive, but very competitive with other tests, is the only test that can be linked to a graphical tool as the PP plot. A graphical tool is always more desirable than a test due to its easier interpretation. However, if the graphical GOF tool can be accompanied by a p-value associated with a GOF test, it is more informative. This is the reason why we have used both GOF tools (Castro-Kuriss et al., 2020).

The Mahalanobis distance is used with data transformed by the Wilson-Hilferty approximation to obtain a normal distribution, which can be verified by the GOF methods as described in Step 2 of Algorithm 20.3. This figure indicates that the FL_5 distribution has a good fit, which is supported by the p-value of 0.8915 from the corresponding KS test. In addition, the presence of possible multivariate atypical data is observed. Multivariate outliers can affect the resulting estimates. Although the detection of outliers in multivariate observations is often based on the Mahalanobis distance (Marchant et al., 2018), sometimes outliers do not have an enough large Mahalanobis distance, which is due to the fact that the estimators based on the

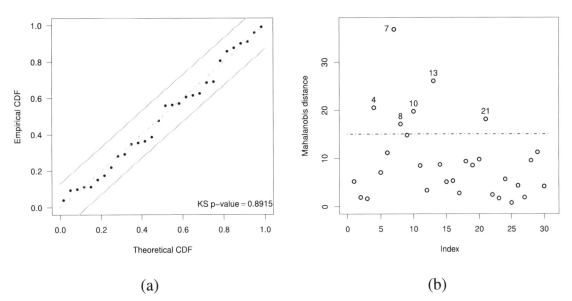

FIGURE 20.3 PP plots with 5% KS acceptance regions for the Mahalanobis distance using the FL_2 distribution (A) and Mahalanobis distance index plot (B) for June 2015 pollution data.

model employed to generate the Mahalanobis distance are nonrobust (Becker and Gather, 1999; Jobe and Pokojovy, 2015). This is named the masking effect and occurs when a group of extreme observations distorts the estimates of the mean vector and/or variance-covariance matrix, producing a small distance from the outlier to the mean. Another aspect to be considered regarding multivariate outliers is the low-dimensional visualization when employing usual scatterplots (2D). This type of visualization is not reliable to identify high-dimensional outliers. There are several outlier identification approaches looking at axis-parallel views or low-dimensional projections (often 2D) which are assumed to indicate high-dimensional outliers (Wilkinson, 2017; Talagala et al., 2021; Ro et al., 2015). Low-dimensional views are risky, as discussed in Wilkinson (2017) and shown by its Fig. 9. The 2D scatterplots fail to reveal 3D outliers, a situation which is even worse in higher dimensions. Then, as mentioned, usual 2D scatterplots may be utilized to support the linear relationships between variables provided by correlation coefficients, such as we did in this application. However, one must be careful when analyzing the 2D scatterplot matrix to detect high-dimensional outliers having in mind such a limitation.

To monitor air quality of June 2015 from the Pudahuel station in Phase II, we employ 5-variate FL charts. We use the LCL and UCL obtained in Phase I summarized in Algorithm 20.3. For the control chart of this month, the number of subgroups and the subgroup size are $r = 30$ days and $n = 24$ h, respectively, giving a total of $N = 744$ observations. Table 20.3 provided descriptive statistics for these variables in Phase II, including central tendency statistics, SD, CV, CS, and CK. Using Algorithm 20.3, a 5-variate FL control chart is constructed for monitoring this process with an FAR fixed at $\varrho = 0.027$. Fig. 20.4 displays the multivariate FL control chart, indicating an out-control state for 13 days, reporting at the same time a very poor air quality in this month of 2015 for the monitoring station located in Pudahuel, Santiago, Chile. Note that this conclusion is coherent with what was obtained with the FL PCI for PM10.

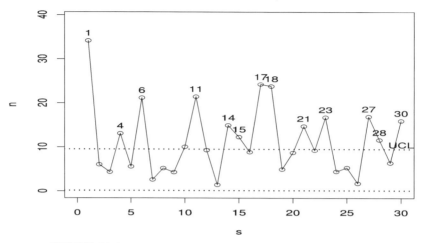

FIGURE 20.4 m-variate fatigue-life control chart for pollution data.

5. Conclusions and future investigation

The fatigue-life distributions, also known as Birnbaum-Saunders distributions, have recently received considerable attention due to its interesting properties and its relationship to the normal distribution. It has now been 50 years since the origins of the fatigue-life distribution, but since then, it went a long way. Its origins were as a model for the cumulative damage describing fatigue and failure of materials. As a unimodal and skewed to the right distribution, it has found widespread use as a life distribution in many applications, even outside material science. Extensions and variations are conducted at multivariate fatigue-life and logarithmic fatigue-life distributions, allowing several other applications. In the last decades, there have been many advances and applications on this statistical distribution. Therefore, the fatigue-life distribution provides another opportunity in modeling. This has already proven to be sharp, giving reliable and insightful results. The study of fatigue-life distributions is now a mature topic, with a substantial literature, which has had significant impact in important and interesting fields, for example, in business and industry, and today in environmental sciences.

In the process control setting, the normal distribution was the most relevant ingredient of statistical process control tools, but nowadays new models are considered, mainly in asymmetric frameworks. In this work, statistical process control tools based on the fatigue-life distribution, such as control charts for attributes and variables, as well as their multivariate versions and capability indexes, have been presented, discussed, and implemented in the R software.

We illustrated the proposed tools with real-world air quality data of Santiago, Chile, one of the most polluted cities around the world. The results of this illustration are worrisome from a public health sense due to many studies that indicated health problems caused by exposure to high levels of air pollution; see, for example, the recent work presented in Cavieres et al. (2020) and references therein. Thus, this illustration showed that our tools are helpful for alerting episodes of extreme air pollution and for preventing adverse effects on human health for the population of Santiago, Chile. Moreover, these alerts are in agreement with the decisions made by the Chilean Environmental Health Authority.

As part of future research, it is of interest to derive the following:

(i) Multivariate fatigue-life process capability indexes;
(ii) Cumulative sum (CUSUM) and the exponentially weighted moving average (EWMA) type control charts for uni- and multivariate fatigue-life distributions;

(iii) Multivariate control charts for percentiles and proportions;

(iv) Control charts with covariates as well as with temporal, spatial, and functional structures, measurement errors, and partial least squares (Huerta et al., 2019; Carrasco et al., 2020; Giraldo et al., 2020; Sánchez et al., 2020);

(v) Traditional robust estimation methods as well as local influence tools (Velasco et al., 2020; Leiva et al., 2020);

(vi) Other applications in the context of multivariate methods are in cluster analysis and principal component analysis, particularly when using principal components to remove the collinearity among variables (Ramirez-Figueroa et al., 2021);

(vii) An interesting field of application is in the statistical learning and neural networks (Aykroyd et al., 2019).

Investigation on these problems is currently in progress and we hope to report these findings in future works.

Acknowledgments

The authors thank the editors and reviewers for their constructive comments on an earlier version of this manuscript. This research was supported partially by project grants "Fondecyt 11190636" (C. Marchant) and "Fondecyt 1200525" (V. Leiva) from the National Agency for Research and Development (ANID) of the Chilean government and by ANID-Millennium Science Initiative Program—NCN17_059 (C. Marchant).

References

Akber, S., 2012. Enhancements to Control Charts for Monitoring Process Dispersion and Location. Ph.D. Thesis. University of Auckland, New Zealand.

Aykroyd, R.G., Leiva, V., Marchant, C., 2018. Multivariate Birnbaum-Saunders distributions: modelling and applications. Risks 6 (1), 1—25. Article 21.

Aykroyd, R.G., Leiva, V., Ruggeri, F., 2019. Recent developments of control charts, identification of big data sources and future trends of current research. Technol. Forecast. Soc. Chang. 144, 221—232.

Becker, C., Gather, U., 1999. The masking breakdown point of multivariate outlier identification rules. J. Am. Stat. Assoc. 4, 947—955.

Birnbaum, Z.W., Saunders, S.C., 1969a. Estimation for a family of life distributions with applications to fatigue. J. Appl. Probab. 6, 328—347.

Birnbaum, Z.W., Saunders, S.C., 1969b. A new family of life distributions. J. Appl. Probab. 6, 319—327.

Carrasco, J.M.F., Figueroa-Zúñiga, J., Leiva, V., Riquelme, M., Aykroyd, R.G., 2020. An errors-in-variables model based on the Birnbaum-Saunders the distribution and its diagnostics with an application to earthquake data. Stoch. Environ. Res. Risk Assess. 34, 369—380.

Castro-Kuriss, C., Huerta, M., Leiva, V., Tapia, A., 2020. On some goodness-of-fit tests and their connection to graphical methods with uncensored and censored data. In: Xu, J., Ahmed, S.E., Duca, G., Cooke, F.L. (Eds.), Management Science and Engineering Management. Springer-Verlag, Berlin, Germany, pp. 157—183.

Cavieres, M.F., Leiva, V., Marchant, C., Rojas, F.A., 2020. A methodology for data-driven decision making in the monitoring of particulate matter environmental contamination in Santiago of Chile. Rev. Environ. Contam. Toxicol. 250, 45—67.

CONAMA, 1998. Establishment of Primary Quality Guideline for PM10 that Regulates Environmental Alerts. Technical Report Decree 59. Ministry of Environment (CONAMA) of the Chilean Government, Santiago, Chile.

Core Team, 2019. R: A Language and Environment for Statistical Computing. R Foundation for Statistical Computing, Vienna, Austria.

Faltin, F., 2007. Control Charts Overview. Wiley Stats Ref., Statistics Reference Online.

Ferreira-Baptista, L., De Miguel, E., 2005. Geochemistry and risk assessment of street dust in Luanda, Angola: a tropical urban environment. Atmos. Environ. 39, 4501—4512.

Giraldo, R., Herrera, L., Leiva, V., 2020. Cokriging prediction using as secondary variable a functional random field with application in environmental pollution. Mathematics 8, 1305.

Grigg, O.A., Farewell, V.T., 2004. A risk-adjusted sets method for monitoring adverse medical outcomes. Stat. Med. 23, 1593—1602.

Huerta, M., Leiva, V., Liu, S., Rodriguez, M., Villegas, D., 2019. On a partial least squares regression model for asymmetric data with a chemical application in mining. Chemometr. Intell. Lab. Syst. 190, 55—68.

Jemayyle, R., Ruhhal, N., 2009. Using of cause-selecting control charts to model and improve service performance of a utilities company. Dirasat: Eng. Sci. 36, 37—50.

Jobe, J.M., Pokojovy, M., 2015. A cluster-based outlier detection scheme for multivariate data. J. Am. Stat. Assoc. 110, 543—1551.

Johnson, N.L., Kotz, S., Balakrishnan, N., 1995. Continuous Univariate Distributions, vol. 2. Wiley, New York.

Kane, V., 1986. Process capability indices. J. Qual. Technol. 18, 41–52.

Kaya, I., Kahraman, C., 2009. Air pollution control using fuzzy process capability indices in the six-sigma approach. Hum. Ecol. Risk Assess. 15, 689–713.

Kumar, K.N., Rao, K., Srinivas, Y., Satyanarayana, C., 2015. Texture segmentation based on multivariate generalized Gaussian mixture model. Comput. Model. Eng. Sci. 107, 201–221.

Kundu, D., 2015. Bivariate log-Birnbaum-Saunders distribution. Statistics 49, 900–917.

Leiva, V., Saunders, S.C., 2015. Cumulative Damage Models. Wiley Stats Ref: Statistics Reference Online, pp. 1–10.

Leiva, V., Sanhueza, A., Sen, P.K., Paula, G.A., 2008. Random number generators for the generalized Birnbaum-Saunders distribution. J. Stat. Comput. Simulat. 78, 1105–1118.

Leiva, V., Marchant, C., Saulo, H., Aslam, M., Rojas, F., 2014. Capability indices for Birnbaum-Saunders processes applied to electronic and food industries. J. Appl. Stat. 41, 1881–1902.

Leiva, V., Marchant, C., Ruggeri, F., Saulo, H., 2015. A criterion for environmental assessment using Birnbaum-Saunders attribute control charts. Environmetrics 26, 463–476.

Leiva, V., Sanchez, L., Galea, M., Saulo, H., 2020. Global and local diagnostic analytics for a geostatistical model based on a new approach to quantile regression. Stoch. Environ. Res. Risk Assess. 34, 1457–1471.

Leiva, V., 2016. The Birnbaum-Saunders Distribution. Academic Press, New York.

Lio, Y.L., Park, C., 2008. A bootstrap control chart for Birnbaum-Saunders percentiles. Qual. Reliab. Eng. Int. 24, 585–600.

Lund, R., Seymour, L., 1999. Assessing temperature anomalies for a geographical region: a control chart approach. Environmetrics 10, 163–177.

Manly, B., Mackenzie, D., 2000. A cumulative sum type of method for environmental monitoring. Environmetrics 11, 151–166.

Marchant, C., Leiva, V., Cavieres, M., Sanhueza, A., 2013. Air contaminant statistical distributions with application to PM10 in Santiago, Chile. Rev. Environ. Contam. Toxicol. 223, 1–31.

Marchant, C., Leiva, V., Cysneiros, F.J.A., Liu, S., 2018. Robust multivariate control charts based on Birnbaum-Saunders distributions. J. Stat. Comput. Simulat. 88, 182–202.

Marchant, C., Leiva, V., Christakos, G., Cavieres, M.F., 2019. Monitoring urban environmental pollution by bivariate control charts: new methodology and case study in Santiago, Chile. Environmetrics 30, env.2551.

MMA, 2011. Establishment of Primary Quality Guideline for Inhalable Fine Particulate Matter PM2.5; Technical Report Decree 12. Ministry of Environment of the Chilean Government, Santiago, Chile.

Morrison, L.W., 2008. The use of control charts to interpret environmental monitoring data. Nat. Area J. 28, 66–73.

Puentes, R., Marchant, C., Leiva, V., Figueroa-Zúñiga, J., Ruggeri, F., 2021. Predicting PM2.5 and PM10 levels during critical episodes management in Santiago, Chile, with a bivariate Birnbaum-Saunders log-linear model. Mathematics 9 (6), 645.

Ramirez-Figueroa, J.A., Martin-Barreiro, C., Nieto, A.B., Leiva, V., Galindo, M.P., 2021. A new principal component analysis by particle swarm optimization with an environmental application for data science. Stoch. Environ. Res. Risk Assess. 35, 1969–1984.

Ro, K., Zou, C., Wang, Z., Yin, G., 2015. Outlier detection for high-dimensional data. Biometrika 102, 589–599.

Sánchez, L., Leiva, V., Galea, M., Saulo, H., 2020. Birnbaum-Saunders quantile regression models with application to spatial data. Mathematics 8, 1000.

Saulo, H., Leiva, V., Ruggeri, F., 2015. Monitoring environmental risk by a methodology based on control charts. In: Kitsos, C., Oliveira, T., Rigas, A., Gulati, S. (Eds.), Theory and Practice of Risk Assessment. Springer, Switzerland, pp. 177–197.

Shewhart, W.A., 1931. Economic Control of Quality of Manufactured Product. D Van Nostrand Company, New York.

Talagala, P.D., Hyndman, R.J., Smith-Miles, K., 2021. Anomaly detection in high-dimensional data. J. Comput. Graph Stat. 30, 360–374.

Tsai, C.C., Chen, C.C., 2006. Making decision to evaluate process capability index c_p with fuzzy numbers. Int. J. Adv. Manuf. Technol. 30, 334–339.

Velasco, H., Laniado, H., Toro, M., Leiva, V., Lio, Y., 2020. Robust three-step regression based on comedian and its performance in cell-wise and case-wise outliers. Mathematics 8, 1259.

Wilkinson, L., 2017. Visualizing big data outliers through distributed aggregation. IEEE Trans. Visual. Comput. Graph. 24, 256–266.

Woodall, W.H., 2006. The use of control charts in health-care and public-health surveillance. J. Qual. Technol. 38, 89–104.

CHAPTER 21

A combined sustainability-reliability approach in geotechnical engineering

Dipanjan Basu, Mina Lee

Department of Civil and Environmental Engineering, University of Waterloo, Waterloo, ON, Canada

1. Introduction

Sustainability is a normative global concept that advocates the survivability, functionality, well-being of the current and future generations and other life forms by living within the carrying capacity of the earth. Rooted in the early environmentalism of the 19th century (Emerson, 1849; Ruskin 1881; Thoreau, 1860), sustainability started gaining wider public attention since the 1980s (Allen, 1980; Brown, 1981, 1984) and now encompasses interconnected economic, ecological, and social aspects like food security, good health and well-being, gender equality, peace, and justice (World Wildlife Fund [WWF], 2016; United Nations [UN], 2015). Brundtland Commission (1987), formed under the auspices of the United Nations, defined sustainable development as "development that meets the needs of the present without compromising the ability of the future generations to meet their own needs." Further, the Brundtland report introduced the concept of triple bottom line which suggests that sustainability should be evaluated with reference to three fundamental pillars related to the environment, economy, and equity

(3 Es). Degradation of earth's life-supporting capacity, growing awareness of different environmental problems, socio-economic issues especially related to poverty and inequality, and impacts of current generation's decision-making on future generations are the key triggers to the idea of sustainability (Gibson, 2006; Hopwood et al., 2005). In fact, according to the *Living Planet Report* (WWF, 2010, 2018), humanity is currently consuming more than 1.5 times the earth's resources. This is a rather grim situation and of particular relevance to civil engineering practices because buildings and construction together account for 36% of global final energy use and 39% of carbon dioxide (CO_2) emissions (International Energy Agency, 2018).

The world we live in today is highly engineered even if we consider the most pristine environment at the remotest corners like the Amazon rainforest or the Himalayan hamlets. The dominance of anthropogenic activities and technological applications is evident in almost every facet of human societies and nature. Thus, sustainability thoughts and practices should inherently consider the contribution of engineering in a system. From that point of

Risk, Reliability and Sustainable Remediation in the Field of Civil and Environmental Engineering
https://doi.org/10.1016/B978-0-323-85698-0.00029-0

© 2022 Elsevier Inc. All rights reserved.

view, sustainability can be thought of as a balance of the four Es—environment, economy, equity, and engineering (Fig. 21.1). Simply put, sustainability can be considered to be a balance between the supplies (capacities) and demands (loads) in a system—as long as the supplies are greater than the demands, the system is sustainable. However, it is the interconnectivity of the different systems that makes the concept and implementation of sustainability rather complex. Achieving sustainability goals in one system may make another connected system unsustainable. For example, technological developments like cars, cell phones, and processed food generate employment and make social life more comfortable, but cause environmental pollution and detrimental impacts on human health. It is important to note that, for a system, supplies, demands, and sustainability objectives may change over time and, therefore, questions related to sustainability should be addressed considering the temporal aspect. In fact, sustainability problems are considered to be similar to *wicked problems*, defined by Horst Rittel in the 1960s (as cited in Buchanan, 1992), with conflicting objectives and without a clear, achievable solution based on linear thinking.

Sustainable development is a quintessential example of practical holism and embodies an ultimate practicality, which is literally meaningless unless implemented in the actual world. Guiding principles such as those advocated by the *Natural Step Framework* are very pertinent in the context of civil engineering and may be adopted for implementing sustainability in practice. The four sustainability principles advocated by the Natural Step are (1) materials from the earth's crust (e.g., heavy metals and fossil fuels) must not be systematically increased in the earth's environment, (2) materials produced by society (e.g., plastics and DDT) must not be systematically increased in the earth's environment, (3) the physical basis for the productivity and diversity of nature must not be systematically diminished (e.g., by over harvesting forests and destroying natural habitats), and (4) there must be fair and efficient use of resources with respect to meeting human needs (e.g., unsafe working condition or insufficient pay for a job must be avoided). Ethical principles and concepts such as *Intergenerational Justice*, *Distributional Equity*, *The Precautionary Principle*, *Protecting the Vulnerable*, and *Respect for Nature* can be used to develop guidelines for sustainable practices. Technical concepts such as *Life Cycle Thinking*, *Life Cycle Management*, *Industrial Ecology*, *By-product Synergy*, and *Biomimicry* can be used to develop a framework within which sustainable practices can be implemented within engineering. Some of these important concepts are summarized in Table 21.1.

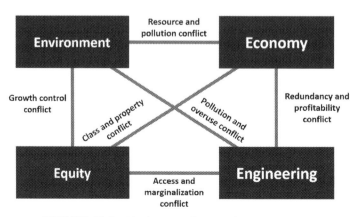

FIGURE 21.1 The four E's of sustainable development.

1. Introduction

381

TABLE 21.1 Ethical and technical concepts underlying sustainability (Kibert et al., 2011; Remmen et al., 2007).

Ethical principles	Description
Intergenerational Justice	The obligation of a generation to provide future generation with a resource base that will enable them to fulfill their own needs.
Distributional Equity	The rights of everyone to an equal distribution of all available resources (e.g., land, water, and fuel) including products and services.
The Precautionary Principle	Taking precautionary measures even if some cause and effect relationships are not fully established scientifically (e.g., deterring people from emitting greenhouse gases even though it is unclear whether human activities actually cause a temperature rise or not).
Protecting the Vulnerable	The obligation to protect the powerless, poverty-stricken people, future generations, and animal world against destruction of ecosystem under the guise of development.
The Reversibility Principle	Decisions taken by the present generation should be such that the effects of the decisions can be undone by the future generation.
Protecting the Rights of Nonhuman World	Protecting and preserving nonhuman world (e.g., plants and animals) and nonliving portion of Earth which are essential to supporting life.
Respect for Nature	An ethics based on biocentrism—integrity of the biosphere would benefit all communities of life including nonhumans.
Land Ethic	Ethical relationship to the land that is based on value in a philosophical sense (e.g., love, respect, and admiration for the land).
Technical concepts	
Life Cycle Thinking	Consideration of environmental implications of decisions over the entire span of the process or product (e.g., design considerations not only on the technical and economic aspects of design and construction but also on the reuse and recyclability of the construction materials).
Life Cycle Management	An integrated framework of concepts and techniques to address environmental, economic, technological, and social aspects of products and organizations from a life cycle point of view.
Industrial Ecology	The study of the physical, chemical, and biological interactions and interrelationships both within and among industrial and ecological systems (e.g., minimization of wastes in industrial processes, redistribution of energy, waste and water with partner companies).
By-product Synergy	Cross-industry collaboration in which information is shared on feed-stock need and unwanted waste products so that a waste of one industry can be used in another.
Closed Loop	The closed loop concept advocates keeping materials in productive cycle even after the end of their designated use (e.g., steel structural elements are in accordance with the concept of closed loop because of the ease in both construction and deconstruction along with its high potential of recyclability).
Construction Ecology	Construction ecology promotes the ideals of closed loop material cycle integrated with eco-industrial and natural systems, depends solely on renewable energy sources, and preserves natural system functions
Design for the Environment	Design philosophy which aims to integrate the environmental considerations into process engineering based on the entire life cycle of the product
Biomimicry	The theory of biomimicry advocates creation of strong, tough, and intelligent materials from naturally occurring materials at ambient temperature using solar energy to run a manufacturing process so that there is no generation of waste

(Continued)

TABLE 21.1 Ethical and technical concepts underlying sustainability (Kibert et al., 2011; Remmen et al., 2007).—cont'd

Ethical principles	Description
Biophilia Hypothesis	The biophilia hypothesis emphasizes that human beings have a genetically based need to affiliate with life and life-like processes
Carrying Capacity	The number of people who can be supported in a given area within natural resource limits, and without degrading the natural, social, cultural, and economic environment for present and future generations.
Ecological Footprint	The land area required to support a certain population or activity (inverse of carrying capacity).
Eco-Efficiency	Reduction of material, energy, and toxic emissions of a process while maximizing material recyclability, sustainable use of renewable resources, product durability, service intensity of goods, and services.
Factor 4	Factor 4 concept visualizes a quadruple increase in capital wealth by double increase in production, using technological innovation, with only half of the resources used.
Factor 10	Factor 10 states that, over a generation, material use should be brought down by a factor of 10 by increasing the technological efficiency 10 times.

The *sustainability science approach* can be the ideal pathway for solving complex sustainable engineering problems because, in this approach, the discipline-specific solution approach to sustainability issues is replaced with a global approach where ideas and concepts are drawn from all related fields of sustainability study and then integrated into a solution for the sustainable engineering problem (Kates et al., 2001; Mihelcic et al., 2003; Seager et al., 2012). For this approach to be successful, engineering curricula have to be revised so that engineers become T-shaped (Guest, 1991) with a broad appreciation of multiple related disciplines and a deep knowledge in the domain of specialization (Fig. 21.2). An example of excellent T-shaped geotechnical engineer who practices sustainable geotechnics would have broad knowledge across multiple disciplines like civil engineering, systems design engineering, management science, environmental science, economics, political science, and sociology, and a deep knowledge in geotechnical engineering. At its present form, the sustainability science is more conceptual and difficult to implement because of which the *systems engineering approach* can be considered a practical pathway forward. In the systems engineering approach, an optimized design solution is sought in which cost minimization is the design objective and the environmental impact is used as a design constraint (Graedel, 1994; Kibert, 2008). A more recent approach to systems engineering is to incorporate the 3 Es—environment, economy and equity—as the design objectives so that cost, environmental impact, and adverse societal impacts are simultaneously minimized (Seager et al., 2012; Slaper and Hall, 2011).

Practicing sustainability in civil and geotechnical engineering is important because civil constructions consume vast amounts of energy, deplete large amounts of stone, rock, and soil reserves, cause significant land, water, and air pollution, and contribute to climate change, ozone depletion, desertification, deforestation, and soil erosion (Kibert, 2008). Civil infrastructure systems provide essential services to societies and play a vital role in maintaining and improving the quality of human life. For example, roads, bridges, dams, levees, power

1. Introduction

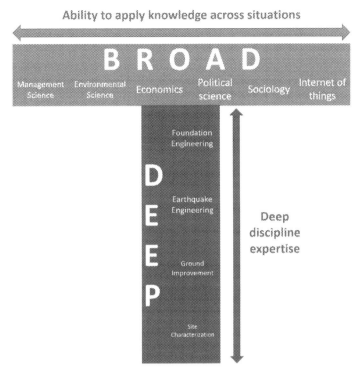

FIGURE 21.2 T-shaped geotechnical engineer.

plants, pipelines, and water and energy utilities allow humans to live in highly populated areas, protect them from natural hazards, provide safety and security, and protect the environment from wastes generated by humans. Large projects involving significant geotechnical construction, e.g., underground metro rail project or a dam construction project, may alter the land use pattern of the surrounding region, change the demographics and economics of the region, and alter social interaction and ethical values of the neighboring communities. Thus, well-planned and executed civil infrastructure may help in promoting sustainability and create important interactions between natural systems and communities. Responsible and ethical engineering practices are therefore necessary to ensure that the least amount of natural resources is used and least amount of wastes is generated. It is important to recognize in this context that no matter where an engineering process is performed, the entire living world lives at the "downstream" of the process and gets affected.

For any infrastructure engineering project, ensuring the safety, serviceability, and reliability of a facility or product is as important as balancing the three Es of sustainability. A design with too much focus on the environmental impact or economy may lead to a marginally safe and "efficient" structure without any scope for redundancy (in this context, efficiency relates to frugal design with minimal use of resources and money, while redundancy refers to extra provisions that may become useful if design loads and stresses are exceeded). Such a design may readily fail under unprecedented and unanticipated external threats (Chateauneuf, 2008) and does not support the sustainability agenda. Unlike structural systems that handle uncertainties related mostly to external loads, geotechnical engineering suffers from significant uncertainties related to soil and rock properties in addition to uncertainties in external loads (Long et al., 2006). The spatial and temporal variabilities in subsurface profiles also lead to difficulties in characterizing the soil properties based on site investigation and mathematical models. Further, material properties in geotechnical design may alter because of a change in the surrounding environment (e.g., a contamination can alter the soil properties and permafrost can get degraded from climate change) and can add to the uncertainties in the system. Therefore, reliability-based design should be an essential part of sustainable geotechnical design. The lack of resilience (the ability to bounce back to normal functionality after experiencing disruptions) against unaccounted external forces and threats is often not acceptable in civil engineering particularly for those structures that are part of critical infrastructures. Failures to critical infrastructure systems may cause severe consequences to the society, such as loss of life, expensive recovery, disrupted public services, and economic crisis. Geo-structures are often important components of critical infrastructures and, hence, thoughts on sustainability in geotechnical engineering should include the reliability and resilience aspects of engineering design. Considering the systems engineering approach, the four Es should be used to develop design objectives such that the reliability and resilience are maximized and the cost, environmental impacts, and societal impacts are minimized simultaneously.

The purpose of this chapter is to introduce a newly developed design framework in geotechnical engineering that considers the environmental, economic, and engineering (reliability) aspects of sustainability. In order to put the framework in perspective, first an overview of the sustainable practices in geotechnical engineering is provided. Subsequently, the life cycle assessment (LCA) procedure is explained for geotechnical problems with examples of pile foundations, which illustrates how the environmental impacts of geotechnical processes and products can be determined. Thereafter, the concepts of reliability and resilience are described in the context of geotechnical problems. In this context, the first-order reliability method (FORM), which captures the uncertainty and reliability of a design in terms of the reliability index, is briefly described. Finally, an integrated design framework for geotechnical engineering is introduced in which the reliability index, cost, and the environmental impacts can be optimized to obtain a sustainable solution. The framework is illustrated using an example of a pile foundation.

2. Sustainable practices in geotechnical engineering

Geotechnical projects involve earth works, ground improvement, building geotechnical structures, and practice of geo-environmental engineering, which require consumption and transportation of both natural resources (e.g., aggregates and soil) and man-made materials (e.g., concrete and steel) over the life cycle of the projects. Although geotechnical engineers cannot solve global sustainability problems, they can contribute significantly toward sustainable development, as illustrated in Fig. 21.3. In geotechnical engineering, the sustainability objectives (i.e., desirable outcomes) can be considered as the following: (1) energy efficiency and carbon reduction, (2) materials and waste reduction, (3) maintaining natural water cycle and enhancing natural watershed, (4) climate change adaptation and resilience, (5) effective land use and management, (6) economic viability and whole life cost, and (7) positive contribution to society (Pantelidou et al., 2012). Basu et al. (2015) considered the following to be part of sustainable geotechnical engineering practice: (1) use of alternate, environment friendly materials and reuse of waste materials in geotechnical

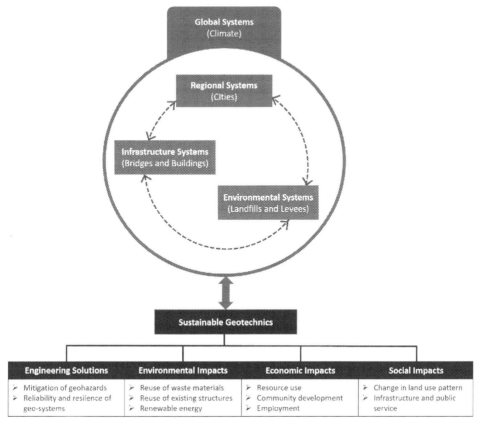

FIGURE 21.3 Scope of sustainable geotechnics.

construction (e.g., use of construction and demolition wastes in pavement subgrade, and appropriate use of geosynthetics), (2) innovative, environment friendly, and energy efficient geotechnical techniques for site investigation, construction, monitoring, retrofitting, ground improvement, and deconstruction (e.g., bioslope engineering and use of natural fiber in soil reinforcement), (3) retrofitting and reuse of foundations and other geotechnical structures, (4) use and reuse of underground space for beneficial purposes like pedestrian pathways, public transit and water distribution system, and for storage of energy, CO_2, and waste products, (5) characterization, analysis, design, monitoring, repairing, and retrofitting techniques in geotechnical engineering that ensure safety, serviceability, and resilience, (6) geotechnical techniques involved in the discovery and recovery of geologic resources like minerals and hydrocarbons, (7) geotechnical techniques for pollution control and redevelopment of brownfields and other marginal sites, (8) mitigation of geohazards (e.g., landslides, earthquakes, heavy rainfall, and blast) that also include the effects of global climate change, (9) practice of energy geotechnics (e.g., geo-facilities for storage and extraction of renewable energy sources like solar, wind, and geothermal energies), (10) environmental and socio-economic impacts from geo-activities, for example, from mining and petroleum extraction, dam construction, and landfill/waste disposal, (11) practice of geoethics and geodiversity, and (12) development of sustainability indicators and assessment tools in geotechnical engineering.

It is important to recognize that sustainability outcomes from a civil engineering solution can be ensured in two ways: by doing the right project and by doing the project right (Institute for Sustainable Infrastructure [ISI], 2018). Choosing the right project often depends on the owner and policy makers, and is often beyond the scope of the civil and geotechnical engineers although they can provide their inputs in the decision-making process. In fact, choosing the right project can have a significantly greater chance of success from a sustainability stand-point than completing the project using good engineering that follows appropriate sustainability guidelines. The earlier the sustainability objectives are considered, the more the choices available and the better the outcome because the availability of sustainable alternatives decreases as a project proceeds from the planning to the execution stage. To achieve the most beneficial outcome from sustainable considerations, it is necessary to incorporate sustainability objectives at the planning and design stages of a project (Misra and Basu, 2011). Geotechnical engineering, being placed at the earlier stage of construction, plays an important role in practicing sustainability in construction projects. For example, in a building foundation project, choices at the planning stage can include types of foundation. At the design stage, the choice is limited to the materials to be used, and at the execution stage, the choice is mostly limited to the machinery used. Decisions at the project level are traditionally made considering the safety, security, and economy. But the environmental impact and even social aspects should also be incorporated, for example, by adopting locally available construction materials and local construction methods that have low carbon footprint, can boost the local economy, and contribute to the well-being of the local community. Basu et al. (2015) outlined the following steps that can be followed at the project level to obtain a sustainable geotechnical solution: (1) involving all the stakeholders (e.g., owner, lawmakers, engineers, architects, users, and members of the affected community) at the planning stage of the project so that a consensus is reached regarding the steps to achieve a sustainable solution (such as control of pollution during and after construction, financial impact on the affected community, choice of environment friendly materials, aesthetic acceptability, and acceptability of the project to the local community), and in subsequent stages to maintain transparent flow of information and to gain consensus on any required change from the initial plan, (2) proper site characterization so that the geologic uncertainties and associated hazards are minimized, (3) robust and reliable analysis, design, and construction that involves minimal financial burden and inconvenience to all the stakeholders, (4) optimal use of materials and energy in planning, design, construction, and maintenance of geotechnical facilities, (5) use of materials and methods that cause minimal negative impact on the ecology and environment, (6) reuse of existing geotechnical elements (e.g., foundations and retaining structures) to minimize wastage, (7) appropriate and adequate instrumentation, monitoring, and maintenance to ensure proper functioning of the facility, and (8) performing adequate checks against resilience (which may include engineering, social, economic, and ecological resilience) and redesigning if necessary. This approach can contribute toward balancing engineering integrity, economic efficiency, environmental quality, and social equity as part of a sustainable geotechnical solution.

In geotechnical engineering, a large number of studies are based on the common notions of sustainability like recycling, reuse and use of alternative materials, technologies, and resources; but, without a life cycle view, it is difficult to understand if these practices are truly sustainable. For example, use of recycled materials may seem a sustainable solution at all times; however, there are cases in which the benefits of recycling are largely offset by the environmental impact of transporting back the recycled materials and may be unsustainable when looked upon with a life cycle point of view. Thus, development and use of a proper sustainability assessment framework is necessary to assess whether sustainable choices are indeed made for a project. Sustainability assessment tools available in geotechnical engineering can be categorized into (1) single criterion—based methods (quantitative), (2) multiple criteria—based methods (qualitative or quantitative or combined), and (3) point-based rating systems (quantitative), as shown in Fig. 21.4.

Quantitative environmental metrics like global warming potential (GWP), carbon, embodied carbon, cumulative energy demand, embodied energy, and a combination of embodied energy and emissions (CO_2, methane (CH_4), nitrous oxide, sulfur oxides, and nitrogen oxides) have been used to compare competing alternatives in geotechnical engineering. But, the assessment of sustainability on the basis of a single metric like embodied carbon or global GWP involves ad hoc assumptions, puts excess emphasis on the environmental aspects, and neglects the technical, financial, and social aspects (Jefferson et al., 2007). According to Carpenter et al. (2007), a combination of life cycle analysis and site- and material-specific factors is more contextual for any decision-making framework than a singular metric. The multidimensional aspect of sustainability has been addressed in geotechnical engineering through qualitative and quantitative methods. Qualitative models are comprised of indicators that are evaluated based on color-coded rose diagrams. Life cycle—based tools have been widely used for quantifying environmental impacts or costs that are expected over the life cycle of geotechnical projects. LCA and life cycle costing (LCC) have been particularly used. LCA focuses on quantifying the environmental impacts and LCC is used to quantify the costs associated with the life cycle of a system (Fig. 21.5). The third approach to sustainability assessment comprises point-based rating systems in which points scored in different relevant categories are used as measures of sustainability of geotechnical projects. Table 21.2 summarizes the past studies that developed or utilized sustainability assessment methods for various geotechnical applications. It is important to note that sustainability assessment frameworks

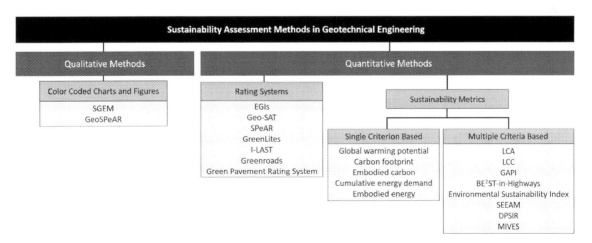

FIGURE 21.4 Existing sustainability assessment frameworks in geotechnical engineering.

in geotechnical engineering should (1) have a life cycle view of the geotechnical processes and products (Dam and Taylor, 2011), (2) incorporate all the four Es of sustainability (Holt et al., 2010a; Steedman 2011), (3) enforce sound engineering design and maintenance, (4) assess the reliability and resilience of the geo-system and offer flexibility to the user to identify site-specific needs, and (5) account for uncertainties in geotechnical works. Many studies, especially the single criterion—based methods, focused on assessing sustainability solely based on environmental impacts and did not incorporate multidimensional perspective by capturing the four Es of sustainability. Further, most studies outlined in Table 21.2 did not consider the impact of engineering design (e.g., reliability and resilience) on sustainability. In the next section, the procedure to conduct LCA is explained, which is an important part in the integrated sustainability framework that will be discussed later.

Life cycle assessment (LCA)

➤ Purpose: to quantify the **environmental impacts** of a product or system throughout its life cycle

➤ Types of environmental impact:
  - **Human health**: ozone depletion, photochemical oxidant and particulate matter formation
  - **Ecosystems**: climate change, acidification, ecotoxicity, land occupation
  - **Resource availability**: fossil fuel, minerals, water depletion

➤ Standards & guidelines: ISO 14040 and 14044

Life cycle costing (LCC)

➤ Purpose: to estimate the overall **cost** of a product or system

➤ Types of cost:
  - **Initial cost**: acquisition and construction cost
  - **Operation cost**: service and energy cost
  - **Maintenance cost**: repair and replacement costs
  - **End-of-life cost**: residual (resale or salvage) values and disposal cost

➤ Standards & guidelines: ISO 15686

FIGURE 21.5 LCA and LCC.

TABLE 21.2 Sustainability assessment methods used in geotechnical engineering.

Sustainability assessment method	Application	Indicator, metric, rating, or model	References
Single criterion —based methods	Concrete retaining walls and bioengineered slopes	Global warming potential	Storesund et al. (2008)
	Ground improvement methods	Carbon footprint	Spaulding et al. (2008)
	Retaining wall systems	Embodied carbon	Inui et al. (2011)
	Ground improvement methods	Embodied carbon	Egan and Slocombe (2010)
	Geosynthetics	Cumulative energy demand	Heerten (2012)
	Retaining wall systems	Embodied energy	Chau et al. (2006) Inui et al. (2011)
	Tunnel	Embodied energy	Chau e t al. (2012)

TABLE 21.2 Sustainability assessment methods used in geotechnical engineering.—cont'd

Sustainability assessment method	Application	Indicator, metric, rating, or model	References
Multiple criteria based methods (qualitative)	Slope stabilization	The Sustainable Geotechnical Evaluation Model (SGEM)	Jimenez (2004)
	Geotechnical projects	Geotechnical Sustainable Project Appraisal Routine (GeoSPeAR)	Holt (2011) Holt et al. (2010b)
Multiple criteria based methods (quantitative)	Pavement design	Life cycle costing (LCC)	Reigle et al. (2002) Praticò et al. (2011) Zhang et al. (2008)
	Airport pavement treatments	Green Airport Pavement Index (GAPI)	Pittenger (2011)
	Recycled materials in pavement	Life cycle assessment (LCA) and LCC	Lee et al. (2010b)
	Highway construction	Building Environmentally and Economically Sustainable Transportation—Infrastructure—Highways (BE^{2}ST-in-Highways)	Lee et al. (2010a)
	Underground mining	Environmental Sustainability Index	Torres and Gama (2006)
	Geotechnical projects	LCA	Misra (2010) Misra and Basu (2012)
	Ground improvement projects	Streamlined Energy and Emissions Assessment Model (SEEAM)	Shillaber et al. (2016a; 2016b)
	Geotechnical projects	Driver—Pressure—State—Impact—Response (DPSIR)	Lee and Basu (2018)
	Treatments for surficial slope	LCA	Das et al. (2018)
	Geotechnical projects	Integrated Value Model for Sustainable Assessment (MIVES)	da S Trentin et al. (2019)
	Retaining walls	LCA and MIVES	Damians et al. (2015, 2016)
Point based rating systems	Ground improvement projects	Environmental Geotechnics Indicators (EGIs)	Jefferson et al. (2007)
	Geotechnical projects	Geotechnical Sustainability Assessment Tool (Geo-SAT)	Raza et al. (2020, 2021)
	Foundation reuse	Sustainable Project Appraisal Routine (SPeAR)	Laefer (2011)
	Highway construction	GreenLites	McVoy et al. (2010)
	Highway construction	Illinois — Livable and Sustainable Transportation (I-LAST)	Knuth and Fortman (2010)
	Highway construction	Greenroads	Muench and Anderson (2009)
	Pavement construction	Green Pavement Rating System	Chan and Tighe (2010)

3. Life cycle assessment

LCA is a methodology for quantification of environmental impacts (e.g., global warming, acidification, ozone depletion, photochemical oxidant formation, particulate matter formation, ecotoxicity, human toxicity, resource depletion, etc.) that are caused throughout the life cycle of a product. The term "product" can be interpreted as primary materials for construction, such as concrete and steel, or an assembled/constructed system like a pile foundation. Life cycle refers to consecutive stages of a product's life from raw material extraction through materials processing, manufacture, transportation, operation, maintenance and repair, and disposal or recycling (International Organization for Standardization [ISO], 2006a). Multiple processes (activities that transform energy and materials into a subproduct) can exist within each stage. For example, in the manufacturing stage, production of concrete and production of steel are two individual processes. The unique stages in the life cycle of building structures are "construction" and "reuse of structure." The "operation" stage may not be applicable for geo-structures unless there exists material and/or energy use during their operations. The existing structures can be reused (e.g., foundation reuse) and this stage should be considered prior to the demolition and disposal stage. Fig. 21.6 shows an example of life cycle of a typical pile foundation.

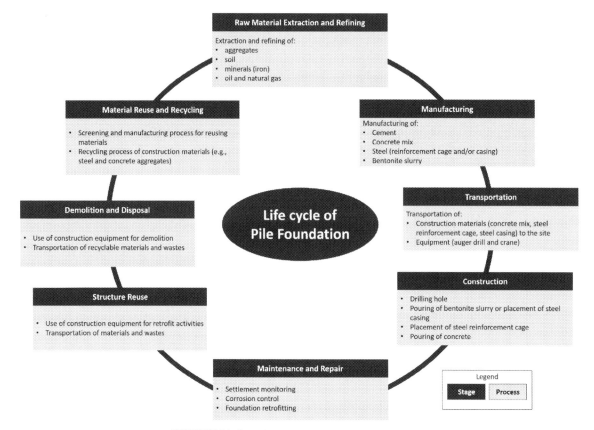

FIGURE 21.6 Life cycle of a typical pile foundation.

The main purpose of LCA is to quantify the inputs (e.g., material and energy), outputs (e.g., pollutants), and environmental impacts associated with the life cycle. LCA can assist in identifying the environmental performance of specific products (e.g., concrete and steel) or a system (e.g., foundation system, retaining structure, and reinforced slope) at various points in their life cycle. It is also useful for comparing design alternatives with respect to environmental sustainability (e.g., drilled shaft vs. driven pile).

LCA is conducted following four phases: (1) goal and scope definition, (2) inventory analysis (IA), (3) environmental impact assessment (EIA), and (4) interpretation (Crawford, 2011; Klöpffer and Grahl, 2014; ISO, 2006a). Fig. 21.7 shows the LCA framework for a built structure showing the four phases with some examples of inputs, life cycle stages, life cycle processes, outputs, environmental impacts, and interpretation.

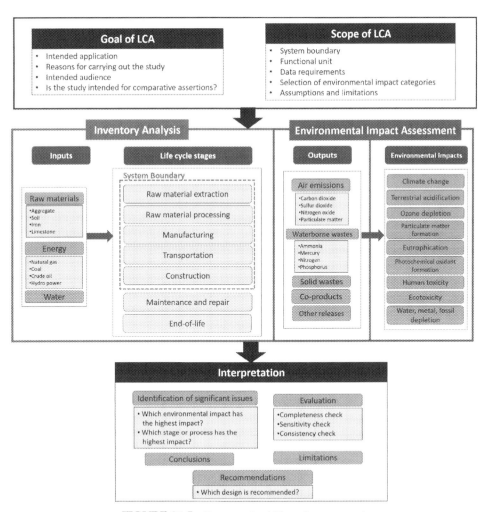

FIGURE 21.7 Framework of life cycle assessment.

3.1 Goal and scope definition

For defining the goal of an LCA study, the following items are considered: (1) the intended application, (2) the reasons for carrying out the study, (3) the intended audience, and (4) whether the results are intended to be used in comparative assertions for public disclosure. The scope of LCA includes defining the system boundary, the functional unit of the system, data requirements, selection of the environmental impact categories, assumptions, and limitations. The system boundary defines the life cycle stages to be included in the system. Examples of typical system boundaries are cradle-to-grave (i.e., full life cycle of a product), cradle-to-gate (from resource extraction to the factory gate), and cradle-to-site (cradle to the end of construction) (ISO, 2006a; Song et al., 2020). Further, the processes associated with each life cycle stage are determined. For example, during the construction (stage) of a drilled shaft foundation, construction activities (processes) include drilling a borehole in the ground, placement of steel reinforcement cage, and concrete pouring. Selection of the system boundary of an LCA depends on the type of geotechnical project and availability of data required for performing the LCA. A functional unit is a quantified description of the performance characteristics to fulfill the primary purpose of the product. Simply put, it describes the intended purpose of the product. For example, the functional unit of a pile foundation can be "supporting the structure load without bearing capacity failure and settlement exceeding 15 mm." All inputs and outputs are relative to the functional unit, as are the resultant environmental impacts. In other words, an increase in the quantity of functional unit will result in an equivalent increase in the associated inputs, outputs, and environmental impacts (Crawford, 2011). For example, if the specified pile settlement tolerance is lower (meaning a higher quality performance requirement), then the dimensions of pile foundation may increase resulting in higher environmental impacts. It is important to clearly define the functional unit when comparing two or more products because the comparison should be made on the basis of equivalent performance (Curran, 2017).

Databases are required for the quantification phases of LCA, i.e., inventory analysis and environmental impact assessment, which can differ depending on the year the database was developed, the geographical coverage, and technology. The data requirements are defined in the scope to ensure that the LCA results are representative of the actual data. For example, databases developed in European countries may not be suitable for products/systems made in North America or Australia. The technology mix can be different from one another depending on how much renewable energy sources are utilized. Based on the defined goal of LCA, the categories of environmental impacts are selected. For example, the goal of an LCA study may be focused only on global warming impact, or may include different types of environmental impacts. Further, any assumptions and limitations pertaining to the LCA are stated in the scope to ensure that comparative assertions are properly made.

3.2 Inventory analysis

The inventory analysis involves compilation and quantification of inputs and outputs of each process in the lifecycle of a product (ISO, 2006a). The analysis is conducted for the processes determined in the system boundary using the databases that comply with the data requirements stated in the scope. All calculations of inputs and outputs are completed in relation to the functional unit. In a life cycle process, inputs (e.g., material and energy) are transformed into outputs (e.g., emissions and by-product). For example, production of concrete requires natural aggregate, soil, limestone, water, and energy

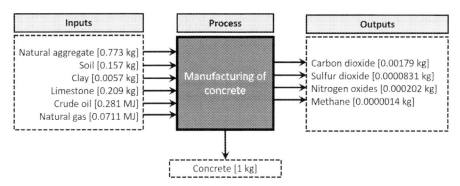

FIGURE 21.8 Sample calculations for inventory analysis.

(inputs), and the production process generates emissions like CO₂, CH₄, sulfur dioxide, and nitrogen oxides (outputs) as environmental consequences. To manufacture 1 kg of concrete, 0.773 kg of natural aggregates is required and 1.79×10^{-3} kg of CO₂ is emitted (see Fig. 21.8). If 100 kg of concrete is required to meet the primary purpose (e.g., concrete shaft for pile foundation) stated in the functional unit, the inputs and outputs, shown in Fig. 21.8, are scaled up by 100. Therefore, the emissions (outputs) are calculated as follows:

$$E_x = m \times (EF)_x \quad (21.1)$$

where E_x is the total emission of the substance x (e.g., CO₂, nitrous oxide, carbon monoxide, CH₄, or nitrogen oxides), m is the mass of subproduct (e.g., concrete), and $(EF)_x$ is the emission factor of the substance x, which indicates the specific emission of substance x per unit mass of the subproduct. For example, as shown in Fig. 21.8, the emission factor of carbon dioxide $(EF)_{CO2} = 0.00179$ kg/kg and the total emission of carbon dioxide $E_{CO2} = 0.179$ kg for 100 kg of concrete manufacturing ($m = 100$ kg). The mass of concrete is determined based on the design results (e.g., pile dimensions). The list of inputs and outputs (i.e., emission factors $(EF)_x$) used in the inventory analysis calculations can be obtained from LCA inventory databases. Hammond and Jones (2008) and National Renewable Energy Laboratory [NREL] (2012) are well-established databases for construction materials that are used in this study.

In summary, the inventory analysis is completed by executing the following steps for each life cycle process in the defined system boundary: (1) determination of the quantity of products for the selected process, (2) finding the inventory database that best represents the selected process, and (3) calculating the total inputs and outputs based on steps (1) and (2).

3.3 Environmental impact assessment

The environmental impact assessment involves associating the emissions (e.g., CO₂, sulfur dioxide, and CH₄), obtained from the inventory analysis, with specific environmental impact categories like global warming, acidification, and ecotoxicity. The emissions, compiled in the inventory analysis, are first classified according to their contributions to the environmental impacts. For example, CH₄ (one of the emissions) is a type of greenhouse gas that contributes to global warming. The total mass of CH₄, quantified in the inventory analysis, is converted into an equivalent measurement that represents global warming (i.e., equivalent mass of CO₂). Fig. 21.9 shows an example of how emissions from concrete are converted into environmental impacts. The details of characterization factors (used to convert the mass of emissions to equivalent quantities of environmental impacts) and

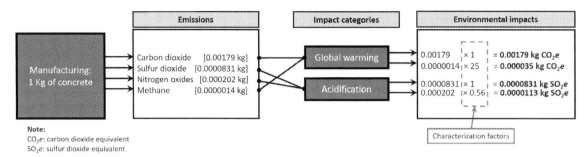

FIGURE 21.9 Sample calculations for environmental impact assessment.

calculation procedure are discussed below with particular reference to the impact category of global warming.

Greenhouse gases (GHGs) such as CH_4 that contribute to global warming may influence the global warming more or less than that of CO_2 because their potencies to trap heat in the atmosphere are different in magnitudes than that of CO_2 (Intergovernmental Panel on Climate Change [IPCC], 1990). Hence, the mass of air emissions of these gases needs to be converted into an equivalent measurement, i.e., GWP or equivalent mass of CO_2, to capture the different contributions toward global warming. This conversion is done using the characterization factor (which is GWP for global warming impacts), as shown in Fig. 21.9. CH_4 has a much higher potency to global warming than CO_2 because of which the mass of CH_4 is converted into an equivalent mass of CO_2 by multiplying with the respective characterization factor of 25 (expressed in terms of GWP) (Fig. 21.9).

GWP of a gas is a measure of how much infrared radiation (Sun's energy or heat radiated by the earth) an emission of 1 kg of the gas will absorb over a given period of time (e.g., 100 years) relative to the instantaneous emission of 1 kg of CO_2. In other words, it is a measurement intended to compare the gases that have different abilities of absorbing energy and the different time periods these gases remain in the atmosphere. The GWP of an emission with respect to CO_2 is calculated as follows:

$$(GWP)_x = \frac{\int_0^{TH} a_x \cdot C_x(t) dt}{\int_0^{TH} a_{CO_2} \cdot C_{CO_2}(t) dt} \quad (21.2)$$

where x is the emission of interest, TH is the time horizon (e.g., 20, 100, or 500 years), a_x is the radiative efficiency of the emission x, $C_x(t)$ is the time-dependent abundance of x (the atmospheric mass of emission x over time starting with a pulse at time $t = 0$ with the assumption of instantaneous release of all the emissions), and a_{CO2} and $C_{CO2}(t)$ are the corresponding quantities for CO_2, which is the reference gas for global warming. The calculated GWP in Eq. (21.2) is also referred to as the characterization factor.

CO_2 has a characterization factor of 1 GWP because all other gases are expressed in terms of equivalent CO_2, and CH_4 has a characterization factor of 25 GWP (indicating 25 kg of CO_2 equivalent) assuming a 100 year of time horizon. The equivalent mass of CO_2 is calculated as follows:

$$CO_2e = E_x \times (GWP)_x \quad (21.3)$$

where CO_2e is the equivalent mass of CO_2, E_x is the emissions of climate forcer x (e.g., GHGs), and $(GWP)_x$ is the 100-year GWP relative to CO_2 of climate forcer x. The characterization factors can be obtained from the ReCiPe database developed by Goedkoop et al. (2014). Similar calculations can be done for other environmental impact categories—see Goedkoop et al. (2013) for detailed explanations.

In summary, the environmental impact assessment is completed by following these steps: (1) compilation of the list of emissions from the inventory analysis, (2) selection of the environmental impact categories of interest, (3) finding the EIA database to obtain characterization factors, and (4) calculation of the environmental impacts in equivalent measurements. A complete list of environmental impact categories is provided in Table 21.3.

3.4 Interpretation of LCA results

The interpretation phase involves reporting the findings from the inventory analysis and the environmental impact assessment. Conclusions are drawn through an iterative process that follows these sequences: (1) identification of significant issues, (2) evaluation of the methodology and results for completeness, sensitivity, and consistency, (3) drawing preliminary conclusions and checking that these are consistent with the goal and scope of the study, data quality requirements, predefined assumptions and values, and limitations, and (4) reporting the conclusions if they are found to be consistent; otherwise, reiteration from steps (1), (2), and (3) as appropriate (ISO, 2006b). Recommendations are given based on the final conclusions and according to the goal of LCA. For example, if the goal of LCA is to compare the global warming impacts of two alternative designs of a pile foundation, the design with the least GWP is recommended.

Significant issues can be identified by investigating the proportional contributions of the different stages or processes to the environmental impacts or to specific emissions of interest. The objective of completeness check is to ensure that all relevant information and data needed for the interpretation are available and complete. The sensitivity check involves evaluating the reliability of the final results based on their sensitivities to the uncertainties in the data. Lastly, the consistency check is completed to determine whether the assumptions, calculation methods, and data are consistent with the goal and scope.

4. Reliability and resilience

4.1 Uncertainty and reliability in geotechnical engineering

An engineering system that provides sufficient resistance (capacity or supply) to the external loads (demand), thereby ensuring adequate safety and serviceability, is said to be reliable. The level of reliability depends on by how much sufficiently the supplies are available to meet the demand requirements. For example, a pile foundation that has a high bearing capacity to withstand the applied loads is said to be reliable, and the greater the bearing capacity for the same applied loads, the higher the reliability. Traditionally, reliability of geo-structures has been achieved through the use of factors or margins of safety and adopting conservative assumptions in the process of design. A common technique is to conservatively assume parameters associated with a minimum capacity that will remain adequate under a maximum loading condition (Ang and Tang, 1984). Uncertainties in geotechnical engineering problems are inevitable because of the natural randomness in subsurface soils. Factor of safety has been used to collectively capture all the uncertainties related to the resistances and loads associated with geo-structures based on past experience with similar structures and types of soil. However, it should be recognized that merely assuming conservative values of parameters and factors of

21. A combined sustainability-reliability approach in geotechnical engineering

TABLE 21.3 Environmental impact categories (Huijbregts et al., 2017).

Impact category	Indicator	Characterization factor	Equivalent measurement unit
Global warming	Infra-red radiative forcing increase	Global Warming Potential (GWP)	[a]kg of CO_2
Ozone depletion	Stratospheric ozone decrease	Ozone Depletion potential (ODP)	[b]kg of CFC-11
Ionizing radiation	Absorbed dose increase	Ionizing Radiation potential (IRP)	[c]kBq of Co-60
Terrestrial acidification	Proton increase in natural soils	Acidification Potential (TAP)	[d]kg of SO_2
Eutrophication	Phosphorous increase in marine or fresh water	Eutrophication potential (EP)	[e]kg of P
Photochemical oxidant formation	Tropospheric ozone increase	Photochemical Oxidant Formation Potential (POFP)	[f]kg of NMVOC
Particulate matter formation	Intake of fine particulate matter	Particulate Matter Formation Potential (PMFP)	[g]kg of PM_{10}
Human toxicity	Risk increase of cancer and noncancer disease incidence	Human Toxicity Potential (HTP)	[h]kg of 1,4-DB
Terrestrial ecotoxicity	Emission of organic substances and chemicals to natural soils	Terrestrial ecotoxicity potential (TETP)	kg of 1,4-DB
Marine ecotoxicity	Emission organic substances and chemicals to marine water	Marine Ecotoxicity Potential (METP)	kg of 1,4-DB
Freshwater ecotoxicity	Emission of organic substances and chemicals to fresh waters	Freshwater Ecotoxicity Potential (FETP)	kg of 1,4-DB
Land use	Occupation and time-integrated land transformation	Agricultural land occupation potential (LOP)	$m^2 \times$ year annual cropland
Water use	Increase of water consumed	Water consumption potential (WCP)	m^3 water consumed
Mineral resource scarcity	Increase of ore extracted	Surplus ore potential (SOP)	[i]kg of Cu
Fossil resource scarcity	Upper heating value	Fossil fuel potential (FFP)	kg of oil

[a] *CO_2: Carbon dioxide.*
[b] *CFC-11: Trichlorofluoromethane.*
[c] *Co-60: Cobalt-60.*
[d] *SO_2: Sulfur dioxide.*
[e] *P: Phosphorous.*
[f] *NMVOC: Nonmethane volatile organic carbon compound.*
[g] *PM_{10}: fine particulate matter.*
[h] *1,4-DB: 1,4 dichlorobenzene.*
[i] *Cu: Copper.*

safety do not properly address the uncertainties, and often lead to other problems like excess use of materials and money that are counterproductive to sustainable designs. Systematic accounting of the different uncertainties using rigorous mathematical tools is necessary, particularly in geotechnical engineering, because transparency is then ensured in the design process so that the design ensures the required quantifiable levels of safety and serviceability without the excess use of materials and money.

Reliability-based design methods provide a framework based on the mathematical theory of probability in which the different uncertainties in design are quantified and satisfactory levels of safety and serviceability, quantified in terms of probabilities of failure, are maintained in the design. The different design parameters, such as the soil friction angle, applied loads, and foundation capacity, are treated as random variables and the uncertainties associated with these parameters (or random variables) are quantified systematically in terms of their respective probability distribution functions. Subsequently, a probabilistic analysis is performed to ensure that the designed capacity (supply) is sufficiently greater than the loads (demand) in a probabilistic sense such that the probability of failure of the system is below an accepted threshold value.

To perform a reliability-based design in geotechnical engineering, it is important to distinguish between the different types of uncertainties associated with geotechnical problems. Two types of uncertainties are usually encountered when characterizing soil properties—aleatory (inherent or natural) and epistemic (lack of knowledge) uncertainties (Fig. 21.10). Aleatory uncertainty represents the natural randomness of a property, and it cannot be reduced or eliminated. Natural variability is associated with the inherent randomness, manifesting as variability over time at a single location (temporal variability) or as variability over space at different locations but at a single time (spatial variability). For any extensive volume

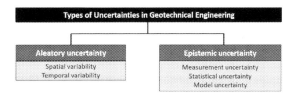

FIGURE 21.10 Types of uncertainties in geotechnical engineering.

of natural soil layer, the properties fluctuate spatially, both laterally and with depth. In geotechnical engineering, three types of epistemic uncertainty can be encountered—measurement uncertainty, statistical uncertainty, and model uncertainty. Measurement uncertainty is related to measurement errors because of imperfections of an instrument or of a method, inconsistency or inhomogeneity of data, data handling and transcription errors, and inadequate representative of data samples because of time and space limitations. Measurement errors can also occur because of bias in the measurement procedures (e.g., effects of excessive sample disturbance). Measurement uncertainty is inevitably encountered in site investigations and laboratory tests. Statistical uncertainty results from the lack of accuracy in assessing parameter values because of limited sets of measurement data. Model uncertainty is related to the limitations of the mathematical models in accurately mimicking the true physical behavior of a system. For example, the discrepancy between pile load test results and design equations is usually caused by model uncertainty. Epistemic uncertainty can be reduced or eliminated by collecting more information, improving the measurement methods, or improving the calculation models (Baecher and Christian, 2003; Christian, 2004; Lacasse and Nadim, 1996).

Uncertainty in applied loads also falls under the categories of aleatory and epistemic uncertainties. The variability in dead loads is fairly small, but minor levels of uncertainty do exist. For example, the statistical uncertainty in defining the unit weight of concrete because of limited availability of data translates into dead load uncertainty. Temporal variability in dead loads can be caused by the change in occupancy of a building over time. Model uncertainty in applied loads is apparent when representing live or transient loads. For example, live loads are usually assumed to be uniformly distributed and winds are modeled as static loads while, in fact, both are random in space and time (aleatory uncertainty). Statistical uncertainty also exists in defining the environmental loads because these are typically defined based on limited data from past recordings (e.g., 100 years or less of data). Another source of uncertainty is the combination of load effects (e.g., primary load effect is caused by occupancy live load and the secondary load effect is caused by snow load). The two load processes need to be examined to determine the worst combined load effect. For example, it is unlikely and not economical to assume that the worst occupancy live load in 50 years occurs at the same with the worst snow load in 50 years (Bulleit, 2008).

Reduction or elimination of uncertainties in geotechnical engineering requires a huge effort—it is almost impossible to gather enough subsurface data to characterize and model the soil as is. Therefore, explicit consideration of uncertainties in the design of geotechnical structures is necessary for establishing reliable safety margins (Lacasse and Nadim 1996). There are several strategies available for dealing with the uncertainties in geotechnical engineering problems: (1) being conservative, (2) use of the observational method, and (3) quantification of uncertainty (Christian, 2004). Being conservative is the simplest method; however, it can be expensive, may drag the project completion timelines, may not be practically feasible, and will lead to excess use of natural resources all of which lead to unsustainable outcomes. The observational method involves adjusting the construction procedures and design details depending upon observations and measurements made as the construction proceeds. However, the observational method requires expensive field measurements, and the engineers may have difficulty in completing the project should a change in design or construction sequence is required mid-project because of limited access to the decision makers (Christian, 2004; Whitman, 2000). Uncertainty quantification seems to be the most rational approach and different techniques are used for quantifying uncertainties in geotechnical engineering. For example, borings for site investigation are usually spaced out to take into account the horizontal spatial variability in subsurface soil. The correlations of soil types with lateral distance and depth are systematically identified using a mathematical technique known as kriging, which involves geostatistical interpolation analysis that considers both the distance and the degree of variation between known data points (autocorrelation) when estimating values in the unknown areas. Kriging is useful in identifying the spatial extents of different soil layers and for estimating scattered values of soil properties (e.g., undrained shear strength) based on site investigation (e.g., cone penetration test) (Lacasse and Nadim, 1996). Accounting for the spatial variability is particularly important for estimating risks of slope failures with long embankments, dispersion of plumes of pollutants, and evaluation of resistance of a site to liquefaction (Whitman, 2000). For embankment design, Christian et al. (1994) used spatial averaging technique to reduce the uncertainty caused by spatial variability and systematic error (e.g., statistical uncertainty and bias in field vane testing) in estimation of soil properties. Model uncertainty can be evaluated from comparisons between model tests and deterministic calculations, pooling of expert opinions, case studies of prototypes or other model tests, results from the literature, and engineering judgment (Lacasse and Nadim 1996, 1998).

The uncertainties and random heterogeneity of the soil properties and the uncertainties in the load are most rigorously modeled by treating these variables (e.g., soil friction angle, soil unit weight, live load, etc.) as stochastic (random) processes defined by their means, variances, and correlation lengths (or times) with associated probability distributions (Fenton and Griffiths, 2003; Popescu et al., 1998). However, more commonly, a simpler approach is often chosen in which these variables are treated as random variables defined by their means and variances with associated probability distributions (e.g., normal, lognormal, and uniform distributions) (Juang et al., 1999; Low, 2005; Low and Tang, 1997; Phoon and Kulhawy, 1999; Vanmarcke, 1980). The coefficient of variation (COV), which is a statistical measure of dispersion relative to the central value, is commonly used to define the characteristics of the random variables, and it is calculated as the ratio of the standard deviation to the mean of a random variable. Lacasse and Nadim (1996) and Lumb (1974) provided the COVs and probability distribution functions of different soil properties (as cited in Baecher and Christian, 2003), and Ellingwood et al. (1980) provided the statistics for different types of loads including the dead load, live load, wind load, snow load, and earthquake load. For example, the friction angle of sand follows a normal distribution with the COV in the range of 0.02—0.05, and the undrained shear strength of clay follows a lognormal distribution with the COV in the range of 0.05—0.2. The dead load follows a normal distribution with the COV in the range of 0.06—0.15, and the maximum wind load follows a Gumbel distribution with the COV in the range of 0.09—0.16 for different areas in the United States.

The importance of reliability-based design and its superiority over conventional, factor of safety—based deterministic design can be illustrated with an example of a pile foundation. A jacket pile installed in 1976 for an offshore structure in Norway was reanalyzed 13 years later (Lacasse and Nadim, 1998) after a new soil investigation and new calculations of the environmental and gravity loads were completed.

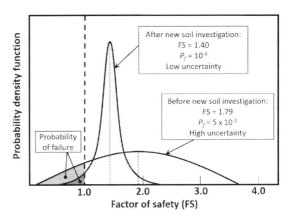

FIGURE 21.11 Probability of failure versus factor of safety (after Lacasse & Nadim, 1996).

Initially, the factor of safety of the pile was calculated to be 1.78, and the new deterministic analysis resulted in a lower factor of safety of 1.40. However, the new information reduced the uncertainties in both the soil and load parameters, which resulted in a safety margin greater than that perceived at the time of first design. The lower uncertainties, illustrated in Fig. 21.11 as a "narrow" probability distribution, led to a reduction in the probability of failure by a factor of 2. This shows that the factor of safety is not a sufficient indicator of safety margin because the uncertainties are not explicitly considered in deterministic calculations. Thus, probabilistic methods like the reliability-based design methods may help reduce excess safety margins that are often used in traditional design methods or in factors of safety prescribed in codes (Lacasse and Nadim, 1996). Naturally, overdesigning is reduced in reliability-based approaches and satisfactory performance of the geotechnical structure is ensured with a quantified probability of failure, which promotes the sustainability goals.

4.2 Reliability-based design

Reliability analysis is related to solving supply—demand problems, in which the supply and demand are considered random variables.

For geotechnical engineering problems, the supply is usually the resistance (capacity) of a system and the demand comprises the applied loads on the system. The difference between the resistance and load is defined as the safety margin:

$$S = R - Q \qquad (21.4)$$

where S is the safety margin, R is the resistance (capacity) of the system, and Q is the load applied to the system. If R and Q are random variables (characterized by probability distribution functions, means, and variances), then S is also a random variable characterized by a probability distribution function, mean, and variance. The probability of failure P_f of the system is estimated as follows:

$$P_f = \int_{-\infty}^{0} f_S(s) ds \qquad (21.5)$$

where P_f is the probability of failure, and $f_S(s)$ is the probability distribution function of S (Fig. 21.12), and s is a realization of the random variable S. Graphically, the probability of failure is the area under $f_S(s)$ for $s < 0$, as indicated by the shaded area in Fig. 21.12. The mean μ_S of the safety margin S is calculated as follows:

$$\mu_S = \mu_R - \mu_Q \qquad (21.6)$$

where μ_R and μ_Q are the means of R and Q, respectively. For statistically independent R and Q, the variance of S is given by the following:

$$\sigma_S^2 = \sigma_R^2 + \sigma_Q^2 \qquad (21.7)$$

where σ_S, σ_R, and σ_Q are the standard deviations of S, R, and Q, respectively. Assuming that the resistance R and load Q follow normal distributions, the probability of failure P_f estimated based on Eq. (21.5) is given by

$$P_f = 1 - \Phi\left(\frac{\mu_S}{\sigma_S}\right) \qquad (21.8)$$

where Φ is the standard normal distribution function. The reliability index β is defined as follows:

$$\beta = \frac{\mu_S}{\sigma_S} \qquad (21.9)$$

and it is related to P_f. Graphically, the reliability index is related to the horizontal shift of the mean of safety margin from $s = 0$, as shown in Fig. 21.12.

The objectives of reliability-based design are to quantify the performance of a system with uncertain information and to minimize the probability of failure (or maximize the reliability index) by adjusting the design parameters or by updating additional information on uncertain variables (e.g., soil properties, loads, and limit state function). Different reliability analysis methods are available such as the first-order second moment (FOSM) method, first-order reliability method (FORM), point estimate method (PEM), and Monte Carlo simulation (MCS) method. The integrated framework described in the next section uses FORM because of which it is briefly described in the following paragraphs.

FORM, also known as the Hasofer–Lind approach, is an optimization method involving the loads and resistances (Q and R) which are the random variables (there can be multiple

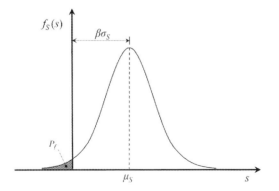

FIGURE 21.12 Probability distribution of safety margin.

loads and resistances) such that the minimum distance between the design state D (corresponding to a particular combination of loads and resistances) and the limit state function $G(R_1, R_2, ..., Q_1, Q_2 ...)$ is determined. In the probability space of the loads and resistances, the limit state function is a surface that separates the feasible design space (within which the design state has to lie) from the unfeasible (unsafe) space. For example, $G(R, Q) = R - Q$ is a possible limit state function and the unsafe (or unfeasible) space consists of all possible combinations of R and Q for which $G < 0$. The minimum distance between D and $G(R_1, ..., Q_1, ...)$ is related to the probability of failure or the reliability index. In the actual calculations, the original probability space of the loads and resistances is mapped to the standard normal space expressed in terms of reduced variables x_i' defined as

$$x_i' = \frac{x_i - \mu_{x_i}}{\sigma_{x_i}} \qquad (21.10)$$

where x_i' is the reduced variable for a set of random variables x_i (with $i = 1, 2, ..., n$) in the original space ($R_1, R_2, ..., Q_1, Q_2, ...$ are the x_i-s), and μ_{xi} and σ_{xi} are the means and standard deviations of x_i, respectively. The problem of finding the minimum distance between D and $G(x_i)$ in the original space translates into finding the minimum distance in the standard normal space between the point of origin and the limit state function (surface) $G(x_i') = 0$ expressed in terms of reduced variables, as shown in Fig. 21.13.

If all the random variables follow normal distribution and are uncorrelated, the reduced variables have a mean of 0 and a unit standard deviation. In the standard normal space (in which the random variables are reduced to standard normal variables), the probability is rotationally symmetric around the origin, illustrated as the gray dashed circles (joint distribution surface of the reduced variables x'_1 and x'_2) in Fig. 21.13.

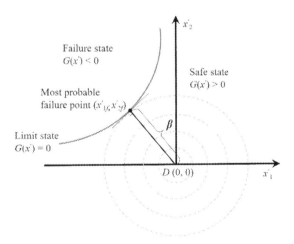

FIGURE 21.13 Illustration of the first-order reliability method (FORM).

The minimum distance from the origin (which indicates the mean values of the reduced variables) to the limit state function $G(x_i') = 0$ is defined as the reliability index. The objective of FORM is to find this minimum distance. As the limit state surface (or failure surface) $G(x_i') = 0$ moves further or closer to the origin, the safe region, $G(x_i') > 0$, increases or decreases, respectively. Therefore, the position of the failure surface relative to the origin of the reduced variates should determine the reliability of the system. The point on the failure surface with minimum distance to the origin is the most probable failure point, as indicated by (x'_{1f}, x'_{2f}) in Fig. 21.13 (Ang and Tang, 1984). Simply put, FORM is a constrained optimization problem with the following formulation:

$$\begin{array}{l} \text{Minimize } \beta \\ \text{Subject to } G(x) = G(x') = 0 \end{array} \qquad (21.11)$$

For multidimensional problems with a number of random variables x_i, the distance from a point $x' = (x'_1, x'_2, ..., x'_n)$ on the failure surface $G(x') = 0$ to the origin is calculated as follows:

$$\beta = \sqrt{(x'_1)^2 + (x'_2)^2 + \cdots + (x'_n)^2} \qquad (21.12)$$

4.3 Need for infrastructure resilience

The underlying assumption in reliability-based design is that the failure states (limit states) can be predicted and the associated probabilities of failure can be estimated, which is the fundamental premise of risk based or *fail-safe* design. In recent times, however, there is a growing consensus that it is not entirely possible to anticipate the likelihood, manifestation, and consequences of all future threats (i.e., failure states). Past experiences, particularly with extreme events, show that indeed there is a serious limitation in our ability to predict unforeseen disasters and resist all surprises. For example, the 2001 World Trade Center disaster (September 11 attacks) killed 2996 people and caused US $3 trillion worth of damage (Carter and Cox, 2011). The collapse of the building caused rupture of water mains and flooding of rail tunnels, a commuter station, and a vault containing all of the cables for one of the largest telecommunication nodes in the world. The recovery from the terror attacks was extremely slow. The 2005 Hurricane Katrina caused two landfalls, generated a large storm surge, and destroyed levees that led to flooding. The total damage reported was US $108 billion, and 1833 direct and indirect fatalities were recorded. After the Hurricane Katrina, a power outage was experienced for more than a month, and the supply of crude oil and petroleum products was interrupted because of the loss of electricity at the pumping stations for major transmission pipelines. Nearly 160 million liters of gasoline production, accounting for 10% of the US supply, was lost per day. In addition, it was reported that the US federal government had spent $120.5 billion on the Gulf Region for the recovery (Ayyub, 2014; Leaning and Guha-Sapir, 2014). Therefore, safety measures are always inadequate and *safe-to-fail* design approaches are necessary (Simoncini, 2011; Vugrin et al., 2011). In the safe-to-fail design approaches, it is accepted that failures to systems can occur, but it is aimed that the

systems lose functionality in controlled ways and remain adaptable to control the consequences by recovering the lost functions or by transforming to serve new purposes. For example, the Indian Bend Wash, an 18-km long greenbelt in Scottsdale, Arizona, U.S.A., designed to manage heavy rainfall events, utilized a bioretention basin in the form of parks and golf courses instead of a traditional concrete channel structure (Kim et al., 2019). The Indian Bend Wash not only serves as a place of recreation for the city residents, but also serves as a storm water management system to mitigate flooding damage during heavy rainfall events. In 2014, the wash was able to accommodate excess runoff caused by a heavy rainstorm and helped drain city streets and neighborhoods. As demonstrated by the Indian Bend Wash case study, it is important that the overconfidence of the fail-safe approach (e.g., designing for oversized or excessively robust infrastructure like the concrete channel structure that was initially recommended) in the current design practice needs to be reconsidered and instead the aim should be on building adaptable and resilient infrastructure systems.

Rogers et al. (2012) pointed out eight categories of possible threats that the physical civil infrastructure may encounter: (1) gradual deterioration from aging, exacerbated by adverse ground conditions (including chemical, biological, and physical threats), (2) damage due to surface loading or stress relief due to open cut interventions, (3) severely increased demand, and ever-changing (different or altered) demands, (4) terrorism, (5) the effects of climate change, (6) the effects of population increase including increasing population density, (7) funding constraints, and (8) severe natural hazards (extreme weather events, earthquakes, landslides, etc.). Although designs should be made flexible and robust to ward off as many threats as possible, it is impossible to design engineering systems that are full proof against all possible threats. Therefore, for systems to be sustainable, it is necessary to ensure that the system is inherently capable of bouncing back to its functionality irrespective of the nature or magnitude of shock or distress it is subjected to. Such systems are called resilient systems.

4.4 Resilience and geo-infrastructure

Sustainability of geo-structures and geo-systems is closely related to resilience (Lee and Basu, 2018). Soil being the weakest of all the civil engineering materials, the vulnerability of the geotechnical components of civil infrastructure against hazards is among the highest. There are real threats of breach of levees and earth dams during hurricanes and tsunamis, breakdown of underground water pipeline network during an earthquake, landslides during heavy rainfall and earthquakes, and disruptions and distress in underground transit systems because of terror attacks (Basu et al., 2013). Resilience is particularly important for interconnected systems where a failure in any part can quickly propagate to other interdependent critical infrastructure systems and can easily trigger a system failure (Park et al., 2011; Rinaldi et al., 2001). For example, closure of a transportation network can affect access to medical care, emergency services, and food and fuel supply from which the impacts propagate to other critical infrastructures such as electric power generation, telecommunications, and water supply facilities (Min et al., 2007). Naturally, public safety, economy, quality of life, and environment are inherently connected to the resilience of the geotechnical infrastructure. Climate change effects also add the need of improving resilience of geotechnical infrastructure systems. The number of natural disasters has substantially increased since 1950, with climate-related events accounting for 80% increase (Leaning and Guha-Sapir, 2014). Rapid sea level rise, increased global surface temperature, warming oceans, shrinking ice sheets, rapid decrease in the extent and thickness of Arctic sea ice, and glacial retreat in mountains are some of the evidence of rapid climate change (National Aeronautics and Space Administration [NASA], 2017). Climate change alters the loads and resistances of physical infrastructure systems that can accelerate the deterioration of the physical structures. Weakened physical infrastructures are more vulnerable to extreme events like terror attacks, earthquakes, floods, and landslides that can severely incapacitate the physical infrastructure. Even if the infrastructure health is maintained at a high level of functionality, extreme events with high severity can damage or destroy it. Therefore, incorporation of resilience thinking is necessary in planning, design, execution, operation, and maintenance of geotechnical infrastructure so that the preparedness and response against disruptive events can be improved. Resilient infrastructure systems are capable of coping with extreme and unexpected events, thereby minimizing the environmental impacts and economic losses, mitigating hazards, ensuring public safety, and continuing to provide the essential services to the public. All these outcomes of resilient infrastructure systems correlate positively with the three elements of sustainability and, therefore, promote sustainable development.

Resilience emphasizes the ability to recover from unforeseen disasters rather than resist all possible disasters, to be prepared so that hazards are mitigated and consequences are minimized, and to continuously adapt and evolve based on the lessons learned from the past. For infrastructure systems, Bruneau et al. (2003) chose four parameters (4 Rs) as measures of community resilience: (1) robustness, as defined by the capacity of the system or its parts to perform its function even under external disturbance, (2) redundancy, as measured by the degree to which an affected part is substitutable, (3) resourcefulness, defined as the ability to identify threats and set up plans for handling such threats, and (4) rapidity with which external disturbances are addressed. Fig. 21.14A shows a typical degradation and recovery of functionality of system over time. Absorption of shocks is reflected

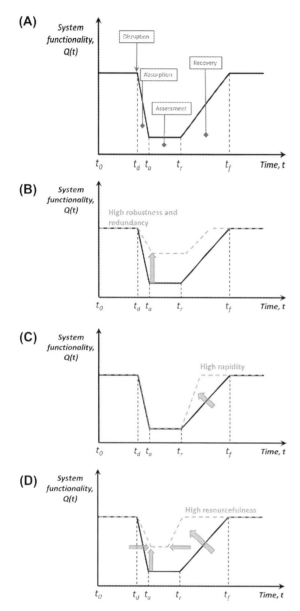

FIGURE 21.14 (A) Engineering resilience expressed in terms of change in system function over time, (B) system with high robustness and redundancy, (C) system with high rapidity, (D) system with high resourcefulness.

by the degradation of the system functionality at the event of a disruption (from time t_d to t_a in Fig. 21.14A). A system with high robustness and redundancy is likely to experience less degradation at the event of disruption, as illustrated in Fig. 21.14A. The recovery efforts can be initiated immediately postdisruption; however, the degraded system functionality may remain unchanged for a certain period of time (from time t_a to t_r in Fig. 21.14A) until adequate resources are collected and response strategies are organized (this is the assessment stage). Ultimately, it is expected that the system functionality recovers to an acceptable level for its normal operation (from time t_r to t_f in Fig. 21.14A). The process of recovery can be hastened if the system has high rapidity (Fig. 21.14C). A system with high resourcefulness can improve its resilience in all aspects (Fig. 21.14D).

From a design point of view, resilience can be improved by ensuring that the engineered system has adequate robustness against external threats. The remaining aspects of resilience (redundancy, resourcefulness, and rapidity) are mostly addressed by proper planning and management and are not entirely part of engineering design. Robustness can be thought as the most important aspect of resilience because it determines the amount of effort needed for assessment and repair. For example, a structure that experienced lower degradation is most likely to require less resources for recovery. To improve the robustness of an engineered system, it needs to be designed such that it is at least safe against possible identified failure states (i.e., the known limit states). The consequences of extreme events (that surpass the known limit states) are mostly handled through strategic planning and management, which are not necessarily related to geotechnical engineering design. Hence, by performing reliability-based design, the robustness aspect of resilience can be taken into account to some extent. Reliability-based design reasonably takes into account the known unknowns in which the uncertainties from external loads and soil properties are considered in a systematic way. To tame the unknown unknowns, however, building resilience in geo-systems is necessary.

5. An integrated sustainability framework

From the foregoing discussion, it is clear that sustainability problems require multidimensional and lifecycle views in a systematic manner to unravel their complexities. In this section, an integrated sustainability assessment framework is introduced which is inspired by the systems engineering approach of optimizing the 3 Es—environment, economy, and equity—simultaneously as design objectives. For geotechnical engineering problems, particularly, engineering design should also be considered as part of the design objectives or constraints (4 Es of sustainability) to ensure safety and serviceability. Sustainable designs are deemed to be lean designs that generate less environmental impacts and require low costs. However, lean designs are not necessarily desirable because they usually have a lower tolerance to disruptions than conservative designs. A balanced design that uses less resources but has sufficiently high reliability should be considered more sustainable from a long-term perspective because the possibility of spending resources for maintenance and repair throughout the life cycle is then relatively less. Therefore, sustainable design entails considering and evaluating multiple criteria that include environmental impact, cost, social implications, and reliability (and/or resilience) over the life cycle of the structure/system.

The main objective of the integrated framework is to optimize geotechnical designs based on reliability and sustainability considerations, which are generally at conflict with each other. The integrated framework involves conducting the following: (1) reliability analysis of the geotechnical structure (e.g., foundation, retaining structure, slope, etc.), (2) LCA of the geotechnical design, and (3) multiobjective optimization considering sustainability and reliability-related objectives (see Fig. 21.15). In this chapter, the procedure of using the integrated sustainability framework is demonstrated with the example of a pile foundation. In the problem definition, the performance functions of geo-structures of interest should be defined. The ultimate limit state and/or serviceability limit state may be used as the performance function depending

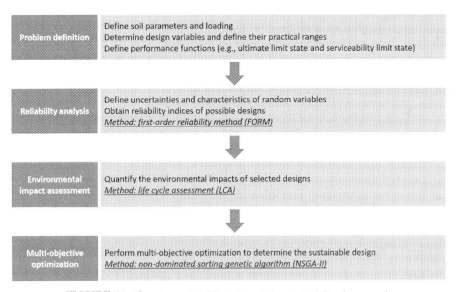

FIGURE 21.15 Elements of the integrated sustainability framework.

on the geo-structure. In this example, the performance function of the pile foundation is defined as the difference between the ultimate bearing capacity and applied axial load. The relevant soil properties, loading conditions, and design variables also have to be defined. Some or all of these parameters can be defined as random variables with appropriate probability distribution functions and characteristics (e.g., mean and standard deviation). The reliability analysis and LCA of the geo-structure are used as tools to formulate the objective functions and constraints for the multiobjective optimization problem. FORM is used to analyze the reliability of the selected geo-structure (pile foundation in this example) to obtain the functions or surfaces of reliability indices βs with respect to practical design dimensions (e.g., pile length and pile diameter). Subsequently, LCA is used to perform the environmental impact assessment of the practical designs considered in the FORM. Multiple environmental impact categories can be selected in the LCA if necessary. In this example, however, only the GWPs of the practical pile designs are quantified using the LCA. The social aspect of sustainability can be considered if quantification is available, but it is omitted in this example problem for simplicity.

The multiobjective optimization of the pile foundation problem can be represented as follows:

Minimize

$$\left\{ \begin{array}{l} \text{Sustainability Objective \#1: GWP} \\ \\ \text{Sustainability Objective \#2: Project cost} \end{array} \right\}$$

subjected to

$$B \in [\text{min}, \text{max}]$$

$$L \in [\text{min}, \text{max}]$$

$$\beta_i \geq \beta_t$$

$$(21.13)$$

where B is the pile diameter (m), L is the pile length (m), β_i is the reliability index of the ith design, and β_t is the target reliability index (2.0 in this example). The algorithm aims at minimizing the multiple sustainability objectives given the practical ranges [min, max] of pile dimensions, and consider only those designs that meet the reliability requirements (target reliability index) as candidates for optimal designs. The lower (minimum) and upper (maximum) bounds of the pile diameter B and length L, denoted by [min, max], are set to ensure that practical designs are only considered and spurious mathematical solutions are eliminated from the feasible solution space. Within the practical range of pile dimensions (i.e., diameter and length), the reliability index β (or probability of failure) of the practical designs are computed using FORM. In the FORM, the uncertainties related to the soil properties, applied loads, design equations, and pile dimensions can be considered. Table 21.4 shows an example of a set of random variables that was used in the example problem of the pile foundation for a particular soil profile. The sustainability objectives (#1 and #2) can be a life cycle environmental impact (e.g., GWP) and project cost; however, other sustainability indicators can be used if relevant data or quantification is available—see Table 21.5 for an example list of sustainability objectives. Fig. 21.16 shows the reliability indices of the corresponding pile foundation designs with different diameters and lengths within the defined minimum and maximum bounds ($B = $ [0.6 m, 1.2 m] and $L = $ [10 m, 20 m]) in Eq. (21.13).

Subsequent to the design calculations, the environmental impacts of the practical designs have to be determined in order to make a connection between the environmental impact (one of sustainability objectives) and reliability. As it is time consuming to conduct LCA for every possible foundation design for which the reliability indices are shown in Fig. 21.16, the response surface methodology is used to

TABLE 21.4 Random variables in FORM of pile foundation design in a soil profile.

	Random variable	Type of distribution	Mean	Coefficient of variation (= Mean/Standard deviation)
γ_b	Total unit weight of soil	Normal	16.4 kN/m^3	0.1
D_R	Relative density	Normal	70%	0.01
φ_c	Critical-state friction angle	Normal	32 degrees	0.02
K_0	At-rest earth pressure coefficient	Uniform	0.45	0.01
DL	Dead load	Normal	500 kN	0.1
LL	Live load	Normal	500 kN	0.18
B	Pile diameter	Normal	0.6–1.2 m	0.02
L	Pile length	Normal	10–20 m	0.02

TABLE 21.5 List of sustainability objectives.

Design objective	Indicator
Environment	Energy use (embodied energy, cumulative energy demand) Environmental impact (Table 21.3)
Economy	Construction cost (material, labor, transportation, equipment) Operation cost (service and energy cost) Maintenance cost (repair, replacement, retrofitting) End-of-life cost (recycling, reuse, demolishing, disposal)
Society	Employment Construction and public safety Improvement of infrastructure service to the public Aesthetics Land use pattern

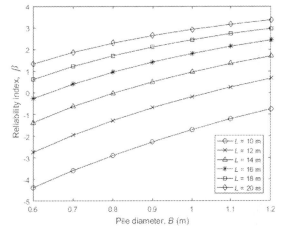

FIGURE 21.16 Reliability index of pile foundation for different pile dimensions.

formulate the environmental impacts of the pile designs as functions of pile dimensions based on a relatively small sample data. For example, using GWP as one of the sustainability objectives (Eq. 21.13), the GWPs for a small set of designs are first calculated, as shown in Table 21.6, and represented as functions of pile diameter (ranging from 0.6 to 1.2 m) and pile length (ranging from 10 to 20 m). The sample data in Table 21.6 are then fitted into a first-order or second-order model given by the following:

$$P = a_0 + a_1 z_1 + a_2 z_2 + a_{12} z_1 z_2 \quad (21.14)$$

$$P = a_0 + a_1 z_1 + a_2 z_2 + a_{11} z_1^2 + a_{22} z_2^2 + a_{12} z_1 z_2 \quad (21.15)$$

where P is the function (surface) to be formulated, a_i are the regression coefficients, and z_i are the design variables. In this study, z_1 represents the pile diameter and z_2 represents the pile length. Eq. (21.14) is a first-order model

TABLE 21.6 GWP sample data for response surface methodology.

	Pile diameter (m)	Pile length (m)	Concrete volume (m^3)	Steel volume (m^3)	GWP (kg of CO_2 equivalent)
Assuming fixed diameter	0.60	26	7.4	0.18	4216
	0.72	22.5	9.2	0.23	5233
	0.84	19.5	10.8	0.27	6157
	0.96	17	12.3	0.31	6999
	1.08	15	13.7	0.34	7676
	1.20	14	15.8	0.40	9139
Assuming fixed length	1.38	10	15.0	0.37	8460
	1.24	12	14.5	0.36	8178
	1.11	14	13.5	0.34	7716
	1.00	16	12.6	0.31	7147
	0.90	18	11.5	0.29	6527
	0.81	20	10.3	0.26	5895

which assumes that the response P (which is the GWP) has a linear relationship with the variables (i.e., design parameters pile length and diameter). Eq. (21.15) considers a second-order variation between GWP and the design variables (pile dimensions). The last terms on the right hand side of in Eqs. (21.14) and (21.15) define the interaction effect of the design parameters on the GWP. The method of least squares is used to estimate the regression coefficients in Eqs. (21.14) and (21.15). Fig. 21.17 shows an example of the GWP surface (response surface), generated using the second-order regression model (Eq. 21.15), for the pile designs considered in the example problem with respect to the defined ranges of pile diameter and length.

The algorithm of the multiobjective optimization (defined in Eq. 21.13) determines the optimized values of the sustainability objectives (e.g., minimum GWP and project cost in this example) within the defined range of the pile dimensions. Fig. 21.18 shows, as an illustration, a set of conceptual results of the optimization

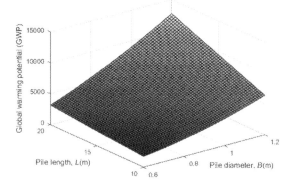

FIGURE 21.17 GWP surface of pile foundation designs formulated using the response surface methodology.

and includes the infeasible designs for explanation purpose. The infeasible designs in Fig. 21.18 have reliability indices lower than the specified target reliability index; hence, these are not considered as candidates for forming a Pareto front (i.e., the set of optimal solutions that cannot be improved without degrading other objective values) by the algorithm of the multiobjective optimization. The designs that

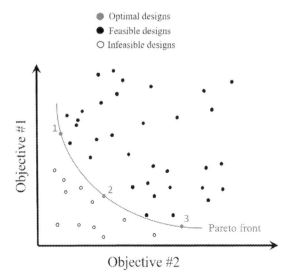

FIGURE 21.18 Optimal designs determined by Pareto front.

form the Pareto front are considered to be the most sustainable and reliable designs, and the corresponding pile dimensions can be extracted from the results of the multiobjective optimization. The choice of the final solution from the Pareto front depends on the preference of the designer based on the importance of the sustainability objectives in a particular project. For example, design 1 in Fig. 21.18 (shown with a red (gray in print version) dot), has a low cost (objective #2) but has the high GWP value (objective #1).

6. Concluding remarks

With increasing risk of resource depletion, environmental pollution, and climate change, there is a growing need for sustainable practices in the modern life. Sustainability advocates taking responsible decisions and optimal use of resources by the current generation such that future generations are not adversely impacted. Sustainability problems are complex and are often discussed, in a simplified way, with respect to three pillars (the 3 Es)—environment, economy, and equity.

Past records indicate that the construction industry is responsible for nearly 40% of the total energy use and carbon emissions in the recent years. It is important to recognize that geotechnical engineering practices are partially responsible for such consequences. Geotechnical constructions are particularly relevant to sustainable development because the construction activities can cause severe environmental impacts through extensive use of natural resources, and carbon and energy intensive materials, and because the activities may potentially influence the economy and social behaviors (e.g., the construction activities may change the land use pattern, provide public safety through hazard mitigation, and provide essential infrastructure services).

In addition to the 3 Es of sustainability, engineering design is an important dimension to be considered in evaluating the sustainability of engineered systems. Reliability and resilience considerations in engineering design are particularly important in geotechnical engineering because geo-structures are often important components of critical infrastructure systems, and failures in such systems can be devastating. The risks of geo-hazards, like earthquakes, landslides, and floods, can be managed efficiently by improving the reliability and resilience of geo-structures. Reliable and resilient systems are expected to resist external loads (e.g., environmental loads and extreme events), resulting in better public safety and reduced use of resources for unexpected repairs which also leads to reduced environmental impacts and economic losses—all of these are in compliance with the sustainability goals. A resilient system is characterized by having high robustness, redundancy, rapidity, and resourcefulness. The robustness aspect of resilience can be maintained by performing reliability-based designs in which the uncertainties related to soil properties and external loads are properly taken into account.

In this chapter, an integrated sustainability framework for geotechnical design is introduced. The framework utilizes a systems design approach and LCA to tackle sustainability problems in geotechnical engineering. The main purpose of the framework is to optimize geotechnical designs considering the reliability and sustainability objectives. The two notions—reliability and sustainability—can be conflicting to each other because a highly reliable geotechnical structure/system may be costly and use larger amount of natural resources, which are not necessarily considered sustainable. Hence, a multiobjective optimization approach is taken to determine geotechnical designs that are considered both sustainable and reliable.

References

Allen, R., 1980. How to Save the World: Strategy for World Conservation. Kogan Page, United Kingdom.

Ang, A., Tang, W., 1984. Probability Concepts in Engineering Planning and Design — Volume II: Decision, Risk, and Reliability. John Wiley and Sons, United States.

Ayyub, B.M., 2014. Systems resilience for multihazard environments: definition, metrics, and valuation for decision making. Risk Anal. 34 (2), 340—355.

Baecher, G., Christian, J., 2003. Reliability and Statistics in Geotechnical Engineering. John Wiley and Sons, West Sussex, England.

Basu, D., Puppala, A.J., Chittoori, B., 2013. Sustainability in geotechnical engineering — general report of TC 307. In: Proceedings of the 18th International Conference on Soil Mechanics and Geotechnical Engineering. Paris, France, pp. 3155—3162.

Basu, D., Misra, A., Puppala, A.J., 2015. Sustainability and geotechnical engineering perspectives and review. Can. Geotech. J. 52 (1), 96—113.

Brown, L.R., 1981. Building a Sustainable Society. The Worldwatch Institute, Norton, NY.

Brown, L.R., 1984. State of the World: A Worldwatch Institute Report on Progress toward a Sustainable Society. The Worldwatch Institute, Norton, NY.

Brundtland, G.H., 1987. Our Common Future. Oxford University Press, United Kingdom.

Bruneau, M., Chang, S., Eguchi, R., Lee, G., O'Rourke, T., Reinhorn, A., Winterfelt, D.von, 2003. A framework to quantitatively assess and enhance the seismic resilience of communities. Earthq. Spectra 19 (4), 733—752.

Buchanan, R., 1992. Wicked problems in design thinking. Des. Issues 8 (2), 5—21.

Bulleit, W.M., 2008. Uncertainty in structural engineering. Pract. Period. Struct. Des. Construct. 13 (1), 24—30.

Carpenter, A.C., Gardner, K.H., Fopiano, J., Benson, C.H., Edil, T.B., 2007. Life cycle based risk assessment of recycled materials in roadway construction. Waste Manag. 27, 1458—1464.

Carter, S., Cox, A., September 8, 2011. One 9/11 Tally: $3.3 Trillion. The New York Times. Retrieved from: https://www.nytimes.com/.

Chan, P., Tighe, S., 2010. Quantifying pavement sustainability in economic and environmental perspective. In: Proceedings of the Transportation Research Board 89th Annual Meeting, Washington, DC, pp. 1—15.

Chateauneuf, A., 2008. Principles of reliability-based design optimization. In: Tsompanakis, Y., Lagaros, N.D., Papadrakakis, M. (Eds.), Structural Design Optimization Considering Uncertainties. Taylor and Francis, London, United Kingdom, pp. 3—30.

Chau, C., Soga, K., Nicholson, D., 2006. Comparison of embodied energy of four different retaining wall systems. In: Butcher, A.P., Powell, J.J.M., Skinner, H.D. (Eds.), Proceedings of International Conference on Reuse of Foundations for Urban Sites. BRE Press, Watford, United Kingdom, pp. 277—286.

Chau, C., Soga, K., O'Riordan, N., Nicholson, D., 2012. Embodied energy evaluation for section of the UK Channel Tunnel rail link. Geotech. Eng. 165 (GE2), 65—81.

Christian, J.T., 2004. Geotechnical engineering reliability: how well do we know what we are doing? J. Geotech. Geoenviron. Eng. 130 (10), 985—1003.

Christian, J.T., Ladd, C.C., Baecher, G.B., 1994. Reliability applied to slope stability analysis. J. Geotech. Eng. 120 (12), 2180—2207.

Crawford, R.H., 2011. Life Cycle Assessment in the Built Environment. Spon Press, New York, NY.

Curran, M.A., 2017. Goal and scope definition in life cycle assessment. In: Klöpffer, W., Curran, M.A. (Eds.), LCA Compendium - the Complete World of Life Cycle Assessment. Springer. https://doi.org/10.1007/978-94-024-0855-3.

da S Trentin, A.W., Reddy, K.R., Kumar, G., Chetri, J.K., Thome, A., 2019. Quantitative assessment of life cycle sustainability (QUALICS): framework and its application to assess electrokinetic remediation. Chemosphere 230, 92—106.

Das, J.T., Puppala, A.J., Bheemasetti, T.V., Walshire, L., Corcoran, M.K., 2018. Sustainability and resilience analyses in slope stabilization. Eng. Sustain. 171 (ES1), 25—36.

Dam, T.V., Taylor, P.C., 2011. Seven principles of sustainable concrete pavements. Concr. Int. 33 (11), 49—52.

Damians, I., Bathurst, R., Adroguer, E., Josa, A., Lloret, A., 2015. Environmental assessment of earth retaining wall structures. Environ. Geotech. 4 (6), 415—431.

Damians, I., Bathurst, R., Adroguer, E., Josa, A., Lloret, A., 2016. Sustainability assessment of earth-retaining wall structures. Environ. Geotech. 5 (4), 187–203.

Egan, D., Slocombe, B.C., 2010. Demonstrating environmental benefits of ground improvement. Ground Improv. 163 (1), 63–69.

Ellingwood, B., Galambos, T., MacGregor, J.G., Cornell, C.A., 1980. Development of a Probability Based Load Criterion for American National Standard A58: Building Code Requirements for Minimum Design Loads in Buildings and Other Structures. (National Bureau of Standards Special Publication 577). U.S. Department of Commerce, Washington, DC.

Emerson, R.W., 1849. Nature. James Munroe and Company, Boston, MA.

Fenton, G., Griffiths, D.V., 2003. Bearing-capacity prediction of spatially random $c - \varphi$ soils. Can. Geotech. J. 40, 54–65.

Gibson, R.B., 2006. Sustainability assessment: basic components of a practical approach. Impact Assess. Proj. Apprais. 24 (3), 170–182.

Goedkoop, M., Heijungs, R., Huijbregts, M., Schryver, A., Struijs, J., van Zelm, R., 2013. ReCiPe 2008. Report I: Characterisation. Den Haag, Netherlands. Retrieved from: https://www.leidenuniv.nl/cml/ssp/publications/recipe_characterisation.pdf. (Accessed 4 May 2020).

Goedkoop, M., Heijungs, R., Huijbregts, M., Schryver, A., Struijs, J., van Zelm, R., 2014. Characterisation and Normalisation Factors. Den Haag, Netherlands. Retrieved from: https://www.rivm.nl/documenten/6recipe111. (Accessed 4 May 2020).

Graedel, T., 1994. Industrial ecology: definition and implementation. In: Socolow, R., Andrews, C., Berkhout, F., Thomas, V. (Eds.), Industrial Ecology and Global Change. Cambridge University Press, Cambridge, United Kingdom, pp. 23–41.

Guest, D., 1991. The Hunt Is on for the Renaissance Man of Computing. The Independent, London, United Kingdom.

Hammond, G., Jones, C., 2008. Inventory of Carbon and Energy (ICE) Version 1.6a. Bath. University of Bath, United Kingdom.

Heerten, G., 2012. Reduction of climate-damaging gases in geotechnical engineering practice using geosynthetics. Geotext. Geomembranes 30, 43–49.

Holt, D.G.A., 2011. Sustainable Assessment for Geotechnical Projects (Doctoral Dissertation). University of Birmingham, Birmingham, United Kingdom.

Holt, D.G.A., Jefferson, I., Braithwaite, P.A., Chapman, D.N., 2010a. Embedding sustainability into geotechnics. Part A: Methodology. Eng. Sustain. 163 (ES3), 127–135.

Holt, D.G.A., Jefferson, I., Braithwaite, P.A., Chapman, D.N., 2010b. Sustainable geotechnical design. In: Proceedings of GeoFlorida 2010: Advances in Analysis, Modeling & Design, Palm Beach, FL, pp. 2925–2932.

Hopwood, B., Mellor, M., O'Brien, G., 2005. Sustainable development: mapping different approaches. Sustain. Dev. 13 (1), 38–52.

Huijbregts, M., Steinmann, Z., Elshout, P., Stam, G., Verones, F., Vieira, M., van Zelm, R., 2017. ReCiPe2016: a harmonized life cycle impact assessment method at midpoint and endpoint level. Int. J. Life Cycle Assess. 22 (2), 138–147.

Institute for Sustainable Infrastructure, 2018. Envision Sustainability Rating System. Retrieved from: https://sustainableinfrastructure.org/wp-content/uploads/EnvisionV3.9.7.2018.pdf. (Accessed 20 March 2021).

Intergovernmental Panel on Climate Change, 1990. Climate Change: The IPCC Scientific Assessment. Press Syndicate of the University of Cambridge, Cambridge, United Kingdom.

International Energy Agency, 2018. Global Status Report 2018: Towards a Zero-Emission, Efficient, and Resilient Buildings and Construction Sector. United Nations Environment Programme. Retrieved from: https://www.worldgbc.org/sites/default/files/2018%20GlobalABC%20Global%20Status%20Report.pdf. (Accessed 16 November 2020).

International Organization for Standardization (ISO), 2006a. ISO 14040: Environmental Management − Life Cycle Assessment − Principles and Framework. ISO, Geneva, Switzerland.

International Organization for Standardization (ISO), 2006b. ISO 14044: Environmental Management - Life Cycle Assessment - Requirements and Guidelines. Geneva, Switzerland.

Inui, T., Chau, C., Soga, K., Nicholson, D., O'Riordan, N., 2011. Embodied energy and gas emissions of retaining wall structures. J. Geotech. Geoenviron. Eng. 137 (10), 958–967.

Jefferson, I., Hunt, D.V.L., Birchall, C.A., Rogers, C.D.F., 2007. Sustainability indicators for environmental geotechnics. Eng. Sustain. 160 (2), 57–78.

Jimenez, M., 2004. Assessment of Geotechnical Process on the Basis of Sustainability Principles. M.Sc. Dissertation. University of Birmingham, Birmingham, United Kingdom.

Juang, C.H., Rosowsky, D.V., Tang, W.H., 1999. Reliability-based method for assessing liquefaction potential of soils. J. Geotech. Geoenviron. Eng. 125 (8), 684–689.

Kates, R.W., Clar, W.C., Corell, R., Hall, J.M., Jaeger, C.C., Lowe, I., Svedin, U., 2001. Sustainability science. Science 292 (5517), 641–642.

Kibert, C.J., 2008. Sustainable Construction, second ed. John Wiley and Sons Inc, Hoboken, NJ.

Kibert, C.J., Monroe, M.C., Peterson, A.L., Plate, R.R., Thiele, L.P., 2011. Working toward Sustainability: Ethical Decision-Making in a Technological World. John Wiley and Sons Inc, Hoboken, NJ.

Kim, Y., Chester, M.V., Eisenberg, D.A., Redman, C.L., 2019. The infrastructure trolley problem: positioning safe-to-fail infrastructure for climate change adaptation. Earth's Future 7 (7), 704—717.

Klöpffer, W., Grahl, B., 2014. Life Cycle Assessment (LCA): A Guide to Best Practice. John Wiley and Sons Inc, Weinheim, Germany.

Knuth, D., Fortmann, J., 2010. The development of I-LAST™ Illinois - Livable and sustainable transportation. In: Proceedings of the Green Streets and Highways 2010 Conference, Denver, CO, pp. 495—503.

Lacasse, S., Nadim, F., 1996. Uncertainties in Characterising Soil Properties. Uncertainty In the Geologic Environment: From Theory to Practice, pp. 49—75.

Lacasse, S., Nadim, F., 1998. Risk and reliability in geotechnical engineering. In: Proceedings of International Conference on Case Histories in Geotechnical Engineering, St. Louis, MO, pp. 1172—1192.

Laefer, D.F., 2011. Quantitative support for a qualitative foundation reuse assessment tool. In: Proceedings of the GeoFrontiers 2011, Dallas, TX, pp. 113—121.

Leaning, J., Guha-Sapir, D., 2014. Natural disasters, armed conflict, and public health. N. Engl. J. Med. 369 (19), 1836—1842.

Lee, M., Basu, D., 2018. An integrated approach for resilience and sustainability in geotechnical engineering. Indian Geotech. J. 48 (2), 207—234.

Lee, J., Edil, T.B., Benson, C.H., Tinjum, J.M., 2010a. Use of BE^{2}ST In-Highways for green highway construction rating in Wisconsin. In: Proceedings of the Green Streets and Highways 2010 Conference, Denver, CO, pp. 480—494.

Lee, J., Edil, T.B., Tinjum, J.M., Benson, C.H., 2010b. Quantitative assessment of environmental and economic benefits of using recycled construction materials in highway construction. Transport. Res. Rec.: J. Transp. Res. Board 2158 (1), 138—142.

Long, J.C.S., Amadei, B., Bardet, J.-P., Christian, J.T., Glaser, S.D., Goodings, D.J., Santamarina, J.C., 2006. Geological and Geotechnical Engineering in the New Millennium: Opportunities for Research and Technological Innovation. The National Academic Press, Washington, DC. Retrieved from: http://www.nap.edu/catalog/11558.html. (Accessed 30 October 2013).

Low, B.K., 2005. Reliability-based design applied to retaining walls. Geotechnique 55 (1), 63—75.

Low, B.K., Tang, W.H., 1997. Reliability analysis of reinforced embankments on soft ground. Can. Geotech. J. 34 (5), 672—685.

Lumb, P., 1974. Application of statistics in soil mechanics. In: Lee, I.K. (Ed.), Soil Mechanics: New Horizons. Butterworths and Company Publishers Limited, London, United Kingdom, pp. 44—112.

McVoy, G.R., Nelson, D.A., Krekeler, P., Kolb, E., Gritsavage, J.S., 2010. Moving towards sustainability: New York state department of transportation's GreenLITES story. In: Proceedings of the Green Streets and Highways 2010 Conference, Denver, CO, pp. 461—479.

Mihelcic, J.R., Crittenden, J.C., Small, M.J., Shonnard, D.R., Hokanson, D.R., Zhang, Q., Schnoor, J.L., 2003. Sustainability science and engineering: the emergence of a new metadiscipline. Environ. Sci. Technol. 37 (23), 5314—5324.

Min, H.S.J., Beyeler, W., Brown, T., Son, Y.J., Jones, A.T., 2007. Toward modeling and simulation of critical national infrastructure interdependencies. IIE Trans. 39 (1), 57—71.

Misra, A., 2010. A Multicriteria Based Quantitative Framework for Assessing the Sustainability of Pile Foundations. University of Connecticut, Storrs, CT (M.Sc. Dissertation).

Misra, A., Basu, D., 2011. Sustainability metrics for pile foundations. Indian Geotech. J. 41 (2), 108—120.

Misra, A., Basu, D., 2012. A quantitative sustainability indicator system for pile foundations. In: Proceedings of the Geocongress 2012, Oakland, CA, pp. 4252—4261.

Muench, T.S., Anderson, J.L., 2009. Greenroads: A Sustainability Performance Metric for Roadway Design and Construction. University of Washington, Seattle, WA. https://www.wsdot.wa.gov/research/reports/fullreports/725.1.pdf.

National Aeronautics and Space Administration, 2017. Climate Change: How Do We Know? [Online]. Retrieved from: https://climate.nasa.gov/evidence/. (Accessed 4 May 2020).

National Renewable Energy Laboratory, 2012. U.S. Life Cycle Inventory Database. Retrieved from: https://www.nrel.gov/lci/. (Accessed 4 May 2020).

Pantelidou, H., Nicholson, D., Gaba, A., 2012. Sustainable geotechnics. In: John, B., Tim, C., Hilary, S., Michael, B. (Eds.), ICE Manual of Geotechnical Engineering, vol. 1. Institute of Civil Engineers, London, United Kingdom.

Park, J., Seager, T.P., Rao, P.S., 2011. Lessons in Risk-versus resilience-based design and management. Integr. Environ. Assess. Manag. 7 (3), 396—399.

Phoon, K., Kulhawy, F.H., 1999. Characterization of geotechnical variability. Can. Geotech. J. 36 (4), 612—624.

Pittenger, D.M., 2011. Evaluating sustainability of selected airport pavement treatments with life- cycle cost, raw material consumption, and greenroads standards. Transport. Res. Rec.: J. Transp. Res. Board 2206, 61—68.

Popescu, R., Prevost, J.H., Deodatis, G., 1998. Spatial variability of soil properties: two case studies. Proc. Geotechn. Earthquake Eng. Soil Dyn. Geotechn. Spec. Publ. 181 (75), 568—579.

Praticò, F., Saride, S., Puppala, A., 2011. Comprehensive life cycle cost analysis for the selection of stabilization alternative for better performance of low volume roads. Transport. Res. Rec.: J. Transp. Res. Board 2204 (1), 120—129.

Raza, F., Alshameri, B., Jamil, M.S., 2020. Engineering aspect of sustainability assessment for geotechnical projects. Environ. Dev. Sustain. 23 (3), 6359–6394. https://doi.org/10.1007/s10668-020-00876-x.

Raza, F., Alshameri, B., Jamil, M.S., 2021. Assessment of triple bottom line of sustainability for geotechnical projects. Environ. Dev. Sustain. 23, 4521–4558.

Reigle, J.A., Zaniewski, J.P., 2002. Risk-based life-cycle cost analysis for project-level pavement management. Transport. Res. Rec.: J. Transp. Res. Board 1816, 34–42.

Remmen, A., Jensen, A.A., Frydendal, J., 2007. Life Cycle Management: A Business Guide to Sustainability. United Nations Environment Programme.

Rinaldi, B., Peerenboom, J., Kelly, T., 2001. Identifying, understanding, and analyzing critical infrastructure interdependencies. IEEE Control Syst. Mag. 21 (6), 11–25.

Rogers, C.D.F., Bouch, C.J., Williams, S., Barber, A.R.G., Baker, C.J., Bryson, J.R., Quinn, A.D., 2012. Resistance and resilience – paradigms for critical local infrastructure. Municip. Eng. 165 (ME2), 73–84.

Ruskin, J., 1881. Unto This Last. Harvard University, Cambridge, MA.

Seager, T., Selinger, E., Wiek, A., 2012. Sustainable engineering science for resolving wicked problems. J. Agric. Environ. Ethics 25, 467–484.

Shillaber, C.M., Mitchell, J.K., Dove, J.E., 2016a. Energy and carbon assessment of ground improvement works. I: definitions and background. J. Geotech. Geoenviron. Eng. 142 (3), 04015083.

Shillaber, C., Mitchell, J.K., Dove, J.E., 2016b. Energy and carbon assessment of ground improvement works. II: working model and Example. J. Geotech. Geoenviron. Eng. 142 (3), 04015084.

Simoncini, L., 2011. Socio-technical complex systems of systems: can we justifiably trust their resilience? In: Jones, C.B., Lloyd, J.L. (Eds.), Dependable and Historic Computing. Springer, Berlin, Germany, pp. 486–497.

Slaper, T.F., Hall, T.J., 2011. The triple bottom line – what is it and how does it work. In: Indiana Business Review. Bloomington. Indiana University.

Song, X., Carlsson, C., Killsgaard, R., Bendz, D., Kennedy, H., 2020. Life cycle assessment of geotechnical works in building construction: a review and recommendations. Sustainability 12 (20), 8442.

Spaulding, C., Massey, F., LaBrozzi, J., 2008. Ground improvement technologies for a sustainable world. In: Proceedings of the GeoCongress 2008, New Orleans, LA, pp. 891–898.

Steedman, R.S., 2011. Geotechnics and society: carbon, a new focus for delivering sustainable geotechnical engineering. In: Iai, S. (Ed.), Geotechnics and Earthquake Geotechnics towards Global Sustainability. Springer, Netherlands, pp. 75–88.

Storesund, R., Messe, J., Kim, Y., 2008. Life cycle impacts for concrete retaining walls vs. bioengineered slopes. In: Proceedings of the GeoCongress 2008, New Orleans, LA, pp. 875–882.

Thoreau, H.D., 1860. The Succession of Forest Trees. New York weekly tribune, New York, NY.

Torres, V.N., da Gama, C.D., 2006. Quantifying the environmental sustainability in underground mining. In: Proceedings of the 15th International Symposium on Mine Planning and Equipment Selection, Torino, Italy, pp. 1–7.

United Nations, 2015. Transforming Our World: The 2030 Agenda for Sustainable Development. Retrieved from: https://sustainabledevelopment.un.org/content/documents/21252030%20Agenda%20for%20Sustainable%20Development%20web.pdf. (Accessed 16 November 2020).

Vanmarcke, E.H., 1980. Probabilistic stability analysis of earth slopes. Eng. Geol. 16 (1–2), 29–50.

Vugrin, E.D., Warren, D.E., Ehlen, M.A., 2011. A Resilience assessment framework for infrastructure and economic systems: quantitative and qualitative resilience analysis of petrochemical supply chains to a hurricane. Process Saf. Prog. 30 (3), 280–290.

Whitman, R.V., 2000. Organizing and evaluating uncertainty in geotechnical engineering. J. Geotech. Geoenviron. Eng. 126 (7), 583–593.

World Wildlife Fund, 2010. Living Planet Report – 2010: Biodiversity, Biocapacity and Development [PDF File]. Retrieved from: https://wwfeu.awsassets.panda.org/downloads/lpr_living_planet_report_2010.pdf. (Accessed 16 November 2020).

World Wildlife Fund, 2016. Living Planet Report – 2016: Risk and Resilience in a New Era [PDF file]. Retrieved from: https://wwfint.awsassets.panda.org/downloads/lpr_2016_full_ report_low_res.pdf. (Accessed 16 November 2020).

World Wildlife Fund, 2018. Living Planet Report – 2018: Aiming Higher [PDF file]. Retrieved from: https://c402277.ssl.cf1.rackcdn.com/publications/1187/files/original/LPR2018_ Full_Report_Spreads.pdf. (Accessed 16 November 2020).

Zhang, H., Keoleian, G.A., Lepech, M.D., 2008. An integrated life cycle assessment and life cycle analysis model for pavement overlay systems. In: Biondini, F., Frangopol, D. (Eds.), Life-Cycle Civil Engineering. Taylor and Francis Group, London, United Kingdom, pp. 907–912.

CHAPTER

22

Safety risks in underground operations: management and assessment techniques

Parthiban Kathirvel

School of Civil Engineering, SASTRA Deemed University, Thanjavur, Tamil Nadu, India

1. Introduction

Due to a rapid growth in the population in the urban areas, the expansion and the effective use of underground space for underground transportation network will be the major outcome leading toward urbanization and rejuvenation in city-intensive regions (Qian, 2016). The urbanization growth is predicted to increase by 66% in 2050 than 54% in 2014 globally (UN, 2015). On account of this tendency, the requirement for transport operations is anticipated to grow continuously worldwide, whereas issues related to urban development, for instance, deficiencies in space for urban construction, severe traffic jam, and the resulting environmental pollution, have confined the urban growth rigorously due to the urban population growth. The expansion and transformation of urban space from aboveground to underground turns out to be an unavoidable tendency for the sustainable improvement of upcoming cities (Broere, 2016). Over the past decade, construction of tunnels has presented a commanding thrust toward speedy monetary growth. Recent investigations reveal that the tunnel transportation is as

safe as the road transport or better than that (Beard and Cope, 2008; Amundsen and Engebretsen, 2009; Naevestad and Meyer, 2014; Kirytopoulos et al., 2017). It is treated to be safer than road transport as it is a closed environment which leads the drivers to drive their vehicles carefully (Kirytopoulos et al., 2017), as well as the restrictions in the availability of junctions, pedestrians, or any advertising boards in the tunnels (Amundsen and Engebretsen, 2009) as these factors are the main causes for accidents or these may tend to aggravate the evolution of accidents. Following the regulatory guidelines strictly by the contemporary tunnels may also result in safer transport operations (EU, 2004; NFPA, 2014). Conversely, the underground construction operation has resulted in serious issues on account of diverse risk factors connected with intricate project atmospheres and breach of safety rules (Liu et al., 2005; Qian and Rong, 2008; Qian, 2014; Wu et al., 2015). The transportation division is considered to be the most important resource-intense division worldwide which consumes around 63% of entire oil utilization (World Energy Council, 2016). Hence, the stakeholders should concentrate heavily on developing proper

Risk, Reliability and Sustainable Remediation in the Field of Civil and Environmental Engineering
https://doi.org/10.1016/B978-0-323-85698-0.00022-8

transportation systems, with advanced transportation expertise which result in effective pollution-free (EC, 2011). Recent studies have shown that the road transportation system holds the premier spot among all urban transport networks even with the recent progress in the underground rail system which is evident from EDT, 2012 which highlighted that around 46% of the passengers and goods are transported by means of road networks in urban areas. Even though the rate of accidents in tunnels is confirmed to be lesser than the open road transportation, but once an accident occurs, the rigorousness will be quite more than the other road networks. In this aspect, an investigation has been carried out in Italy which portrays that severe accidents are more recurrent in tunnel road network which lies between 9.13 and 20.45 crashes/108 veh.km, whereas this lies in the range of 8.62 and 10.14 crashes/108 veh.km for the associated motorways (Caliendo and De Guglielmo, 2012). One more investigation was carried out in Norway, the nation where 1000 tunnels were built over a stretch of 800 km results in an average tunnel fire accidents, was recorded to be 21.25 per year per 1000 tunnels between the year 2008 and 2011 (Naevestad and Meyer, 2014). Human beings are not getting affected with all these accidents, yet the statistics reveals that there is a severe threat for the tunnel operation personnel. Hence, it requires more attention as it is one of the most critical networks for the day-to-day function in urban areas. Due to the immense technical complexity and huge investment in the resource, the utilization of underground space in the existing scenario typically has been restricted with shallow depths rather than deep underground space. Nevertheless, in some of the cities in China, for instance, Chongqing, the utilization of deep underground space is certainly unavoidable due to the constraint that the lands were mountainous and exceptionally centralized urban development and found no issues related to technical and structural integrity during the process of deep underground space utilization

(Xie et al., 2020). Various underground metro construction projects were undertaken worldwide as a result of speedy growth in the urbanization of the nation. The growth may result in rapid progress in the associated industrial sectors with the increase in employment opportunities. On the other hand, accidents in underground metro construction are also recorded frequently due to volatile environmental and hydrological conditions, deprived supervision of insecure actions, and hazardous circumstance of construction equipment and working atmosphere (Ding and Zhou, 2013). Many underground construction accidents were recorded which include collapse of Xi'an metro line 3 tunnel shaft on course of excavating the tunnel which results in five fatalities. The failure could be owing to the ignorance in breaking into a hazardous area in the tunnel during excavation in addition to the deprived geological circumstances and too-much quarrying. Overall, there are five varieties of collapses that may be encountered in the underground tunnel construction such as failure of daylight, failure of subsurface, explosion of rocks, penetration of water, and failure of portal (Guglielmetti et al., 2008). During the course of daylight failure, the ground is ragged to the surface and the disseminating collapse of the surface may be exceedingly fast. But, the principal source is deprived safety management (Cao et al., 2012; Mahdevari et al., 2014; Wu et al., 2014). Hence, there is a requirement to increase the workers awareness toward impending risk in the construction site.

2. Potential risks in underground operation

Owing to the rapid increase in the population and shifting toward urbanization, the utilization of underground space has widened with the features of huge, deep, rapid, and dense with most likely the construction of high-rise buildings and tunneling more often involving deep excavation. The excavation process involves certain risks as a result of its reduced duration for construction, complication in the soil layer (Juang et al., 2018), and adjoining buildings (Houhou et al., 2019a,b) in addition to the variety of process and units involved in the construction. The underlying risks associated with the underground operations can be assessed with the aid of engineering characteristics which involve excavation collapse, settlement of ground and pipeline (Ni and Mangalathu, 2018), and the potential hazards to the adjoining structures (Zhang et al., 2018). Various types of accidents associated with the excavation and their distribution (Zhou et al., 2009) are shown in Fig. 22.1, wherein the event of seepage failure (Zhou et al., 2009; Xu et al., 2019; Wang et al., 2019) is found to be predominant with a maximum distribution rate of 62% followed by anchorage instability and pit landslide sharing 13% each with further distribution by means of mechanical injury (8%) and kick damage and inrush damage which accounts for 2% each. It is noted that the occurrence of seepage failure is mainly owing to the inferior characteristics of the waterproof outcome of the walls enclosing the excavation under deep condition (Wu et al., 2019, 2020).

The risks associated with the deep excavation are mainly related with the factors which may trigger the recurrent risks of deep excavation that are classified into three phases namely the design, construction, and hydro-geological factors. These factors reveal the critical need for establishing the system for forecasting the risks and to prevent risks associated with excavation as well as alleviate the damage as a result of deep excavation failure during the course of construction. Selection of concrete strength, thickness and insertion ratio in the design of wall, depth and width of excavation scheme, and number and form of support design are associated with the design factors. Construction factors emphasize the speed of excavation, steel and concrete support reliability, timeliness, and specification of support erection. Confined water distribution, earth pressure, groundwater level, and shear strength of soil are associated with hydro-geological factors.

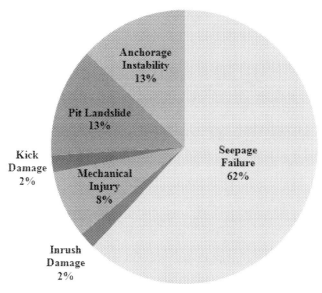

FIGURE 22.1 Types of accidents and their distribution.

3. Identifying the risks in underground construction

International Tunnel Association has released the first edition of *"Guidelines for tunneling risk management,"* during the year 2004 which signifies the serious consideration of risk identification in underground construction (Eskesen et al., 2004). In order to develop the management of tunnel construction, European Parliament has summoned *"Assessment of the Safety of Tunnels"* in the year 2008 (Beard and Cope, 2008), which has led to the implementation of proper guidance in identifying the risk and how to accomplish successful risk management for the engineers during the construction of tunnels. The vulnerability index was computed with the aid of conventional geotechnical parameters using interaction matrix (Fattahi and Moradi, 2017). Despite the fact that the vulnerability index categorization could perceive risky zones in underground tunnel constructions (Benekos, 2017), it merely paid attention on geotechnical dangers. The factors that influence the risk signify the estimation of size and occurrence of risk susceptibility events was developed by Park and Kim (2013) with the drawback that this model can be applicable only for open-cut form of underground construction and the efficiency of other types was not verified. The development of Risk Management Software (TRM1.0) for shield tunneling operations with the aid of risk records was implemented by Huang et al. (2006) for evaluating the factors that influence the risks at various stages of underground tunnel construction. Digitalized Tunnel Face Mapping System (DiTFAMS) was introduced by Sagong et al. (2006) derived from the technology of PDA and WLAN which may possibly transfer the geological data rapidly, for instance, pictures of the tunnel face to the decision makers. Similarly, an IT-based system of managing the tunneling risk (TURISK) was developed by Yoo et al. (2006) to evaluate the risks on the contiguous environment resulting from the impact of tunnel construction using ArcGIS software and to suggest suitable assistance in the design of tunnel (Perlman and Barak, 2014). Zhang et al. (2014) has listed some common methods for recognizing the risks, which were then advanced with the hybrid approach (Zhang et al., 2017) that combines Monte Carlo simulation technique, fuzzy matter element technique, and Dempster—Shafer evidence theory to identify the size of the risk for the tunnel-induced building damage at an early construction phase.

The surrounding environment in terms of buildings and water bodies has major influence as a result of underground operations. The effective way of minimizing the impending harmful impacts on surrounding environments owing to excavation has been the study on research in recent years (Houhou et al., 2019a,b). It is obvious that a particular category of failure on the adjacent structures possibly results in settlement (Azadi et al., 2013; Hashash et al., 2010). For instance, Zhang et al. (2019) noticed that the failure risk of surrounding structures can be well assessed with the prediction of the settlement arising in the piles of the structures (Zhang et al., 2019; Li et al., 2019). In fact, a plenty of techniques consist of the analysis by quantitative and qualitative methods that have been established to comprehend the performance of adjacent structures as a result of settlement risks (Aye et al., 2006; Ou and Hsieh, 2011). The provision of underground road tunnels meets up definite environmental constraints with the achievement of connecting and accessing inaccessible urban places, avoiding the complexity occurs due to the available inner city built environment and hence alleviating noise and environmental pollution. In spite of the advantages mentioned earlier, the development of underground road tunnel covers momentous limitation results in the occurrence of the accident's severity. There might be a momentous unfavorable consequence as far as loss of human life and the devastation of structures owing to the accidents in underground tunnels under high traffic volume (Beard and Carvel, 2012). Taking in to consideration of the past accidents, the aforementioned impacts tend to augment if the accident on account of fire (AADT, 1999; Ntzeremes et al., 2016). The evaluation of risk is treated to be a precious instrument in order to address the accidents due to fire (PIARC, 2008; PIARC, 2013). However, the ambiguity with respect to the significant factors arises during the process of system and is measured to be noteworthy challenges in the assessment of risks (PIARC, 2016). Regardless of that, the present methodologies continue to treat the tunnel system factors deterministically such as, fire characteristics, volume of daily traffic, and activity of the workers trapped in the tunnel by overlooking their surrounding uncertainty. But this deterministic approach tends to affect the process involved in assessing the risk and hence the precision of the results is questionable.

During the recent past, various researchers were working on the safety assessment of the underground cavern construction and achieved the utilization of real-time screening data to estimate the contiguous rock's stability and support during their construction (Oreste, 2005; Perras et al., 2005; Wang et al., 2011; Garg and Jaiswal, 2016). The construction of underground caverns seems to be complex in natural geological medium, and hence, it may affect the surrounding rock's stability by means of various parameters such as geological configuration, characteristics of rock mechanics, stress developed in the soil medium, and hydrological circumstances which are usually indefinite during the process of design and execution (Han et al., 2009; Renani et al., 2016). Consequently, the analysis results of the cavern construction are mostly deviated from the results obtained in the site which may result in various safety risks (Chen et al., 2001; Ni et al., 2013). Hence, the safety and accomplishment of the cavern construction mainly depend

on the capacity to envisage precisely the contiguous rock's safety in consistent with the real-time screening data, which makes the theme widely studied in engineering and academic spheres (Zhu et al., 2010). The distribution of stress in soil gets altered during the excavation operation in constructing underground cavern, which may lead to failure in the contiguous rock mass (Chan et al., 2011). The most widely recognized modes of failure in the contiguous rock are volatile deformation and rock subsiding (Luo et al., 2008) in which the volatile deformation is mainly influenced by the additional and divergence deformation in the contiguous rock which is in general a typical trend with steady modification (Jia et al., 2011). The subsidence of the rock mass is primarily attributed to the reduction in the mass of the contiguous rock, which in general described by the transformation that take place at the top or the side wall of the cavern (Yi et al., 2011).

4. Development and progression of safety management system

The safety management system has been instituted and enhanced in the progression of enduring safety production practices in order to offer the support for guiding enterprises safety management. On the whole, the system of safety management has gone on in three stages namely the stage of inferior, middle, and advanced. It has started with the inferior stage, where the management of accidents was modeled based on the analysis of accident causation theory during the initial phase of human business growth. Hence, this model has been treated as accident management model which utilize the accidents as management entity based on the experience postaccidents. Secondly, the hidden danger management model which is considered to be the middle stage was developed based on the assumption of danger analysis, which employs hidden

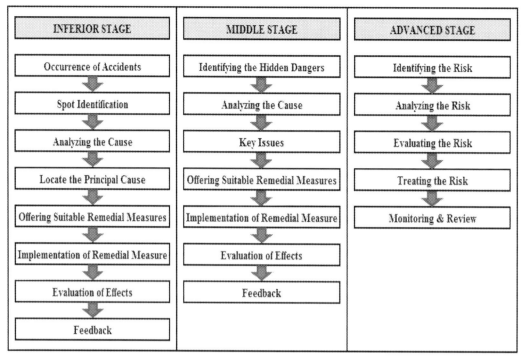

FIGURE 22.2 Systematic procedures in the system of safety management.

dangers as management entity which is characterized on the course of accidents based on institutions, i.e., standardization management. The final stage is the risk management model which is termed to be the advanced stage in the safety management system which was developed based on the assumption of risk control which employs risk as the management entity and this model has been characterized by the systematic management before the occurrence of accidents. The systematic procedures followed in all the stages are shown in Fig. 22.2.

The organization implementing the safety management has come up with two major progresses under the assistance of various stages of theory based on safety management. The first out of which is the progression from experience to institution management and the second one is from institution to risk management. It was recorded that in China, most of the enterprises are in the stage of progressing to institution management from experience management with few higher organizations that are in the level of progressing to risk management from institution management.

5. Approaches in assessing the safety risks

Initially, the assessment of risks in engineering was done by the domain experts and the resulting evaluation data are highly subjective owing to the inadequate information and the manners of the experts as individual on risks. In recent years, owing to the rapid development in the information technology and the available sophisticated equipment which lead to the development of various intelligent algorithms that helps in obtaining enormous quantity of monitored data. These algorithms are valuable and capable of dealing with the information measured in highly nonlinear correlations and the estimated results are highly intended and consistent, the principal benefit of employing machine learning methods in the assessment of risks.

Variety of investigations established risk-based assessment so as to evade numerous losses in properties and human life that has basically classified into two techniques: qualitative and quantitative risk assessments (Smith et al., 2009). Qualitative techniques essentially paid attention on utilizing approaches namely fault tree analysis (FTA), fuzzy set theory, safety check list or sensitivity analysis, while the latter utilizes neural networks (NNs), decision trees, influence diagrams, job risk analysis, and others in order to investigate the mechanism of settlement threat for safety management in complex engineering projects significantly (Li et al., 2018; Alfredo, 2002; Piniella et al., 2009). Nevertheless, these techniques have the limitations of applying it only to static control and management (Alaeddini and Dogan, 2011). For instance, Nezarat et al. (2015) recognized the potential risk factors resulting in accidents due to settlement by the application of FTA and subsequently positioned them by chances. The ground surface settlement of adjacent buildings was predicted by Leu and Lo (2004) with the aid of ANN-based regression model. Khakzad et al. (2011) illustrates FTA inappropriate for intricate problems as a result of its inadequacy in clearly signifying the dependency of events, revise prospects, and manages with ambiguity. When the related constraints, for instance, geological, design, and construction constraints are varied, the abovementioned techniques cannot precisely describe the restructured characteristic of dynamic environments while the growth of construction persists. Since the constraints are not up-to-date, the provision of real-time implication nor except support may not be feasible. However, the application of qualitative techniques exclusively offers theoretical outcomes exclusive of utilizing monitoring records, which will results in restricted practical applications in

terms of controlling the settlement (Chen et al., 2017). By distinction, quantitative techniques are rather utilized in the analysis of settlement of adjacent structures (Fang et al., 2014; Vahdatirad et al., 2010; Zhou et al., 2018). So far, the undesirable effect on surrounding tunnels induced by the excavation was evaluated with the aid of 2D/3D finite element analysis (Li et al., 2019).

Regardless of their prosperity, the abovementioned techniques actually have limitations, as they principally rely upon relationship between observed/expected results and essentials of national standards (Fang et al., 2011; Ding et al., 2013). To begin with, the estimation of the existing threshold has to be estimated with the knowledge of various experts which is relatively subjective (Chheng and Likitlersuang, 2018). Furthermore, the assessment of risk has been estimated with the consideration of excavation that will in general be displayed as the entire framework actually, as well as the interface and the significance of every monitoring point (Ou and Hsieh, 2011). For example, a few insignificant points which are treated as high risk while the entire excavation proved to be safe (Zhou et al., 2018). Obviously, the acceptable limit of settlement varies with varying point of consideration attributable to varying hydrogeological nature as well as their role toward excavation system (Yu et al., 2014). Therefore, the provision of risk assessment is absolutely inadequate by evaluating using unified standards. Finally, the existing threshold-based technique overlooked the systematic impacts and dynamic characteristics of unsafe components. Also, it neglects to accomplish the overall risk evaluation of building location which includes excavation and adjoining site and reveal the advancement standards. Therefore, the aforementioned issues need to be addressed immediately by evaluating the risk related to settlement for the adjoining environments using a consistent system of systematic risk assessment wherein all the observing points are considered as a whole. On the whole, in order to investigate the relationship between various observing points and the track dynamic attributes of settlement characteristics, a variety of settlement time sequences were employed for back analysis. But, these observing information are high dimensional, nonlinear, and nonisometric, so as to establish a substantial challenge for system dynamic analysis of settlement characteristics.

5.1 Building information modeling

Over the recent past, "Building Information Modeling" (BIM) has turned out to be the prime platform for the construction management in the construction sector owing to its fitting multidimensional prophecy, interactivity, and better sharing role. The application of BIM-Cloud has enabled the sharing and storing of engineering information which has led to the advancement of slowly replacing the conventional 2-D CAD. In addition, the use of BIM also enables to pull-out the internal and external engineering information with the aid of programming in nature as a result of interface provision function. For instance, the application of BIM models not only offers the data related to internal static structural elements, but it has the capacity to afford data related to external environment. Ultimately, a system, which could mine and distribute engineering data from BIM models on its own, is capable to be recognized (Volk et al., 2014). The recognition of various safety risks in construction and the mode of reducing the causalities are much significant at the project preparation phase. Combining existing requirement for recognizing the safety risk effectively as well as the development of future trend in BIM leads to a basic technical problem of developing a technique for recognizing the risk based on BIM, which could accomplish appropriate and precise identification of safety hazards in underground construction at the preconstruction stage.

BIM technology is considered to be the second revolution in the construction sector due to its 3D models and could be an unavoidable option in the near future (Eastman et al., 2011). Statistics also reveals that almost 50% of the construction industries in North America are employing BIM or associated tools, whereas in Germany, Norway, and Finland nearly 70% of their construction projects are dealt with BIM (Travaglini et al., 2014). In China, Hong Kong leads the path with 20 out of 28 underground metro stations modeled using BIM technology (Ma, 2017). Over 500 collision points were observed with the aid of collision detection technique using the platform of BIM by means of integrating pipeline and infrastructure carried out by China Railway Design Institute (Zhang et al., 2016). In similar way, the quality of the Shanghai underground constructions was greatly improved with the application of BIM by the installation of large-scale apparatus (Qian and Lin, 2016). On the other hand, it was observed that the application of BIM technology was primarily employed in large-scale structures, and was firmly limited to a particular phase. The process of applying BIM technology in underground construction is insufficient apart from building simulation and the analysis of pipeline collision (Min and Miya, 2016). Apparently, the role of BIM was not completely applied in the underground construction projects. Concurrently, the cases of underground construction projects derived from BIM technology are constantly rising. Li et al. (2018) instituted a system that could accomplish a timely and precise identification of risks in the underground construction at the preconstruction stage using BIM and categorized the overall risk detection process in three parts namely risk knowledge database, relation analysis of engineering information, and risk recognition mechanism as shown in Fig. 22.3.

By applying the merits of BIM expertise, it is viable to perform valuable safety risk assessment for underground operations. The precise and

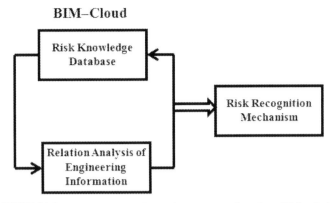

FIGURE 22.3 Structure for safety risk recognition based on BIM technique.

implied information regarding safety risk was acquired and categorized so as to create a complete safety risk knowledge database. The categories were arranged in the form of risks related to technical, geological, and environment. Subsequently, the factors related to safety hazards are articulated by SQL database and stored in BIM cloud. The models developed from BIM are used to carter the engineering information and the majority of the safety risks linked with engineering information are classified into four categories: characteristics of project, techniques adopted in construction, atmosphere in the construction field, and geology hydrological information. Hence, the relationship among them was analyzed and the engineering information was made in such a way that link the database related to safety awareness in the BIM network. Finally, the mechanism related to automated safety risk recognition was clarified and concluded that the level of confidence becomes a bridge to connect information related to project with the knowledge of risk.

5.2 Structural health monitoring

With the intention of overcoming the challenges faced in the underground construction operations, novel techniques, approaches, and tools for safety construction management were employed in the aforementioned analytical methods. In the safety management of underground construction, a complete instrumentation of structures and environment "structural health monitoring" (SHM) is extensively acknowledged as a critical factor (Bhalla et al., 2005). SHM system is projected to envisage the risks associated with the structural and environmental instability, which are often encountered in subway construction (Bhalla et al., 2005; Chai et al., 2011; Lin et al., 2014). It turns to be feasible to automatically and wisely examine and forecast the characteristics of underground structures in real time with the installation of highly durable and robust sensors (Khoury and Kamat, 2009). Geo-DATA organization in Italy has developed an information management system called Geodata Master System (GDMS) to manage the risks in underground engineering. GDMS has the tendency to offer comprehensive risk management plans and hence it is extensively applied in the subway construction projects in Russia and Italy and not in the case of China owing to the variation in the monitoring techniques and standards pertaining to construction management. Extensive study has been performed on examining the tracking expertise and their relevance which can meet diverse necessities in underground construction practices (Lin et al., 2014). These advanced techniques can be utilized in identifying and preventing human error and characteristic risks in underground construction. As on date, variety of tracking techniques were established such as ZigBee and indoor GPS (Ergen et al., 2007), radio frequency identification device (RFID) (Khoury and Kamat, 2009; Tu et al., 2009; Seco et al., 2010; Rao and Chandran, 2013), ultrawide band (UWB) (Carbonari et al., 2011), global positioning system (GPS), and wireless local area network (WLAN or WiFi) (Jiang et al., 2015). These techniques are viable to cover an extensive array of space and provide comparatively precise results.

5.3 Fuzzy set theory

Fuzzy set theory was developed by Zadeh in the year 1965, which offered an efficient tool in dealing with ambiguity and uncertainty as a result of the development of member function grading and the flowchart representing the fuzzy computing model is shown in Fig. 22.4 (Falcone et al., 2020). Variety of sets comprise more than one membership criterion such as interval-valued fuzzy set (Chen, 2000), Pythagorean fuzzy set (Ren et al., 2016), and intuitionistic fuzzy sets (Bao et al., 2014; Pan, 2009; Zou and Li, 2010; Wei et al., 2020; Sun, 2010) as a result of a steady course of conversion to

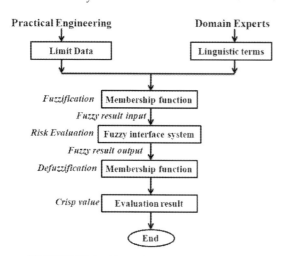

FIGURE 22.4 Hierarchy of fuzzy computation.

nonmembership from membership. The level of judgment in deep excavation has been formulated using fuzzy synthetic evaluation method with the intention of constructing the judgment matrix and to propose quantitative computations (Bao et al., 2014). Various techniques are recommended or developed in the area of fuzzy computing which are derived from fuzzy set theory. For instance, TOPSIS technique was developed in the year 1981 by Hwang and Yoon (1981) which is a valuable device in resolving multicriteria decision-making (MCDM) issues. The objective of TOPSIS technique is to pick the finest contestant having diminutive distance from the positive best resolution and extreme distance from the negative ideal solution. Diverse extensions of TOPSIS with fuzzy sets contain diverse geometric area. The linguistic expressions need to be converted into fuzzy number in the assessment of risk.

The common processes involved in fuzzy computing for the applying in the evaluation of risks involved in excavation can in general be explained by the subsequent steps: (1) recapitulate the different impending risks derived from the analysis of abundant accidents as a result of deep excavation and realistic engineering; (2) encourage field specialists to apple linguistic expressions in order to estimate the entries derived from their skill and converting the linguistic expressions into fuzzy numbers; (3) institute a risk evaluation matrix for the deep excavation; and (4) apply the fuzzy complete evaluation technique to determine the level of risk in deep excavation quantitatively. Nonetheless, definite impediments are available in the estimation of risks in deep excavation using fuzzy computing with the conditions of restricted information. For instance, under the circumstances of inadequate information, not all impact aspects are taken during the analysis of the data that impacts the level of safety in deep excavation. Additionally, the results based on the assessment still demonstrate definite subjectivity while necessitating field specialists

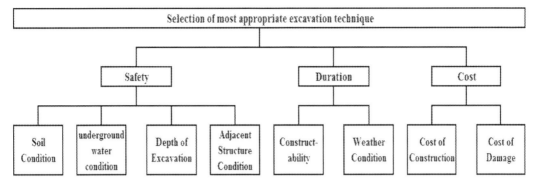

FIGURE 22.5 Hierarchy of selecting the most appropriate excavation technique.

to implement linguistic expressions in the assessment of items.

A technique derived from fuzzy TOPSIS was proposed by Mahdevari et al. (2014) in order to evaluate the risks connected with human health in underground coal mines so as to supervise control measures and support decision-making, which could afford the exact poise among various problems, namely safety and costs. Based on the information collected from three hazardous coal mines in Iran, it was identified with 86 hazards and was classified under eight categories: geo-mechanical, geochemical, electrical, mechanical, chemical, environmental, personal, and social, cultural, and managerial risks. It was concluded that the proposed model can be largely designed to recognize the impending risks and facilitate to take suitable measures so as to reduce or remove the risks before the occurrence of accidents. Pan (2009) constructed a four-level hierarchical network to select the most appropriate excavation technique as shown in Fig. 22.5 and employed the triangular fuzzy number and fuzzy number transaction set to replace the prospect of incidence and to infer a relationship algorithm of fuzzy accident tree which helps in computing the probability of support structure failure in deep excavation.

The degree of ambiguity in deep excavation has been verified with the triangular and trapezoidal fuzzy numbers in order to challenge the expert's intrinsic subjectivity and indistinct decision to precise arithmetic values (Zou and Li, 2010). Fuzzy evidential reasoning approach was adopted by Sun (2010) to estimate the overall level of risks associated with the excavation by developing a combined probability mass with the reassigning technique. To estimate the hazards tha are related to rock bursting during the excavation of tunnels using hesitant fuzzy sets, TOPSIS and VIKOR techniques were developed in resolving multi-criteria decision making issues. Joshi (2018) addressed the issues related to the combination of Pythagorean fuzzy values by developing a novel technique of incorporating the thought of the general factors of the domain experts in the Pythagorean fuzzy environment and suggests a framework for estimating the reliability of the data in the usual Pythagorean fuzzy set to eliminate any deviation in the inclination of the domain expert.

5.4 Artificial neural network

An artificial neural network (ANN) is an artificial intelligent technique intended to excite the activity of human brain to deal with processing complicated data. An ANN structure basically consists of input layer and output layer along with variety of hidden layers which is depicted in Fig. 22.6. The introduction of the renowned back-propagation algorithm made the pathway of thorough study and widespread application of ANN (Leu and Lo, 2004; Rumelhart et al.,

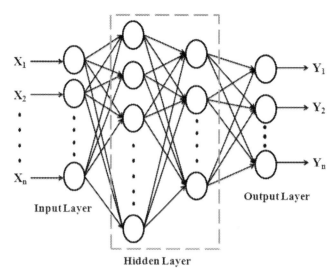

FIGURE 22.6 Structure of ANN with two hidden layers.

1986). The performance of ANN mainly depends on the neurons that are utilized in the mathematical model analogous to the role of biological neuron. The data that flow from the input information to the output result through one or more hidden nodes in the forward direction are the principal characteristics of ANN. The number of hidden layers/nodes purely depends on the training data, whereas the input node is constant with the number of basic risk-inducing factors (Jan et al., 2002).

The benefits of applying ANN comprise the fact that the knowledge of source and its consequence is not mandatory. The use of ANN modeling essentially paid attention on forecasting the deflection of the diaphragm wall as well as the settlement of the ground derived from the observed data in real-time engineering application. All the indices have the ability to offer material to forecast the risks associated with the deep excavation technique. For instance, the size, position, and the limiting rate of deflection associated with the diaphragm wall has been forecasted effectively based on the approach proposed by Jan et al. (2002) as shown in Fig. 22.7.

In addition, the settlement of ground surface as a result of deep excavations has been precisely estimated by the model proposed by Leu and Lo (2004), which as well acquired the size, position, and the limiting rate of deflection realistically. In this, in order to predict the settlement, the application of ANN-NDC models was developed separately for design and construction stage in which the design stage focused mainly on the information received from the analysis and the construction stage focused mainly on the monitoring data as shown in Fig. 22.8.

5.5 Bayesian network

Bayesian network (BN) (also referred to as Bayes network, belief network, or decision network) is categorized under probabilistic graphical model that is applied to construct a model derived from the data and/or expert opinion. It comprises probability and graph theory which is termed to as linked joint probability distribution and a directed acyclic graph (Zhu and Deshmukh, 2003). BNs are perfect for capturing an occurrence and forecasting the chances that any one of the numerous potential

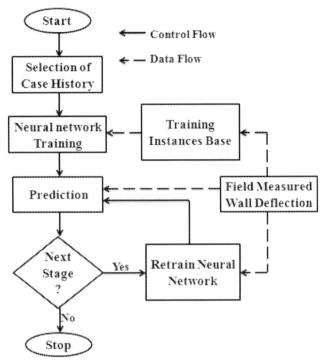

FIGURE 22.7   Flowchart of verification.

identified source was the contributing feature. The configuration of BN and conditional probability table of the node is represented in directed lines and a set of nodes in which the nodes resemble the system parameter, whereas the lines symbolize the relationship between cause and effect or the dependencies between the parameters in the directed acyclic graph. This makes it feasible to estimate the prospect allocation of parameters evidently (Wang and Chen, 2017; Cooper, 1990; Weber et al., 2012), derived from the preceding knowledge and the study of parameters in BNs. The foremost benefit of applying BN is that it can augment the competence and consistency of a given form and endorse its relevance in managing risk in real-time engineering, derived from experimental proof, enduring practices, and chronological statistics (Zhou et al., 2018; Zhou and Zhang, 2011). It is recognized that the skills of civil engineering are momentous in excavating deep foundation and supplementary infrastructures, the outcome being a prospective technique for the risk estimation sector. Fuzzy comprehensive BN technique was adopted to evaluate the total risk involved in the metro construction and to make a decision with the systematic approach as shown in Fig. 22.9 (Wang and Chen, 2017).

BN is extensively applied in the field of engineering for the evaluation of risk involved in excavation operation. For instance, Zhou and Zhang (2011) projected a fuzzy complete assessment technique derived from the BN, which augments the modeling competence and consistency of the outcomes related to risk evaluation. Wang et al. (2014) suggested a fusion process of integrating procedures of the BN, which can be applied to illustrate accident circumstances, and evaluate their prospect of rigorousness and occasion. Zhou et al. (2018) recommended a risk

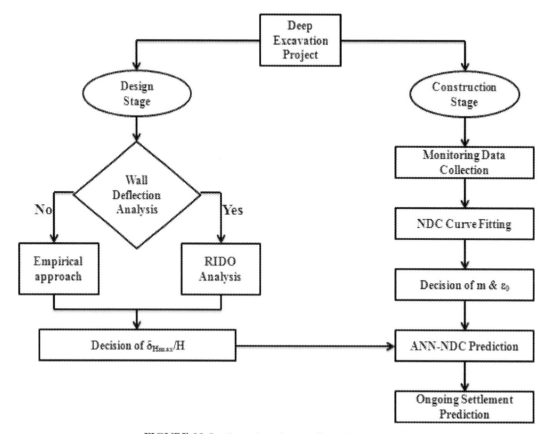

FIGURE 22.8 Typical application flow of ANN-NDC.

investigation model for forecasting the deflection in diaphragm wall with the uses of BN, which could recognize risks in vague atmosphere and appreciate dynamic control to make sure safety in the metro construction that requires fours steps as shown in Fig. 22.10.

Zhou et al. (2020) proposed a flexible assessment technique in evaluating the risk associated with the sewer pipelines with the aid of BN combined with D-S evidence theory. BN has been framed to exhibit the intricate instrument that causes disaster in sewer pipelines under utility tunnels; and the diverse professional awareness has been processed using D-S evidence theory. It was concluded that the framework based on BN is an efficient technique in assessing the complicated hazards and disasters under dynamic circumstances.

5.6 Other techniques

Support vector machine (SVM) was proposed by Cortes and Vapnik in the year 1995 (Cortes and Vapnik, 1995), derived from the theory of minimizing the structural risk and theory of statistical learning. The problems associated with the categorization and regression can be resolved with the aid of SVM, which in addition augment the exact forecasting as well as avoiding the issue of overfitting. Similar to the ANN, the configuration of SVM model also comprises three layers namely input, hidden, and output layers as depicted in Fig. 22.11, derived from definite classification standards.

To monitor the convergence mode of the tunnel, Mahdevari et al. (2013) developed a dynamic model which relies on SVM algorithm

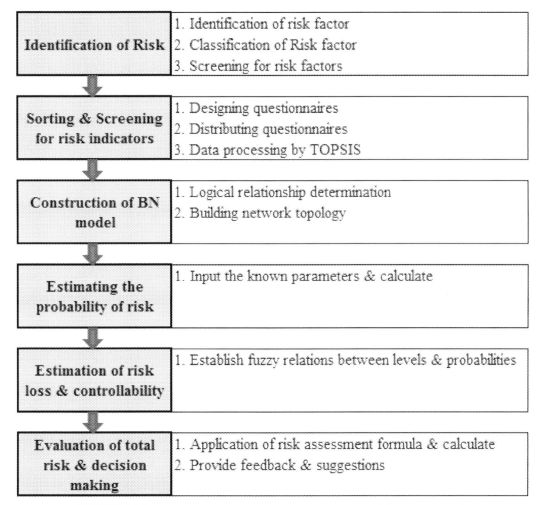

FIGURE 22.9 Step-by-step procedure involved in FCBN-based decision approach.

and the proposed approach architecture is shown in Fig. 22.12. In this process, the data affecting the geo-mechanical factors and monitored displacements in various segments of the tunnel were initiated as a training set in SVM model to evaluate the indefinite nonlinear relationship between the soil characteristics and convergence of tunnel. From the results obtained, it was observed that the values predicted were well correlated with the in situ results and concluded that the development of SVM can forecast the tunnel convergence during excavation in addition to the unexcavated areas.

The application of Fisher discriminant analysis and SVM techniques was proposed by Zhou et al. (2011) to estimate the stability of pillars in underground mines with the aid of mechanical and index characteristics such as height, width, ratio of width to height, uniaxial compressive strength of the rock, and the stress induced in the pillars derived from various coal mines. Comparing to Fisher discriminant analysis, SVM exhibits the best results and concluded

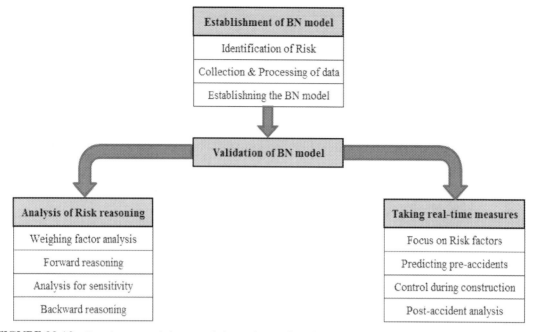

FIGURE 22.10 Bayesian network framework for analyzing the safety risk of diaphragm wall in metro construction.

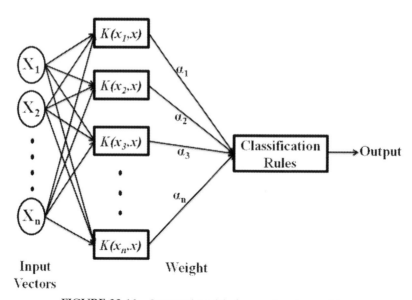

FIGURE 22.11 Structural model of support vector machine.

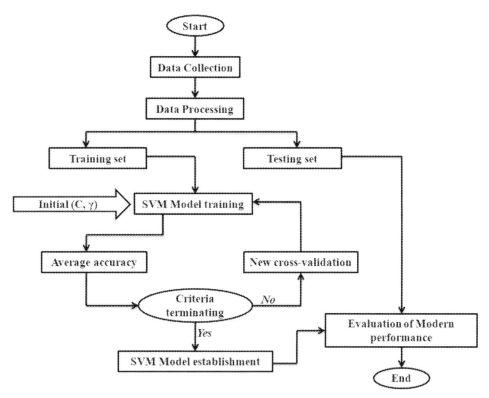

FIGURE 22.12 Proposed approach architecture using SVM algorithm.

that the application of SVM in underground mines reliable than the Fisher discriminant analysis to evaluate the stability of pillars.

Random forest is a technique developed by Breiman by merging randomized node optimization, enhancing aggregation, and a classification and regression tree toward prevailing over the demerits of less forecasting and likelihood of overfitting with a decision tree. Besides, use of Random forest does not involve in processing the data, whereas the processing of numerical and categorical information can function well with large dataset. A strategy of forecasting the injuries caused in construction process has been developed by Tixier et al. (2016) by utilizing the process of machine learning and stochastic gradient tree boosting. The unstructured datasets pertaining to the details of injury were utilized in this model and observed various safety implications with respect to construction injuries by the application of text mining system. With similar analysis, Goh and Ubeynarayana (2017) also addressed the issue of categorizing the information pertaining to unstructured construction by text mining. Poh et al. (2018) applied the various machine learning techniques for evaluating the factors resulting in safety accidents and observed that the application of random forest technique resulting in most accurate results compared to all other machine learning techniques. Kang and Ryu (2019) forecast the various construction site occupational accident types in Korea by constructing a model to obtain significant characteristics as well as verifying the precision in forecasting. It was suggested that the key management features will result in offering considerable contribution toward work of both practitioners and researchers in the construction sector.

6. Conclusions

The increase in the population and a transformation toward urbanization leads to a rapid increase in the infrastructural development. Due to the lack of land resources, there is an increase in the underground construction for both structures and transportation purpose in addition to the continuous mining operations. Even with large technological development in the construction sector, the risk associated with the underground construction has a serious threat to the human being. The inaccuracy in the identification of the potential risk, improper management, and limitations in the assessment techniques has led to various accidents in the underground construction. This has led to the increase in the awareness of the stakeholders and the employees in overcoming these safety hazards. This chapter overviews the various accidents associated with the underground construction, how to identify the potential hazards in construction, factors involving in hazards and their management techniques, and the various techniques developed in assessing and forecasting the potential risks in underground construction operations.

References

AADT, 1999. Task Force for Technical Investigation of the March 1999 Fire in the Mont Blanc Vehicular Tunnel. Ministry of Equipment, Transportation and Housing, Paris.

Alaeddini, A., Dogan, I., 2011. Using Bayesian networks for root cause analysis in statistical process control. Expert Syst. Appl. 38 (9), 11230–11243.

Alfredo, D.C., 2002. Integrated methodology for project risk management. J. Construct. Eng. Manag. 128 (6), 473–485.

Amundsen, F.H., Engebretsen, A., 2009. Studies on Norwegian Road Tunnels II. An Analysis on Traffic Accidents in Road Tunnels 2001–2006, Oslo: Vegdirektoratet, Roads and Traffic Department, Traffic Safety Section, Rapport Nr: TS4.

Aye, Z.Z., Karki, D., Schulz, C., 2006. Ground movement prediction and building damage risk assessment for the deep excavations and tunneling works in Bangkok subsoil. In: International Symposium on Underground Excavation and Tunelling Urban Tunnel Construction for Protection of Environment.2-4 February 2006, Bangkok.

Azadi, M., Pourakbar, S., Kashfi, A., 2013. Assessment of optimum settlement of structure adjacent urban tunnel by using neural network methods. Tunn. Undergr. Space Technol. 37, 1–9.

Bao, X.H., Fu, Y.B., Huang, H.W., 2014. Case study of risk assessment for safe grade of deep excavations. Chin. J. Geotech. Eng. 36 (zk1), 192–197.

Beard, A., Carvel, R., 2012. Road Tunnel Fire Safety, second ed. Thomas Telford, London.

Beard, A., Cope, D., 2008. Assessment of the Safety of Tunnels. European Parliament, Brussels.

Benekos, D.D., 2017. On risk assessment and risk acceptance of dangerous goods transportation through road tunnels in Greece. Saf. Sci. 91, 1–10.

Bhalla, S., Yang, Y.W., Zhao, J., Soh, C.K., 2005. Structural health monitoring of underground facilities: technological issues and challenges. Tunn. Undergr. Space Technol. 20 (5), 487–500.

Broere, W., 2016. Urban underground space: solving the problems of today's cities. Tunn. Undergr. Space Technol. 55, 245–248.

Caliendo, C., De Guglielmo, M.L., 2012. Accident rates in road tunnel and social costs evaluation. In: SIIV — 5th International Congress — Sustainability of Road Tunnels Infrastructures. Rome, in Procedia — Social and Behavioural Sciences, vol. 53, pp. 166–177.

Cao, Q.G., Li, K., Liu, Y.J., 2012. etc. Risk management and workers' safety behavior control in coal mine. Saf. Sci. 50 (4), 909–913.

Carbonari, A., Giretti, A., Naticchia, B., 2011. A proactive system for real-time safety management in construction sites. Autom. ConStruct. 20 (6), 686–698.

Chai, J., Liu, J.X., Qiu, B., Li, Y., Zhu, L., Wei, S.M., Wang, Z.P., Zhang, G.W., Yang, J.H., 2011. Detecting deformations in uncompacted strata by fiber Bragg grating sensors incorporated into GFRP. Tunn. Undergr. Space Technol. 26 (1), 92–99.

Chan, X.L., Yu, S.C., Ma, G., et al., 2011. Particle swarm optimization based on particle migration and its application to geotechnical engineering. Chinese J. Rock Soil Mech. 32 (4), 1077–1082.

Chen, S.H., Chen, S.F., Shahrour, I., et al., 2001. The feedback analysis of excavated rock slope. Rock Mech. Rock Eng. 34 (1), 39–56.

Chen, T., Deng, J., Sitar, N., Zheng, J., Liu, T., Liu, A., Zheng, L., 2017. Stability investigation and stabilization of a heavily fractured and loosened rock slope during

References

construction of a strategic hydropower station in China. Eng. Geol. 221, 70−81.

Chen, C.T., 2000. Extensions of the TOPSIS for group decision-making under fuzzy environment. Fuzzy Set Syst. 114 (1), 1−9.

Chheng, C., Likitlersuang, S., 2018. Underground excavation behaviour in Bangkok using three-dimensional finite element method. Comput. Geotech. 95, 68−81.

Cooper, G.F., 1990. The computational complexity of probabilistic inference using bayesian belief networks. Artif. Intell. 42 (2−3), 393−405.

Cortes, C., Vapnik, V., 1995. Support-vector networks. Mach. Learn. 20 (3), 273−297.

Ding, L.Y., Zhou, C., 2013. Development of web-based system for safety risk early warning in urban metro construction. Autom. ConStruct. 34, 45−55.

Ding, L.Y., Zhou, C., Deng, Q.X., Luo, H.B., Ye, X.W., Ni, Y.Q., Guo, P., 2013. Real-time safety early warning system for cross passage construction in Yangtze Riverbed Metro Tunnel based on the internet of things. Autom. ConStruct. 36, 25−37.

Eastman, C.M., Eastman, C., Teicholz, P., Sacks, R., Liston, K., 2011. BIM Handbook: A Guide to Building Information Modeling for Owners, Managers, Designers, Engineers and Contractors. John Wiley & Sons (ISBN: 0470541377, 9780470541371).

EC, 2011. White Paper: Roadmap to a Single European Transport Area - Towards a Competitive and Resource Efficient Transport System. European Commisssion, Brussels.

EDT, 2012. Road Transport: A Change of Gear. European Department of Transportation, European Commission, Brussels.

Ergen, E., Akinci, B., Sacks, R., 2007. Tracking and locating components in a precast storage yard utilizing radio frequency identification technology and GPS. Autom. ConStruct. 16 (3), 354−367.

Eskesen, S.D., Tengborg, P., Kampmann, J., Veicherts, T.H., 2004. Guidelines for tunneling risk management: international tunneling association working group No. 2. Tunn. Undergr. Space Technol. 19 (3), 217−237.

EU, 2004. Minimum Safety Requirements for Tunnels in the Trans-european Road Network, Directive 2004/54/EC of the European Parliament and of the Council. European Union, Brussels.

Falcone, R., Lima, C., Martinelli, E., 2020. Soft computing techniques in structural and earthquake engineering: a literature review. Eng. Struct. 207 (110269), 0141−0296.

Fang, Q., Zhang, D., Wong, L.N.Y., 2011. Environmental risk management for a cross interchange subway station construction in China. Tunn. Undergr. Space Technol. 26, 750−763.

Fang, Y.-S., Wu, C.-T., Chen, S.-F., Liu, C., 2014. An estimation of subsurface settlement due to shield tunneling. Tunn. Undergr. Space Technol. 44, 121−129.

Fattahi, H., Moradi, A., 2017. Risk assessment and estimation of TBM penetration rate using RES-based model. Geotech. Geol. Eng. 35 (1), 365−376.

Garg, P., Jaiswal, A., 2016. Estimation of modulus of the caved rock for underground coal mines by back analysis using numerical modelling. J. Inst. Eng.: Series D 97 (2), 269−273.

Goh, Y.M., Ubeynarayana, C.U., 2017. Construction accident narrative classification: an evaluation of text mining techniques. Accid. Anal. Prev. 108, 122−130.

Guglielmetti, V., Grasso, P., Mahtab, P., Xu, S.L., 2008. Mechanized Tunneling in Urban Areas: Design Methodology and Construction Control. Taylor & Francis, London, UK.

Han, R.R., Zhang, J.H., Zhang, X., et al., 2009. Study of prediction on feedback calculation of excavated monitor to the underground powerhouse of Xiluodu. Chinese J. Shandong Univ. (Eng. Sci.) 39 (4), 140−144.

Hashash, Y.M.A., Levasseur, S., Osouli, A., Finno, R., Malecot, Y., 2010. Comparison of two inverse analysis techniques for learning deep excavation response. Comput. Geotech. 37, 323−333.

Houhou, M.N., Emeriault, F., Belounar, A., 2019a. Three-dimensional numerical back-analysis of a monitored deep excavation retained by strutted diaphragm walls. Tunn. Undergr. Space Technol. 83, 153−164.

Houhou, M.N., Emeriault, F., Belounar, A., 2019b. Three-dimensional numerical backanalysis of a monitored deep excavation retained by strutted diaphragm walls. Tunn. Undergr. Space Technol. 83, 153−164.

Huang, H.W., Zeng, M., Chen, L., Hu, Q., 2006. Risk management software (TRM1. 0) based on risk database for shield tunneling. Chin. J. Undergr. Space Eng. 2 (1), 36−41.

Hwang, C.L., Yoon, K., 1981. Multiple Attributes Decision Making Methods and Applications. Springer, Berlin Heidelberg. https://doi.org/10.1007/978-3-642-48318-9.

Jan, J.C., Hung, S.L., Chi, S.Y., Chern, J.C., 2002. Neural network forecast model in deep excavation. J. Comput. Civ. Eng. 16 (1), 59−65.

Jia, S.P., Wu, G.J., Chen, W.Z., et al., 2011. Application of finite element inverse model based on improved particle swarm optimization and mixed penalty function. Chinese J. Rock Soil Mech. 32 (S2), 598−603.

Jiang, H., Lin, P., Qiang, M., Fan, Q.X., 2015. A labor consumption measurement system based on real-time tracking technology for dam construction site. Autom. ConStruct. 52, 1−15.

Joshi, B.P., 2018. Pythagorean fuzzy average aggregation operators based on generalized and group-generalized parameter with application in MCDM problems. Int. J. Intell. Syst. 34, 895−919.

Juang, C.H., Gong, W., Martin, J.R., Chen, Q., 2018. Model selection in geological and geotechnical engineering in the face of uncertainty - does a complex model always outperform a simple model? Eng. Geol. 242, 184−196.

Kang, K., Ryu, H., 2019. Predicting types of occupational accidents at construction sites in Korea using random forest model. Saf. Sci. 120, 226–236.

Khakzad, N., Khan, F., Amyotte, P., 2011. Safety analysis in process facilities: comparison of fault tree and Bayesian network approaches. Reliab. Eng. Syst. Saf. 96 (8), 925–932.

Khoury, H.M., Kamat, V.R., 2009. Evaluation of position tracking technologies for user localization in indoor construction environments. Autom. ConStruct. 18 (4), 444–457.

Kirytopoulos, K., Kazaras, K., Papapavlou, P., Ntzeremes, P., Tatsiopoulos, I., 2017. Exploring driving habits and safety critical behavioural intentions among road tunnel users: a questionnaire survey in Greece. Tunn. Undergr. Space Technol. 63 (3), 244–251.

Leu, S.S., Lo, H.C., 2004. Neural-network-based regression model of ground surface settlement induced by deep excavation. Autom. ConStruct. 13, 279–289.

Li, M., Yu, H., Liu, P., 2018. An automated safety risk recognition mechanism for underground construction at the pre-construction stage based on BIM. Autom. ConStruct. 91, 284–292.

Li, M.-G., Xiao, X., Wang, J.-H., Chen, J.-J., 2019. Numerical study on responses of an existing metro line to staged deep excavations. Tunn. Undergr. Space Technol. 85, 268–281.

Lin, P., Li, Q.B., Fan, Q.X., Gao, X.Y., Hu, S.Y., 2014. A real time location-based services system using Wi-Fi fingerprinting algorithm for safety risk assessment of workers in tunnels. Math. Probl Eng. 371456.

Liu, T.M., Zhong, M.H., Xing, J.J., 2005. Industrial accidents: challenges for China's economic and social development. Saf. Sci. 43 (8), 503–522.

Luo, R.L., Ruan, H.N., Huan, Y.Z., et al., 2008. Particle swarm optimization inversion method of initial ground stress and implementation in FLAC3D. Chinese J. Yangtze River Sci. Res. Inst. 25 (4), 73–76.

Ma, X., 2017. Application of BIM in engineering construction, building, building materials. Decoration 107, 12.

Mahdevari, S., Haghighat, H.S., Torabia, S.R., 2013. A dynamically approach based on SVM algorithm for prediction of tunnel convergence during excavation. Tunn. Undergr. Space Technol. 38, 59–68.

Mahdevari, S., Shahriar, K., Esfahanipour, A., 2014. Human health and safety risks management in underground coal mines using fuzzy TOPSIS. Sci. Total Environ. 488 (1), 85–89.

Min, X., Miya, S., 2016. Analysis of the application value and Barriers of BIM. In: Metro Station Project, pp. 41–44.

Naevestad, T.O., Meyer, S., 2014. A survey of vehicle fires in Norwegian road tunnels 2008–2011. Tunn. Undergr. Space Technol. 41 (3), 104–112.

Nezarat, H., Sereshki, F., Ataei, M., 2015. Ranking of geological risks in mechanized tunneling by using Fuzzy Analytical Hierarchy Process (FAHP). Tunn. Undergr. Space Technol. 50, 358–364.

NFPA, 2014. Standards for Road Tunnels, Bridges, and Other Limited Access Highways. National Fire Protection Association, New York.

Ni, P., Mangalathu, S., 2018. Fragility analysis of gray iron pipelines subjected to tunneling induced ground settlement. Tunn. Undergr. Space Technol. 76, 133–144.

Ni, S.H., Xiao, M., He, S.H., et al., 2013. Back analysis in underground engineering based on parallel computing and optimization algorithm and its verification. Chin. J. Rock Mech. Eng. 32 (3), 501–511.

Ntzeremes, P., Kirytopoulos, K., Tatsiopoulos, I., 2016. Management of infrastructure's safety under the influence of normative provisions. WIT Trans. Built Environ. 64, 51–60.

Oreste, P., 2005. Back-analysis techniques for the improvement of the understanding of rock in underground constructions. Tunn. Undergr. Space Technol. 20 (1), 7–21.

Ou, C.Y., Hsieh, P.G., 2011. A simplified method for predicting ground settlement profiles induced by excavation in soft clay. Comput. Geotech. 38, 987–997.

Pan, N.F., 2009. Selecting an appropriate excavation construction method based on qualitative assessments. Expert Syst. Appl. 36 (3), 5481–5490.

Park, C.S., Kim, H.J., 2013. A framework for construction safety management and visualization system. Autom. ConStruct. 33, 95–103.

Perlman, S.R., Barak, R., 2014. Hazard recognition and risk perception in construction. Saf. Sci. 64, 22–31.

Perras, M.A., Wannenmacher, H., Diederichs, M.S., 2005. Underground excavation behaviour of the queenston formation: tunnel back analysis for application to shaft damage dimension prediction. Rock Mech. Rock Eng. 48 (4), 1647–1671.

PIARC, 2008. Risk Analysis for Road Tunnels, Paris: Ref: 2008R02EN. ISBN: 2-84060-202-4.

PIARC, 2013. Current Practice for Risk Evaluation for Road Tunnels, Paris: World Road Association, Technical Committee 3.3, Road Tunnel Operation, ISBN 978-2-84060-290-3.

PIARC, 2016. Road Tunnels: Complex Underground Road Networks, Paris: World Road Association, Technical Committee 3.3, Road Tunnel Operation, ISBN 978-2-84060-404-4.

Piniella, F., Fern, A., Ndez-Engo, M.A., 2009. Towards system for the management of safety on board artisanal fishing vessels: proposal for check-lists and their application. Saf. Sci. 47 (2), 265–276.

Poh, C.Q., Ubeynarayana, C.U., Goh, Y.M., 2018. Safety leading indicators for construction sites: a machine learning approach. Autom. ConStruct. 93, 375–386.

Qian, Q., Lin, P., 2016. Safety risk management of underground engineering in China: progress, challenges and strategies. J. Rock Mech. Geotech. Eng. 8 (4), 423–442.

Qian, Q.H., Rong, X.L., 2008. State, issues and relevant recommendations for security risk management of China's underground engineering. Chin. J. Rock Mech. Eng. 27 (4), 649–655.

Qian, Q.H., 2014. Report on the Strategy and Countermeasure of Safety Risk Management System for Civil Engineering in China. Consulting Research Project of Chinese Academy of Engineering.

Qian, Q., 2016. Present state, problems and development trends of urban underground space in China. Tunn. Undergr. Space Technol. 55, 280–289.

Rao, K.S., Chandran, K.R., 2013. Mining of customer walking path sequence from RFID supermarket data. Electron. Govern. 10 (1), 34–55.

Ren, P., Xu, Z., Gou, X., 2016. Pythagorean fuzzy TODIM approach to multi-criteria decision making. Appl. Soft Comput. 42, 246–259.

Renani, H.R., Martin, C.D., Hudson, R., 2016. Back analysis of rock mass displacements around a deep shaft using two- and three- dimensional continuum modeling. Rock Mech. Rock Eng. 49 (4), 1313–1327.

Rumelhart, D.E., Hinton, G.E., Williams, R.J., 1986. Learning representations by backpropagating errors. Nature 323 (6088), 533–536.

Sagong, M., Lee, J.S., You, K., Kim, J.G., 2006. Digitalized tunnel face mapping system (DiTFAMS) using PDA and wireless network. Tunn. Undergr. Space Technol. 21, 390.

Seco, F., Plagemann, C., Jimenez, A.R., Burgard, W., 2010. Improving RFID-based indoor positioning accuracy using Gaussian processes. In: Proceedings of the International Conference on Indoor Positioning and Indoor Navigation (IPIN). IEEE, pp. 1–8.

Smith, N.J., Merna, T., Jobling, P., 2009. Managing Risk in Construction Projects. Wiley-Blackwell, Hoboken, New Jersey, USA.

Sun, F., 2010. SVM in predicting the deformation of deep foundation pit in soft soil area. In: 2010 Int. Conf. Mach. Vis. Human-Machine Interface, vol. 10, pp. 761–763.

Tixier, A.J.P., Hallowell, M.R., Rajagopalan, B., Bowman, D., 2016. Application of machine learning to construction injury prediction. Autom. ConStruct. 69, 102–114.

Travaglini, A., Radujković, M., Mancini, M., 2014. Building information modelling (BIM) and project management: a stakeholders perspective, organization. Technol. Manag. 6 (2), 1001–1008.

Tu, Y.J., Zhou, W., Piramuthu, S., 2009. Identifying RFID-embedded objects in pervasive healthcare applications. Decis. Support Syst. 46 (2), 586–593.

UN, 2015. World Urbanization Prospects: The 2014 Revision, New York: Department of Economic and Social Affairs, United Nations.

Vahdatirad, M.J., Ghodrat, H., Firouzian, S., Barari, A., 2010. Analysis of an underground structure settlement risk due to tunneling - a case study from Tabriz. Iran. Songklanakarin J. Sci. Technol. 32, 145–152.

Volk, R., Stengel, J., Schultmann, F., 2014. Building Information Modeling (BIM) for existing buildings—literature review and future needs. Autom. ConStruct. 38, 109–127.

Wang, Z.Z., Chen, C., 2017. Fuzzy comprehensive Bayesian network-based safety risk assessment for metro construction projects. Tunn. Undergr. Space Technol. 70, 330–342.

Wang, G., Jiang, Y.J., Li, S.C., 2011. Rapid feedback analysis method for underground caverns during construction. Chinese J. Shandong Univ. (Eng. Sci.) 41 (4), 133–136.

Wang, F., Ding, L.Y., Luo, H.B., Love, P.E.D., 2014. Probabilistic risk assessment of tunneling-induced damage to existing properties. Expert Syst. Appl. 41, 951–961.

Wang, X.W., Yang, T.L., Xu, Y.S., Shen, S.L., 2019. Evaluation of optimized depth of waterproof curtain to mitigate negative impacts during dewatering. J. Hydrol. 577, 123969.

Weber, P., Medina-Oliva, G., Simon, C., Iung, B., 2012. Overview on Bayesian networks applications for dependability, risk analysis and maintenance areas. Eng. Appl. Artif. Intell. 25 (4), 671–682.

Wei, D.J., Xu, D.S., Zhang, Y., 2020. A fuzzy evidential reasoning-based approach for risk assessment of deep foundation pit. Tunn. Undergr. Space Technol. 97, 103232.

World Energy Council, 2016. World Energy Resources. World Energy Council, London.

Wu, B., Xu, Z.D., Zhou, Y., 2014. etc. Study on coal mine safety management system based on "Hazard, Latent Danger and Emergency Responses". Procedia Eng. 84, 172–177.

Wu, X.G., Liu, H.T., Zhang, L.M., Skibniewski, M.J., Deng, Q.L., Teng, J.Y., 2015. A dynamic Bayesian network based approach to safety decision support in tunnel construction. Reliab. Eng. Syst. Saf. 134, 157–168.

Wu, Y.X., Lyu, H.M., Han, J., Shen, S.L., 2019. Dewatering-induced building settlement around a deep excavation in soft deposit in Tianjin, China. J. Geotech. Geoenviron. Eng. ASCE 145 (5), 05019003.

Wu, Y.X., Shen, S.L., Lyu, H.M., Zhou, A.N., 2020. Analyses of leakage effect of waterproof curtain during excavation dewatering. J. Hydrol. 583, 124582.

Xie, R., Pan, Y., Zhou, T., Ye, W., 2020. Smart safety design for fire stairways in underground space based on the ascending evacuation speed and BMI. Saf. Sci. 125, 104619.

Xu, Y.S., Yan, X.X., Shen, S.L., Zhou, A.N., 2019. Experimental investigation on the blocking of groundwater seepage from a waterproof curtain during pumped dewatering in an excavation. Hydrogeol. J. 27 (7), 2659–2672.

Yi, D., Chen, S.H., Ge, X.R., 2011. A methodology combining genetic algorithm and finite element method for back analysis of initial stress field of rock masses. Chinese J. Rock Soil Mech. 32 (S2), 598–603.

Yoo, C., Jeon, Y.W., Choi, B.S., 2006. IT-based tunnelling risk management system (ITTURISK)–development and implementation. Tunn. Undergr. Space Technol. 21 (2), 190–202.

Yu, Q.Z., Ding, L.Y., Zhou, C., Luo, H.B., 2014. Analysis of factors influencing safety management for metro construction in China. Accid. Anal. Prev. 68, 131–138.

Zhang, L., Skibniewski, M.J., Wu, X., Chen, Y., Deng, Q., 2014. A probabilistic approach for safety risk analysis in metro construction. Saf. Sci. 63, 8–17.

Zhang, L., Wu, X., Ding, L., Skibniewski, M.J., Lu, Y., 2016. BIM-based risk identification system in tunnel construction. J. Civ. Eng. Manag. 22 (4), 529–539.

Zhang, L., Ding, L., Wu, X., Skibniewski, M.J., 2017. An improved Dempster-Shafer approach to construction safety risk perception. Knowl.-Based Syst. 132, 30–46.

Zhang, X., Yang, J., Zhang, Y., Gao, Y., 2018. Cause investigation of damages in existing building adjacent to foundation pit in construction. Eng. Fail. Anal. 83, 117–124.

Zhang, Z., Huang, M., Zhang, C., Jiang, K., Lu, M., 2019. Time-domain analyses for pile deformation induced by adjacent excavation considering influences of viscoelastic mechanism. Tunn. Undergr. Space Technol. 85, 392–405.

Zhou, H.B., Zhang, H., 2011. Risk assessment methodology for a deep foundation pit construction project in Shanghai, China. J. Construct. Eng. Manag. 137 (12), 1185–1194.

Zhou, H.B., Cai, L.B., Gao, W.J., 2009. Statistical analysis of the accidents of foundation pit of the urban mass rail transit station. Hydrogeol. Eng. Geol. 36 (2), 67–71.

Zhou, J., Xi-bing, L., Xiu-zhi, S., Wei, W., Bang-biao, W., 2011. Predicting pillar stability for underground mine using Fisher discriminant analysis and SVM methods. Trans. Nonferrous Metals Soc. China 21, 2734–2743.

Zhou, C., Ding, L., Zhou, Y., Luo, H., 2018. Topological mapping and assessment of multiple settlement time series in deep excavation: a complex network perspective. Adv. Eng. Inf. 36, 1–19.

Zhou, Y., Li, C., Zhou, C., Luo, H., 2018. Using Bayesian network for safety risk analysis of diaphragm wall deflection based on field data. Reliab. Eng. Syst. Saf. 180, 152–167.

Zhou, R., Fang, W., Wu, J., 2020. A risk assessment model of a sewer pipeline in an underground utility tunnel based on a Bayesian network. Tunn. Undergr. Space Technol. 103, 103473.

Zhu, J.Y., Deshmukh, A., 2003. Application of Bayesian decision networks to life cycle engineering in green design and manufacturing. Eng. Appl. Artif. Intell. 16 (2), 91–103.

Zhu, W.S., Li, X.J., Zhang, Q.B., et al., 2010. A study on sidewall displacement prediction and stability evaluations for large underground power station caverns. Int. J. Rock Mech. Min. Sci. 47 (7), 1055–1062.

Zou, P.X.W., Li, J., 2010. Risk identification and assessment in subway projects: case study of Nanjing Subway line 2, Constr. Manag. Econ. 28 (12), 1219–1238.

CHAPTER 23

Sustainability: a comprehensive approach to developing environmental technologies and conserving natural resources

Hosam M. Saleh, Amal I. Hassan

Radioisotopes Department, Nuclear Research Center, Egyptian Atomic Energy Authority, Giza, Egypt

1. Introduction

Environmental degradation, much of it over the past century, refers to systems that deplete nonrenewable resources, and that exploit renewable resources to a greater degree than their viability. It alters the chemistry of the Earth and distorts its ecosystems, generating irreversible damage to the land, water, and air (Criekemans, 2018). Therefore, the excessive exploitation and destruction accompanying development are a product of modern industrial society, and in particular its system of values, beliefs, and political construction. Although the system of modernity has many achievements, it also has its dark side represented in the corruption of the environment. However, most people are so immersed in this model of modernity that they are unable to realize that the structures and processes on which daily life is based are the cause of environmental devastation.

Sustainability is the study of how natural systems function, diversity, and production of everything the natural environment needs to remain balanced (Schlosberg and Coles, 2016). Sustainability also recognizes that human civilization provides resources for the sustainability of our contemporary way of life. There are many examples throughout human history where civilizations destroyed their environment and seriously affected their chances of survival (Flint, 2013). Sustainability takes into account how we live in harmony with the natural world and protect it from destruction.

Sustainable living should focus only on people who live in cities. Rather, improvements must be made everywhere (it is estimated that we consume about 40% more resources annually than we can) and that this needs to make changes to maintain the sustainability of these resources (Hume, 2010).

Sustainability and sustainable development focus on the balance between calculating needs, our need to use technology economically, and the need to protect the environments in which we live. Sustainability is not related to the environment only, but rather it is related to the health of societies and to ensure that people are not exposed to suffering due to environmental legislation, with the need to examine the long-term effects of human actions and ask questions about how can the situation be improved (Moldan et al., 2012). Many specialists and researchers see that sustainable development after it has become circulating, multiuse and diverse in meanings, is rich in components and economic, social, and environmental components and dimensions that are interconnected, interacting, and balanced, to achieve sustainable development as an ethical vision (Saleh et al., 2020a), commensurate with the interests and priorities of the new world order that is still under formation, and whose agenda is currently being arranged, as well as a method for errors and changes of previous Western development models in their relationship with the environment, in all its economic, social, and natural aspects, as a new frame of reference, whose emergence coincides with the emergence of the postmodern stage, and as a conscious environmental management issue, and new planning for the exploitation of natural resources, to achieve human needs in the present and future (Winder and Le Heron, 2017; Saleh et al., 2020b).

Sustainable agriculture consists of environmentally friendly farming methods that allow crops or livestock to be produced without harming human or natural systems. It involves preventing the harmful effects of soil, water, biodiversity, ocean, or downstream resources, as well as for those who work or live on the farm or in the vicinity (Machovina et al., 2015). The concept of sustainable agriculture extends between generations, passing based on natural resources, conservation, and economic vitality that is saved or improved rather than being depleted or polluted (Rands et al., 2010). Elements of sustainable agriculture include permaculture, agroforestry, mixed farming, multiple harvesting, and crop rotation (Altieri et al., 2017). They include agricultural methods that do not undermine the environment and smart agricultural technologies that promote a good environment for people to thrive and restore and transform deserts into farmland. Sustainable development is based on conjecturing the integrated and continuous relationship between evolution and the environment, to satisfy the needs of the population and taking into account environmental considerations, which is known as sustainability (Phillips, 2011). Sustainable development is based on proper planning, based on data that balance the real needs of the population and the available societal capabilities, and the realizing utilization of these human and physical capabilities that can be produced in light of agreed priorities, and taking into account the balance between the interest of the individual and society alike, which is achieved (Bibri, 2018). The process of evaluating projects and sustainable development programs, to identify areas of weakness and work to avoid them, and aspects of strength and working on their development, provided that this process is completed in all stages of planning, implementation, and follow-up, using it. Using the method of subsystems and their integration preserve the life of society, through a concern in economic, social, and environmental aspects, in a way that guarantees the balance of the global system without adverse consequences. Because environmental problems are linked to patterns of economic development, and the agricultural policies applied in many countries of the world are directly responsible for the deterioration of soil and the uprooting of forests (Sharifi and Murayama, 2013), leading to the rapid flow of surface water and acid rain, to the destruction of forests and water bodies, in particular, closed one. Sustainable development consists of three important elements: human and financial wealth that use modern technology, and natural wealth interacting with each other (Gude, 2016). Sustainable development requires coordination of capabilities and high efficiency to lead the crucial elements that aim at human well-being. Since the human being is the goal and means of developing nonhuman resources, his development is considered a fundamental pillar that leads to more benefit than he can obtain from his nonhuman resources, by raising the efficiency and productivity of human resources and continuing their capacity, whether through education or adequate and balanced food (Lisca, 2014), training, counseling and experience, health, and social care. The participation of individuals in development efforts and planning for decision-making to achieve sustainable development plays an essential role in economic and social life. The development needs of everyone and opportunities for all are shared, which requires the distribution of the growth product. Providing value to the uses of resources and energy forms the basis of life (Porter and Kramer, 2019). They have always been used as free resources such as air, water, and genetic diversity, especially among wildlife, and a strategy was followed in their use in a way that allows their consumption at rates not exceeding their rates of renewal, in a manner that preserves them from depletion and leaves future generations the opportunity to enjoy them as a resource (Sikes et al., 2016). Work to find alternatives to these resources and adjust their consumption rates to provide the appropriate technology and begin to address the problems of resource depletion and environmental stress, and their effects on the world (Omer, 2008). The science and knowledge base and its accumulations are used to create new technology that reduces pressure on the consumption of natural resources, increases the efficiency of its use, and reduces the quantities of energy used in production, taking into account the long-term effects of the uses of new technologies (Saleh et al., 2019), as well as taking into account the deliberate intervention in ecosystems and the environment (Omer, 2009). For example, the use of genetic engineering to change the biological properties and functions of living organisms, through the technique in which scientists were able to separate the thaumatin gene (a sugary substance found in the fruit of an African plant) to produce it in the laboratory in commercial quantities in bacterial laboratories. The size of this scientific laboratory cultivation on the production of sugar cane and beet in various parts of the world, and the effect of this on the employment in the farms, and on the land was exploited in this regard.

2. Sustainable development goals

Sustainable development, through its contents, seeks to achieve a set of goals that include achieving a better quality of life for the population by focusing on relationships in the activities of the population and the environment (Biggs et al., 2015). It also deals with environmental systems preserving the quality of the environment and reform and preparing and implementing some development projects and programs at all economic, political, social, and cultural levels. In addition, sustainability focuses on the quality of development and achieving justice in the present and future and the exploitation of environmental resources, taking into account the rights of individuals and groups to satisfy the appropriate needs of their needs in the future (Omer, 2009).

Achieving technical—economical growth preserves the natural capital, which includes natural resources and the environment, and this, in turn, requires institutions, infrastructure, and appropriate management of risks and fluctuations, to ensure equality in sharing wealth between successive generations (Hammond and Xie, 2020). Providing and revitalizing global partnership opportunities for development and popular, governmental, and private sector participation in activating education, training, awareness, evaluation, and environmental trends, to stimulate creativity, search for new ways of thinking, use knowledge, unleash and develop human energies and establish the concept of environmental citizenship, and to protect the environment from the problems it threatens (Laven et al., 2010). Analyzing the social, most economical, political, administrative, and environmental conditions as a holistic and integrative vision is primarily based on the harmony of the entire surroundings, and the interconnectedness of its subsystems. Sustainability contributes to developing present environmental resources and rationally using them, without extravagance and waste, through their protection and the invention of recent investments (Olsson and Jerneck, 2018). Employing present-day technology in a way serves the goals of society through the great of the population, and the significance of its goals, without resulting in environmental dangers or being managed (Butzer, 2005).

In the absence of increased efficiency and the reuse and recycling of waste, the volume of waste will continue to rise, and the world will rapidly produce 2.2 billion tons of global, in addition to pollution of soil, water, and air. The World Bank estimates waste by 2025, double its current 1.3 billion tons. Air, water, and soil pollution is an ongoing problem of local and global proportions that harm human health and ecosystems (Kumar et al., 2017; Kaza et al., 2018). According to the Organization for Economic Cooperation and Development, there is already significant exposure to hazardous chemicals worldwide and is likely to increase in the coming decades, particularly in emerging economies and developing countries. Pollutant concentrations are now above safe levels in some cities (Reisch et al., 2013; Saleh et al., 2020c, 2021). The unabated increase in pollution is likely to double the number of premature deaths from airborne particles in urban areas, to 3.6 million deaths per year by 2050, knowing that most deaths will occur in China and India at the same time. The benefit-to-cost ratio of pollution control may be as high as 10 in 1 in emerging economies due to pollution. Respiratory problems can also increase, especially in urban areas (Markandya et al., 2018). At the same time, indoor air pollution from burning biomass, coal, and kerosene cause at least 1.5−2 million premature deaths every year, and most of the victims are from women and children. Pollution trends in general have the potential to exacerbate existing inequalities and vulnerabilities among the poor (Outwater et al., 2013). Freshwater is already scarce in many parts of the world. The water distress is expected to increase, with water supplies expected to meet only 60% of global needs within 20 years (Boretti and Rosa, 2019). The Organization for Economic Cooperation and Development predicts, in its Environmental Outlook to 2050 report, that the number of people living in areas suffering from acute water distress will increase by 2.3 billion people, exceeding 40% of the world's total population in 2050 (Case and Deaton, 2017). The existing water shortage will impede the growth of many economic activities. The sectors of industry, energy generation, human consumption, and agriculture will be increasingly competing with each other for water, which will have serious implications for food security (Popp et al., 2014). The wealth of plant and animal species provides the basis for food production and the provision of raw materials for a range of basic commodities and products, from textiles and construction materials to paper and pharmaceuticals. The number and diversity of species are crucial to the stability of ecosystems (Popa, 2018). Its mortality rate is now 100−1000 times higher than what can be considered normal. Approximately 30% of all mammal, bird, and amphibious species will be at risk of extinction during this century. While the main drivers of biodiversity loss are land use and management change as well as pollution, climate change is expected to become the fastest-growing factor in biodiversity loss by 2050 (Powers and Jetz, 2019). Despite the vital function of ecosystem services, biodiversity and its immense value are often overlooked. These values and the costs of losing them are not systematically reflected in national accounts. Remember as the market signals in the business decision-making process (Salles, 2011).

Continued biodiversity loss and ecosystem degradation reduce the ability of biodiversity and ecosystems to provide essential life-sustaining services (Wang et al., 2019).

One of the most dangerous global threats, and one that exacerbates other environmental concerns such as water scarcity and loss of biodiversity, is climate change. Climate change, in the medium and long term, leads to an increase in global average temperatures, changes in rainfall regimes, and an increase in sea levels. In the short term, the impacts of climate change are attributable to variable weather patterns and more severe weather events (Hanjra and Qureshi, 2010). The cause of climate change is primarily the increase in greenhouse gas concentrations in the atmosphere, mainly caused by the burning of fossil fuels, biomass, livestock, irrigation of paddy fields, and the use of nitrogenous fertilizers. These greenhouse gases trap more of the energy the Earth receives from the sun, giving an effect similar to that of a greenhouse (Wang et al., 2012; Tian et al., 2016). Energy supply and agriculture-related activities together account for about 57% of total emissions. Industrial activity (i.e., manufacturing) and transportation are the two main sources of emissions, with 19% and 13%, respectively (Gavrilova and Vilu, 2012).

High-income countries remain the largest emitters of greenhouse gases per capitates in 2011. Increased emissions, under the future scenarios, could increase the concentration of greenhouse gases in the atmosphere from 390.5 ppm to 685 ppm of 25 greenhouse gases in the atmosphere, with a temperature increase likely of between 3 and 6°C by 2050 (Agustina et al., 2019; Ramanathan and Feng, 2009). This concentration and the accompanying successive rise in temperatures would result in dire consequences that may be irreversible and would go far beyond the internationally agreed maximum rates in the range of 450 parts per million and 2°C. Because greenhouse gases remain active for long periods in the atmosphere, global temperatures and sea levels will continue to rise for centuries even after stabilizing the level of greenhouse gas concentrations (Price et al., 2013). Hence, the future challenge is to reduce emissions drastically and rapidly, as well as to adapt to climate change that has become a fait accompli and will last for decades due to the already released greenhouse gas emissions (Price et al., 2013).

The phenomenon of drought, desertification, and scarcity of lands suitable for the exploitation of agricultural activities that afflict many developing countries, as a result of harsh climatic conditions, such as low rain rates, high temperatures, and evaporation rates, and the acute shortage and pollution of water resources (Akinmoladun et al., 2019). In addition, the weakness of the human capacity capable of dealing with knowledge, technologies to benefit from advanced research in this field, and the delay of educational and research institutions, especially in those countries, to keep pace with scientific and technical progress in the world (Akinmoladun et al., 2019; Yasmin et al., 2019).

3. Recent environmental technologies to reach sustainability

The purposes of the environment are achieved through the appropriate selection of plants in the appropriate locations for them (Poff et al., 2010), as well as taking into account other design elements, which made the design work to reduce pollution, protect from wind and unwanted solar radiation, control the degree of humidity, regulate air movement, reduce noise, and provide unspoiled air while minimizing erosion factors while positively affecting the characteristics of the existing ecosystem. Plants are the main component of the design of the green area, and after studying and knowing fully the nature of their growth and the characteristics of each one, they are chosen (Cuellar and Idir, 2014). These plants are placed in the appropriate place for it to fulfill the required purpose of its cultivation and use, whether it is placed singly in the middle of green spaces or groups or as background scenes for the identification or in groups adjacent to any element to show its surroundings, or to achieve diversity (Rosol, 2010). The factors that determine the choice of plants are divided into primary and secondary elements, where the primary elements include the type of plants used, whether they are trees, shrubs, ground plants or ground cover, the height and width of the plants used, the distinctive texture and shape of the plant, seasonal characteristics, and color of the plants (Olson et al., 2018). The secondary factors include the plant's ability to withstand drought, plant resistance to diseases and insects, as well as the ability of the plant to adapt to the soil and the characteristics of the land such as irregular inclinations and depressions, stones, rocks, and cornices, the plant's tolerance of the sun and shade, the plant's tolerance of moisture, and exposure to wind and salinity (Qasem, 2015). The process of appropriate selection of these elements is considered extremely important for the sustainability of green areas, as they must be put in place and they must all be taken into consideration without ignoring any of them so that the appropriate selection and organization of suitable plants in the appropriate sites for them to achieve the desired purpose of it (Barbosa et al., 2012).

Environmental sustainability is a state of positive interaction between humans and the components of the environment to ensure its continuity. Sustainability aims to conserve resources and protect all components of the environment. The plant is one of the basic elements of human and animal life, and scientists classify it within the product series, and it is the first in the group of living chains, which depend on each other, and the products provide food for themselves and other organisms that are known as consumers, such as humans and animals (Brown et al., 2019). Agriculture is an essential pillar in the process of preserving the environment. Trees purify the air from pollution, release oxygen gas into the atmosphere, absorb carbon, and give humans many advantages and benefits. The urban population of the world is 3.9 billion. More than half of them live in "small" cities with a population of no more than 500,000, while about 12% reside in major cities (more than 10 million people) (Hill, 2020). By 2050, an estimated two-thirds of the world's population will live in urban centers or about 6.2 billion people. Interestingly, the fastest-growing urban settlements are not the megacities that often hit the headlines, but the medium and small cities of less than 1 million people (Verhezen et al., 2016). By 2025, megacities will account for only 10% of global urban growth. Medium and large cities will contribute more than half of the global growth, followed by small cities. Most medium and small cities will be in low- and middle-income countries, and they will often face different sustainability challenges than large cities. In general, poverty rates—for example—may be higher, and challenges may relate more to the efficiency of basic services rather than their availability (Rob and Talukder, 2013).

Urban growth is often supported by natural population growth and rural-to-urban migration (Doan and Oduro, 2012). In the first place, migration is driven primarily by economic incentives such as trade, as well as the drive to improve the quality of life. In addition to the pressures arising from an increasing population, cities face numerous environmental, social, and economic challenges.

To meet these challenges, some cities in the developing world are transforming, and are pioneering innovative planning, integrative design, and technology use. Whether they are called sustainable cities, eco-friendly cities, low-carbon cities, "smart" cities, or zero-energy cities, there is a striving for a safe and healthy environment for all residents (De Jong et al., 2013). These cities share the basic characteristics of sustainable development, such as reduced energy consumption, minimal encroachment on environmental spaces, reduced use of harmful building materials, or expansion of closed systems for waste treatment (Warner, 2010). Some cities aim to reduce carbon emissions, either by developing high-tech systems for storing greenhouse gases to keep track of emissions or by creating "green" or "walkable" neighborhoods through participatory planning (Wachsmuth and Angelo, 2018). Societies use modern technologies to map the poor areas, helping planners and politicians know how to provide better services. Nature simulation represents a new scientific approach to designing buildings that use nature as a model, scale, and teacher in solving complex human problems (Wachsmuth and Angelo, 2018). Technology has transformed the means for monitoring, managing, and designing urban systems, such as those for traffic and water. The most successful technologies are combined with programs that encourage their use, providing tools that enhance sustainability. But these systems can sometimes be costly, and they may not necessarily be suitable for rapidly urbanizing medium cities (Dhakal and Chevalier, 2017).

4. Mechanisms for activating solar energy applications

The issue of using new and renewable energy, especially concentrated solar energy, has emerged in the global arena as one of the strategic options to meet the local and global future energy requirements. Perhaps it is well established that there is a strong correlation between the success of development and the energy saved as the main driver for it (Dubash and Florini, 2011). This added a very important dimension that is evident with the beginning of the depletion of the traditional sources of energy during the next 30 years, as well as another issue related to the contribution of the process of saving energy based on the traditional sources

of energy to the steady rise of the global energy centers (Kellens et al., 2017). Accordingly, efforts seek to identify the appropriate mechanisms to activate the systems of using solar energy in our societies to achieve and activate the principles of sustainability in the development process (Bocken et al., 2014). The most important of these were research and development, partnership and financing, awareness and motivation, legislation and law, inclusion in development plans for regions, and urban communities (Trencher et al., 2014). Because of the close association between environmental pressures and high pollution rates on the one hand, and the increasing urban development, on the other hand, efforts supporting sustainability adopt a set of mechanisms to reduce the negative impacts of increasing urbanization on the environment. The trend toward the use of renewable energy sources, especially those generated from focusing the sun's rays at all levels, by monitoring the effects of the use of traditional energy on the steady increase in pollution rates in urban centers, is based on many experiences, also reviewing the most important experiences in the application of energy technologies generated from concentrating sunlight in urban centers and their effects in reducing pollution. In addition to extracting the most important positive effects of using this technology in protecting the environment and achieving the principles of sustainability (Daramola and Ibem, 2010). Also, to identify the appropriate mechanisms to activate the systems of using solar energy applications and reaching sustainable cities (Daramola and Ibem, 2010). The sun is the main source of many energy sources in nature. Solar energy is used directly in many applications, the most important of which are heating, lighting, water heating, cooling, steam production, seawater desalination, and thermal electricity generation. By 2025, solar thermal systems will boost electricity by 130 gigawatts to achieve a carbon-neutral grid and produce clean fuels (Sahoo, 2016).

Global interest in the environment and achieving its sustainability has led his current interests in sustainable clean energy sources of all kinds (wind, sunlight concentration, bioenergy, geothermal heat, hydropower) as a strategic option to provide the future requirements for energy development, especially with the presence of many of the most important sources of energy (Abbasi and Abbasi, 2012). The relationship between this and the climate changes resulting from global warming is one of the most important causes of which is the use of traditional sources to save energy, such as oil and gas, which contribute about 26% to emissions (Abbasi and Abbasi, 2012). Given the importance of the issue and its effects at global levels, there have been numerous studies and research that dealt with it, and it has recommended the necessity to reduce carbon dioxide emissions worldwide until the middle of the current century by about 30% so that the concentration in the atmosphere can be stabilized at 450 parts per million (Bilgen, 2014). There are many fields of energy use generated from the concentration of sunlight, and the levels of their applications vary. Indeed, the matter now goes beyond the fields of use until it is presented as one of the strategic axes of the developmental visions of urban centers in many experiences at the global level (Oyedepo, 2012). This is in pursuit of the principles of sustainability and preserving the environment from degradation as a global goal that states will unite toward achieving. For example, a thermal turbine system using concentrated solar energy. It consists of a group of mirrors reflecting the sun rays distributed in arrays according to the total area required to generate the required volume of thermal energy (Oyedepo, 2012). This system relies on its work on concentrating the largest possible amount of sun rays on a tank or tube containing a brine solution, which leads to heating this solution to degrees of use to turn it into steam that drives a turbine. It is worth noting that there are different types of this system, according to the

mechanism of generation and storage using separate units of photovoltaic cells to generate the electricity needed for outdoor lighting works. In addition to the use of separate units of cells, solar thermal water heaters generate the electricity required for water heating works in public and private facilities in urban centers, by supplying the roofs of buildings with water heaters powered by solar thermal energy (Herrando and Markides, 2016). The establishment of large solar plants within urban areas and major economic areas such as industrial parks, craft areas, and shopping centers will develop manufacturing techniques and reduce costs (Herrando and Markides, 2016; Boiko, 2017).

Many studies have confirmed the existence of correlations between the elements of green innovation and the promotion of environmental sustainability. Green innovation means creating products or production processes aimed at addressing the environmental problems resulting in the product life cycle (Boiko, 2017). It is innovation related to green products and processes, including innovation in technologies that contribute to energy savings, pollution prevention, waste recycling, and green product designs. The importance of green innovation is highlighted by its interest in reducing pollution, improving environmental performance, improving resource productivity, increasing energy efficiency, and reducing waste, as well as reducing the costs of produced materials (Zailani et al., 2015). Green innovation is a green product, green process, and organizational innovation (Zhang and Zhu, 2019). A green product is a process of modifying the use of natural resources and raw materials in line with environmental requirements and standards, and modifying existing production processes mainly to reduce spoilage through production processes and reduce pollution levels to the lowest possible degree as well as production processes and reduce the possibility of benefiting again from their waste by recollecting, treating, and classifying them. Regarding the goal of the green product, Burger Chakraborty et al., (2013), Zhang and Zhu, (2019) indicated that the green product aims to conserve energy and various natural resources, which include products or services that conserve energy and reduce the use of fuels and materials (Burger Chakraborty et al., 2013; Zhang and Zhu, 2019), reducing pollution, which includes products or services that provide clean energy, or preventing, treating, reducing or controlling it, or measuring environmental damage, eliminating the effects of transportation or storage (Burger Chakraborty et al., 2013) Green process means eliminating waste by redefining the existing production process or the existing system so that the same thing is reached at the end of the production line as it is recycled and used again, and this concept contributes to addressing the social and environmental impacts of the pollution process as well as controlling the work environment and reduce the costs involved due to imbalance in production methods (Lombardi and Laybourn, 2012).

5. Conclusion

Sustainability is one of the most prominent modern concepts that have emerged recently and everyone began to research it and hold seminars and workshops related to it. Sustainability in general is a continuous, dynamic, and evolving way of life, and it is not a specific and understandable result that can be achieved after a specific time. It is mainly a process and a political approach rather than a design problem that needs specific magic solutions. The problem stems mainly in the absence of designs designed to define, prepare, and prepare the application of sustainability concepts locally. Land use policies can exacerbate environmental problems, as the fumes and gases rising and the high temperatures on the surface of the earthwork to raise the sea level and thus cause the destruction of many countries, and also the threat posed by climate change and the effects of air pollution. Therefore, sustainable planning policies must achieve sound urban growth along with the development of neglected urban lands and reuse of the existing land through redevelopment, by increasing the oxides of carbon, chlorofluorocarbons, and nitrogen oxides. Methane, and others, which all work to increase the earth's temperature, affects the weather and increases sea levels. The disappearance of forests due to climate and environmental changes and thus the disappearance of their role. Trees, woods, and forests absorb carbon oxides and thus reduce the effect of global heat.

Exhaustion of nonrenewable sources of fuel, minerals, rocks, sand, and others. Therefore, there is a need to replace these energy sources with solar panels and those generated from the wind. Therefore, many trends appeared in an attempt to avoid such problems, such as the various environmental protection groups in an attempt to reduce the negative effects, conserve resources, and not exhaust them as much as possible. Among the most prominent of these ideas was sustainability, which is a development that does not destroy the environment, and that increases the city's ability to be economically and socially sustainable. The idea of sustainable development emerged as a response to environmental regression. Sustainability does not only aim to protect the environment, it also seeks to reduce the consumption and demand for raw materials. Consequently, this leads to its continuation for a longer time for future generations. No doubt, implementing sustainability programs requires a great collective effort. It is quite clear that we are living in an unsustainable way. This requires us to think seriously about changing how we reuse natural resources into a more sustainable methodology. One of the most important human impacts on earth systems is the destruction of natural resources. Therefore, more stringent practices in managing these resources must be applied to several sectors such as manufacturing, agriculture, and others.

The culture of individual sustainability must also be promoted and encouraged by changing the lifestyle of individuals and societies by reducing the use of natural land resources.

References

Abbasi, T., Abbasi, S.A., 2012. Is the use of renewable energy sources an answer to the problems of global warming and pollution? Crit. Rev. Environ. Sci. Technol. 42, 99–154.

Agustina, R., Dartanto, T., Sitompul, R., et al., 2019. Universal health coverage in Indonesia: concept, progress, and challenges. Lancet 393, 75–102.

Akinmoladun, O.F., Muchenje, V., Fon, F.N., 2019. Small ruminants: farmers' hope in a world threatened by water scarcity. Animals 9, 456.

Altieri, M.A., Nicholls, C.I., Montalba, R., 2017. Technological approaches to sustainable agriculture at a crossroads: an agroecological perspective. Sustainability 9, 349.

Barbosa, A.E., Fernandes, J.N., David, L.M., 2012. Key issues for sustainable urban stormwater management. Water Res. 46, 6787–6798.

Bibri, S.E., 2018. The IoT for smart sustainable cities of the future: an analytical framework for sensor-based big data applications for environmental sustainability. Sustain. Cities Soc. 38, 230–253.

Biggs, E.M., Bruce, E., Boruff, B., et al., 2015. Sustainable development and the water–energy–food nexus: a perspective on livelihoods. Environ. Sci. Pol. 54, 389–397.

Bilgen, S., 2014. Structure and environmental impact of global energy consumption. Renew. Sustain. Energy Rev. 38, 890–902.

Bocken, N.M.P., Short, S.W., Rana, P., Evans, S., 2014. A literature and practice review to develop sustainable business model archetypes. J. Clean. Prod. 65, 42–56.

Boiko, V., 2017. Diversification of business activity in rural areas as a risk minimization tool of economic security. Manag. Theor. Stud. Rural Bus. Infrastruct. Dev. 39, 19–32.

Boretti, A., Rosa, L., 2019. Reassessing the projections of the world water development report. NPJ Clean Water 2, 1–6.

Brown, K., Adger, W.N., Devine-Wright, P., et al., 2019. Empathy, place and identity interactions for sustainability. Global Environ. Change 56, 11–17.

Burger Chakraborty, L., Qureshi, A., Vadenbo, C., Hellweg, S., 2013. Anthropogenic mercury flows in India and impacts of emission controls. Environ. Sci. Technol. 47, 8105–8113. https://doi.org/10.1021/es401006k.

Butzer, K.W., 2005. Environmental history in the Mediterranean world: cross-disciplinary investigation of cause-and-effect for degradation and soil erosion. J. Archaeol. Sci. 32, 1773–1800.

Case, A., Deaton, A., 2017. Mortality and morbidity in the 21st century. Brookings Pap. Econ. Act 2017 397–476.

Criekemans, D., 2018. Geopolitics of the renewable energy game and its potential impact upon global power relations. In: The Geopolitics of Renewables. Springer, pp. 37–73.

Cuellar, C., Idir, N., 2014. EMI filter design methodology taking into account the static converter impedance. In: 2014 16th European Conference on Power Electronics and Applications. IEEE, pp. 1–10.

Daramola, A., Ibem, E.O., 2010. Urban environmental problems in Nigeria: implications for sustainable development. J. Sustain. Dev. Afr. 12, 124–145.

De Jong, M., Wang, D., Yu, C., 2013. Exploring the relevance of the eco-city concept in China: the case of Shenzhen Sino-Dutch low carbon city. J. Urban Technol. 20, 95–113.

Dhakal, K.P., Chevalier, L.R., 2017. Managing urban stormwater for urban sustainability: barriers and policy solutions for green infrastructure application. J. Environ. Manag. 203, 171–181.

Doan, P., Oduro, C.Y., 2012. Patterns of population growth in peri-urban Accra, Ghana. Int. J. Urban Reg. Res. 36, 1306–1325.

Dubash, N.K., Florini, A., 2011. Mapping global energy governance. Glob. Policy 2, 6–18.

Flint, R.W., 2013. Basics of sustainable development. In: Practice of Sustainable Community Development. Springer, pp. 25–54.

Gavrilova, O., Vilu, R., 2012. Production-based and consumption-based national greenhouse gas inventories: an implication for Estonia. Ecol. Econ. 75, 161–173.

Gude, V.G., 2016. Desalination and sustainability—an appraisal and current perspective. Water Res. 89, 87–106.

Hammond, M.J., Xie, J., 2020. Towards Climate Resilient Environmental and Natural Resources Management in the Lake Victoria Basin.

Hanjra, M.A., Qureshi, M.E., 2010. Global water crisis and future food security in an era of climate change. Food Pol. 35, 365–377.

Herrando, M., Markides, C.N., 2016. Hybrid PV and solar-thermal systems for domestic heat and power provision in the UK: techno-economic considerations. Appl. Energy 161, 512–532. https://doi.org/10.1016/j.apenergy.2015.09.025.

Hill, M.K., 2020. Understanding Environmental Pollution. Cambridge University Press.

Hume, M., 2010. Compassion without action: examining the young consumers consumption and attitude to sustainable consumption. J. World Bus. 45, 385–394.

Kaza, S., Yao, L., Bhada-Tata, P., Van Woerden, F., 2018. What a Waste 2.0: A Global Snapshot of Solid Waste Management to 2050. The World Bank.

Kellens, K., Baumers, M., Gutowski, T.G., et al., 2017. Environmental dimensions of additive manufacturing: mapping application domains and their environmental implications. J. Ind. Ecol. 21, S49–S68.

Kumar, A., Holuszko, M., Espinosa, D.C.R., 2017. E-waste: an overview on generation, collection, legislation and recycling practices. Resour. Conserv. Recycl. 122, 32–42.

Laven, D., Ventriss, C., Manning, R., Mitchell, N., 2010. Evaluating US national heritage areas: theory, methods, and application. Environ. Manage. 46, 195–212.

Lisca, C.F., 2014. Criteria, standards and quality assurance mechanisms of the educational management system in Romania. Valahian J Econ. Stud. 5, 49.

Lombardi, D.R., Laybourn, P., 2012. Redefining industrial symbiosis: crossing academic–practitioner boundaries. J. Ind. Ecol. 16, 28–37.

Machovina, B., Feeley, K.J., Ripple, W.J., 2015. Biodiversity conservation: the key is reducing meat consumption. Sci. Total Environ. 536, 419–431.

Markandya, A., Sampedro, J., Smith, S.J., et al., 2018. Health co-benefits from air pollution and mitigation costs of the Paris Agreement: a modelling study. Lancet Planet Heal. 2 e126–e133.

Moldan, B., Janoušková, S., Hák, T., 2012. How to understand and measure environmental sustainability: indicators and targets. Ecol. Indicat. 17, 4–13.

Olson, M.E., Soriano, D., Rosell, J.A., et al., 2018. Plant height and hydraulic vulnerability to drought and cold. Proc. Natl. Acad. Sci. U.S.A. 115, 7551–7556.

Olsson, L., Jerneck, A., 2018. Social fields and natural systems. Ecol. Soc. 23.

Omer, A.M., 2008. Energy, environment and sustainable development. Renew. Sustain. Energy Rev. 12, 2265–2300.

Omer, A.M., 2009. Energy use and environmental impacts: a general review. J. Renew. Sustain. Energy 1, 53101.

Outwater, A.H., Ismail, H., Mgalilwa, L., et al., 2013. Burns in Tanzania: morbidity and mortality, causes and risk factors: a review. Int. J. Burns Trauma. 3, 18.

Oyedepo, S.O., 2012. On energy for sustainable development in Nigeria. Renew. Sustain. Energy Rev. 16, 2583–2598.

Phillips, J., 2011. The conceptual development of a geocybernetic relationship between sustainable development and environmental impact assessment. Appl. Geogr. 31, 969–979.

Poff, N.L., Richter, B.D., Arthington, A.H., et al., 2010. The ecological limits of hydrologic alteration (ELOHA): a new framework for developing regional environmental flow standards. Freshw. Biol. 55, 147–170.

Popa, V.I., 2018. Biomass for fuels and biomaterials. In: Biomass as Renewable Raw Material to Obtain Bioproducts of High-Tech Value. Elsevier, pp. 1–37.

Popp, J., Lakner, Z., Harangi-Rakos, M., Fari, M., 2014. The effect of bioenergy expansion: food, energy, and environment. Renew. Sustain. Energy Rev. 32, 559–578.

Porter, M.E., Kramer, M.R., 2019. Creating shared value. In: Managing Sustainable Business. Springer, pp. 323–346.

Powers, R.P., Jetz, W., 2019. Global habitat loss and extinction risk of terrestrial vertebrates under future land-use-change scenarios. Nat. Clim. Change 9, 323–329.

Price, D.T., Alfaro, R.I., Brown, K.J., et al., 2013. Anticipating the consequences of climate change for Canada's boreal forest ecosystems. Environ. Rev. 21, 322–365.

Qasem, J.R., 2015. Prospects of wild medicinal and industrial plants of saline habitats in the Jordan valley. Pakistan J. Bot. 47, 551–570.

Ramanathan, V., Feng, Y., 2009. Air pollution, greenhouse gases and climate change: global and regional perspectives. Atmos. Environ. 43, 37–50.

Rands, M.R.W., Adams, W.M., Bennun, L., et al., 2010. Biodiversity conservation: challenges beyond 2010. Science 329 (80-), 1298–1303.

Reisch, L., Eberle, U., Lorek, S., 2013. Sustainable food consumption: an overview of contemporary issues and policies. Sustain. Sci. Pract. Pol. 9, 7–25.

Rob, U., Talukder, M.N., 2013. Urbanization prospects in Asia: a six-country comparison. Int. Q Community Health Educ. 33, 23–37.

Rosol, M., 2010. Public participation in post-Fordist urban green space governance: the case of community gardens in Berlin. Int. J. Urban Reg. Res. 34, 548–563.

Sahoo, S.K., 2016. Renewable and sustainable energy reviews solar photovoltaic energy progress in India: a review. Renew. Sustain. Energy Rev. 59, 927–939.

Saleh, H.M., Bondouk, I.I., Salama, E., Esawii, H.A., 2021. Consistency and shielding efficiency of cement-bitumen composite for use as gamma-radiation shielding material. Prog. Nucl. Energy 137, 103764. https://doi.org/10.1016/j.pnucene.2021.103764.

Saleh, H.M., El-Sheikh, S.M., Elshereafy, E.E., Essa, A.K., 2019. Performance of cement-slag-titanate nanofibers composite immobilized radioactive waste solution through frost and flooding events. Constr. Build. Mater. 223, 221–232. https://doi.org/10.1016/j.conbuildmat.2019.06.219.

Saleh, H.M., Moussa, H.R., El-Saied, F.A., Dawoud, M., Bayoumi, T.A., Wahed, R.S.A., 2020a. Mechanical and physicochemical evaluation of solidified dried submerged plants subjected to extreme climatic conditions to achieve an optimum waste containment. Prog. Nucl. Energy 122, 103285. https://doi.org/10.1016/j.pnucene.2020.103285.

Saleh, H.M., Refaat, F.A., Hazem, H.M., 2020b. Qualification of corroborated real phytoremeediated radioactive wastes under leaching and other weathering parameters. Prog. Nucl. Energy 119, 103178.

Saleh, H.M., Salman, A.A., Faheim, A.A., El-Sayed, A.M., 2020c. Sustainable composite of improved lightweight concrete from cement kiln dust with grated poly (styrene). J. Clean. Prod. 277, 123491. https://doi.org/10.1016/j.jclepro.2020.123491.

Salles, J.-M., 2011. Valuing biodiversity and ecosystem services: why put economic values on Nature? C R Biol. 334, 469−482.

Schlosberg, D., Coles, R., 2016. The new environmentalism of everyday life: sustainability, material flows and movements. Contemp. Polit. Theor. 15, 160−181.

Sharifi, A., Murayama, A., 2013. A critical review of seven selected neighborhood sustainability assessment tools. Environ. Impact Assess. Rev. 38, 73−87.

Sikes, R.S., Mammalogists, A.C., UC of the AS of, 2016. 2016 Guidelines of the American Society of Mammalogists for the use of wild mammals in research and education. J. Mammal. 97, 663−688.

Tian, H., Lu, C., Ciais, P., et al., 2016. The terrestrial biosphere as a net source of greenhouse gases to the atmosphere. Nature 531, 225−228.

Trencher, G., Bai, X., Evans, J., et al., 2014. University partnerships for co-designing and co-producing urban sustainability. Global Environ. Change 28, 153−165.

Verhezen, P., Williamson, I., Crosby, M., Soebagjo, N., 2016. Doing Business in ASEAN Markets: Leadership Challenges and Governance Solutions across Asian Borders. Springer.

Wachsmuth, D., Angelo, H., 2018. Green and gray: new ideologies of nature in urban sustainability policy. Ann. Assoc. Am. Geogr. 108, 1038−1056.

Wang, M., Han, J., Dunn, J.B., et al., 2012. Well-to-wheels energy use and greenhouse gas emissions of ethanol from corn, sugarcane and cellulosic biomass for US use. Environ. Res. Lett. 7, 45905.

Wang, J., Zhai, T., Lin, Y., et al., 2019. Spatial imbalance and changes in supply and demand of ecosystem services in China. Sci. Total Environ. 657, 781−791.

Warner, K., 2010. Global environmental change and migration: governance challenges. Global Environ. Change 20, 402−413.

Winder, G.M., Le Heron, R., 2017. Assembling a Blue Economy moment? Geographic engagement with globalizing biological-economic relations in multi-use marine environments. Dialogues Hum. Geogr. 7, 3−26.

Yasmin, H., Nosheen, A., Naz, R., et al., 2019. Regulatory role of rhizobacteria to induce drought and salt stress tolerance in plants. In: Field Crops: Sustainable Management by PGPR. Springer, pp. 279−335.

Zailani, S., Govindan, K., Iranmanesh, M., et al., 2015. Green innovation adoption in automotive supply chain: the Malaysian case. J. Clean. Prod. 108, 1115−1122.

Zhang, F., Zhu, L., 2019. Enhancing corporate sustainable development: stakeholder pressures, organizational learning, and green innovation. Bus. Strat. Environ. 28, 1012−1026. https://doi.org/10.1002/bse.2298.

CHAPTER

24

Effectiveness and efficiency of nano kaolin clay as bitumen modifier: part A

Ramadhansyah Putra Jaya[1], Haryati Yaacob[2], Norhidayah Abdul Hassan[2], Zaid Hazim Al-Saffar[3]

[1]Department of Civil Engineering, College of Engineering, University of Malaysia Pahang, Kuantan, Pahang, Malaysia; [2]Faculty of Engineering, School of Civil Engineering, University of Technology Malaysia, Skudai, Malaysia; [3]Building and Construction Eng. Technical College of Mosul, Northern Technical University, Iraq

1. Introduction

Increasing traffic volumes and rising cost of bitumen make it necessary to improve the engineering properties and performance of binder through bitumen modification (Liu and Wu, 2011). Various types of modifiers have been employed as bitumen binders in order to improve the properties of the bitumen mixture (Kaloush et al., 2010; Ye et al., 2009), particularly with regard to its resistance to aging, cracks due to fatigue and thermal conditions, moisture-induced damage, and permanent deformation. However, fibers and polymers are common materials used in bitumen modification (Anurag et al., 2009; Liu et al., 2009).

Currently, nonmetallic mineral resources such as clay, silica, and so on are widely used as a constituent part or substitution element in the production of modified asphalt binder to modify conventional binder properties (Golestani et al.,

2015). These mineral resources normally are formed through the weathering process and can be found abundantly in tropical climate country. These additives come in solid and powder forms that are desirable for modifying and upgrading the properties of construction material. In order to improve the performance of pavement against distress and deterioration of service of life, the bituminous layer should be strengthened with high quality of mineral resources materials to improve its mechanical properties in terms of permanent deformation resistance, fatigue, and so on. The utilization of mineral resources has proved that its fulfilled the sustainable development requirement which is help toward increasing the application of green technology materials.

In the last decade or so, nanomaterials have emerged as the potential solution to greatly

Risk, Reliability and Sustainable Remediation in the Field of Civil and Environmental Engineering
https://doi.org/10.1016/B978-0-323-85698-0.00019-8

© 2022 Elsevier Inc. All rights reserved.

enhance the properties of binder with the purpose of improving the performance of asphalt mixture (Li et al., 2017). Nanomaterials are defined as materials with at least one dimension that falls in the length scale of 1–100 nm. Due to the small size and high surface area, the property of nanomaterials is much different from the macroscale and microscale size materials (Yao et al., 2013). Besides that, nanomaterial also can be considered as the economically interesting material because conventional materials are usually added in percentages varying between 20% and 40%, while nanomaterial at a typical quantity may be between 2% and 5%, which shows it is a light weight material (Ray and Okamoto, 2003; Paul and Robeson, 2008; Jahromi and Khodaii, 2009). This is also supported by Roy et al. (2015), which reported that only a small weight percentage of nanomaterial is sufficient to enhance the specimens.

One of the common and widely used nanomaterials in bitumen is nanoclays (Farias et al., 2016). Nanoclays are nanoparticles of layered mineral silicates in which the class of nanoclay is depending on chemical composition and nanoparticle morphology. Nowadays, researchers start to apply nanoclay in bitumen modification due to significant improvements in mechanical, thermal, and barrier properties.

The mechanical performance of nanoclay as a modifier was evaluated by Juraidah et al. (2019). The mechanical properties findings indicate that nanoclay increased the stiffness, indirect tensile strength, resilient modulus, Marshall stability, and improved the rutting resistance. This mechanical property result coincides with the study conducted by Goh et al. (2011), wherein an enhancement of tensile strength was achieved and moisture susceptibility was improved. Therefore, this study is to address the effectiveness and efficiency of nano kaolin clay (NKC) as bitumen modifier.

2. Preparation of NKC

Kaolin Clay (KC) was used as a modifier in bitumen through the study. A ball mill machine (Fig. 24.1) was used as grinding machine to produce the NKC. There is no calcination process involved because the KC powder that used in this study was from pure natural clay. Calcination, also known as thermal treatment process, is defined as a process of heating the material to the high temperatures in air or oxygen in which the temperature applied was in the range of 500–700°C for clay within a controlled atmosphere. In addition, there is also no crushing

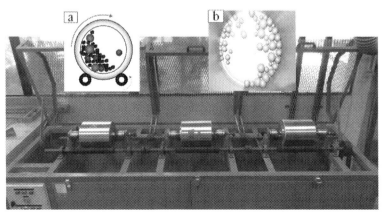

FIGURE 24.1 Bowl ball-mill machine (A) rotation the bowl and shifts and (B) different sizes of balls in the bowl.

3. Chemical analysis

FIGURE 24.2 Pure kaolin clay powder.

process in the preparation of NKC as the KC was already in powdery state (Fig. 24.2).

Approximately 500 g of KC powder was sieved using a sieve size of 0.075 mm to remove agglomerate and impurity materials. The KC powder passing through 0.075 mm sieve size was placed in the drum together with the steel balls to produce the collision force for the sample to be crushed efficiently (Fig. 24.1). The KC powder was ground into five different durations namely 0, 5, 10, 15, and 20 h to obtain the NKC. After completing the grinding time, the KC powder underwent a Transmission Electron Microscopy (TEM) to measure the particle size of KC powder.

3. Chemical analysis

3.1 X-ray fluorescence

The X-ray Fluorescence (XRF) test was conducted in order to determine the elemental composition of the KC powder. The chemical

TABLE 24.1 XRF analysis of raw KC powder.

Oxides	tWt. (%)
SiO_2	53.52
Al_2O_3	45.20
TiO_2	0.47
Fe_2O_3	0.46
CaO	0.12
K_2O	0.11
MgO	0.03
ZrO_2	0.03
P_2O_5	0.02
SrO	0.01
Na_2O	0.01

composition of the KC powder sample was estimated through the quantitative determination of its constitution oxides. Table 24.1 tabulated the XRF analysis of the raw KC powder. It is observed that the major constituents in the raw KC are silicon dioxide [SiO_2] and aluminum oxide [Al_2O_3]. The higher SiO_2 and lower Al_2O_3 content are mainly due to the predominance of clay mineral as well as the presence of considerable amounts of quartz (SiO_2) and feldspar (alkali alumina-silicates) as nonclay minerals (Nurulain et al., 2017; Jaya et al., 2019).

3.2 X-ray diffraction

X-ray diffraction (XRD) was performed to analyze the phase identification of a crystalline material and at the same time provide information on unit cell dimensions. Raw KC powder with 200–250 nm average size particle was used to perform the XRD test. The graph pattern of the raw KC powder generated from the XRD is presented in Fig. 24.3. Based on the XRD graph, it can be noticed that the presence of a large peak was observed for raw KC powder. The sharp peak appeared at the atoms that contained mullite, with the chemical formula $Al_2(Al_{2.5}Si_{1.5})O_{9.75}$, followed by Bismuthpentafluoride $[BiF_5(SbF_5)_3]$ and Zeolite X $[Ca_{43.3}Al_{76.8}Si_{115.2}O_{384}]$. According to Assaedi et al. (2017), mullite is the main crystalline content of the nanoclay; hence, it is stable and unreactive in the alkaline environment. The measurement of the highest peak was $2\theta = 21$ degrees with $d = 4.26$. Generally, the value of "d" represented the peak width, which indicated the crystalline size. The presence of peak indicated that the atom was arranged in the periodic array and was thereby recognized as the crystalline structure. Therefore, this sample was identified as a perfect crystalline structure due to the existence of peak being dominant in the raw KC sample.

3.3 FESEM

Fig. 24.4 illustrated the Field Emission Scanning Electron Microscopy (FESEM) image for NKC with the average particle size being approximately 40–50 nm. The image is observed at 30,000 times magnification with 1.0 μm used as a scale reference in this observation. The figure shows that the formation of the morphology structure was visibly clustered, overlapping and in jointed form. The clay platelets are stacked together in a disordered pattern due to the agglomerates process. Agglomerates happen due to the very fine particles of NKC, which implies that severely curled or crumpled structures are formed much more easily. The image pattern obtained in this study was similar with the findings from Liu et al. (2011), which concluded that the internal structure of nanoclay was massive, aggregated morphology with some bulky flakes and curve plates. This was agreed by Abdullah et al. (2016), who stated that due to a lack of regularity and uniformity in size, the specific surface area of the modified binder becomes larger, which can contribute to a better fatigue resistance.

In addition, a porous structure could also be observed from the FESEM image. This indicated

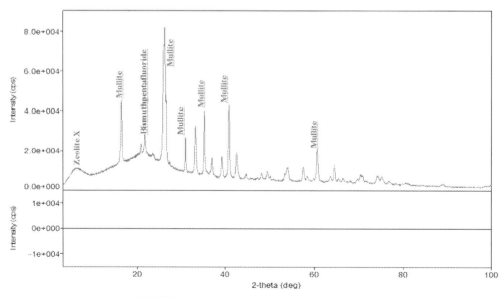

FIGURE 24.3　XRD analysis of raw KC powder.

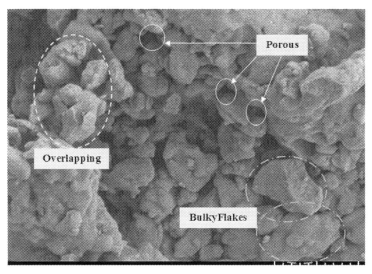

FIGURE 24.4 FESEM image for NKC.

that NKC has high absorption properties. Due to the porous structure of NKC, it can be said that the amount of asphalt binder was reduced when NKC was used as a modifier. Therefore, the design binder content for modified asphalt binder declines from 4.6% to approximately 4.2% when NKC is added in control asphalt binder as discovered in Chapter 6. At the same time, the porous and fibrous structure of NKC may help build a strong chemical bonding that improves the properties and performance of the modified binder.

3.4 EDX

The elemental composition of NKC is illustrated in Fig. 24.5. The Energy Dispersive X-ray (EDX) investigation was considered as an important analysis since it discovered the correlation between the elemental compositions that affected the morphological changes. According to the elemental information generated from the EDX, for the most part, the element composition in the NKC, constituted by oxygen, was denoted as O and recorded about 46.6%. Other elements detected in NKC were Silicon (Si), Aluminum (Al), and Carbon (C). These elements were represented as 22.3% (Si), 21.1% (Al), and 10.0% (C), respectively, in the bituminous mixture constituent.

4. Determination size of NKC

In order to classify as nanomaterials, the particle size should be less than 100 nm or otherwise it can be categorized as ultrafine particles (particle size less than 500 nm).

From Fig. 24.6A and B, it can be observed that most of the raw KC powder before grinding has a particle size ranging from 200 to 250 nm and some particles exceeded 500 nm. However, no nano particle size was found in this sample. Therefore, KC powder before grinding was classified as ultrafine particles. Fig. 24.7A and B shows that the average particle size distribution for 5 h of grinding was in the range of 140–160 nm. A particle size of 250 nm and above was also found in this sample. Even though this sample was produced after 5 h of

454   24. Effectiveness and efficiency of nano kaolin clay as bitumen modifier: part A

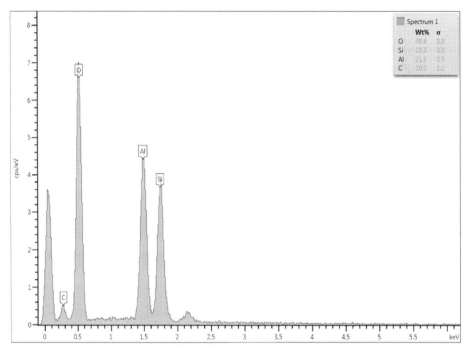

FIGURE 24.5   EDX analysis for NKC.

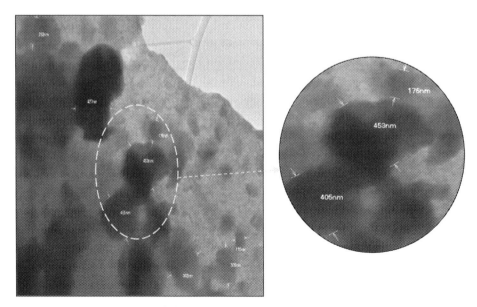

FIGURE 24.6(A)   TEM image of KC powder before grinding.

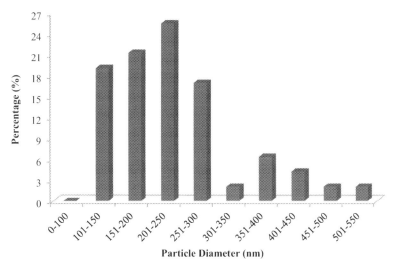

FIGURE 24.6(B) Particle size distribution for KC powder before grinding.

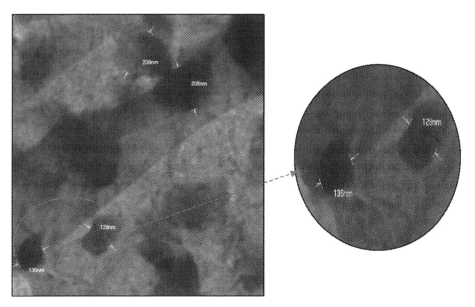

FIGURE 24.7(A) TEM image of KC powder after 5 h grinding.

grinding, no nano particle size was observed. Hence, it can be concluded that 5 h of grinding was insufficient to produce the NKC.

Fig. 24.8A and B indicates the particle size distribution for raw KC powder after 10 h of grinding. Based on the figure, the average particle size distribution was in the range of 70–80 nm. Even though the average particle size distribution was less than 100 nm, some particle sizes were spotted above 100 nm. For that reason, this sample cannot be classified as fully nanomaterial.

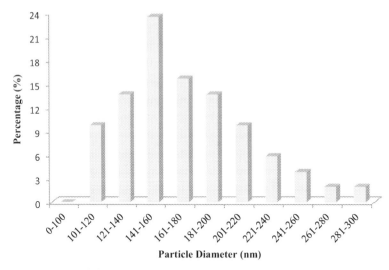

FIGURE 24.7(B) Particle size distribution for KC powder after 5 h grinding.

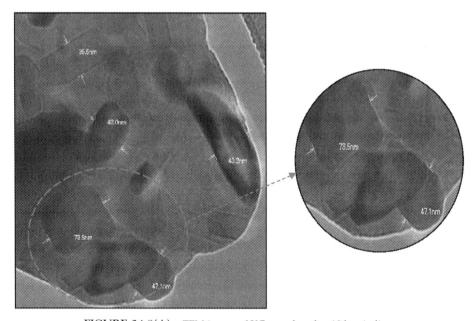

FIGURE 24.8(A) TEM image of KC powder after 10 h grinding.

Figs. 24.9A,B and 24.10A,B show that the particle size distribution ranged was from 40 to 50 nm and 30–40 nm for the raw KC powder after 15 and 20 h of grinding, respectively. Both ranges were within the limit of nanomaterial where the overall particle size was less than 100 nm. Furthermore, no particle size above 100 nm was observed in both samples.

4. Determination size of NKC

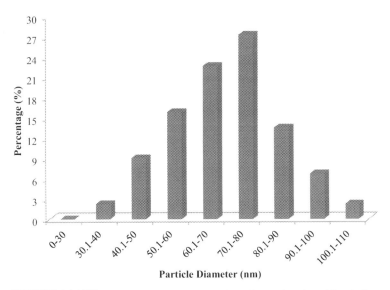

FIGURE 24.8(B) Particle size distribution for KC powder after 10 h grinding.

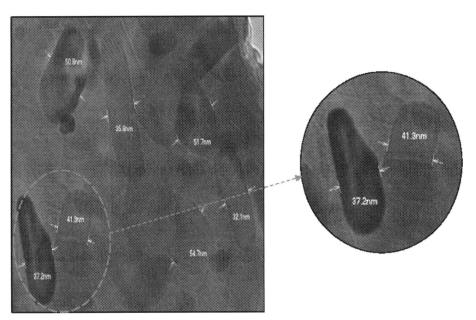

FIGURE 24.9(A) TEM image of KC powder after 15 h of grinding.

Therefore, both samples can be considered as nanomaterials. However, in consideration of cost effectiveness and energy consumption for the production of NKC, the KC powder after 15 h grinding was selected as the optimum grinding.

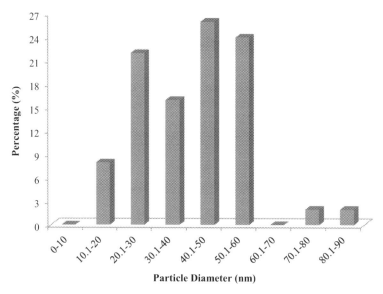

FIGURE 24.9(B)   Particle size distribution for KC powder after 15 h of grinding.

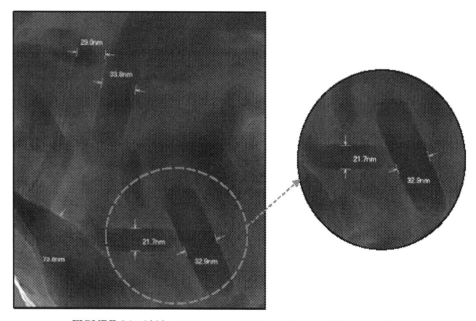

FIGURE 24.10(A)   TEM images of KC powder after 20 h of grinding.

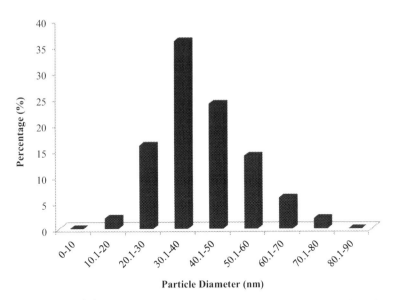

FIGURE 24.10(B) Particle size distribution for KC powder after 20 h of grinding.

5. Conclusions

The study of the morphology structure and chemical composition was extensively observed and evaluated in order to clarify the effectiveness and efficiency of NKC as a modifier before it can be applied in asphalt binder and mixture. The most critical issue in this study is the production of NKC. This is because the previous research works had used the clay that undergoes modification during the manufacturing process. Some of the nanoclay might have been mixed with other materials before it can produce the nanoparticle size. Consequently, no guideline or reference was found in the production of NKC because pure KC powder was used as a modifier in this research. The TEM findings indicated that the 15 h of grinding was sufficient to produce the particle sizes from 40 to 50 nm, and of less than 100 nm. Therefore, this material can be considered as a nanomaterial. The chemical composition of KC powder was also being observed in this study through the XRF analysis. The major chemical compound in KC powder was silicon oxide [SiO_2], also known as silica and aluminum oxide [Al_2O_3], a chemical compound of aluminum and oxygen. In addition, the XRD findings illustrated that NKC was a perfect crystalline structure due to the existence of a dominant sharp peak in the sample. This indicates that NKC was stable and unreactive in the alkaline environment. The results from the FESEM analysis indicated that the internal structure of the NKC was clustered, overlapping, and in jointed form. Agglomerates occurred, which suggests that severely curled or crumpled structures are formed much more easily. This indicated that NKC has a better fatigue resistance due to the irregularity in size and shape. Furthermore, a porous structure was also observed in the internal structure of NKC. This result implied that NKC has high absorption properties, which can reduce the amount of asphalt binder when NKC is used as a modifier. Thus, this can imply that the 15 h of grinding time is sufficient to produce NKC, which can help enhance the properties of asphalt binder and mixture.

Acknowledgments

This study was supported by the Malaysian Ministry of Higher Education, Universiti Malaysia Pahang, and Universiti Teknologi Malaysia in the form of research grant number Q.J130000.2451.09G26 and FRGS/1/2018/TK01/UTM/02/10 and is highly appreciated.

References

Abdullah, M.E., Hainin, M.R., Md Yusoff, N.I., Zamhari, K.A., Hassan, N.A., 2016. Laboratory evaluation on the characteristics and pollutant emissions of nanoclay and chemical warm mix asphalt modified binders. Construct. Build. Mater. 113, 488–497.

Anurag, K., Xiao, F., Amirkhanian, S.N., 2009. Laboratory investigation of indirect tensile strength using roofing polyester waste fibers in hot mix asphalt. Construct. Build. Mater. 23 (5), 2035–2040.

Assaedi, H., Shaikh, F.U.A., Low, I.M., 2017. Effect of nanoclay on durability and mechanical properties of flax fabric reinforced geopolymer composites. J. Asian Ceram. Soc. 5 (1), 62–70.

Farias, L.G.A.T., Leitinho, J.L., Amoni, B.C., Bastos, J.B.S., Soares, J.B., Soares, S.A., Sant'Ana, H.B., 2016. Effects of nanoclay and nanocomposites on bitumen rheological properties. Construct. Build. Mater. 125, 873–883.

Goh, S.W., Akin, M., You, Z., Shi, X., 2011. Effect of deicing solutions on the tensile strength of micro or nano-modified asphalt mixture. Construct. Build. Mater. 25 (1), 195–200.

Golestani, B., Nam, B.H., Nejad, F.M., Fallah, S., 2015. Nanoclay application to asphalt concrete: characterization of polymer and linear nanocomposite-modified asphalt binder and mixture. Construct. Build. Mater. 91, 32–38.

Jahromi, S.G., Khodaii, A., 2009. Effects of nanoclay on rheological properties of bitumen binder. Construct. Build. Mater. 23 (8), 2894–2904.

Jaya, R.P., Che'Mat, N., Hassan, N.A., Mashros, N., Yaacob, H., Hainin, M.R., Satar, M.K.I.M., Ali, M.I., August 29, 2019. Effects of Nano-kaolin clay on the rutting resistance of asphalt binder. AIP Conf. Proc. 2151. Article number 020023.

Juraidah, A., Ramadhansyah, P.J., Azman, M.K., Nurulain, C.M., Naqiuddin, M.W.M., Norhidayah, A.H.,

Haryati, Y., Nordiana, M., 2019. Performance of Nano kaolin clay as modified binder in porous asphalt mixture. IOP Conf. Ser. Earth Environ. Sci. 244 (1) article no. 012036.

Kaloush, K.E., Biligiri, K.P., Zeiada, W.A., Rodezno, M.C., Reed, J.X., 2010. Evaluation of fiber-reinforced asphalt mixtures using advanced material characterization tests. J. Test. Eval. 38 (4), 400–411.

Li, R., Xiao, F., Amirkhanian, S., You, Z., Huang, J., 2017. Development of nano materials and technologies on asphalt materials – a review. Construct. Build. Mater. 143, 633–664.

Liu, X., Wu, S., 2011. Study on the graphite and Carbon fiber modified asphalt concrete. Construct. Build. Mater. 25 (4), 1807–1811.

Liu, D., Luo, Q., Wang, H., Chen, J., 2009. Direct synthesis of micro-coiled Carbon fibers on graphite substrate using Co-electrodeposition of nickel and sulfur as catalysts. Mater. Des. 30 (3), 649–652.

Liu, B., Wang, X., Yang, B., Sun, R., 2011. Rapid modification of montmorillonite with novel cationic gemini surfactants and its adsorption for methyl orange. Mater. Chem. Phys. 130 (3), 1220–1226.

Nurulain, C.M., Ramadhansyah, P.J., Haryati, Y., Norhidayah, A.H., Abdullah, M.E., Wan Ibrahim, M.H., 2017. Performance of macro clay on the porous asphalt mixture properties. IOP Conf. Ser. Mater. Sci. Eng. 271 (1) art. no. 012050.

Paul, D.R., Robeson, L.M., 2008. Polymer nanotechnology: nanocomposites. Polymer 49 (15), 3187–3204.

Ray, S.S., Okamoto, M., 2003. Polymer/layered silicate nanocomposites: a review from preparation to processing. Prog. Polym. Sci. 28 (11), 1539–1641.

Roy, S., Kar, S., Bagchi, B., Das, S., 2015. Development of transition metal oxide-kaolin composite pigments for potential application in paint systems. Appl. Clay Sci. 107, 205–212.

Yao, H., You, Z., Li, L., Goh, S.W., Lee, C.H., Yap, Y.K., 2013. Rheological properties and chemical analysis of nanoclay and Carbon microfiber modified asphalt with fourier transform infrared spectroscopy. Construct. Build. Mater. 38, 327–337.

Ye, Q., Wu, S., Li, N., 2009. Investigation of the dynamic and fatigue properties of fiber-modified asphalt mixtures. Int. J. Fatig. 31 (10), 1598–1602.

CHAPTER

25

Nano kaolin clay as bitumen modifier for sustainable development: part B

Ramadhansyah Putra Jaya[1], Khairil Azman Masri[1], Sri Wiwoho Mudjanarko[2], Muhammad Ikhsan Setiawan[2]

[1]Department of Civil Engineering, College of Engineering, University of Malaysia Pahang, Kuantan, Pahang, Malaysia; [2]Narotama University, Sukolilo, Surabaya, Indonesia

1. Introduction

This chapter presented the continued investigation outlined in *Part A: Effectiveness and Efficiency of Nano Kaolin Clay as Bitumen Modifier*. The asphalt binder incorporating nanoclay was tested in the unaged and aged condition to simulate the pavement condition during service life.

Among the potential nanoparticle, nanoclay received much attention as a modifier due to the expectation to strengthen and enhance the physical and rheological properties of the asphalt binder. Nanoclay is a layered silicate that has a layer thickness between 1 and 100 nm and widely used in the binder modification to improve mechanical and thermal properties. In addition, nanoclay is the most commonly used nanomaterial to modify bituminous material due to their low cost of production and abundance in nature. There are four main groups of clays which are kaolinite, montmorillonite, illite, and chloride. Generally, there are two main categories of nanoclay used in asphalt binder which are nonmodified nanoclay (NMN) and polymer modified nanoclay (PMN). NMN is the most frequently used with 2-to-1 layered structure clay consisting of one octahedral alumina sheet sandwiched between two tetrahedral silica (Ray and Okamoto, 2003).

It has been proved that the utilization of nanoclay in bituminous material shows positive improvement in modified asphalt binder. For instance, Ezzat et al. (2016) studied the effect of heating, cooling, and reheating the binder after mixing with 3%, 5%, and 7% of nanoclay by weight of the binder for 1 hour at a temperature of $145 \pm 5°C$. In order to conduct these studies, samples were prepared and tested for penetration, softening point, and viscosity at times of 1, 5, and 10 days from the mixing time. No significant influence was observed on the test results up to 10 days from the first testing day. However, the penetration for modified binder decreased from 55.5 dmm down to 33.33 dmm at a

Risk, Reliability and Sustainable Remediation in the Field of Civil and Environmental Engineering
https://doi.org/10.1016/B978-0-323-85698-0.00006-X

nanoclay content of 3% by weight of binder. On the other hand, it then increased the penetration up to 67 dmm at 7% nanoclay by weight of the asphalt binder. In addition, the nanoclay increased the viscosity at 135°C by more than 230% at 3%, and decreased the viscosity to 153% at 7% nanoclay content. The nanoclay results indicated that higher percentages of nanoclay material tended to cluster and form batches; therefore, the optimum percentage of 3% nanoclay was recommended. Farias et al. (2016) investigated the effect of two different types of nanoclay (organo-clay montmorillonite (OMMT) and Cloisite 20A) with asphalt binder 50/70 in terms of the physical parameter correlated to pavement performance. They found that the softening point changed depending on the type of nanoclay; it is probably an indication that different nanoclay could be dispersed in a different way in asphalt binder. Cloisite seemed to present a higher effect than OMMT. This could be due to better dispersion and compatibility in the binder.

In addition, Abdullah et al. (2016) evaluated the physical properties of the unmodified and modified asphalt binders using the penetration test, softening point test, penetration index (PI), and penetration-viscosity number (PVN). Their research consists of two different surface-modified montmorillonite types of clay, referred to as Nanoclay A and Nanoclay B. The finding noticed that the addition of nanoclay modified binder A (NCMB A) and nanoclay modified binder B (NCMB B) reduced penetration and the increased softening point value; however, the effect produced by NCMB A was larger compared with the effect of NCMB B. Furthermore, the temperature susceptibility of the modified asphalt binder samples was determined based on the value of PI and PVN. Higher values of PI and PVN represent lower temperature susceptibility, which provides better resistance to the cracking and rutting of asphalt mixture (Tasdemir, 2009; Lee et al., 2009). The PI values increased for both NCMB A and NCMB B. Additionally, the similar trend as PI was observed for PVN value. This indicated that asphalt binder containing nanoclay yields higher PI and PVN values, which could cause the mixtures to be more resistant to thermal cracking and rutting.

2. Rheological properties of the asphalt binder incorporating nanoclay

The rheological properties of unmodified and modified binder were determined by using a Dynamic Shear Rheometer (DSR). The DSR is commonly used to characterize the viscous and elastic behaviors of asphalt by measuring the complex shear modulus (G^*), shear storage modulus (G'), shear loss modulus (G''), and the phase angle (δ). Both G^* and δ provide a measure of rutting resistance to deformation (Paul and Robeson, 2008).

Goh et al. (2011), You et al. (2011), and Xiao et al. (2011) stated in their research that the complex shear modulus of the nano-modified binder increased relative to the control binder, as well as the failure temperature and high temperature performance grade. The rutting resistance performance of nanoclay modified binder would be enhanced. This statement supported the research conducted by You et al. (2011). In their research, they found that when the percentages of nanoclay added in the binder increased, the complex shear modulus significantly increased. An interesting feature in the shear complex moduli results of the DSR test is their convergence as the frequency increases from 0.01 to 100 Hz. It is believed that as the frequency increases to 100 Hz (or traffic loading time decreases), the nanoclay molecules play a lesser role in bearing the shearing load compared to low frequency or increased traffic loading where the action of the nanoclay molecules starts to become prominent. Therefore, it can be concluded that nanoclay presents more positive improvement in low frequency rather than high frequency or lower traffic loading.

3. Asphalt binder characterization

Asphalt binder is one of the main components in the porous asphalt mixture. It acts as a binder to create adhesion and cohesion bonding between the aggregate and asphalt binder. Therefore, the properties of asphalt binder are considered critical parameters, which significantly affect the performance and mechanical properties of the porous asphalt mixture. In this study, different percentages, i.e., 0%, 3%, 5%, 7%, and 9% of the NKC content were tested. Table 25.1 indicates the basic specification of asphalt 60/70.

TABLE 25.1 Specification of asphalt binder 60/70.

Parameter	Specification
Specific gravity	1.01–1.06
Ductility at 25°C and 5 cm/min (cm)	>100
Flash point (°C)	>250
Loss of heating (%)	<0.50
Penetration at 25°C and 100 g for 5 s (mm)	60–70
Softening point (°C)	49–56
Penetration index	−1 to +1

4. Penetration

In order to measure the consistency of the asphalt, a penetration test was conducted at 25°C to stimulate the average yearly service temperature. According to Fig. 25.1, the penetration values of unaged modified binder slightly decreased by 6.9%–13.5% compared to the unaged control binder. The same situation occurs for aged modified binder, where the penetration value declined correspondingly with the addition of NKC by 10.0%–27.8% compared to the aged control binder. Lower penetration values indicate that the asphalt binder is stiff and hard, which can lead to cracking issues (Hussein et al., 2017). Therefore, in order to ensure the long service life of pavement, the asphalt binder used should be as soft as possible without reducing stability below the minimum required to prevent displacement under traffic.

5. Softening point

Fig. 25.2 illustrates the relationship of the softening point value and NKC content. The result

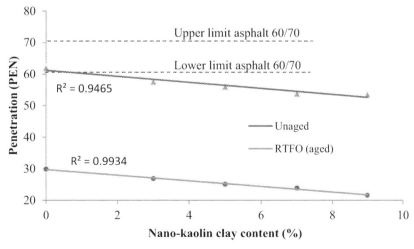

FIGURE 25.1 Relationship between penetration value and NKC content.

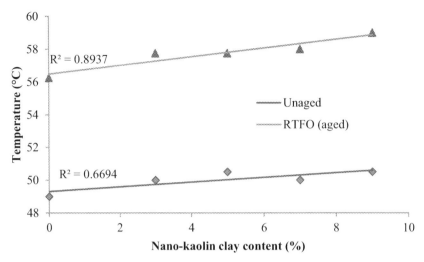

FIGURE 25.2 Relationship between softening point value and NKC content.

shows that the temperatures of the asphalt binder increased significantly with the increasing of NKC content. This finding is identical to the penetration test result, where the softening point value is getting higher if a lower penetration value is observed. From Fig. 25.2, the softening point for the unaged modified binder increased from about 1 to 3°C as compared to the unaged control binder. The similar trend was also observed at softening point for aged modified binder, where the softening point value increased from 2 to 3°C as compared to the aged control binder. The high softening point value in NKC content showed the low temperature susceptibility properties, which had high resistance to temperature exposure. The high softening point performance was desirable since this characteristic can withstand high temperature effect thereby minimizing the presence of permanent deformation deterioration in pavement. Approximately 5% of NKC content was identified as an optimum binder based on the highest softening point.

6. Storage stability

The storage stability test was conducted to ensure that the blending process was effective and the modified binders would remain stable during storage. Stability storage is a critical property for modified binders because of the differences in the solubility parameter and density between the modifier and binder; phase separation would take place during storage at elevated temperatures (Jeffry et al., 2018). Fig. 25.3 shows a comparison between the control and modified binders in terms of their storage stability. In general, increasing the percentage of modifiers would increase the difference in the softening point values between the top and bottom sections. From the result, the difference between the top and bottom sections was less than 2.2°C for all percentages of modified binder. Therefore, it can be said that NKC modified binder improved the storage stable blend.

7. Rutting resistance

The DSR test was used to define the rutting resistance of asphalt binder at high service temperature (Airey et al., 2008 and Xiao et al., 2011). Figs. 25.4 and 25.5 show the rutting resistance for the unaged and aged sample, respectively. In Fig. 25.4, the rutting factor values of modified asphalt are higher than the control binder. A similar trend was observed in

7. Rutting resistance

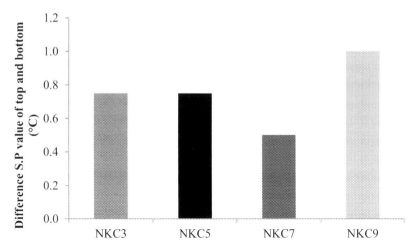

FIGURE 25.3 Storage stability test result for NKC modified binders.

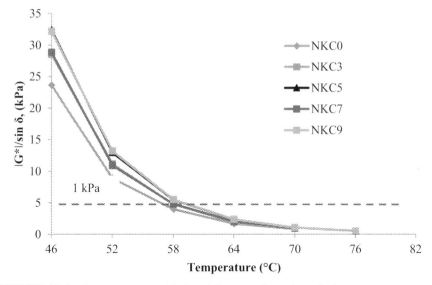

FIGURE 25.4 Rutting resistance (G*/sin δ) for unaged samples at high service temperature.

Fig. 25.5 for the aged condition. However, for the unaged condition, the highest rutting factor occurs at 5% and 9% NKC content, while for the aged condition the highest rutting factor shows at 3% NKC content. This implied that the tendency of the permanent deformation was minimized in modified binder with NKC when the high failure temperature was achieved. This improvement can be attributed to the stiffening effect caused by the addition of the modifiers and is found to be consistent with the viscosity analyses at high service temperatures. According to the ASTM D7175 (ASTM, 2015) and BS EN 14770 (BS EN, 2012) specifications, these modified binders reduce the rutting susceptibility, and also illustrate the improvement on

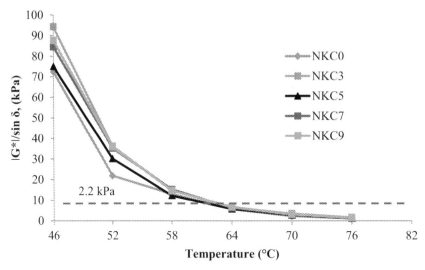

FIGURE 25.5 Rutting resistance (G*/sin δ) for aging samples at high service temperature.

permanent deformation resistance at high temperature compared to the control binder. In addition, as predicted, a higher rutting resistance was achieved at aged conditions as compared to the unaged condition due to aging. Theoretically, aging caused the mechanical properties of binder to become harder and more elastic. The hard binder with high stiffness possessed in aged binder sample was able to withstand against permanent deformation at high temperature exposure, and thereby had increased the rutting resistance performance. Therefore, it can be observed that the higher rutting resistance was achieved by aged sample as compared to the unaged sample.

8. Failure temperature

The failure temperature of asphalt binders was determined based on the data of the rutting resistance test for the unaged and aged condition. Based on Lee et al. (2009), the higher failure temperature values showed that binders were less susceptible to permanent deformation at high service temperatures. Failure temperature is a critical temperature recorded when the rutting resistance (G*/sin δ) is less than 1 kPa for an unaged sample, while for aged sample is it less than 2.2 kPa based on the ASTM specification. It is related to permanent deformation occurrence; that is, the higher the failure temperature, the higher is the resistance to permanent deformation (Hassan et al., 2015). Failure temperature provides an indication of an inclination to temperature resistance and sensitivity toward the temperature alteration (Poovaneshvaran et al., 2020). A lower failure temperature results in a low resistance performance against rutting problem. Fig. 25.6 and 25.7 show the failure temperature at 70°C of the control and modified binders in both unaged and aged conditions, respectively. Based on Fig. 25.6, the unaged control binder shows the lowest complex shear modulus at a temperature of 70°C with 0.816 kPa. The differences in increment from control to modified binders were recorded as 18.5%, 31.7%, 19.1%, and 33.9% for 3%, 5%, 7%, and 9% of NKC replacement. It can be noticed that a higher difference in increment was observed at 5% and 9% of NKC content. This is because the elastic property was enhanced in

8. Failure temperature

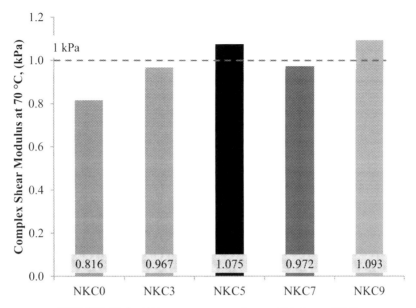

FIGURE 25.6 Failure temperature of unaged binders at 70°C.

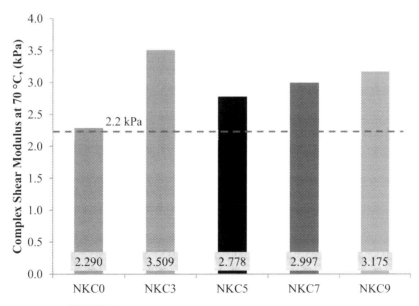

FIGURE 25.7 Failure temperature of aged binders at 70°C.

modified binder with NKC which resulted in high recoverable deformation, and thereby increased rutting resistance performance at high failure temperature. This finding was consistent with the results reported by Nazzal et al. (2012) and You et al. (2011), which indicated that rutting resistance was improved with addition of nanoclay in binder. In addition, the amount of nanoclay may also be one of the parameters that influenced the rutting performance of asphalt binder. A failure temperature increment was noticed after aging condition as compared to the unaged condition, especially for modified binder containing 3% of NKC content as illustrates in Fig. 25.7. However, the difference in increment from control to modified binder was recorded higher as 53.2%, 21.3%, 30.9%, and 38.6% for 3%, 5%, 7%, and 9% of NKC replacement at 70°C failure temperature. The increase in failure temperature, especially at 3% of NKC content, was noticeable due to aging during which the modified binder was hardened and thereby could withstand and increase the rutting resistance at high failure temperature. As a conclusion, the modification of binder with NKC has improved the failure temperature for better resistance performance against permanent deformation and can reflect more energy when loading is applied as compared to the control binder.

9. Phase angle

The phase angle (δ) is defined as the time lag between strain and stress under traffic loading and is highly dependent on the temperature and frequency of loading. Therefore, the measurement of δ is commonly considered to be more sensitive to the chemical and physical structure than the complex modulus for the modification of binder (Ouyang et al., 2006). Under normal pavement temperatures and traffic loading, asphalt binders act in both characteristics, which are viscous liquids and elastic solids. For that reason, δ can be used as an indicator of the viscosity and elasticity of binders. The small phase angle indicated that the binder has high elastic recovery for better resistance to permanent deformation. During the test, δ was recorded automatically at a temperature ranging between 46 and 76°C. Figs. 25.8 and 25.9 show the relationship between δ and temperature for the unaged and aged condition, respectively. Based on the graph, the δ increased linearly as the test temperature increased.

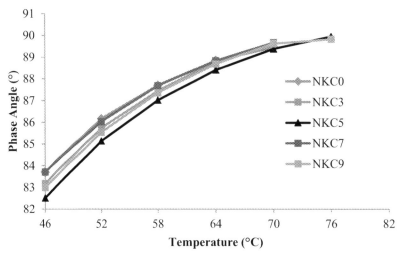

FIGURE 25.8   Changes of phase angle versus temperature for unaged binder.

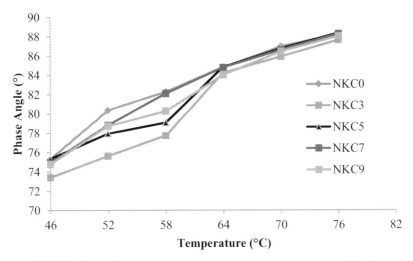

FIGURE 25.9 Changes of phase angle versus temperature for aged binder.

Based on Fig. 25.8, it can be noticed that 5% of the NKC content had the lowest δ values compared to control binder and other percentages of NKC replacement. The lower δ was recorded as 82.4, 85.1, 87.0, 88.4, 89.2, and 89.7 degrees at the failure temperatures of 46, 52, 58, 64, 70, and 76°C, respectively. This result shows that asphalt with 5% NKC possesses a better performance specifically at high temperatures and at the same time approaching the behavior of the ideal binder. Furthermore, a smaller δ value can also indicate that the elasticity recovery performance of the binder was greatly improved. The increasing elasticity characteristic due to the reduction in phase angle improved the complex modulus, thereby exhibiting better rutting resistance performance in modified binder with NKC as compared to control binder.

Fig. 25.9 illustrates a similar trend with unaged sample in which modified aged binder presents a lower δ as comparison to aged control binder. Based on the graph, the increase of δ was linearly recorded as the test temperature increased. It can be noticed that all δ values of the modified binder demonstrate lower δ in comparison to the control binder. However, 3% of the NKC content shows the lowest δ values compared to other percentages in the aged condition. The lower δ was recorded as 73.8, 75.6, 77.9, 84.0, 85.7, and 87.2 degrees at the failure temperatures of 46, 52, 58, 64, 70, and 76°C, respectively. During the aging process, binders become stiff compared to unaged binder, which leads to the alteration of physical characteristic from viscous to elastic property. As a result, it can be concluded that higher stiffness indicated an increase of elastic property in aged asphalt binder containing NKC. The elastic characteristic had high potential for recoverable deformation which in turn improved the rutting resistance performance of modified binder incorporated with NKC.

10. Fatigue resistance

Fatigue resistance ($|G^*|\sin \delta$) for control and modified binder after conditioning (RTFO + PAV residual) was determined using a DSR machine. The specification sets a maximum limit of 5000 kPa, as shown in Fig. 25.10 for the fatigue cracking index at intermediate service temperature. Based on the

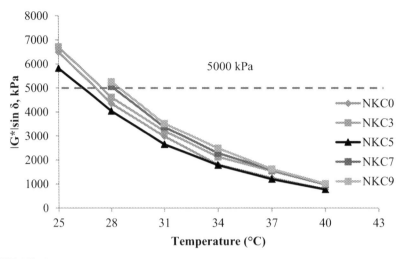

FIGURE 25.10 Fatigue resistance ($|G^*|\sin \delta$) for aged binder at intermediate service temperature.

graph, 5% of NKC content presents the lowest fatigue cracking index compared to the control binder and other percentages of NKC. The lower G*sin δ was recorded as 776.5, 1202.0, 1785.0, 2650, 4040.0, and 5815 kPa at the failure temperatures of 40, 37, 34, 31, 28, and 25°C, respectively. According to Lee et al. (2009) and Yu et al. (2009), the lower $|G^*|\sin \delta$ values are considered desirable from the fatigue cracking resistance point of view. Therefore, it can be said that the utilization of NKC can enhance the fatigue resistance of the asphalt binder. However, Fig. 25.10 also shows that when a high percentage of NKC was replaced in binder, $|G^*|\sin \delta$ decreased compared to the control binder. It can be noticed that modified binder with 7% and 9% of NKC replacement presents earlier failure temperature, at which the samples failed at the temperature of 28°C as compared to control binder. This occurred because the high amount of NKC increases the stiffness of the binder and reduced the elastic property, which can cause the binder to break easily during the test. The elastic property has nonrecoverable deformation when subjected to a lower temperature which thereby had reduced the fatigue resistance at intermediate failure temperature. As a conclusion, the modification of binder with NKC has improved the fatigue resistance as compared to the control binder when appropriate amount of NKC was replaced in the binder.

11. AFM analysis

The atomic force microscopy (AFM) test was conducted in order to measure the average surface roughness of the asphalt binder. In this investigation, two different asphalt binders consisting of the control binder (NKC0) and the modified binder (NKC5) were investigated. According to ASTM D907-15 (ASTM, 2015), adherence can be defined as the chemical–physical interaction that occurs when two different surfaces are held together and come into contact by interlocking forces. The presence of interlocking forces produced the attraction between the molecules for sharing the electrons, thus creating a chemical bonding between materials. The adhesion between the binder and aggregates mainly occurred at their interface; therefore, the surface microstructure of the materials contributed as an important factor in the effectiveness of the adhesion mechanism that needed to be observed. According to Xu et al. (2016), the adhesion rate between the binder and aggregate in the mixture was affected by the binder microstructure surface roughness to attach and coat the aggregates. The higher the surface roughness, the lower the adhesion bonding rate between the contacted materials and vice versa. The high adhesiveness could increase the interlocking bonding and thus improved the strength performance of the asphalt mixture. However in this study, the three distinct phases were observed in the AFM image, which can be divided into a) catana phase ("bee" structure); b) peri phase, which is peripheral to the catana phase; c) and para phase, which is adjacent to the peri phase. The *bumble bees* term was used to name the ripples microstructure, which resembles the yellow black strip (Wan Azahar et al., 2016). According to Jager et al. (2004), the "bees" part, which is the bright area, represents the lager stiffness and lower adhesive force. Meanwhile, the dark area shows a lower stiffness and high adhesive force.

Figs. 25.11 and 25.12 illustrate the AFM images for NKC0 and NKC5, respectively. NKC0 presents the abundance of the catana phase, while a low catana phase was observed in the NKC5. The average surface roughness for NKC0 was 3.062 nm, while the average surface roughness of NKC5 was 2.873 nm, as tabulated in Table 25.2. This indicates that NKC5 has a lower stiffness with a high adhesive force compared to NKC0. From the AFM image observation, the decrease of the catana phase minimized the availability of the high part "bees," which reduced the surface roughness tendency of the modified binder with NKC. This evidence shows that the modified binder produced a low stiffness surface and high adhesive force. Furthermore, the aggregates surface is also significant in the coatability performance, in which a smoother surface can coat easily as compared to a rough surface. Therefore, the smooth and low surface roughness was required to improve the greater adhesion performance.

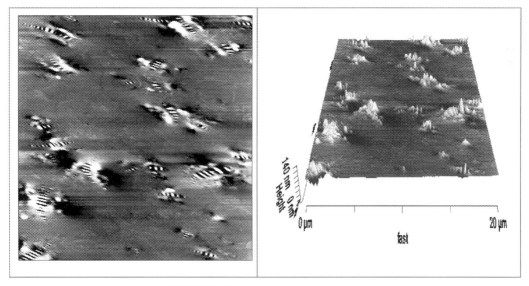

FIGURE 25.11 AFM image for NKC0.

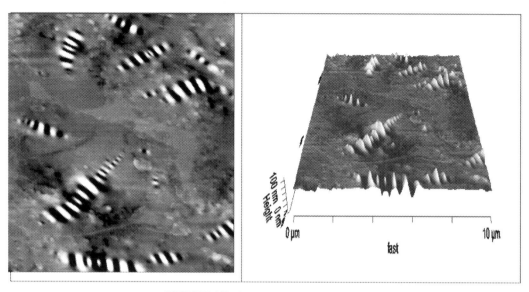

FIGURE 25.12   AFM image for NKC5.

TABLE 25.2   Average surface roughness.

Designation	Average surface roughness (nm)
NKC0	3.062
NKC5	2.873

12. XRD analysis

The XRD test for the control binder (NKC0) with penetration grade 60/70 was conducted in order to investigate the crystalline structure in the sample. The XRD graph pattern was illustrated in Fig. 25.13. In the NKC0 sample, the graph pattern generated from this test interpreted data on the arrangement of atoms inside the material. According to the graph, no sharp peak was observed in this sample. This indicated that the control binder was perfectly amorphous. The amorphous broad phase generated between $2\theta = 15$ degrees and 25 degrees for this sample revealed the reactivity of binder. The highest peak appeared at $2\theta = 19$ degrees with $d = 4.48$. Fig. 25.14 illustrates the trend of the graph for the modified binder incorporating 5% NKC from the XRD test. In the NKC5 sample, the graph pattern interpreted two types of data, which were the arrangement of the atom inside the material and the homogeneity of NKC in the modified binder. A similar graph pattern with the control binder was observed in the modified binder containing NKC, which shows that the sample maintained the amorphous structure. However, small peaks were also observed in this sample, which indicates that the sample was not perfectly amorphous. The crystalline phase found in this sample was Sillimanite [Al_2SiO_5] and Cristobalite [SiO_2], also known as quartz. The highest peak was generated at 20 degrees = 2θ with $d = 4.43$.

By comparing the X-ray diffractograms of NKC0 and NKC5, it was found that a slight shift in the position of 2θ planes (2θ changed from 19–20 degrees) took place, indicating that an increment in the basal spacing of these planes occurred. The result implied that NKC5 is more efficient than NKC0. According to Bakharev (2006), the amorphicity degree remarkably influences the mechanical properties of the modifier. Therefore, when the amorphous content is

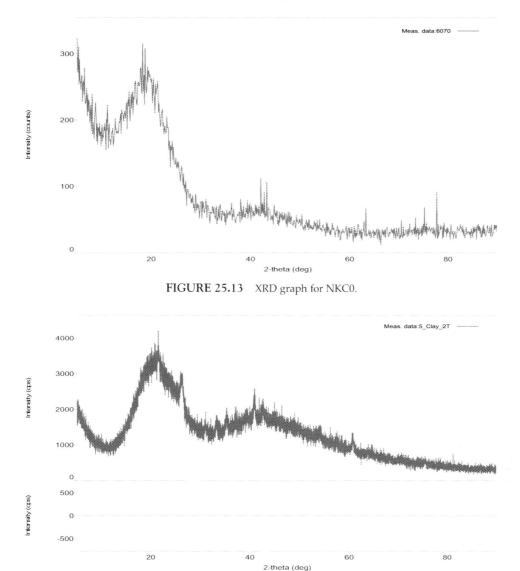

FIGURE 25.13 XRD graph for NKC0.

FIGURE 25.14 XRD graph for NKC5.

higher, the strength of the NKC is also higher. Furthermore, the trend of the graph obtained in this study was similar with the findings from Assaedi et al. (2016), who concluded that the addition of nanoclay particles in asphalt binder increased the amorphous content, resulting in superior mechanical performance. In addition, the flat pattern of the XRD graph for the modified binder also showed that the NKC was uniformly dispersed and homogenously mixed with the binder. This result was supported by the findings from You et al. (2011), which can be concluded that the asphalt binder and NKC were well blended during the mixing process.

13. Summary

From the results, it can be said that 5% NKC was a suitable percentage to be incorporated into the porous asphalt mixture. However, to validate these results, all percentages of NKC (0%, 3%, 5%, 7%, and 9%) by weight of binder were tested for the engineering properties and performance of the porous asphalt mixture. Generally, the superior strength of the asphalt mixture depended on the adhesion bonding in the bituminous mixture. The adhesion performance in binder–aggregates interactions was affected by the mechanical and chemical factors. The mechanical theory suggested that the microstructure surface roughness influenced the adhesiveness properties between the aggregates and binder as the binding process occurred at the interface of the materials. The AFM image demonstrated that the highest surface roughness occurred at the binder incorporating 5% of NKC. This evidence shows that the modified asphalt binder produced a low stiffness surface and high adhesive force in asphalt binder. Additionally, structural information was revealed on the chemical composition, crystalline structure, crystallite size, preferred orientation, and layer thickness. The XRD findings illustrated that the XRD graph for the modified binder incorporating 5% NKC was considered amorphous in structure due to the small peaks being more dominant, which is the same as the control binder (NKC0) by comparison. This result implied that NKC was uniformly dispersed and homogenously mixed with the binder. In summary, the addition of NKC in the binder proved outstanding and promising better physical and rheological properties of the modified binder in both unaged and aged condition.

Acknowledgments

This study was supported by the Malaysian Ministry of Higher Education and Universiti Malaysia Pahang in the form of research grant that is highly appreciated.

References

Abdullah, M.E., Zamhari, K.A., Hainin, M.R., Oluwasola, E.A., Hassan, N.A., Md Yusoff, N.I., 2016. Engineering properties of asphalt binders containing nanoclay and chemical warm-mix asphalt additives. Construct. Build. Mater. 112, 232–240.

Airey, G.D., Mohammed, M.H., Fichter, C., 2008. Rheological characteristics of synthetic road binders. Fuel 87 (10–11), 1763–1775.

Assaedi, H., Shaikh, F.U.A., Low, I.M., 2016. Effect of nanoclay on mechanical and thermal properties of geopolymer. J. Asian Ceramic Soc. 4 (1), 19–28.

ASTM D7175, 2015. Standard Test Method for Determining the Rheological Properties of Asphalt Binder Using a Dynamic Shear Rheometer. ASTM International, West Conshohocken, PA.

Bakharev, T., 2006. Thermal behaviour of geopolymers prepared using class F fly ash and elevated temperature curing. Cement Concr. Res. 36 (6), 1134–1147.

BS EN 14770, 2012. Bitumen and bituminous binders. Determination of complex shear modulus and phase angle. Dynamic Shear Rheometer (DSR). British Standard Institution.

Ezzat, H., El-Badawy, S., Gabr, A., Zaki, E.I., Breakah, T., 2016. Evaluation of asphalt binders modified with nanoclay and nanosilica. Procedia Eng. 143, 1260–1267.

Farias, L.G.A.T., Leitinho, J.L., Amoni, B.C., Bastos, J.B.S., Soares, J.B., Soares, S.A., Sant'Ana, H.B., 2016. Effects of nanoclay and nanocomposites on bitumen rheological properties. Construct. Build. Mater. 125, 873–883.

Goh, S.W., Akin, M., You, Z., Shi, X., 2011. Effect of deicing solutions on the tensile strength of micro or nano-modified asphalt mixture. Construct. Build. Mater. 25 (1), 195–200.

Hassan, N.H., Gordon, D.A., Nur Izzi, M.Y., Mohd Rosli, H., Ramadhansyah, P.J., Mohd Ezree, A., Maniruzzaman, A.A., 2015. Microstructural characterisation of dry mixed rubberised asphalt mixtures. Construct. Build. Mater. 82, 173–183.

Hussein, A.A., Ramadhansyah, P.J., Norhidayah, A.H., Haryati, Y., Ghasan Fahim, H., Mohd Haziman, W.I., 2017. Performance of nanoceramic powder on the chemical and physical properties of bitumen. Construct. Build. Mater. 156, 496–505.

Jager, A., Lackner, R., Eisenmenger-Sittner, C., Blab, R., 2004. Identification of four material phases in bitumen by atomic force microscopy. J. Road Mater. Pavement Des. 5, 9–24.

Jeffry, S.N.A., Ramadhansyah, P.J., Norhidayah, A.H., Haryati, Y., Jahangir, M., Siti Hasyyati, D., 2018. Effects of nanocharcoal coconut-shell ash on the physical and rheological properties of bitumen. Construct. Build. Mater. 158, 1–10.

Lee, S.J., Amirkhanian, S.N., Park, N.W., Kim, K.W., 2009. Characterization of warm mix asphalt binders containing artificially long-term aged binders. Construct. Build. Mater. 23 (6), 2371–2379.

Nazzal, M., Kaya, S., Gunay, T., Ahmedzade, P., 2012. Fundamental characterization of asphalt clay nanocomposites. J. Nanomech.Micromech. 3 (1), 1–8.

Ouyang, C., Wang, S., Zhang, Y., Zhang, Y., 2006. Thermorheological properties and storage stability of SEBS/kaolinite clay compound modified asphalts. Eur. Polym. J. 42 (2), 446–457.

Paul, D.R., Robeson, L.M., 2008. Polymer nanotechnology: nanocomposites. Polymer 49 (15), 3187–3204.

Poovaneshvaran, S., Mohd Rosli, M.H., Ramadhansyah, P.J., 2020. Impacts of recycled crumb rubber powder and natural rubber latex on the modified asphalt rheological behaviour, bonding, and resistance to shear. Construct. Build. Mater. 234, 1–19.

Ray, S.S., Okamoto, M., 2003. Polymer/layered silicate nanocomposites: a review from preparation to processing. Prog. Polym. Sci. 28 (11), 1539–1641.

Tasdemir, Y., 2009. High temperatures properties of wax modified binders and asphalt mixtures. Construct. Build. Mater. 23 (10), 3220–3224.

Wan Azahar, W.N.A., Ramadhansyah, P.J., Mohd Rosli, H., Mastura, B., Norzita, N., 2016. Chemical modification of waste cooking oil to improve the physical and rheological properties of asphalt binder. Construct. Build. Mater. 125, 218–226.

Xiao, F., Amirkhanian, A.N., Amirkhanian, S.N., 2011. Influence of carbon nano-particles on the rheological characteristics of short-term aged asphalt binders. J. Mater. Civ. Eng. 23 (4), 423–431.

Xu, B., Chen, J., Li, M., Cao, D., Ping, S., Zhang, Y., Wang, W., 2016. Experimental investigation of preventive maintenance materials of porous asphalt mixture based on high viscosity modified bitumen. Construct. Build. Mater. 124, 681–689.

You, Z., Mills-Beale, J., Foley, J.M., Roy, S., Odegrad, G.M., Dai, Q., Goh, S.W., 2011. Nanoclay-modified asphalt materials: preparation and characterization. Construct. Build. Mater. 25 (2), 1072–1078.

Yu, J.Y., Feng, P.C., Zhang, H.L., Wu, S.P., 2009. Effect of organo-montmorillonite on aging properties of asphalt. Construct. Build. Mater. 23 (7), 2636–2640.

CHAPTER

26

Prediction of rutting resistance of porous asphalt mixture incorporating nanosilica

Khairil Azman Masri[1], Ramadhansyah Putra Jaya[1], Chin Siew Choo[1], Mohd Haziman Wan Ibrahim[2]

[1]Department of Civil Engineering, College of Engineering, University of Malaysia Pahang, Kuantan, Pahang, Malaysia; [2]Faculty of Civil Engineering and Built Environment, Tun Hussein Onn University of Malaysia, Batu Pahat, Johor Bahru, Malaysia

1. Introduction

Porous asphalt (PA) has been well known for its advantages in improving skid resistance of pavement during rain, reducing splashing effects, and producing lower riding noise (Wu et al., 2020; Tang et al., 2013; Yao et al., 2011). These criteria exist due to the high porosity possessed by PA layer which allows for high drainage capability of surface run-off. According to the Public Works Department of Malaysia (JKR/SPJ/2008, 2008), PA should have a total percentage of voids between 20% and 25% which is relatively high compared to conventional hot mixed asphalt (Yao et al., 2013). The high voids content in PA have been enabled through the use open-graded type of aggregates (Yao et al., 2013). The gradation of PA consists mainly of coarse aggregates with dimension size larger than 2.36 mm (No. 10 sieve) together with small amount of fine aggregates (not more than 15%) and also mineral filler not exceeding 5% of the total aggregate weight

(BS812-103.1, 1985). Hence, this type of gradation produces a relatively high interconnected air voids after compaction.

PA is generally considered as a nonstructural layer of flexible pavement. However, it should possess sufficient strength in bearing the external loads imposed by vehicular traffic. Some mechanical properties owned by conventional asphalt layer such as dynamic modulus, rutting resistance, stripping potential, resilient modulus, indirect tensile strength, and stability should also be evaluated for PA (Frigio et al., 2016). This is important since PA forms the uppermost layer of flexible pavement, thus receiving the loads from moving traffic directly. The mechanical properties of PA greatly depend on several factors and one of them is related to the binder used (Yi et al., 2014; Chen and Wong, 2013).

Nanotechnology is a promising and creative area in the material industry. It has also been widely applied to various fields nowadays including in asphalt modification. An example of the nanomaterial used for asphalt modification

Risk, Reliability and Sustainable Remediation in the Field of Civil and Environmental Engineering
https://doi.org/10.1016/B978-0-323-85698-0.00018-6

© 2022 Elsevier Inc. All rights reserved.

is carbon nanofiber (George and Mallery, 2010). Other than nanofiber, nanosilica (NS) is also a type of nanoparticle that is very common for construction material industry (Zhang et al., 2020; Slebi et al., 2020; Zafari et al., 2014). NS is a relatively new inorganic material that is used due to its potentially beneficial properties such as large surface area, strong adsorption, good dispersal ability, high chemical purity, and excellent stability (Hu et al., 2019; Aggarwal et al., 2015; Storch et al., 2020). In this study, the use of NS is introduced to further enhance the performance of PA while at the same time preserving the drainage capabilities of PA.

2. Nanosilica and mixing process

The binder mixer is used for the mixing process of the NS with the bitumen. Firstly, chunks of bitumen that has been placed in a steel cup are melted by putting it into the oven at the temperature of 140°C for about 1 h 30 min. Next the bitumen is weighed for about 500 g and the NS is prepared by the interval of 1–5 g. The binder mixer machine is conditioned at 160°C priors before mixing and 1800 RPM is used for mixing. The NS is added gradually for the first 30 min and continues to be mix for another 60 min for the mix to be homogenous. Fig. 26.1 shows the binder mixer used through the study.

3. Porous asphalt mix design

There are three main processes to obtain Design Binder Content (DBC) for PA, which are air void determination, lower limit determination, and upper limit determination (JKR/SPJ/2008, 2008). Air void determination was calculated based on the volumetric properties of PA. Among the volumetric properties are Theoretical Maximum Density (G_{mm}) and Bulk Specific Gravity (G_{mb}), Void in total Mix (VTM), Void in Mineral Aggregate (VMA), and

FIGURE 26.1 Binder mixer.

Void filled with Asphalt (VFA). Lower Limit and Upper Limit value of DBC are based on Cantabro Loss Test and Binder Draindown Test. The Marshall Mix design was used to determine the volume of aggregates and binder in order to produce a mixture with the desired properties. The volumetric properties that include in Marshall Mix design were theoretical maximum specific gravity (G_{mm}), the bulk specific gravity of the mix (G_{mb}), percentage air void (VTM), voids in mineral aggregate (VMA), and voids filled with asphalt (VFA).

4. Rutting resistance

The rutting resistance of NS modified porous was evaluated using the wheel-tracking device, which is known as Asphalt Pavement Analyzer (APA). The APA was used to analyze the potential of various amounts of NS in reducing the rutting occurrence of PA. The amounts of NS used were 0%, 2%, and 4% by weight of the binder. Approximately, 2500 g of combined aggregate with different size was used to prepare sample. At the laboratory, weighted aggregate, Superpave mold, tray, aggregate, and asphalt binder were placed in the oven for 2 h with temperature of 110°C. After 2 h, the aggregate was transferred into pan and the modified binder was added according to the required weight in order to obtain the desired binder content. The aggregate and binder were mixed until all the aggregate was well coated with binder. After that, the loose mix was poured into the Superpave mold (paper disc was placed at the bottom of mold). The mixtures were taped for about 15 times around the perimeter and 10 times in the inner surface using a spatula to distribute the loose mix evenly. After that, a paper disc was placed at the top surface of the sample. Then, the mixture was compacted using a Gyratory Compactor machine. The machine was set to terminate when the sample has reached the desired height which was 75 mm. After gyration process, the sample was extruded from the mold and the paper disc at the bottom and the top of the specimen was removed. The sample was allowed to cool at room temperature. The processes were repeated for the whole samples preparation. Fig. 26.2A–C show the APA machine, laboratory oven, and Superpave mold respectively.

After the preparation of the whole specimens completed, rutting resistance test was performed. The APA machine only has four slots of samples per testing; thus, this test was conducted in nine sets. Each set took about 5 h of testing where the first 3 h for conditioning process of specimen in testing temperature and the remaining 2 h were taken for loading cycles to complete. The specimens were subjected to 8000 cycles with testing temperature of 45°C. In accordance to BS598-110 (BS 598-110, 1998), rutting test for PA was conducted at a temperature of 45°C to avoid the PA samples failing due to high temperature. A hose pressure of 100 psi and vertical wheel load of 100 lbs were used for the testing. Fig. 26.3A–C show the rutting samples process.

5. Results and discussions

5.1 Descriptive statistic of the rutting data

An overview of the rut depth collected data can be summarized in descriptive statistics which contain all variables information such as the value of mean, median, minimum, maximum value, skewness, kurtosis, and standard deviation. From the information provided by descriptive statistic, extreme values can be detected during screening process. The Rut Depth data after screening are tabulated in Table 26.1.

5.2 Correlation analysis for the rutting parameters

Correlation analysis is the procedure applied to identify the possible relationship between the dependent and independent variables later will be used to develop regression models. The hypothesis for the correlation analysis test can be stated as follows:

H_0 = There is no correlation between two variables.
H_1 = There is a correlation between two variables.

All possible variable combinations were analyzed using correlation analysis and the result for overall variables is shown in the correlation matrix in Table 26.2. Each column denotes both Pearson Correlation value and P-value, where Pearson Correlation value is presented as top row, meanwhile P-value presented at bottom row.

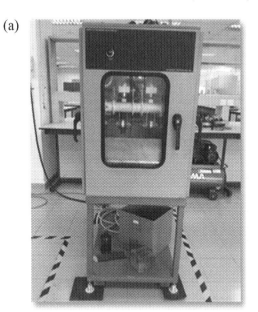
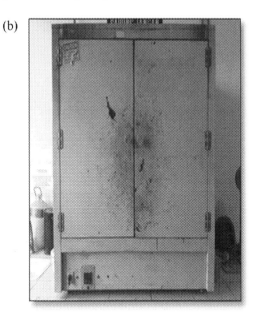

FIGURE 26.2 (A) Asphalt pavement analyzer, (B) Laboratory oven, and (C) Superpave sample mold.

5.3 Multiple linear regression for the rutting parameters

The first model of multiple linear regressions was carried out with the rutting resistance of asphalt mixture as the response, while nominal maximum aggregate size, amount of NS, temperature, rate of rutting, rutting index of asphalt binder and Cantabro loss were the predictors. Table 26.3 lists the value of the constant and the six predictors.

The hypothesis for the final model for estimating this model can be stated as follows:

H_0 = the predictor cannot be used for predicting in the Rut* model.

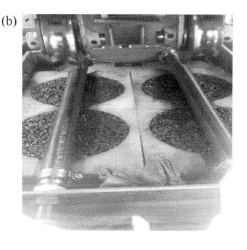
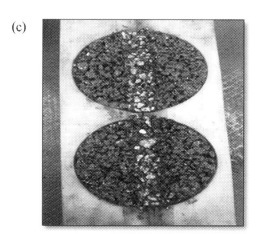

FIGURE 26.3 (A) Rutting samples before testing, (B) Rutting samples during testing, and (C) Rutting samples after testing.

H_1 = the predictor can be used for predicting in the Rut* model.

According to Table 26.3, it is described that certain variables for the Rut* model were significant independent variables for predicting Rut* where the P-value of the multiple linear regression was less than 0.05 (P-value $<$ .05), which mean the null hypothesis (H_0) was rejected and the alternative hypothesis (H_1) was accepted. Hence, certain predictors can be included in the model for predicting the Rut*. The standard error coefficient for the constant value for all variables was very small (less than 1.0). Therefore, the standard errors for each variable have small values indicating that it is reliable in estimating the population parameter (Zhao et al., 2019; Goh and You, 2012). The square root of the

TABLE 26.1 Descriptive statistics for rutting resistance.

Variable	Mean	St. Dev	Minimum	Median	Maximum	Skewness	Kurtosis
Rut*	2.595	1.056	1.641	2.297	4.749	1.24	0.25
NMAS	12	2.014	10	12	14	0.00	−2.06
NS	2.5	1.72	0	2.5	5	−0.00	−1.27
Temp	61	10.32	46	61	76	0.00	−1.27
RR	0.4372	0.41	0.011	0.337	1.264	1.15	0.19
G*/Sinδ	54.41	73.78	0.64	18.18	305.6	1.86	2.92
CL	14.782	4.179	9.37	14.805	20.77	0.07	−1.56

CL, Cantabro Loss, %; *G*/Sinδ*, Rutting Parameter of Asphalt Binder; *NMAS*, nominal maximum aggregate size, mm; *NS*, Amount of NS, %; *RR*, Rate of Rutting, mm/h; *Rut**, Rut depth of mixture, mm; *Temp*, Temperature, °C.

TABLE 26.2 Correlation matrix among variables.

	Rut*	NMAS	NS	Temp	RR	G*/Sinδ
NMAS	−0.756 0.000					
NS	−0.440 0.000	0.000 1.000				
Temp	0.000 1.000	0.000 1.000	0.000 1.000			
RR	0.964 0.000	−0.696 0.000	−0.438 0.000	0.000 1.000		
G*/Sinδ	−0.100 0.403	0.000 1.000	0.110 0.359	−0.752 0.000	−0.161 0.177	
CL	0.670 0.000	−0.339 0.004	−0.887 0.000	0.000 1.000	0.646 0.000	−0.097 0.419

mean square error (S) and R-squared (R^2) is used to indicate how well the model fits the data. In addition, R^2 is one of the criteria that is used to determine whether a linear relationship between the response and the predictor fits the data well. From Table 26.3, the variables used for the model were 95.46% of the variations. Then, the analysis of variance (ANOVA) portion of the output is as shown in Table 26.4. The hypothesis for the ANOVA test can be stated as follows:

H_0 = The Rut* model cannot be used for prediction.

H_1 = The Rut* model can be used for prediction.

Table 26.4 shows that the *P*-value was less than the α-level of 0.05, thus rejecting the H_0 and accepting the H_1. Hence, the regression model was significant and thus could be used to explain or predict the Rut* if the empirical data of selected independent variables were used. In conclusion, the equations on predicting this model were developed as shown in Eq. (26.1) as the regression equation for the Rut*.

5. Results and discussions

TABLE 26.3 Multi linear regression model for rutting.

Term	Coef	SE Coef	T	P
Constant	2.2475	0.750227	2.9958	.004
NMAS	−0.09847	0.025968	−3.7921	.000
NS	−0.02947	0.049233	−0.5987	.551
Temp	0.00673	0.004261	1.5789	.119
RR	2.02092	0.126718	15.9482	.000
G*/sinδ	0.00125	0.000612	2.0428	.045
CL	0.0163	0.021491	0.7584	.451

R-Sq, 95.46%.

TABLE 26.4 Analysis of variance for rutting model.

Source	DF	SS	MS	F	P
Regression	4	74.5076	18.6269	269.879	.0000
Error	67	4.6243	0.0690		
Total	71	79.1319			

$$Rut^* = 0.390247 + 0.0887033\ NS + 2.28589\ RR$$
$$+ 0.0592326 + CL + 0.0824978\ Log\ G^*/sin\delta$$

$$(26.1)$$

where

Rut* = Rut Depth of Mixture, mm
NS = Amount of NS,
RR = Rate of Rutting, mm/hr
G*/Sinδ = Rutting Parameter of Asphalt Binder
CL = Cantabro Loss, %

The equation showed that the coefficient for all important variables for this model had a positive sign implying that an increment of all independent variables would lead to an increment value of the Rut*.

5.4 Normality test for residuals

Normality test is used to determine whether a data set is modeled for normal distribution.

Many statistical functions require that a distribution be normal or nearly normal. There are both graphical and statistical methods of evaluating normality. Graphical methods include the histogram and normality plot. As for this study, two numerical measures of shape which are Skewness and Kurtosis together with Kolmogorov−Smirnov (KS) are used to test for normality. Skewness gives a measure of how symmetric the observations are about the mean. For a normal distribution, the skewness is 0. A distribution skewed to the right has positive skewness and a distribution skewed to the left has negative skewness, while Kurtosis gives a measure of the thickness in the tails of a probability density function. According to George and Mallery (George and Mallery, 2010) it is stated that the values for asymmetry and kurtosis between −2 and +2 are considered acceptable in order to prove normal univariate distribution. Table 26.5 shows the skewness and kurtosis values of the Rut* residual.

Table 26.5 represents the skewness value for Rut* is −0.34, while Kurtosis value is −0.71 which is closed to 0, that means the data can be accepted for model development. Other than that, Normality Test is also evaluated using Kolmogorov−Smirnov (KS) Test, where the P-value should more than 0.150 to indicate the normality of developed model. At first, the P-value was less, only 0.036 which did not pass the requirement; thus, the initial data for model development were transform. After data transformation process, the P-value was greater than 0.150. Figs. 26.4 and 26.5 show the probability plot for residual before and after data transformation.

TABLE 26.5 Skewness and Kurtosis values for Rut* Residual.

Variable	Skewness	Kurtosis
RES 27- Rut*	−0.34	−0.71

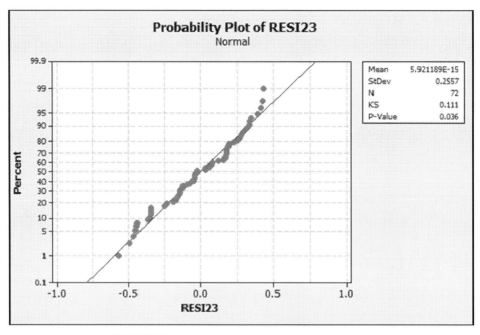

FIGURE 26.4 Probability plot before data transform.

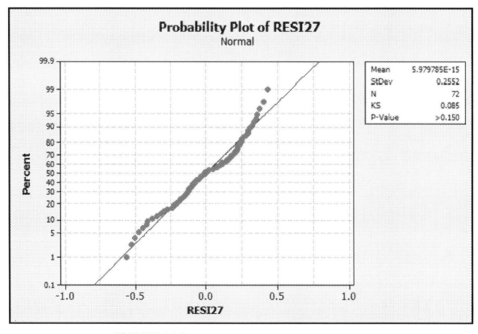

FIGURE 26.5 Probability plot after data transform.

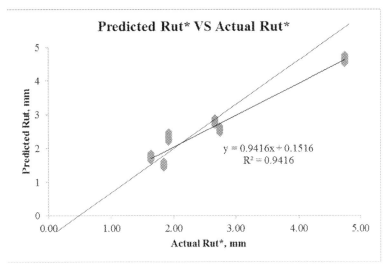

FIGURE 26.6 Predicted Rut* versus empirical Rut*.

5.5 Scatter plot of the Rut* model

Fig. 26.6 shows the relationship between the empirical Rut* and predicted Rut* from the model developed in this study.

5.6 RMSE, MAE, and MAPE of Rut* model

Table 26.6 summarizes the comparisons of RMSE, MAE, and MAPE of the Rut* model. Based on the result summary, the RMSE deviation from the empirical value of Rut* was 0.25 mm. The MAE deviation from the empirical value of Rut* was 0.21 mm. The MAPE for Rut* from the empirical value was 9.96%. Thus, it can be concluded that the Rut* model can be accepted to predict the rutting resistance of mixture due to a small value of the RMSE, MAE, and MAPE.

TABLE 26.6 RMSE, MAE, and MAPE for Rut* model.

Model	RMSE (mm)	MAE (mm)	MAPE (%)
Rut*	0.25	0.21	9.96

5.7 Paired T-test of Rut* model

A comparison of the means was made between the predicted Rut* and actual Rut* from the observed data by using the validation data set as shown in Table 26.7. A hypothesis was set up under the following null and alternative hypotheses:

H_0 = The mean difference of Rut* equals to zero.
H_1 = The mean difference of Rut* does not equal to zero.

From Table 26.7, the *P*-value was greater than 0.05. Therefore, the null hypothesis (H_0) was not rejected at the 5% level of significance. This indicated that the Rut* predicted model did not differ much from the Rut* empirical values.

TABLE 26.7 Validation analysis result for Rut* model.

Test	Value
t-value	−0.73
P-value	0.438

6. Summary

By analyzing all the values from the developed model, it could be seen that the addition of NS to modify the binder can enhance the performance of PA in terms of rutting resistance. This finding proved the advantage of NS, where theoretically it is known to have a huge surface area for interaction and can provide a good bonding in terms of cohesion and adhesion between mixtures in PA and very stable at high temperatures. Thus, this may result in reduced percentage of bonds between binder and aggregate particles that are susceptible to break. Finally, by addition of a little amount of NS, it can enhance the rutting resistance performance of PA that can lead to increase the service life and quality of PA.

Acknowledgments

The authors would to acknowledge the Fundamental Research Grant Scheme-Racer, Ministry of Education Malaysia, with grant number RACER/1/2019/TK06/UMP//1 and Universiti Malaysia Pahang grant number RDU192604 for funding this research.

References

Aggarwal, P., Singh, R.P., Aggarwal, Y., 2015. Use of nanosilica in cement based materials—a review. Cog. Eng. 2 (1), 1078018.

BS 598-110, 1998. Sampling and Examination of Bituminous Mixtures for Roads and Other Paved Areas. Methods of Test for the Determination of Wheel-Tracking Rate and Depth. BSI Group.

BS812-103.1, 1985. Testing Aggregates. Method for Determination of Particle Size Distribution. Sieve Tests, British Standards.

Chen, M.J., Wong, Y.D., 2013. Porous asphalt mixture with 100% recycled concrete aggregate. Road Mater. Pavement Des. 14, 921—932.

Frigio, F., Raschia, S., Steiner, D., Hofko, B., Canestrari, F., 2016. Aging effects on recycled WMA porous asphalt mixtures. Construct. Build. Mater. 123, 712—718.

George, D., Mallery, P., 2010. SPSS for Windows Step by Step: A Simple Guide and Reference 17.0 Update, tenth ed. Pearson, Boston.

Goh, S.W., You, Z., 2012. Mechanical properties of porous asphalt pavement materials with warm mix asphalt and RAP. J. Transport. Eng. 138 (1), 90—97.

Hu, J., Qian, Z., Liu, P., Wang, D., Oeser, M., 2019. Investigation on the permeability of porous asphalt concrete based on microstructure analysis. Int. J. Pavement Eng. 21 (13), 1683—1693.

JKR/SPJ/2008, 2008. Standard Specification for Road Works. Public Works Department, Kuala Lumpur.

Slebi, C.J., Gonzalez, P.L., Vega, I.I., Fresno, D.C., 2020. Laboratory assessment of porous asphalt mixtures reinforced with synthetic fibers. Construct. Build. Mater. 234, 117224.

Storch, I.S., Souza, L.R., Franchi, L., Souza, T.A.J., 2020. Application of Green Nanosilica in Civil Engineering, Green Synthesis of Nanoparticles: Applications and Prospects. Springer, Singapore, pp. 301—316.

Tang, B.Y., Yao, H., Feng, Y., Hu, X.D., 2013. Performance of nanomodified asphalt binder and mixture. Adv. Mater. Res. 721, 219—223.

Wu, J., Wang, Y., Liu, Q., Wang, Y., Ago, C., Oeser, M., 2020. Investigation on mechanical performance of porous asphalt mixtures treated with laboratory aging and moisture actions. Construct. Build. Mater. 238, 117694.

Yao, H., Li, L., Xie, H., Dan, H., Yang, X., 2011. Microstructure and performance analysis of nanomaterials modified asphalt. Road Mater. New Innov. Pav. Eng. 223, 220—228.

Yao, H., You, Z., Li, L., Goh, S.W., Mills-Beale, J., Shi, X., et al., 2013. Evaluation of asphalt blended with low percentage of carbon micro-fiber and nanoclay. J. Test. Eval. 41 (2), 278—288.

Yao, H., You, Z., Li, L., Lee, C.H., Wingard, D., Yap, Y.K., et al., 2013. Rheological properties and chemical bonding of asphalt modified with nanosilica. J. Mater. Civ. Eng. 25 (11), 1619—1630.

Yi, J., Shen, S., Muhunthan, B., Feng, D., 2014. Viscoelastic—plastic damage model for porous asphalt mixtures: application to uniaxial compression and freeze—thaw damage. Mech. Mater. 70, 67—75.

Zafari, F., Rahi, M., Moshtagh, N., Nazockdast, H., 2014. The improvement of bitumen properties by adding nanosilica. Study Civil Eng. Arch. 3, 62—69.

Zhang, Z., Sha, A., Liu, X., Luan, B., Gao, J., Jiang, W., Ma, F., 2020. State-of-the-art of porous asphalt pavement: experience and considerations of mixture design. Construct. Build. Mater. 262, 119998.

Zhao, Y., Tong, L., Zhu, Y., 2019. Investigation on the properties and distribution of air voids in porous asphalt with relevance to the Pb(II) removal performance. Adv. Mater. Sci. Eng. https://doi.org/10.1155/2019/4136295.

CHAPTER 27

Policy options for sustainable urban transportation: a quadrant analysis approach

Anish Kumar[1,2], Sanjeev Sinha[2]

[1]Department of Civil Engineering, Rajkiya Engineering College, Azamgarh, Uttar Pradesh, India;
[2]Department of Civil Engineering, National Institute of Technology, Patna, Bihar, India

1. Introduction

Like many other mid-sized urban cities of India, Patna too suffers the common problems related to urban transportation. Some of the contemporary issues related to transportation can be enlisted as Congestion, Capacity Reduction, Road Safety, Demand Supply Imbalance, Nonmotorized and Motorized Traffic Interaction, Current Scenario in Urban India, Poor Vehicle Quality and Maintenance, Increased Use of Low-Capacity Vehicles, etc. The study was performed in a mid-sized urban city Patna, the capital city of Bihar, India.

1.1 Sustainability and sustainable development

Sustainable development refers to such developments through which the needs of the present generation are fulfilled without compromising the needs of the coming generations (World Commission on Environment and Development, 1987). Sustainability and Sustainable development have been in discussions (Newman and Kenworthy, 1999; Nijkamp, 1999; Cervero, 1998; Haq, 1997) from quite a long and is also referred to be the blueprint of survival (Haq, 1997). Daly (1991) defined Sustainable Development as one in which rate of utilization of renewable and nonrenewable resources should not exceed the rate at which they regenerate. It was highlighted by Daly (1991) that the rate of increase of pollution should not exceed the rate which the environment naturally treats it. Beatley (1995) stated that due to wide applications of sustainability and sustainable development, no common definition can be formulated which is universally acceptable in every context.

1.2 Sustainable transportation

Sustainability in transportation basically refers to the balance between the social,

Risk, Reliability and Sustainable Remediation in the Field of Civil and Environmental Engineering
https://doi.org/10.1016/B978-0-323-85698-0.00001-0

© 2022 Elsevier Inc. All rights reserved.

economic, and environmental factors (OECD, 1996; Ruckelhaus, 1989; Litman, 2003; WCED, 1987). Mid-sized cities in India and in almost all the developing countries are now facing several problems which is not only degrading the present infrastructure but also degrades the standard of people living in those cities. One of the major problems of these cities is related to transportation. The mid-sized cities in most of the developing countries feel more or less the same kind of problems related to transportation like over congestion, air and sound pollution, etc. Patna like other cities of developing countries suffers from acute difficulties caused by transportation, such as air pollution, traffic congestion, open land depletion, etc. According to reports, the transportation sector consumes the largest amount of fossil fuel (International Energy Agency, 2019b; Taghvaee and Hajiani, 2014). It is evident that the natural and economic resources are scarce, and sustainable development of the transportation systems in large city areas is extremely important in order to maintain future existence as well as quality of life. There is a lack of an efficient transport system in Patna which is degrading the standard of living in Patna. The lack of efficient sustainable transport system in a historic city like Patna compelled us to carry out such work.

1.3 Delphi method

Delphi method is based on the philosophical assumption that more heads are better than one (Dalkey, 1972). Linstone and Turoff (2011) commented that the Delphi method structures the communication of individuals in a way that allows a group of individuals to deal with complex problem. Hallowell (2008) proposed the guidelines along with the basic flow of steps which can produce a sound and effective survey. The Delphi technique is a feedback-based questionnaire tool for organizing and expressing opinions (Bardecki 1984; Käpplinger and Lichte, 2020). Okoli and Pawlowski (2004) gave a

detailed comparison of Delphi method and conventional research survey. They also suggested that the minimum sample size for conducting a Delphi survey is 10. Goldfisher (1992) proposed that the Delphi method was primarily introduced for market research and forecasting sales. The Delphi method consists of identification of relevant experts, formulation of questionnaire, and analysis of their responses (Cabanis 2001; Outhred 2001). The selection of sample size and the number of rounds depend upon the experts as different researchers have used different sample sizes in their studies. The details of the sample size used in the current study is presented in Table 27.1.

1.4 Quadrant analysis

Quadrant Analysis is a simple graphical technique that is used to correlate data for decision-making. A 2×2 quadrant analysis can be carried out to help one make decisions in a variety of situations. It's as simple as choosing two criteria with which you would like to evaluate your choice, and placing them on their respective axis. Most commonly, this type of analysis is used in business situations, to determine which endeavors should be pursued, and compare certain potential opportunities. In this chapter, the whole chart area was divided into four quadrants. *The X axis represents Applicability and Y axis represents Desirability*. Both the axes were then divided by a line which demarcates the transition between high and low (Fig. 27.1 and Fig. 27.2).

2. Data and methods

Data were collected by means of a three round Delphi survey. The experts from various backgrounds were contacted for the current study viz. working professionals, academia, researchers, scientists, transport planners, and administrative personnel (Table 27.2).

2. Data and methods

TABLE 27.1 Sample size of the study.

Total questionnaires distributed	No. of filled questionnaires received	No. of valid responses	Turn up %
378	329	295	78.04

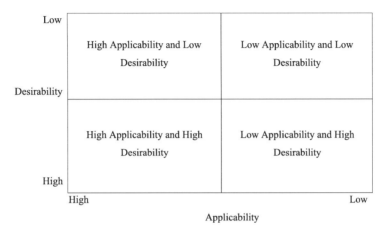

FIGURE 27.1 Template for Quadrant Analysis used in the study.

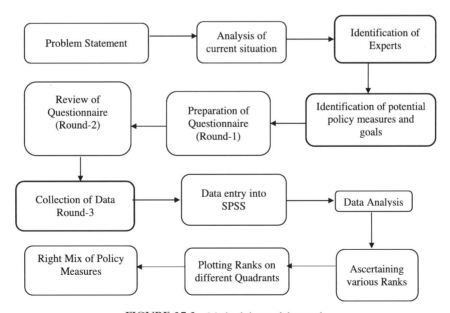

FIGURE 27.2 Methodology of the work.

2.1 Round-1 Delphi: *formulation of the questionnaire (sample Size: 100)*

The first round was dedicated for the formulation of the questionnaire. Several experts were asked to share their views regarding different aspects of sustainable urban transport system and the desirable parameters which can be studied to assess the effectiveness of a sustainable transport system. After contacting 100 experts, going through several sample questionnaires, studying the 17 sustainable development goals as stipulated by United Nations and reviewing past research works (Shiftan et al., 2003), the policies were grouped under 5 categories with a total of 23 policy measures and a format of questionnaire containing 23 × 9 matrix was finalized, where 23 refers to 23 policy measures and 9 refers to the probability of implementation (Applicability) of these policy measures and 8 desirable goals of an efficient sustainable transportation. On the recommendation of experts, it was decided to rate the 23 policy measures on a scale of 0–10 (0 being no contribution, 1 least contribution, and 10 being highest contribution) as per their contributions to achieve the 8 sustainable transport system goals. The policy measures were also rated on the basis of their probability of implementation (Applicability) on the scale of (0%–100%). The 23 policy measures were categorized into the following groups: (1) Government Policy Measures, (2) Social Policy Measures, (3) Economic Policy Measures, (4) Spatial Policy Measures, and (5) Technological Policy measures. The 23 policy measures so identified are listed in Table 27.3.

The 17 SDGs are as follows: *(1) No Poverty, (2) Zero Hunger, (3) Good Health and Well-being, (4) Quality Education, (5) Gender Equality, (6) Clean Water and Sanitation, (7) Affordable and Clean Energy, (8) Decent Work and Economic Growth, (9) Industry, Innovation and Infrastructure, (10) Reducing Inequality, (11) Sustainable Cities and Communities, (12) Responsible Consumption and Production, (13) Climate Action, (14) Life Below Water, (15) Life On Land, (16) Peace, Justice, and Strong Institutions, and (17) Partnerships for the Goals* (United Nations, 2015).

TABLE 27.2 Nature of respondents.

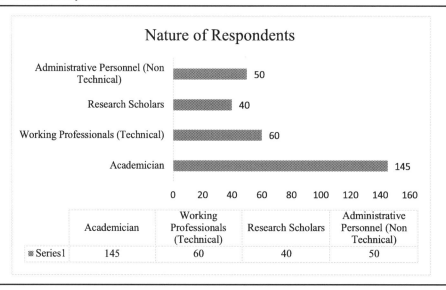

TABLE 27.3 Categories of policy measures.

Government policy measures	Social policy measures	Economic policy measures	Spatial policy measures	Technological policy measures
Continuous (24 × 7) operation of public transport	Promotion of ridership and car pooling	Economic evaluation of present and future projects	Land use policy based on neighborhood principle	Introduction of monorail/metro in the core area
Identification and recovery of encroached spaces along the public transport corridors	Effective public information system	Subsidization of public transport fares	Introduction of car restricted and pedestrian friendly zones	Promotion of nonmotorized vehicles
Reducing the development of new roads and focusing on the development of existing road	Trip management workshops	Introduction of congestion pricing	Reducing parking spaces in the areas well served by public transport services	Development of an effective intelligent transport system
Ensuring proper facilities at public transport stations	Information about the negative effects of transportation system	High parking Fees	Increased parking spaces near the public transport system to facilitate park and ride	A high-quality public transport system based on buses
Effective traffic policing		Taxes on more than one vehicle per household		

The experts designated eight different goals of an efficient sustainable transport system. The Cronbach Alpha was estimated for applicability as well as desirability of these policy measures. In both the cases the Cronbach Alpha value exceeded 0.7, which falls under acceptable range (Taber, 2018). The eight goals identified were the following: *(1) Decreased sound and air pollution (Mapped to SDG 13), (2) Optimal use of open spaces (Mapped to SDG 15,12), (3) Minimal infrastructure cost (Mapped to SDG 8), (4) Reduced Travel time (Mapped to SDG 3), (5) Biodiversity Conservation (Mapped to SDG 13), (6) Accessibility (Mapped to SDG 15), (7) Reduced Energy usage (Mapped to SDG 7), and (8) Safety (Mapped to SDG 3).* The reliability of expert opinions was assessed by Cronbach's Alpha Reliability Test.

2.2 Round-2 Delphi: *adequacy of the questionnaire (sample Size: 100)*

The comprehensive questionnaire was further put to test for its accuracy and adequacy. The same experts were asked to pen down their responses on the given questionnaire. The experts gave some suggestions which were successfully incorporated in the final questionnaire. In the interest of research and with a majority in affirmation, the questionnaire was confirmed and Round-3 of the Delphi was conducted.

2.3 Round-3 Delphi: *data collection (sample size: 295)*

The effective sample size third round Delphi was 295 (valid responses); the selection of such sample size was justified on the basis of the recommendations of experts and past studies made using Delphi. It is to be noted that only those questionnaires were termed valid in which all the entries were made by the respondents. Out of the 378 questionnaires sent to the experts for getting their valuable responses, 295 responses were found to be complete and were used in the current research. The data so obtained in this round are used for the analysis.

3. Data analysis

After collecting the data, it was fed into SPSS in which data analysis was done. Data analysis included descriptive analysis, quadrant Analysis, etc. After analyzing the data, two rankings of the potential policy measures were obtained, namely (i) on the basis of probability of implementation of potential policy measures (applicability) (Table 27.4) and (ii) on the basis of desirability (performance on the basis of fulfilling eight sustainable transport goals) (Table 27.5), respectively. Now quadrant analysis was performed to find out the right mix of policy measures by combining both the ranks based on probability as well as desirability of different policy measures. A template was developed to find out the right mix of policy measures which is suitable on both the scales of probability as well as desirability.

4. Results and discussion

The answers of the experts in the Delphi round were analyzed as follows. A frequency analysis of the probabilities and desirability of the 23 policy measures was performed. Quadrant analysis is used to combine both the rankings. It will give us the right mix of policy measures.

4.1 Ranks on the basis of probability of implementation (applicability) of different policy measures

The experts were asked to rate the policy measures on the basis of probability of their implementation. The ranks varied from 1 to 23 (1 being best performer and 23 being worst performer). The ranks on the basis of probability of implementation of various policy measures (applicability) are enumerated in Table 27.4. On the basis of recommendations of several experts, the Rank 12 was assumed to be to the threshold rank to decide the transition from high probability to low probability of implementation of different policy measures. The policies ranked from 1 to 12 were the policies with high probability of implementation and the polices ranked between 13 and 23 were the policies with low probability.

4.2 Ranks on the basis of desirability of implementation of different policy measures

The experts were asked to rate the policy measures against each of the eight policy measures and then the mean score of the responses given by the experts was calculated. The mean score was then used to rank the policy measures on the basis of their desirability of implementation. On the basis of recommendations of several experts, the Rank 12 was assumed to be to the threshold rank to decide the transition from high desirability to low desirability of implementation of different policy measures, i.e., the policies ranked from 1 to 12 were the policies with high desirability of implementation.

4.3 Right mix of policy measures

The ranks obtained on the basis of probability of implementation and desirability of implementation are plotted on template given

4. Results and discussion

TABLE 27.4 Rank of Policy Measures on the basis of probability of implementation (applicability).

Rank	Policy measures	Mean
1	A high-quality public transport system based on buses	78.67
2	24 × 7 operation of public transport	78.00
3	Effective traffic policing	77.50
4	Subsidization of public transport fares	77.33
5	Economic evaluation of present and future projects	76.83
6	Promotion of nonmotorized vehicles	74.83
7	Effective public information system	74.00
8	Ensuring Proper Facilities at public transport Stations	73.83
9	Introduction to Monorail/Metro in the core area	73.50
10	Promotion of ridership and carpooling	71.67
11	Development of an effective intelligent transport system	71.33
12	Introduction of congestion pricing (*Transition Level*)	70.17
13	Increased parking spaces near the public transport Stations to Facilitate Park and Ride	69.83
14	Reducing the development of new roads and focusing on the development of the existing roads	61.50
15	Promotion of zero emission vehicles	60.83
16	Identification and recovery of encroached spaces along the public transport corridors	60.33
17	Introduction of car restricted and pedestrian friendly zones	59.50
18	Taxes on more than one vehicle per household	58.17
19	Land use policy based on neighborhood principle	57.50
20	Trip management workshops	56.67
21	Reducing parking spaces in the areas well served by public transport services	55.67
22	High parking fees	55.00
23	Information about the negative effects of transportation system	53.50

in Fig. 27.3 and Fig. 27.4 shows the performance of 23 policy measures to achieve eight Sustainable Transport Criteria (Goals). After plotting the implementation ranks and desirability ranks on X and Y axis respectively, four mixes of policy measures are obtained:

1. Policies with high applicability and high desirability
2. Policies with low applicability and high desirability.
3. Policies with high applicability and low desirability

TABLE 27.5 Rank of Policy Measures on the basis of Desirability.

Rank	Policy measures	Mean
1	Economic evaluation of present and future projects	7.533
2	Introduction of monorail/metro in the core area	7.287
3	Promotion of nonmotorized vehicles	7.037
4	Development of an effective intelligent transport system	6.95
5	Subsidization of public transport fares	6.675
6	24×7 operation of public transport system	6.571
7	Introduction of congestion pricing	6.471
8	Promotion of ridership and car pooling	6.442
9	A high-quality public transport system based on buses	6.425
10	Identification and recovery of encroached spaces along the public transport corridors	6.396
11	Promotion of zero emission vehicles	6.392
12	Reducing the development of new roads and focusing on the development of the existing roads (Transition Level)	6.375
13	Ensuring proper facilities at public transport stations	6.337
14	Effective public information system	6.292
15	Effective traffic policing	6.192
16	Land use policy based on neighborhood principle	6.037
17	Introduction of car restricted and pedestrian friendly zones	5.925
18	Reducing parking spaces in the areas well served by public transport services	5.883
19	Increased parking spaces near public transport station to facilitate park and ride	5.879
20	Trip management workshop	5.85
21	Information about the negative effects of transportation services	5.762
22	High parking fees	5.683
23	Taxes on more than one vehicle per house hold	5.671

4. Policies with low applicability and low desirability.

The right mix of policy measures is given by the policies which fall under "policies with high probability as well as high desirability". The right mix of policy measures is as follows:

1. A high-quality public transport system based on buses.
2. 24×7 operation of public transport.
3. Subsidization of public transport fares.
4. Economic evaluation of present and future projects.

4. Results and discussion

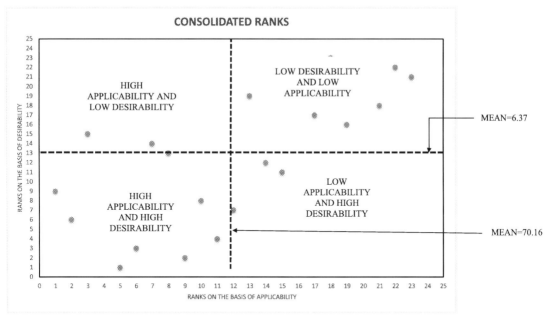

FIGURE 27.3 Consolidated ranks: Plot of ranks on quadrant analysis template.

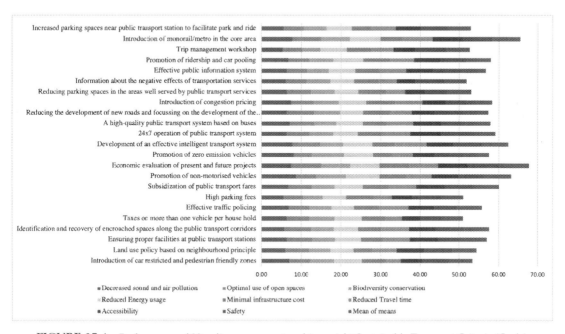

FIGURE 27.4 Performance of 23 policy measures to achieve eight Sustainable Transport Criteria (Goals).

5. Promotion of nonmotorized vehicles.
6. Introduction of monorail/metro in the core area.
7. Promotion of ridership and carpooling.
8. Development of an effective intelligent transport system.
9. Introduction of congestion pricing.
10. Promotion of zero emission vehicles.

5. Conclusions

This research has substantial inferences both from the policy point of view and from a methodological point of view. Taking account of the policy point of view, it is evident that for Patna, a right mix of policy measures is indeed required to create a sustainable transport system. The policies related to land use and technical policy measures were the substantial part of the desired scenario of Patna. It was very much surprising to know the economic policies which talked about decreasing fares got better results as compared to the policies which advocated increase in parking fees. The social policy measured was the least performers as the people must have thought they do not have time for awareness programs and other social happenings. There are some policy measures which are dependent on each other such as the policy related with high parking fees can only be implemented when there is availability of public transport system in that area. Educational policy measures will not be as effective as it was expected to be because of the lack of the awareness about the negative effects of transportation.

Quadrant analysis which can be called as a slight deviation from conventional gap analysis was used to find out the right mix of policy measures. Using quadrant analysis, various policies were grouped as policies with high desirability and high probability, high desirability and low probability, low desirability and high probability, low desirability, and low probability.

Funding information

All the funding related to this research has been made by the Authors only.

References

Bardecki, M., 1984. Participants' response to the Delphi method: an attitudinal perspective. Technol. Forecast. Soc. Change 25 (3), 281–292. https://doi.org/10.1016/0040-1625(84)90006-4.

Beatley, T., 1995. The many meanings of sustainability: introduction to a special issue of JPL. J. Plann. Lit. 9 (4), 339–342. https://doi.org/10.1177/088541229500900401.

Cabanis, K., 2001. Counseling and Computer Technology in the New Millennium—An Internet Delphi Study [Online dissertation]. http://scholar.lib.vt.edu/theses/available/etd-03072001-175713/. (Accessed 12 October 2019).

Cervero, R., 1998. The Transit Metropolis-A Global Inquiry. Island Press, Washington DC.

Daly, H.E., 1991. Steady State Economics. Island Press, Washington, DC.

Dalkey, N.C., 1972. The Delphi method: an experimental study of group opinion. In: Dalkey, N.C., Rourke, D.L., Lewis, R., Snyder, D. (Eds.), Studies in the Quality of Life: Delphi and Decision-Making. Lexington Books, Lexington, MA, pp. 13–54.

Goldfisher, K., 1992. Modified Delphi: a concept for product forecasting. J. Bus. Forec. Winter 1992–93.

Hallowell, M., 2008. A Formal Model of Construction Safety and Health Risk Management. Ph.D. dissertation. Oregon State Univ., Corvallis, Ore.

Haq, G., 1997. Towards Sustainable Transport Planning: A Comparison between Britain and the Netherlands. Ashgate, Aldershot England.

International Energy Agency, 2019. Website of IEA. www.iea.org/sankey/#?c=Islamic%20Republic%20of%20Iran&s=Balance. (Accessed 12 October 2019).

Käpplinger, B., Lichte, N., 2020. "The lockdown of physical co-operation touches the heart of adult education": a Delphi study on immediate and expected effects of COVID-19. Int. Rev. Educ. 66, 777–795. https://doi.org/10.1007/s11159-020-09871-w.

Linstone, H.A., Turoff, M., 2011. Delphi: a brief look backward and forward. Technol. Forecast. Soc. Change 78 (9), 1712–1719. https://doi.org/10.1016/j.techfore.2010.09.011.

Litman, T., 2003. Sustainable transportation indicators. Victoria Transport Policy Institute, Victoria, BC, Canada. Available from: www.vtpi.org. Accessed 7/9/2020.

Newman, P., Kenworthy, J., 1999. Sustainability and Cities-Overcoming Automobile Dependence. Island Press, Washington DC.

Nijkamp, P., 1999. Sustainable transport: new research and policy challenge for the next millennium. Eur. Rev. 7 (4), 551–563. https://doi.org/10.1017/S1062798700004476.

OECD, 1996. Towards Sustainable Transportation, the Vancouver Conference. OECD Publications, Paris. http://www.oecd.org/greengrowth/greening-transport/2396815.pdf. Accessed 22/4/2021.

Okoli, C., Pawlowski, S.D., 2004. The Delphi method as a research tool: an example, design considerations and applications. Inf. Manag. 42 (1), 15–29. https://doi.org/10.1016/j.im.2003.11.002.

Outhred, G.P., 2001. The Delphi method: a demonstration of its use for specific research types. In: Proceeding of the RICS Foundation, Construction & Building Research Conference, 3rd-5th Sept,2001, School of Built & Environment, Glasgow Caledonian University, vol. 1. Department of Building & Construction Economics, RMIT, Melbourne, Australia.

Ruckelhaus, W.D., 1989. Toward a sustainable world. Sci. Am. 114–120. https://doi.org/10.1038/scientificamerican0989-166. September.

Shiftan, Y., Kaplan, S., Hakkert, S., 2003. Scenario building as a tool for sustainable transportation system. Transport. Res. Transport Environ. 8 (5), 323–342. https://doi.org/10.1016/s1361-9209(03)00020-8.

Taber, K.S., 2018. The use of Cronbach's Alpha when developing and reporting research instruments in science education. Res. Sci. Educ. 48, 1273–1296. https://doi.org/10.1007/s11165-016-9602-2.

Taghvaee, V.M., Hajiani, P., 2014. Price and income elasticities of gasoline demand in Iran: using static, ECM, and dynamic models in short, intermediate, and long run. Mod. Econ. 05 (09), 939–950. https://doi.org/10.4236/me.2014.59087.

United Nations, 2015. Resolution Adopted by the General Assembly on 25 September 2015, Transforming Our World: The 2030 Agenda for Sustainable Development. https://sdgs.un.org/2030agenda. Accessed 23/4/2021.

WCED (World Commission on Environment and Development), 1987. Our Common Future. Oxford University Press, Oxford. https://sustainabledevelopment.un.org/content/documents/5987our-common-future.pdf. Accessed 23/04/2021.

CHAPTER

28

Pavement structure: optimal and reliability-based design

Primož Jelušič, Bojan Žlender

University of Maribor, Faculty of Civil Engineering, Transportation Engineering and Architecture, Slovenia

1. Introduction

The period of the last 100 years is characterized by intensive population growth and at the same time the growth of transport infrastructure. Nowadays, any urban environment in the world has a road network with a large total lengths and surface of pavements (Road Statistics Yearbook, 2017). To achieve the appropriate quality of roads, it is necessary to build or restore at least 6% of the length of the entire network annually. This produces large amounts of built-in material, energy consumption, and high costs. Pavements are an important element of any road, which is most depended on for the safety of roads, their sustainability, and the standards of driving and account up to 40% of the road investment costs and about 70% of the road maintenance management.

A pavement structure is a geometrically simple, multilayered structure. The upper layers are formed by bound materials, such as asphalt and concrete, while the lower base layer and subbase layer consists of unbound stone aggregate mixtures. The subgrade layer under the

pavement structure is an embankment or natural ground, with features created by geological processes. The possible use of geosynthetic reinforced soil in pavement can restrain the lateral movement of the unbound material, restrain the stiffness and shear strength of the unbound material, improve the load distribution on the subgrade, and reduce the shear stress of the subgrade (Giroud and Han, 2004). The asphalt layer is subject to fatigue due to repeated traffic loads and is also taken into account. Fatigue is given by the maximum number of cyclical repetitions of traffic load, specified as a function of deformation (Papagiannakis and Masad, 2008). Quality criteria for base and subbase layer and procedures for the investigation of mixtures of stone grains, which are intended for unbound bearing layers, are provided by European standards (EN 1097, 2011; European Standard EN 1744, 2003; EN 932, 1999; EN 933-1, 2012; European Standard EN 13286-7, 2004). As a rule, field investigation of subgrade includes boreholes and excavation pits, sampling and stratification inventory, groundwater level measurement, and field tests. The laboratory tests

Risk, Reliability and Sustainable Remediation in the Field of Civil and Environmental Engineering
https://doi.org/10.1016/B978-0-323-85698-0.00005-8

© 2022 Elsevier Inc. All rights reserved.

according to European standard (European Standard EN 13286-7, 2004) contain basic tests to determine the physical properties of the soil. The result of the CBR test (BS 1377 Part 9, 1990) is essential in the dimensioning of the pavement structure.

The basic parameters in empirical procedure for determining the dimensions of pavement structures are service life of the pavement structure, ground load-bearing capacity (CBR), relevant daily traffic load, climatic and hydrological conditions, characteristics of the pavement structure materials, and usability of the road surface at the end of the life cycle. Different methods can be used for designing pavement (Papagiannakis and Masad, 2008); among them, mechanistic empirical method, based on multilayer elastic system analysis (Huang, 2004), and the AASHTO guides (AASHTO, 2015) or semiempirical method (AASHTO, 1993) are used in practice.

The modern road network with quality pavements definitely enables high quality of transport. On the other hand, the need for energy sources, environmental pollution, and related climate change is becoming increasingly pressing issues. Today, there is a general scientific consensus that current climate change is mainly due to anthropogenic causes, i.e., increasing greenhouse gas emissions (IPCC Climate Change report, 2007). Therefore, new approaches to the use of energy for transport are being intensively explored. Solutions are also being intensively sought for more optimal use of materials and energy use for construction works, which is most relevant in the construction of pavement structures. Finding alternative ways to design pavement structures is therefore a permanent engineering challenge and has been the subject of numerous studies.

Nowadays, it is necessary to design geotechnical structures in such a way that they are economically efficient and have a low probability of failure. For this purpose, a number of optimization algorithms are used, such as mixed-integer nonlinear programming (Jelušič and Žlender, 2018a,b; Jelušič and Žlender, 2020a; Jelušič et al., 2016) and evolutionary algorithms (Jelušič and Žlender, 2020b; Stajkowski et al., 2020). The reliability-based design method is used to analyze the failure probability of geotechnical structures (Kumar and Samui, 2020; Ray et al., 2021; Umar et al., 2018; Ching et al., 2020; Tang and Phoon, 2020; Bozorgzadeh et al., 2020; Bathurst et al., 2019). From the rare literature, it can be concluded that the optimization of pavement structures is rarely performed, nor is the reliability-based design of the pavement structure often considered (Bueno et al., 2020; Sanchez-Silva et al., 2005; Dilip et al., 2013).

In this chapter, an optimization model for the pavement design is proposed and the optimal solution for the failure probability is investigated. The applicability of the optimization model is shown in a case study, for which the optimal design of the pavement structure is determined. In this case study, the probability that the pavement structure will perform its function in the expected lifetime is estimated.

2. Optimizing algorithm

The evolution algorithms are especially used to solve optimization problems where functions are not linear or smooth nonlinear. The most widely used evolution algorithm is genetic algorithm. Integer constraints have many important applications in structural engineering, but the presence of even one such constraint in an optimization model makes the problem an integer programming problem, which may be much more difficult to solve than a similar problem without the integer constraint. However, the genetic algorithm is also able to solve mixed-integer nonlinear problems where both continuous and integer variables are included in optimization problem.

3. Optimization model for pavement structure

To perform the optimization by using the evolutionary algorithm, the design problem of flexural pavement structure was translated into standard formulation of mixed optimization problem. The optimization was performed by using Excel's Built-In Solver [ref]. The proposed optimization model (PAVEOPT) includes input data (constants), variables, and the structure's cost objective function, which is subjected to geotechnical analysis and dimensioning constraints. This PAVEOPT model consists of variables, input data, and cost objective functions of a pavement structure and constraints. The input data represent the specified project requirements and site conditions together with the economic data for the optimal design. The following geometric variables were used in the optimization model PAVEOPT (see Fig. 28.1): thickness of asphalt surface layer d_{as} (m), thickness of asphalt base layer d_{ab} (m), thickness of unbound base layer d_b (m), and pavement construction costs $COST$ (EUR).

3.1 Cost objective function of pavement structure

The cost objective function includes the construction costs of pavement structure (EUR/km), see Eq. (28.1):

$$\min: COST = C_{exc} + C_{gc} + C_{fill,b} + C_{as,sub} + C_{ab} + C_{as} \tag{28.1}$$

where $COST$ designates the construction costs per unit of the pavement structure. The denotations $C_{exc,re}$, C_{gc}, $C_{fill,b}$, $C_{as,sub}$, C_{ab}, and C_{as} designate the material and labor cost items that are included in the cost objective function (see, Table 28.1).

The unit costs c_{exc}, c_{gc}, $c_{fill,b}$, $c_{as,sub}$, c_{ab}, c_{as}, and geometrical properties are given in Table 28.2.

3.2 Design constraints build in optimization model

In presented analyses, the semiempirical AASHTO method is used for pavement design, based on the number of equivalent single axle

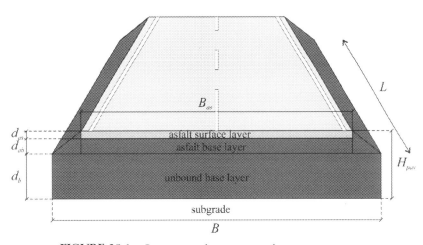

FIGURE 28.1 Geometry and parameters of pavement structure.

TABLE 28.1 Cost items of flexure pavement structure.

Cost items		
Ground excavation	$C_{exc} = [L \cdot H_{pav} \cdot B] \cdot c_{exc}$	(28.2)
Ground leveling and compaction	$C_{gc} = [L \cdot B] \cdot c_{gc}$	(28.3)
Base layer fill	$C_{fill.b} = [L \cdot B \cdot d_b] \cdot c_{fill.b}$	(28.4)
Asphalt fine substrate	$C_{as.sub} = [L \cdot B] \cdot c_{as.sub}$	(28.5)
Asphalt base layer	$C_{ab} = [L \cdot B_{as} \cdot d_{ab}] \cdot c_{ab}$	(28.6)
Asphalt surface layer	$C_{as} = [L \cdot B_{as} \cdot d_{as}] \cdot c_{as}$	(28.7)

TABLE 28.2 Factors increasing traffic load depending on the planned annual growth rate of traffic and the planned duration.

Lifetime of pavement	Annual traffic growth rate										
	0%	1%	2%	3%	4%	5%	6%	7%	8%	9%	10%
	Factor of increasing traffic load f_{it}										
5	5	5	5	5	6	6	6	6	6	7	7
10	10	11	11	12	12	13	14	15	16	17	17
15	15	16	18	19	21	23	25	27	29	32	35
20	20	22	25	28	31	35	39	44	49	56	63

loads (ESALs), caused by traveling on the road pavement in the design life. It is based on the following experimentally obtained parameters such as annual average daily traffic in the first year of the design life, factor considering the division of traffic into directions and lanes, overall conversion factor of the total passages of different heavy vehicle axles into ESALs, heavy vehicle annual growth rate, and design life in years. In this chapter, ESAL was calculated using local technical specifications for roads, which based on AASHTO and European standards.

The geotechnical analysis and dimensioning constraints restrain the cost objective function.

This is done to make sure that the conditions for design are in accordance with the required recommendations. Six different conditions (Eqs. 28.8–28.19) have been defined in accordance with the recommendations, which were put into the optimization model PAVEOPT as geotechnical analysis and dimensioning constraints.

The proposed optimization model (see, Fig. 28.1) includes input data (constants). The input data are comprised of the width of pavement structure covered with asphalt layer B_{as} (m), the width of pavement structure B (m), the depth of frost penetration H_{fr} (m), the factor for hydrological conditions f_{fr} (−), total number of axle load transitions N (−), and the California Bearing Ratio of the subgrade CBR (%). Once all of the above input data are defined, the design constrains are checked.

Constrain 1: The total thickness of frost-resistant layers H_{pav} (m) is greater than the depth of frost h_{min}, see Eq. (28.8):

$$H_{pav} \geq h_{min} \qquad (28.8)$$

$$H_{pav} = d_{as} + d_{ab} + d_b \qquad (28.9)$$

$$h_{min} = f_{fr} \cdot H_{fr} \qquad (28.10)$$

Constrain 2: The minimum thickness of unbound base layer d_b (m) must be provided, see Eq. (28.11):

$$d_b \geq 0.3 \; m \qquad (28.11)$$

Constrain 3: The thickness of unbound base layer d_b (m) must be greater that required thickness of unbound base layer $d_{b,req}$ (m), see Eq. (28.12):

$$d_b \geq d_{b,req} \qquad (28.12)$$

$$d_{b,req} = ((6.6502 - 0.4454 \cdot CBR) \cdot \ln(N)$$
$$+ 2.5586 \cdot CBR - 29.936)/100$$
$$(28.13)$$

Constrain 4: The total thickness of asphalt layer $(d_{ab} + d_{as})$ must be greater that required thickness of asphalt layer $d_{ab \, + \, as,req}$ (m), see Eq. (28.14). Note that the required thickness of asphalt layer is in relation with the total number of axle load transitions N (−).

$$d_{ab} + d_{as} \geq d_{ab+as,req} \qquad (28.14)$$

$$d_{ab+as,req} = \left(0.657 \cdot N^{-0.217}\right)/100 \qquad (28.15)$$

Constrain 5: The minimum thickness of asphalt base layer d_{ab} (m) must be provided due to technical specifications, see Eq. (28.16):

$$d_{ab} \geq d_{ab,min} \qquad (28.16)$$

$$d_{ab,min} = 0.06 \text{ m} \qquad (28.17)$$

Constrain 6: The minimum thickness of asphalt surface layer d_{as} (m) must be provided for pavement smoothness, see Eq. (28.18):

$$d_{as} \geq d_{as,min} \qquad (28.18)$$

$$d_{as,min} = 0.04 \text{ m} \qquad (28.19)$$

4. Application of PAVEOPT model—case study

To illustrate the usefulness of the optimization model presented in this chapter, an example is given for determining the cheapest possible pavement structure and finding an optimal design for the given design parameters. Fig. 28.2 shows the mean and characteristic values of the CBR for a 3 km of road section. The subgrade strength was determined from 15 CBR tests.

The total daily equivalent traffic load in the cross-section of the pavement T_d is determined on the basis of the planned average daily number of motor vehicles in the first year of pavement use according to Eq. (28.20):

$$T_d = \sum (FE_v \cdot n_v) \qquad (28.20)$$

where

FE_v is equivalence factor of a representative motor vehicle and n_v is the number of motor vehicles of a certain type (representative) per day at the beginning of pavement use. The total number of axle load transitions N must be determined by Eq. (28.21):

$$N = 365 \cdot T_d \cdot f_{cs} \cdot f_{lw} \cdot f_{li} \cdot f_{di} \cdot f_{it} \qquad (28.21)$$

where

f_{cs} is a pavement cross-sectional factor, f_{lw} is lane width factor, f_{li} is longitudinal inclination factor, f_{di} is factor of additional dynamic impacts, and f_{it} is factor of increasing traffic load due to traffic growth over the lifespan. Among all involved factors, the factor of increasing traffic load due to traffic growth over the lifespan f_{it} is the most uncertain (see, Table 28.2).

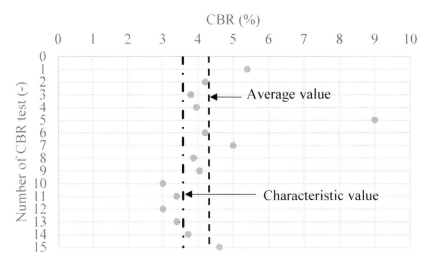

FIGURE 28.2 Average and characteristic value of the subgrade CBR.

Fig. 28.3 shows the average and the characteristic values of the total number of axle load transitions N, for the selected lifetime of pavement 20 years and the expected annual traffic growth rate between 0% and 5%.

The Schneider method (Orr and Farrell, 1999) is used to calculate the characteristic values of the subgrade CBR and total number of axle load transitions N:

$$CBR = CBR_{av} - 0.5 \cdot \sigma_{CBR} = 4.3\% - 0.5 \cdot 1.46\%$$

(28.22)

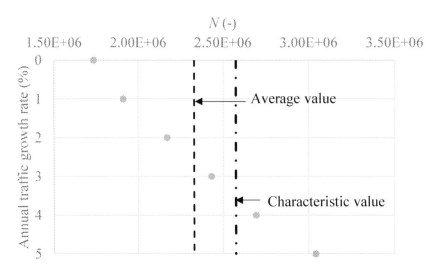

FIGURE 28.3 Prediction based on design data, past measured transitions, and expected growth rate of transitions.

$$N = N_{av} + 0.5 \cdot \sigma_N = 2.33 \cdot 10^6 + 0.5 \cdot 4.90 \cdot 10^5$$
$$= 2.58 \cdot 10^6$$

$$(28.23)$$

The design data is comprised of the characteristic California Bearing Ratio $CBR = 3.6\%$, the total number of axle load transitions $N = 2.58 \cdot 10^6$, the depth of frost penetration $H_{fr} = 0.8$ m, the width of pavement structure covered with asphalt layer $B_{as} = 8$ m, the width of pavement structure $B = 9$ m, and the length of road section $L = 1$ km. All the data involved in the optimization model PAVEOPT are presented in Table 28.3.

The results show that the optimal pavement structure for the given project data leads to self-manufacturing cost of about 576.52 €/m. The degree of utilization of each condition shows (see, Table 28.4) that for the optimum design of the pavement structure for the selected example all constraints are fully utilized.

5. Failure probability of an optimally designed pavement structure—case study

Once optimal design of the pavement structure was obtained, the failure probability is estimated due to the uncertain CBR, total number of

TABLE 28.3 Input data for PAVEOPT model.

B_{as}	**Width of pavement structure covered with asphalt layer**	**8 m**
B	Width of pavement structure	9 m
L	Length of road section	1000 m
H_{fr}	Depth of frost penetration	0.8 m
f_{fr}	Factor for hydrological conditions	0.8
T_d	Total daily equivalent traffic load	300
f_{cs}	Pavement cross-sectional factor	0.5
f_{lw}	Lane width factor	1.4
f_{li}	Longitudinal inclination factor	1.05
f_{di}	Factor of additional dynamic impacts	1.08
c_{exc}	Unit cost for ground excavation	9 €/m^3
c_{gc}	Unit cost for ground leveling and compaction	2.5 €/m^2
$c_{fill,b}$	Unit cost of base layer construction	36 €/m^3
$c_{as,sub}$	Unit cost of asphalt fine substrate	1.5 €/m^2
c_{ab}	Unit cost of asphalt base layer	200 €/m^3
c_{as}	Unit cost of asphalt surface layer	300 €/m^3
CBR	Characteristic California bearing ratio of subgrade	3.6%
N	Total number of axle load transitions	$2.58 \cdot 10^6$

TABLE 28.4 Optimal design of a pavement structure—case study.

Optimal design obtained with the PAVEOPT optimization model	
d_b	0.55 m
d_{ab}	0.13 m
d_{as}	0.04 m
Constrain 1 $H_{pav} \geq h_{min}$	0.72 m $\geq$ 0.64 m (89%)
Constrain 2 $d_b \geq 0.3$ m	0.55 m $\geq$ 0.30 m (55%)
Constrain 3 $d_b \geq d_{b,req}$	0.55 m $\geq$ 0.54 m (98%)
Constrain 4 $d_{ab} + d_{as} \geq d_{ab + as,req}$	0.17 m $\geq$ 0.162 m (95%)
Constrain 5 $d_{ab} \geq d_{ab,min}$	0.13 m $\geq$ 0.06 m (46%)
Constrain 6 $d_{as} \geq d_{as,min}$	0.04 m $\geq$ 0.04 m (100%)
COST	576,520 €
COST/m	576.52 €/m

TABLE 28.5 Uncertainty model, pavement structure case study.

Random variable	Distribution type	Mean	Standard deviation
CBR	Normal	4.31	1.46
N	Normal	$2.33 \cdot 10^6$	$4.90 \cdot 10^5$
H_{fr}	Normal	0.8	0.05

axle load transitions N, and depth of frost penetration H_{fr}. In order to assess probability failure of a pavement structure, three main tasks were performed. The first task is to develop a deterministic model that includes a computational model of the response variable of interest (e.g., six constraints in optimization model) for a given set of input parameter values such as subgrade properties and traffic load. The deterministic model includes equations that were presented in optimization model PAVEOPT. In the second task, the uncertain parameters are selected and their statistical parameters (mean value, standard deviation, distribution type) are determined. The second task refers to us as uncertainty modeling and the model includes all information needed to generate a random sample of uncertain parameters. The uncertainty model for the case study of pavement structure is presented in Table 28.5.

The uncertainty model includes three random variables such as California Bearing Ratio of the subgrade CBR (%), total number of axle load transitions N (−), and depth of frost penetration H_{fr}. Once the deterministic and uncertainty modeling have been carried out, they are connected to create a probabilistic model of the problem. In the third task, the response variable of interest in deterministic model is calculated according to the uncertain parameters defined in uncertainty model. For that purpose, the subset simulation method (Au and Wang, 2014) was applied in order to estimate the complementary cumulative distribution functions (CCDFs) of three response (Constrain 1, Constrain 3, and Constrain 4) quantities of interest that are subjected to uncertainties. The CCDFs are then used for estimating the failure probability. After the Subset Simulation (Au and Wang, 2014), the CCDF and the corresponding values of the driving variables were obtained. Fig. 28.4 shows the CDF plot for three driving variables.

The results show that for the optimal design of pavement structure for selected case study there is 3% probability that the damage in pavement structure will occur due insufficient total thickness of frost-resistant layers H_{pav}. However, there is large probability $P(d_b - d_{b,req} < 0) = 0.178 = 17.8\%$ that the thickness of unbound base layer d_b will be insufficient to sustain traffic load for 20 years. By increasing the unbound base layer from 0.55 to 0.58 m, the construction cost of pavement will increase from 576,520 € to 588,670 €, but the probability of exceeding the Constrain 3 will diminish. The constrain 4 will be exceed for selected design data with the small probability of 3%.

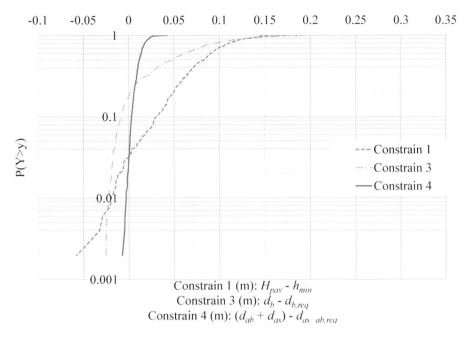

FIGURE 28.4   CDF plot of Constrain 1, Constrain 3, and Constrain 4.

6. Conclusion

In this chapter, an optimization model for the pavement design is proposed and the optimal solution for the failure probability is investigated. The applicability of the optimization model is shown in a case study for which the optimal design of the pavement structure is determined. In this case study, the probability that the pavement structure will perform its function during the expected service life is estimated. The optimization problem was solved by using the genetic algorithm which was able to solve mixed-integer nonlinear problems where both continuous and integer variables are included in optimization problem. The cost objective function and the geotechnical/design constraint were developed and included in the optimization model. The presented case study shows how an optimal solution for adapted project data can be achieved. Once the optimal solution was achieved, the reliability-based design of the pavement structure was carried out. Furthermore, the results of the failure probability allow us to assess which important elements of the pavement structure should be modified in order to reduce the failure probability. The results of the case study show that there is large probability that the thickness of unbound base layer will be insufficient to sustain traffic load for 20 years. However, there is a small probability that the thickness of the asphalt layer and the total thickness of frost-resistant layers will be insufficient for expected lifetime.

Acknowledgment

The authors acknowledge financial support from the Slovenian Research Agency; research core funding No. P2-0268.

References

AASHTO, 1993. Guide for Design of Pavement Structures. American Association of State Highway and Transportation Officials, Washington, DC, USA.

AASHTO, 2015. Mechanistic-Empirical Pavement Design Guide — A Manual of Practice. American Association of State Highway and Transportation Officials, Washington, DC.

Au, S.K., Wang, Y., 2014. Engineering Risk Assessment with Subset Simulation. https://doi.org/10.1002/9781118398050.

Bathurst, R.J., Lin, P., Allen, T., 2019. Reliability-based design of internal limit states for mechanically stabilized earth walls using geosynthetic reinforcement. Can. Geotech. J. 56, 774—788. https://doi.org/10.1139/cgj-2018-0074.

Bozorgzadeh, N., Bathurst, R.J., Allen, T.M., 2020. Influence of corrosion on reliability-based design of steel grid MSE walls. Struct. Saf. 84, 101914. https://doi.org/10.1016/j.strusafe.2019.101914.

BS 1377 Part 9, 1990. Methods for Test for Soils for Civil Engineering Purposes. In-Situ Tests.

Bueno, L.D., Schuster, S.L., Specht, L.P., Pereira, D. da S., do, N.L.A.H., Kim, Y.R., et al., 2020. Asphalt pavement design optimisation: a case study using viscoelastic continuum damage theory. Int. J. Pavement Eng. 1—13. https://doi.org/10.1080/10298436.2020.1788030.

Ching, J., Phoon, K.-K., Khan, Z., Zhang, D., Huang, H., 2020. Role of municipal database in constructing site-specific multivariate probability distribution. Comput. Geotech. 124, 103623. https://doi.org/10.1016/j.compgeo.2020.103623.

Dilip, D.M., Ravi, P., Babu, G.L.S., 2013. System reliability analysis of flexible pavements. J. Transport. Eng. 139, 1001—1009. https://doi.org/10.1061/(ASCE)TE.1943-5436.0000578.

EN 1097, 2011. Tests for Mechanical and Physical Properties of Aggregates - Part 1: Determination of the Resistance to Wear (Micro - Deval), Part 2: Methods for Determination of Resistance to Fragmentation, Part 5: Determination of the Water Content by Drying in a Ventilated Oven, Part 6: Determination of Particle Density and Water Absorption.

EN 932, 1999. Tests for General Properties of Aggregates - Part 1: Methods for Sampling, Part 2: Methods for Reducing Laboratory Samples, Part 3: Procedure and Terminology for Simplified Petrographic Description.

EN 933-1, 2012. Tests for Geometrical Properties of Aggregates-Part 1: Determination of Particle Size Distribution - Sieving Method, Part 4: Determination of Particle Shape - Shape Index, Part 8: Assessment of Fines - Sand Equivalent Test, Part 9: Assessment of Fines - Methylene Blue Test.

European Standard, EN 13286-7, 2004. EN 13286-47:2012. Unbound and Hydraulically Bound Mixtures - Part 7 Repeated Load Triaxial Test for Unbound Mixtures, Part 47 Unbound and Hydraulically Bound Mixtures. Test Method for the Determination of California Bearing Ratio, Immediate Bearing Index and Linear Swelling.

European Standard EN 1744, 2003. Tests for Mechanical and Physical Properties of Aggregates.

Giroud, J., Han, J., 2004. Design method for geogrid-reinforced unpaved roads. I. Development of design method. J. Geotech. Geoenviron. Eng. 130 (8), 775—786.

Huang, Y.H., 2004. Pavement Analysis and Design, second ed. Prentice Hall Inc., Upper Saddle River, NJ, USA.

IPCC Climate Change report, 2007. The Physical Science Basis. Contribution of Working Group I to the Fourth Assessment Report of the Intergovernmental Panel on Climate Change. Cambridge University Press, Cambridge.

Jelušič, P., Žlender, B., 2018a. Optimal design of piled embankments with basal reinforcement. Geosynth. Int. 25, 150—163. https://doi.org/10.1680/jgein.17.00039.

Jelušič, P., Žlender, B., 2018b. Optimal design of pad footing based on a MINLP optimization. Soils Found. 58 (2), 277—289. https://doi.org/10.1016/j.sandf.2018.02.002.

Jelušič, P., Žlender, B., 2020a. Determining optimal designs for conventional and geothermal energy piles. Renew. Energy 147 (March), 2633—2642. https://doi.org/10.1016/j.renene.2018.08.016.

Jelušič, P., Žlender, B., 2020b. Determining optimal designs for geosynthetic-reinforced soil bridge abutments. Soft Comput. 24 (5), 3601—3614. https://doi.org/10.1007/s00500-019-04127-8.

Jelušič, P., Žlender, B., Dolinar, B., 2016. NLP optimization model as a failure mechanism for geosynthetic reinforced slopes subjected to pore-water pressure. Int. J. GeoMech. 16 (5).

Kumar, M., Samui, P., 2020. Reliability analysis of settlement of pile group in clay using LSSVM, GMDH, GPR. Geotech. Geol. Eng. 38 (6), 6717—6730. https://doi.org/10.1007/s10706-020-01464-6.

Orr, T.L.L., Farrell, E.R., 1999. Geotechnical Design to Eurocode 7. https://doi.org/10.1007/978-1-4471-0803-0.

Papagiannakis, A.T., Masad, E.A., 2008. Pavement Design and Materials. John Wiley & Sons, pp. 376–379.

Ray, R., Kumar, D., Samui, P., Roy, L.B., Goh, A.T.C., Zhang, W., 2021. Application of soft computing techniques for shallow foundation reliability in geotechnical engineering. Geosci. Front. 12 (1), 375–383. https://doi.org/10.1016/j.gsf.2020.05.003.

Road Statistics Yearbook, 2017. European Union Road Federation. http://www.erf.be/wp-content/uploads/2018/01/Roadstatistics2017.pdf.

Sanchez-Silva, M., Arroyo, O., Junca, M., Caro, S., Caicedo, B., 2005. Reliability based design optimization of asphalt pavements. Int. J. Pavement Eng. 6, 281–294. https://doi.org/10.1080/10298430500445506.

Stajkowski, S., Kumar, D., Samui, P., Bonakdari, H., Gharabaghi, B., 2020. Genetic-algorithm-optimized sequential model for water temperature prediction. Sustainability 12, 5374. https://doi.org/10.3390/su12135374.

Tang, C., Phoon, K.-K., 2020. Statistical evaluation of model factors in reliability calibration of high-displacement helical piles under axial loading. Can. Geotech. J. 57, 246–262. https://doi.org/10.1139/cgj-2018-0754.

Umar, S.K., Samui, P., Kumari, S., 2018. Reliability analysis of liquefaction for some regions of Bihar. Int. J. Geotech. Earthq. Eng. 9, 23–37. https://doi.org/10.4018/IJGEE.2018070102.

CHAPTER

29

Assessment of factors affecting time and cost overruns in construction projects

J. Vijayalaxmi, Umair Khan

School of Planning and Architecture, Vijayawada, Andhra Pradesh, India

1. Introduction

The construction sector assumes a critical part in contributing to the economy and the overall improvement of the country. The construction sector additionally influences the GDP development rate and work in the country and thus it is viewed as an asset for a country's financial development. Along these lines, this specific industry needs a large quantity of materials than other industry inside the country. The construction industry is currently experiencing massive cost overruns due to inadequate cost and time management. Poor cost control and overruns are a major problem and a significant concern in both developed and developing countries when it comes to project costs. The complexity of construction projects, as well as the environment in which they are built, extends more pressure on the management to complete projects timely, within estimated cost, and with superior quality. A well-known reality is that various projects in India are getting delayed due to a variety of factors. As a result, it's important to identify the cost and time overrun factors so that they can be systematically addressed to prevent and minimize problems.

1.1 Indian construction sector

The Indian construction industry, being one of the economic boosters, needs to minimize the delays as well as cost overruns. The higher the factor of self-dependence, the more energy, manpower, materials, equipment, and capital market exchange are given from within the financial system.

Most infrastructure projects are delayed on a daily basis around the world due to stringent permits, land acquisition issues, lack of qualified personnel, inadequate conflict resolution mechanisms, and geological challenges. However, if the project leader anticipates these issues, he or she will be able to efficiently organize all and validate that the project is well managed.

According to a research by India's Ministry of Statistics and Program Implementation (MOSPI, 2019), cost overruns in the public sector are a

Risk, Reliability and Sustainable Remediation in the Field of Civil and Environmental Engineering
https://doi.org/10.1016/B978-0-323-85698-0.00028-9

© 2022 Elsevier Inc. All rights reserved.

severe problem in India for a variety of reasons. A total of 1634 projects in India were selected for assessment. There were 373 ongoing projects with cost overruns and 552 projects running with time overruns, according to the findings. Delays and price increase are common in projects all over the world. These are, however, more severe in poor countries. According to MOSPI reality findings, overall cost overrun is 19.85% of original cost, while the average time overrun is 45.63 months. On the other hand, it has set an ambitious aim of INR 50 lakh crore in infrastructure spending for the years 2019—23. In order to meet India's infrastructure development needs, annual infrastructure expenditures must be substantial. A few studies on cost overruns in construction projects have been conducted in India. However, these studies did not cover the entire country or include causes of suffering of high valued project.

2. Literature review

This study by Narayanan et al. (2019) found that big infrastructure projects in India experience considerable time and expense overruns. In general, in large infrastructure projects, a rise in time overrun leads to an increase in cost overrun. Projects of a longer life are more susceptible to scope revisions, climatic changes, and inflation. As a result, there will be time and expense overruns. Contrary to popular belief, it is discovered that projects with a longer length have fewer time overrun. This research aids in determining the impact of project duration and budget on time and cost overruns, as well as the relationship between time and cost overruns. Because of the small sample size, the findings of this study may not be generalizable. Further study with a large number of projects from various sectors and locations is required to validate the findings.

Harsha et al. (2020) conducted a study with 500 civil workers with a questionnaire which was analyzed using RII method of analysis. This study found that almost all the projects get delayed because of cost overrun issues. It was found that the factors affecting cost problems were material-related factors such as cost fluctuation of materials, material shortage, late delivery of materials, and specification changes in material; machinery-related factors like insufficient no of equipment's, equipment failure, costly machines, and maintenance; and manpower-related factors like shortage of technical personnel, costly labor, shortage of workers, severe overtime, labor absenteeism, project management, and communication-related factors. RII method used after questionnaire was found to be giving accurate causes of construction delays and cost overruns.

A study on value management was done by Laila et al. (2019) to attain cost optimization. This paper found that the cost increment in residential construction projects varied from 21% to 55%, where value engineering can be implemented to minimize the cost increment. Two cases were taken to analyze the application of value engineering to achieve cost optimization and to know the cost variation before and after the application of value engineering. This paper found that the application of value management helped in achieving the cost savings from 15% to 40% due to cost saving from specific items. Further research was done to know the perception of stakeholders with respect to value engineering in residential projects.

Ametepey et al. (2018) conducted a study following two-stage approach where the first stage involved the exploratory method to find out the causes of construction delays and second stage involved the questionnaire survey and the analysis of data gathered from it. The study listed 10 major causes of delay and a major contribution from the perspective of importance is that it derived the empirical relation between the delay causes and its effects. It helped the practitioners to understand the project management dynamics and efforts can be made to reduce delay incidents. The results of this study were useful in designing the questionnaire for the survey to be done in Indian context.

Tsegay et al. (2017) conducted an empirical study to analyze the delay impacts on construction projects. The survey results show a high level of agreement among the respondents, indicating that there is a considerable difference in the effects of building delays. The comparison of the three construction stages and the overall resulted in the conclusion that the overall is highly similar with all stages of delay causes. As a result, the top three most significant reasons for the Ethiopian construction project's delay have been determined. Corruption, a lack of utilities on site, material inflation or price increases, a lack of quality materials, late design, and design documents, sluggish material delivery, late approval and acceptance of completed project work, poor site management and performance, late budget/funds release, and ineffective project planning and scheduling were all placed in that order.

According to a study by Abbas Niazi et al. (2017), project cost overruns are a major problem for the Afghan construction industry. 69 causes of building cost overrun were discovered after a thorough analysis of the literature. In Afghanistan, data were collected using a standardized questionnaire survey. 75 questionnaire sets were distributed to 517 consultants, clients, and contractors. To rate the causes of cost overrun, an RII approach was used. Corruption, progress payment delays by clients, difficulties in funding projects by contractors, defense, and change orders by clients during the construction process were discovered to be the most significant causes of cost overruns in the Afghan construction industry.

Taking these factors into account, this paper aims to investigate the reasons why most building construction projects in India are delayed beyond their completion dates, necessitating additional expenditure above, and beyond the initial contract costs. The investigation looks into how these problems can be significantly minimized. A literature review is undertaken to come up with a list of items that are considered to be the most common and frequently occurring causes of time and cost overrun in residential building construction projects in order to achieve these objectives and identify the factors that cause time and cost overrun in residential building construction projects.

3. Methods and materials

The investigation of the practical problem of time and cost overruns in construction projects is based on the observation of construction projects in India. To better grasp stakeholder perceptions of time and cost overruns, quantitative analysis is used.

The method involves conducting research to determine the most common causes of cost and time overruns, as well as identifying feasible and realistic solutions to reduce overruns in residential building construction projects. These goals are accomplished by the use of research methodologies such as literature reviews and questionnaire surveys.

The literature review draws on similar past studies and assists with factor and category recognition, research methods, and data analysis. 36 causes of delay and time overrun (as shown below in Table 29.1) have been established and categorized into major classes based on the origins of delay and time overrun. The value of the identified causes is measured by compiling and structuring these causes into questions. A questionnaire survey was used to collect the first data used in this analysis. The residential building construction stakeholders is given a form created from an in-depth literature analysis of various causes of cost and time overruns in construction projects, as well as secondary information sources. The designed questionnaire is distributed to solicit perception

TABLE 29.1 Factors of time and cost overruns.

No.	Causes of time and cost overruns		
	Client-related factors	19	Conflicts among consultant with other parties
1	Client delaying progress payments		Labor-related factors
2	Delay in delivering the site to the contractor	20	Shortage of technical staff
3	Frequent change orders during construction	21	High cost of labor
4	Delay in approval of design documents		Material and machinery related factors
5	Unrealistic contract duration	22	Delay in material delivery
6	Delay in decision making process	23	Improper procurement planning
7	Owner interference	24	Material shortage in the market
8	Delay in change orders by client	25	Construction equipment shortage
9	Poor financial control mechanism	26	Equipment failure
	Contractor-related factors		External factors
10	Financial difficulties by contractors	27	Corruption
11	Delay in site mobilization	28	Security
12	Mistakes during construction by contractors	29	Permits from related authorities
13	Ineffective planning and scheduling	30	Natural disasters (COVID-19)
14	Poor site management by contractors	31	Laws and regulations changes
	Consultant-related factors	32	Lengthy dispute settlement
15	Delay in reviewing the design documents	33	Market inflation
16	Poor inspection plan by consultants	34	Legal disputes between parties
17	Project design complexity	35	Bidding for projects that are inappropriate
18	Project document errors and contradictions	36	Contract documents errors and discrepancies

of professionals in residential building construction industries (project managers, clients, consultant, and contractors) concerned in residential building construction projects. Statistical procedure (RII), visual analysis, tabulating, and categorizing are used to analyze the details collected from the questionnaire. The observations and conclusions will be understood and discussed after the collected data have been analyzed. Finally, the study findings will be presented, which will serve as the foundation for making recommendations to prevent time overruns which cause a loss to the project.

4. Factors causing cost and time overruns

These are the factors which are related to cost and time overruns and they will be used for questionnaire preparation:
- Labor
- Material and Machinery
- Client
- Contractor
- Consultant
- Additional

5. Method of analysis

For this research, the quantitative method is adopted. A questionnaire is used to determine the major factors affecting cost and delay overruns in construction projects in India. The Likert scale is a five-point ordinal scale that ranges from 1 to 5 depending on the degree of contribution where

- 1 strongly disagree,
- 2 disagree,
- 3 neutral,
- 4 agree, and
- 5 strongly agree.

The five-point scale is adopted and transformed to Relative Importance Indices (RIIs) for each factor as follows:

$$RII = \sum W/(A \times N)$$

where W is the respondents' weighting of each factor (from 1 to 5), A is the highest weight (in this case, 5), and N is the total number of respondents. The greater the RII value, the more significant the cause of delays.

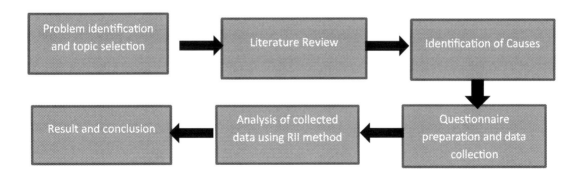

6. Results and discussion

The designed questionnaire survey was collected from 50 professionals in construction industry. 58% of the respondents were professionals with degree in civil engineering, 36% of the professionals had master's degree, and more than 50% of the professionals had experience of up to 5 years as shown in Figs. 29.1 and 29.2. These professionals also included academicians, client's representative, client's project manager, contractor's representative, contractor's site manager, contractor's estimator, consultant, manager, probationary officer, planning engineer, and site engineer as shown in Fig. 29.3. These professionals were

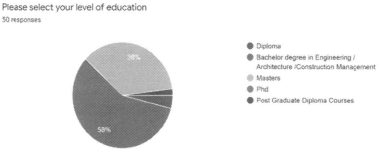

FIGURE 29.1   Education level of participants.

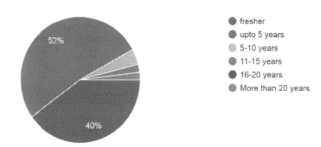

FIGURE 29.2   Work experience of participants.

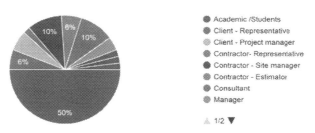

FIGURE 29.3   Profession of participants.

from different reputed organizations such as AECOM, BSPTCL, Cushman & Wakefield, Tata projects, Tracecost, L&T, etc.

7. Conclusion

The results of the study shows that the most important factor affecting cost and schedule overrun is progress payment delays by clients. This situation is due to the current situation (COVID-19) which caused the construction industry to suffer. Reduction in investment in real estate leads to the financial constraints for the contractors to complete the projects. Contractors encounter financial challenges and cash flow issues at the job site, which are compounded by owner delays in making progress payments to the contractor factor, payment methods factor, lack of financial management at the job site factor, and delayed payment to subcontractors. Another key component was discovered to be long dispute resolutions, which caused construction projects to be halted. These disputes are caused due to the contract conditions, design deficiency, construction process, consumer reaction, etc.

The responses received were analyzed by RII method as shown below in Table 29.2. Top

TABLE 29.2 Relative Importance Index of the causes of time and cost overruns.

How much in agreement or disagreement are you with each of the following causes of time and cost overruns	Strongly agree (5)	Agree (4)	Neutral(3)	Disagree(2)	Strongly Disagree(1)	Total	Total number (N)	A × N	RII
Delay in progress payments by client	85	92	21	4	1	203	50	250	0.812
Lengthy dispute settlement	80	84	30	4	1	199	50	250	0.796
Financial difficulties by contractors	65	96	30	4	1	196	50	250	0.784
Ineffective planning and scheduling	90	64	36	4	2	196	50	250	0.784
Permits from related authorities	70	92	24	6	2	194	50	250	0.776
Market inflation	80	64	42	4	2	192	50	250	0.768
Poor site management by contractors	75	76	30	8	2	191	50	250	0.764

Continued

518

29. Assessment of factors affecting time and cost overruns in construction projects

TABLE 29.2 Relative Importance Index of the causes of time and cost overruns.—cont'd

How much in agreement or disagreement are you with each of the following causes of time and cost overruns	Strongly agree (5)	Agree (4)	Neutral(3)	Disagree(2)	Strongly Disagree(1)	Total	Total number (N)	A × N	RII
Poor procurement programming of material	60	88	39	0	3	190	50	250	0.76
Frequent change orders during construction by client	65	80	33	10	1	189	50	250	0.756
Mistakes and discrepancies in contract documents	50	104	24	10	1	189	50	250	0.756
Delay in change orders by client	60	88	27	12	1	188	50	250	0.752
Equipment failure	50	92	39	6	1	188	50	250	0.752
Natural disasters	75	72	24	14	2	187	50	250	0.748
Delay in delivering the site to the contractor	50	92	33	10	1	186	50	250	0.744
Construction equipment shortage	65	68	42	10	1	186	50	250	0.744
Poor inspection plan by consultants	45	104	24	10	2	185	50	250	0.74
Mistakes and discrepancies in design documents	45	96	33	10	1	185	50	250	0.74
Conflicts among consultant with other parties	40	92	45	6	1	184	50	250	0.736

7. Conclusion

TABLE 29.2 Relative Importance Index of the causes of time and cost overruns.—cont'd

How much in agreement or disagreement are you with each of the following causes of time and cost overruns	Strongly agree (5)	Agree (4)	Neutral(3)	Disagree(2)	Strongly Disagree(1)	Total	Total number (N)	A × N	RII
Mistakes during construction	50	88	33	10	2	183	50	250	0.732
Project design complexity	55	68	51	8	1	183	50	250	0.732
Inappropriate type of project bidding and award	40	96	36	10	1	183	50	250	0.732
Delay in decision making process	50	88	30	12	2	182	50	250	0.728
Delay in site mobilization	60	68	45	6	3	182	50	250	0.728
Late in approving design documents	55	84	24	16	2	181	50	250	0.724
Laws and regulations changes	65	52	54	8	2	181	50	250	0.724
Owner interference	45	72	57	4	2	180	50	250	0.72
Delay in reviewing the design documents	50	72	45	12	1	180	50	250	0.72
Corruption	60	60	51	6	3	180	50	250	0.72
Delay in material delivery	55	60	51	10	2	178	50	250	0.712
Shortage of technical staff	55	72	30	18	2	177	50	250	0.708
Poor financial control mechanism	60	68	39	2	7	176	50	250	0.704

Continued

TABLE 29.2 Relative Importance Index of the causes of time and cost overruns.—cont'd

How much in agreement or disagreement are you with each of the following causes of time and cost overruns	Strongly agree (5)	Agree (4)	Neutral(3)	Disagree(2)	Strongly Disagree(1)	Total	Total number (N)	A × N	RII
Shortage of materials in the market	40	72	45	16	1	174	50	250	0.696
High cost of labor	30	72	57	10	2	171	50	250	0.684
Unrealistic contract duration	35	84	33	14	4	170	50	250	0.68
Security	20	76	57	12	2	167	50	250	0.668

RII $= \sum W/(A \times N)$ where W is the weighting given to each factor by the respondents (ranging from 1 to 5), A is the highest weight (i.e., 5 in this case), and N is the total number of respondents. Higher the value of RII, more important was the cause of delays.

factors affecting cost and schedule overrun are as follows:

1. Delay in progress payments by client
2. Lengthy dispute settlement
3. Financial difficulties by contractors
4. Ineffective planning and scheduling
5. Difficulty in obtaining permits
6. Market inflation
7. Poor site management by contractors
8. Poor procurement programming of material
9. Client's frequent change orders throughout construction
10. Errors and inconsistencies in contract documents

The factors found have major influence on the construction projects, so in order to complete the projects within stipulated time and within budget, various key points are to be taken care of while handling such large projects.

- Progress payment should be on time.
- More communication and coordination between project participants during all project phases
- Training courses and workshops should be conducted to improve managerial skills of project participants
- Sufficient time should be given for preparing feasibility studies, planning, design, information documentation, and tender submission. This helps avoiding or minimizing late changes.

The objective of this study was to identify factors influencing time and cost overruns in a construction project. This study revealed the most severe causes of time and cost overruns; the most common factors influencing the time and cost overrun were identified by relative importance index method. Because the project involves a large number of components and participants, each one has its own set of causes. However, key stakeholders such as the owner, contractor, and consultant have a greater impact on project outcomes. As a result, the causes of these participants are examined, which will aid in improving project delivery time and cost efficiency. Most severe factors affecting time and cost overrun found are delay in progress payments by client, lengthy dispute settlement, financial difficulties by contractors, ineffective planning and scheduling, difficulty in obtaining permits, market inflation, poor site management by contractors, poor procurement programming of material, frequent change orders during construction by client, and mistakes and discrepancies in contract documents. The report provides project managers and contractors with information and statistics to help them focus on the most significant cause of cost overruns in building projects. The findings highlight the importance of enhancing the construction industry's skill and performance in order to achieve better cost performance and avoid repeated project failure.

References

Ametepey, S.O., Gyadu-Asiedu, W., Assah-Kissiedu, M., 2018. Causes-Effects Relationship of Construction Project Delays in Ghana: Focusing on Local Government Project's. Springer International Publishing Journals. https://doi.org/10.1007/978-3-319-60450-3_9.

Harsha, M., Balaji, V., Venugopal, P., jan-2020. Impact of cost overrun factors in constructing apartments in Tamil nadu. J. Eng. Des. Technol. 9 (3), 456—459. https://doi.org/10.35940/ijitee.C8092.019320. ISSN: 2278-3075.

Laila, M.K., Ghandour, A.El, feb-2019. Examining the role of value management in controlling cost overrun (application on residential construction projects in Egypt). Ain Shams Eng. J. 471—479.

MOSPI (Ministry of Statistics and Programme Implementation, Govt, of India), 2019. Infrastructure Statistics.

Narayanan, S., Kure, A.M., Palaniappan, S., 2019. Study on time and cost overruns in mega infrastructure projects in India. J. Inst. Eng. India Ser. A 100, 139—145. https://doi.org/10.1007/s40030-018-0328-1.

Niazi, G.A., Noel, P., march-2017. Significant factors causing cost overruns in the construction industry in Afghanistan. Procedia Eng. 182, 510—517. https://doi.org/10.1016/j.proeng.2017.03.145, 7th International Conference on Engineering, Project, and Production Management.

Tsegay, G., Luo, H., 2017. Analysis of delay impact on construction project based on RII and correlation coefficient: empirical study. Procedia Eng. 196, 366—374. Creative Construction Conference 2017, CCC 2017, 19-22 June 2017, Primosten, Croati.

Index

Note: 'Page numbers followed by "f" indicate figures those followed by "t" indicate tables and 'b' indicate boxes.'

A

Active Vulnerability Indices (AVIs), 133–134, 139–140
Adaptive Kriging Monte Carlo Simulation (AK-MCS), 126–127
  advantage, 127
  cost-effective flexible pavement structures
    asphalt layer coefficient of variation (COV), 129–130
    asphalt surface course, 128–129
    conventional and alternative pavement structure, 129f
    HMA density, 128–129
    metamodels, 130–131
    pavement thickness levels, 129
    recycled plastics incorporation, 128–129
    target reliability, 130f
  design parameters, 127
  metamodel, 126
  Multi-Layer Elastic Analysis model, 127
  U-function, 126–127
  validation, 128f
Advanced first-order second-moment (AFOSM) reliability analysis, 184
Afghan construction industry, 513
Aggregate Crushing Value, 325
Aggregated index, 155–156
Aggregates, 74–75, 75t
Air quality, 250
Akaike information criteria (AIC), 81–82
Alkali Silica Reaction (ASR), 276, 277t
ALPRIFT vulnerability indices
  data layers, 150t
  Interferometric Synthetic Aperture Radar (InSAR) values, 138–139

Passive Vulnerability Indices (PVIs), 138
Subsidence Vulnerability Indices (SVIs), 138
Analysis of variance, rutting model, 483t
Andesite
  body geometry, 271
  dry unit weight, 270t
  porosities, 271t
  properties, 270t
  quarry in Cambodia, 274
  quarry, Java, Indonesia
    body geometry, 271
    geology, 268–271
    quality, 272–273, 273t
    rock mass rating (RMR), 271–272, 273t
    structural geology, 271–272
  thickness, 269
  uniaxial compressive strength, 272t
Ant Colony Optimization (ACO) algorithm, 80
Aquifer management
  artificial neural networks (ANN), 169, 173–175
  data fusion technique, 168–169, 173, 175f
  Exclusionary Multiple Modeling (EMM) practices, 156
  Groundwater Quality Index (GQI), 156, 163–166, 171–173
  Groundwater Quality Index (GWQI), 156, 166–167, 171–173
  Gultepe-Zarinabad subbasin
    agricultural activities, 159
    geological map, 162f
    hydrogeology, 160, 161f
    location, 159–160
    pliocene sediments, 160
    pollution data availability, 160–161

quaternary deposits, 160
temperate climate and cold winters, 159
volcanic rocks, 160
human health risk assessment (HHRA), 156, 167–168
inclusive multiple modeling (IMM)
  artificial neural network (ANN), 156–159
  data fusion, 170
  dimensions, 156
  index values calculation, 170
  modeling strategy, 156–159
  performance metrics and banding spatial results, 171
  unsupervised data fusion technique, 156–159
innovatory features, 155–156
methodology, 161–171, 165f
normalization of indices, 169–170
spatial modeling, 161–163, 175–176
study area, 159
USEPA indices, human health risk, 160
Artificial neural network (ANN), 183–184
  ANN-NDC models, 426, 428f
  aquifer management, 169
  hidden layers/nodes, 425–426
  liquefaction hazard mitigation
    FOS and F_L trend, 190f
    multifaceted network mapping, 187
    multilayer perceptron (MLP) neural network, 187
    statical performance parameters, 189t
    training and testing data, 187–188
  performance, 425–426
  structure, 425–426, 426f
  verification, 426, 427f
Asphalt binder modification

524

Index

Asphalt binder modification (*Continued*)
atomic force microscopy (AFM) test, 470–471
characterization, 463
failure temperature, 466–468, 467f
fatigue resistance, 469–470, 470f
penetration test, 463
phase angle, 468–469, 469f
rheological properties
complex shear modulus, 462
dynamic shear rheometer (DSR), 462
rutting resistance, 462
rutting resistance, 464–466, 465f
softening point, 463–464, 464f
specification, 463t
storage stability test, 464, 465f
XRD analysis, 472–473
Asphalt pavement analyzer (APA), 478–479, 480f
Associated Soapstone Distributing Company Private Limited (ASDC), 308–309
Atomic force microscopy (AFM) test
asphalt binders
adherence, 470–471
adhesion mechanism, 470–471
average surface roughness, 472t
NKC0 and NKC5, 471, 472f
Autocorrelation function (ACF), 81
Autoregressive Integrated Moving Average (ARIMA), 80–81
Autoregressive Integrated Moving Average with external input (ARIMAX), 81–82
Autoregressive model (AR)
residual plots, 88f
scatter plots, 86, 87f
Taylor diagram, 85–86, 86f
Average run length (ARL), 361–362

B

Bamboo reinforcement concrete
compressive strength test, 77t–78t
construction principles, 74
flexural strength test, 74
hollow tubular structure, 73–74
preparation, 76, 77f
tensile property, 73–74
Bangladesh Medical Waste Management and Processing Rules 2008, 46

Basic sustainability framework (BSF)
decision-making, 2, 16–23
dimensions, 1
goal orientation, 2, 12–16
governance, 1–2, 6–11
sustainable development, 14–16
Bayesian Information Criterion (BIC), 81–82
Bayesian network (BN)
benefit, 426–427
configuration, 426–427
flexible assessment technique, 428
Fuzzy comprehensive, 426–427
integrating procedures, 427–428
metro construction, 430f
probability and graph theory, 426–427
Bias correction, 117
Biochemical oxygen demand (BOD), 94–95
Biomimicry, sustainability, 381t–382t
Bitumen binder modification
modifiers, 449
nano kaolin clay (NKC)
energy dispersive X-ray (EDX) investigation, 453, 454f
field emission scanning electron microscopy (FESEM), 452–453
particle size distribution, 455f–459f
preparation, 450–451, 450f
TEM image, 453–455, 454f–458f
X-ray diffraction (XRD), 451–452
X-ray fluorescence (XRF) test, 451
Blastability index
blast design, 259
computational techniques, 257
density parameter, 256–257
discontinuity spacing ratio (DSR), 257, 257t
joint plane orientation (JPO), 257
joint plane spacing (JPS), 257t
prediction model, 259
specific gravity influence, 256–257
Blast Danger Zone (BDZ), 233
Blast-induced flyrock. *See also* Flyrock
basic blast design, 210, 211f
civil engineering projects, 209
drill diameter, 211
excavation types, 210f
large opencast mines, 209
mechanical excavations, 209

mine-mill fragmentation system and optimization, 210, 210f
residual explosives energy, 209
rockmass and explosives, 210
Bottom ash, 304
Box-Jenkins model, 81
Building Construction Authority (BCA), 276
Building foundation project, 386
Building Information Modeling (BIM) technology, 422–423, 423f
Built environment and climate change, 252–253
By-product Synergy, sustainability, 381t–382t

C

Calcination, 450–451
California Bearing Ratio (CBR), 505
Capacity spectrum method (CSM), 62
Carbon dioxide (CO_2) emission prediction
Autoregressive Integrated Moving Average (ARIMA), 80–81
Autoregressive Integrated Moving Average with external input (ARIMAX), 81–82
data and study area, 84, 85t
discrete gray prediction model, 80–81
empirical results
ARIMA group models, 84–85
examined models, 85, 85t
residual plots, 86, 88f
scatter diagrams, 86, 87f
Taylor diagram, 85–86, 86f
unit root test, 84–85, 85t
unrestricted cointegration rank test, 85, 85t
Gaussian Process Regression (GPR), 82
global warming, 79
GP-ARX model, 83–84, 86–87
gray model, 80
Iran, 79–80, 89f
Machine Learning (ML) models, 80
nonlinear gray Bernoulli models (NGBM), 80
performance metrics, 84
prediction model, 80
reduction policy and evaluation, 80
support vector machine (SVM), 83

in Taiwan, 80
Central Indo-Gangetic Plain (CIGP), seismic risk assessment
  building inventory details
    household distribution, 58t—60t
    model building type, 60t, 66f
  disaster mitigation, 69—70
  earthquake loss estimation, 54
  economic losses, 69f
  fragility functions
    buildings typologies, 57
    damage probability, 62
    economic losses, 62—63
    fragility curves, 63f
    HAZUS-MH database, 56—57
    human casualties, 63—64
    seismic risk, 58—64
    SELENA, 57
  geo-unit Allahabad, 64—65
  hazard estimations, 54
  human losses, 68f—69f
  life and assets damage, 53
  National Disaster Management Authority report, 54
  population density, 56, 58f
  probabilistic concept, 53—54
  seismic demand, 55—56, 56f
  seismic hazard distribution, 64, 65f
  seismicity parameters, 56t
  seismic zoning map, 54
  sensitivity analysis, 65—66
  shear wave velocity, 57f
  site-specific seismic hazard, 55, 64
  vernacular residential buildings, 55
Change Intensity (CI), soil erosion, 112
Chemical decomposition, 255—256
Chilean Metropolitan Environmental Health Authority, 371
Chronic Daily Intake (CDI), 168
Civil engineers, 253
Civil infrastructure systems, 382—383
Climate change, 79
  adverse events, 79
  greenhouse gas emissions, 79
  greenhouse gases, 441
Cloisite 20A, 461—462
Coal mines, overburden materials, 327
Coarse aggregates, 74—75, 325
Complementary cumulative distribution functions (CCDFs), 506

Complexity science, 6
Compressive strength test
  bamboo reinforcement concrete, 77t
  steel reinforcement concrete, 76t
Comte's Theory of Science, 25
Concrete
  advantages, 73—74
  normal vs. bamboo reinforcement, 73—74
    aggregates, 74—75, 75t
    bamboo, 75
    compressive strength, 77f
    construction principles, 74
    different species, 74
    experimentation, 75—76
    flexural strength test, 74, 78f, 78t
    portland slag cement (PSC), 74t
    superplasticizer, 75
    tensile strength, 73—74
  steel reinforcement, 73—74
Construction Ecology, sustainability, 381t—382t
Construction projects
  complexity, 511
  delays and cost overrun issue
    Afghan construction industry, 513
    empirical study, 513
    RII method of analysis, 512
    two-stage approach, 512—513
    value management, 512
  Indian construction sector. See Indian construction sector
  large infrastructure projects, 512
Copper slag, 304
Cost objective function, pavement structure, 501
Critical mass, 4
Crop management factor (C), 116
Crushed sand, 324
Crushing process, manufactured sand, 326
Cumulative distribution function (CDF), 362—363
Cyclic resistance ratio (CRR), 187
Cyclic Stress Ratio (CSR), 185

D

Data fusion technique, 168—169
Decision-making
  decision-making tools, 18—23
    precautionary principle, 22—23
    probability, 19
    risk-based decisions, 19—21

    system reliability, 18—19
    uncertainty, 21—22
  perceptive model, 16
  SDGs, 16
  sustainability appraisals and indices, 16—18
Design Basis Earthquake (DBE), 54
Design binder content (DBC), 478
Differential Global Positioning System (DGPS) survey, 318
Digital elevation model (DEM), 112—113
Digitalized Tunnel Face Mapping System (DiTFAMS), 418
Discipline-specific solution approach, 382
Discontinuity spacing ratio (DSR), 257, 257t
Distributional Equity, sustainability, 381t—382t
District Level Sand Committee (DLSC), 317—318
Dose—response assessment, 168
DPSIR framework, 25—26
DRASTIC vulnerability indices, 134
  anthropogenic impacts, 138
  data layers, 148t
  parameters, 149f
Dynamic Shear Rheometer (DSR), 462

E

Ecological Modernization theory (EMT)
  development stages, 40
  ecological reform, 40
  economic networks, 41
  empirical studies, 40—41
  environmental sociology, 40
  intergenerational solidarity, 41
  market dynamics and economic agents, 41
  technological innovations, 40—41
  triad network approach, 41—42
Ecological restructuring of industrial society, 40
Economic growth, 79
Economic networks, 47—48, 47f
Element equilibrium conditions, 335
Energy dispersive X-ray (EDX) investigation
  nano kaolin clay (NKC), 453, 454f
Engineer's role, sustainable development, 253

Environmental degradation, 437
Environmental impact, life cycle
    assessment (LCA)
  categories, 396t
  emissions, 393–394, 394f
  global warming potential (GWP),
    394–395
  greenhouse gases (GHGs), 394
Environmental pollution risk
  control charts, 359–362
  fatigue-life distribution, 362
  Santiago, Chile
    Chilean guideline values, 371t
    data description, 371
    exploratory data analysis,
      371–372, 372f
    fatigue-life np charts, 372–373,
      373f
    fatigue-life process capability
      indexes, 374
    multivariate fatigue-life control
      charts, 374–375, 376f
    pollution and meteorological data,
      373t
    random variables, 371
Environmental sustainability
  climate change, 250
  construction
    air quality, 250
    energy efficiency, 250–251
    engineers role, 253
    recyclable materials, 249
    water quality, 251–252
  economic and social development,
    249–250
Environmental technologies
  agriculture, 442
  nature simulation, 442–443
  plants choice, 441–442
  sustainability, 442
  sustainable cities, 442–443
  urban growth, 442
Equation-based empirical
    approaches, 183–184
Equivalent single axle loads (ESALs),
    501–502
Ethical and technical concepts,
    sustainability, 381t–382t
Exclusionary Multiple Modeling
    (EMM) practices, 140, 156
Exposure assessment, Human Health
    Risk Assessment, 168
Exposure rock mass (ERM), 266–267

F

Factor of Safety (FS), 197
Fast Lagrangian Analysis of Continua
    (FLAC), 339
Fatigue-life distribution
  log-fatigue-life distribution, 363–364
  multivariate, 364–365
  multivariate log, 365, 365b
  statistical process control
    multivariate monitoring, 369–370
    np charts, 366–368
    process capability index (PCI),
      368–369
  univariate, 362–363
Fault tree analysis (FTA), 421
Field emission scanning electron
    microscopy (FESEM)
  nano kaolin clay (NKC), 452–453,
    453f
Fine aggregate, sand, 324
First and second-order reliability
    method (FORM and SORM),
    124
First-order reliability method
    (FORM), 198, 384, 400–401
  formulation, 401
  loads and resistances, 400–401
Fisher discriminant analysis, 429–431
Flexible pavement analysis
  aleatory uncertainties, 124
  deep structural rutting, 123–124
  epistemic uncertainties, 124
  fatigue and rutting failure criterion,
    123–124
  fatigue cracking, 123–124
  probabilistic techniques, 124
  reliability-based design
    Adaptive Kriging Monte Carlo
      Simulation (AK-MCS). See
      Adaptive Kriging Monte Carlo
      Simulation (AK-MCS)
    allowable load repetitions, 125
    axle-load applications, 125
    definition, 124–125
    failure probability, 125–126
    fatigue cracking and rutting,
      124–125
    fatigue failure, 125
  reliability-based design setting, 124
  uncertainties, 124
Flyrock
  accidents, 213, 214t
    operational causes, 215

    safety, 231f
    statistics, 214t
  adverse impact, 215
  causes and control measures, 240t
  classical works, 212
  controllable and noncontrollable
    variables, 214t
  data analysis
    bench height frequency, 215, 219f
    burden frequency, 218f
    drill diameters, 215, 218f
    flyrock distance, 220f
    published literature, 216t–217t
    rock density, 220f
    specific charge, 221f
    statistical analysis, ANOVA, 222t
    stemming length frequency, 215,
      219f
  definitions, 211
  distance
    computational and intelligent
      methods, 226, 227t–230t
    definitions, 212, 213f
    kinematic and empirical models,
      225t
    prediction models, 223–226
  geology and associated risk
    damaged front row burden,
      223f–224f
    explosive column, 223f
    geological discontinuities,
      221–222
    incompetent strata, competent
      rockmass, 222f
    irregularities, 222
    too less stemming, 223f
    voids, 222f
  incidents, 213
  input parameters, 213–214
  intelligent techniques, 226
  mechanisms, 212
  pit configuration and habitat, 212f
  postrelease behavior, 224–226
  primary causes, 214–215
  risk and management measures
    Blast Danger Zone, 233–235
    fuzzy set theory, 226
    ground vibrations and air
      overpressure, 226
    published literature, 231t–232t
    wild flyrock, 234
  shortcomings, 224
  trajectory and angle, 212

Index

527

Forest and Climate Change (MoEF), 45–46
Foundry sand, 304
Fuzzy comprehensive Bayesian network (BN) technique, 426–427
Fuzzy set theory
    excavation, 424–425
    fuzzy computation hierarchy, 424f
    Fuzzy evidential reasoning approach, 425
    membership criterion, 424
    TOPSIS technique, 424

G

Gaussian Process (GP) models, 82
Gaussian Process Regression (GPR), 82
    residual plots, 88f
    scatter plots, 86, 87f
    Taylor diagram, 85–86, 86f
General circulation models (GCMs), 111–112
Genetic engineering, 438–439
Geodata Master System (GDMS), 423–424
Geological strength index (GSI), 259–260
Geotechnical engineering system
    geo-structures, 383–384
    infrastructure resilience, 402–403
    integrated sustainability framework
        balanced design, 405
        engineering design, 405
        environmental impact, 406–408
        GWP sample data, response surface methodology, 408t
        multiobjective optimization, 406, 408–409
        objective, 405–406
        pile foundation, 405–406
        random variables, 407t
        reliability analysis and LCA, 405–406
        reliability indices, 406, 407f
        sustainability objectives, 407t
    life cycle assessment (LCA)
        data requirements, 392
        environmental impact assessment, 393–395
        environmental impacts quanti? cation, 390
        environmental performance, 391

functional unit, 392
        goal and scope, 392
        interpretation phase, 395
        inventory analysis, 392–393, 393f
        manufacturing stage, 390
        operation stage, 390
        phases, 391, 391f
        product's life, 390
        typical pile foundation, 390f
        typical system boundaries, 392
    reliability-based design
        demand problems, 399–400
        first-order reliability method (FORM), 400–401
        objectives, 400
        probability of failure, 399–400
        reliability index, 399–400
        safety margin, 399–400
        variance, 399–400
    resilience and geo-infrastructure, 403–404
    sustainable practices
        building foundation project, 386
        civil engineering solution, 386
        geotechnical projects, 384–386
        life cycle analysis, 387–388
        materials and energy use, 386
        multiple criteria based methods, 388t–389t
        point-based rating systems, 387–388, 388t–389t
        project level, 386
        proper site characterization, 386
        quantitative environmental metrics, 387–388
        recycled materials, 386–387
        single criterion based methods, 388t–389t
        stakeholders, 386
        sustainability assessment tools, 386–387
        sustainability objectives, 384–386
    uncertainty and reliability
        applied loads, 397–398
        bearing capacity, 395–397
        coefficient of variation (COV), 398–399
        epistemic uncertainty, 397
        geostatistical interpolation analysis, 398
        kriging, 398
        measurement uncertainty, 397
        model uncertainty, 397–398

natural variability, 397
        observational method, 398
        overdesigning, 399
        reduction/elimination, 398
        reliability-based design methods, 397
        reliability level, 395–397
        safety factor, 395–397, 399
        spatial variability, 398
        statistical uncertainty, 397–398
        stochastic (random) process, 398–399
        strategies available, 398
Global Seismic Hazard Assessment Program, 53–54
Global warming potential (GWP), 387–388, 394–395
Goal orientation
    different feedback loops, 12–13, 13f
    feedforward loops, 14
    negative feedback loops, 13
    systems science emergence, 12
Governance
    new paradigm, 8
    past political orders
        Anglo-American political theories, 7–8
        humanism and enlightenment, 7
        liberalism, 7
        modernity, 6–7
        policymaking and decision-making, 7
    policymaking and planning
        central governments, 10
        critical research, 9
        intelligent authorities, 10
        interpretivist, 9
        local governments, 10
        moral philosophies, 9
        philosophy-driven doctrines, 11
        positivist, 9
        social democracy and liberal democracy, 9
        social-learning, 9
        utilitarianism, 8–9
GP-ARX model
    residual plots, 88f
    scatter plots, 86, 87f
    Taylor diagram, 85–86, 86f
Gramhouse, 48
Granite
    body geometry, 271
    dry unit weight, 270t

Granite (*Continued*)
porosities, 271t
properties, 270t
Thailand
compressive strength, 273–274
exploration data, 273–274
fresh granite, 274t
geology, 273
thickness, 269
uniaxial compressive strength, 272t
Granulated blast furnace slag, 304
Gray model, 80
Green Concrete, 302
Greenhouse gas emissions, 79
Greenhouse gases (GHGs), 394
Green innovation, 444–445
Green product, 444–445
Groundwater Quality Index (GQI),
157t–158t, 163–166
Groundwater Quality Index (GWQI),
156, 166–167

H

HasofereLind–reliability index
method, 188–189
Hasofer–Lind approach, 400–401
Hazard identification, 168
Hazardous waste, 38
Healthcare establishments (HCEs), 37
Healthcare waste (HCW)
developed countries, 37–38
developing countries, 37–38
Dhaka city discharge, 38
hazardous waste, 38
healthcare activities, 38
HIV infection, 38
Healthcare waste management,
Bangladesh
Dhaka city, 38
inadequate, 39–40
institutional transformation, 50–51
Mymensingh Medical College
Hospital (MMCH)
current HCWM practice, 47f
data collection, 43–44
feasible measures, 49–50, 50f
healthcare waste transportation,
44–45, 46f
triad network model analysis,
45–49
waste collection, 44, 45f
waste generation, 44
Mymensingh municipal area, 39

network analysis, 49, 49f
network arrangement, 38
objectives, 40
principles, 39
research design, 42–43
service delivery mechanisms, 38
small and medium-sized cities, 39
theoretical framework, 40–42
Heuristics, 26
Hidden danger management model,
419–420
Hotelling statistics, 369
Human health risk assessment
(HHRA), 156, 167–168
USEPA indices, 160
Hydraulic analysis program
(HEC-RAS), 330

I

Improved displacement coefficient
method (I-DCM), 62
Inclusive multiple modeling (IMM)
artificial neural network (ANN),
156–159
data fusion, 170
dimensions, 156
index values calculation, 170
modeling strategy, 156–159
performance metrics and banding
spatial results, 171
unsupervised data fusion technique,
156–159
Indian construction sector
cost and time overruns
factors causing, 514t, 515
literature review, 513–515
method of analysis, 515
questionnaire survey, 513–517,
516f
Relative Importance Index,
517t–520t
MOSPI reality findings, 511–512
India's Ministry of Statistics and
Program Implementation
(MOSPI), 511–512
Indus Ganga plains (IGPs), 53
Industrial Ecology, sustainability,
381t–382t
Infrastructure engineering project,
383–384
Intergenerational Justice,
sustainability, 381t–382t
Interpretivist, 9

Inventory analysis, life cycle
assessment (LCA), 392–393,
393f
IT-based system of managing the
tunneling risk (TURISK), 418

J

Joint plane orientation (JPO)
parameters, 257, 257t
Joint plane spacing (JPS) parameters,
257, 257t

K

Kallada River basin, precipitation
variability
principal component analysis (PCA)
Eigen values, 33, 33t
methodology, 32
principal components, 32–33, 34f,
34t, 35f
seasonal dataset, 33t
study area details, 32f
variance percentage, 33t
temporal variation, 31
Kolmogorov–Smirnov (KS) test, 374,
483
Kosi river basin, soil erosion. *See* Soil
erosion, climate change
Kriging metamodel, 124

L

Land and slope factor (LS), 116
Land Ethics, sustainability,
381t–382t
Latin Hypercube Sampling (LHS)
scheme, 126
Least squares support vector machine
(LSVM) model, 80–81
Leopold matrix analysis, 96, 105
Liberal democracy, 9
Liberalism, 7
Life cycle assessment (LCA), 384,
387–388, 388f
geotechnical engineering system. *See*
Geotechnical engineering
system
Life cycle costing (LCC), 387–388,
388f
Life Cycle Management,
sustainability, 381t–382t
Likert scale, 43
Linear-elastic pavement analysis
software, 124
Liquefaction hazard mitigation

advanced first-order second-moment (AFOSM) reliability analysis, 184, 191–193

Artificial Neural Network (ANN) model, 183–184

data processing and analysis
min-max normalization technique, 189
normalization, 189
statical performance parameters, 189–190
training and testing phases, 189

empirical and computational model
artificial neural network (ANN), 187–188
cyclic resistance ratio (CRR), 187
Cyclic Stress Ratio (CSR), 185
drawback, 190–191
factor of safety (FOS), 187, 191–193, 191f, 193f–194f
stress reduction coef?cient, 185–187

geologic and geotechnical data, 183–184

liquefaction potential, soil deposit, 183–184

probability of liquefaction, 192f–194f

reliability-based design methods, 184

soil failure, 183–184

stress and energy dissipation, 183–184

study area and data collection
Muzaffarpur district, 184–185
statical details, 185t

Log-fatigue-life distribution, 363–364

Lower bound finite element limit analysis, 335–339

M

Machine Learning (ML) models, 80

Mahalanobis distance, 365, 374–375

Malaysia, river sand mining practices
excavation method, 292f
dirty sand stack, digging, 294f
RB excavator, 293f
sand washing bar, 294f
sand washing process, 295f
gravel pump method, 297t
constructed trough size, 293
sand extraction, 295f, 297f
sand processing area, 293
sand processing bar, 296f
sand wash bar, 296f
suction strength, 293
operating site preparation
cabin site and weight bridge, 291
entrance exit, 291
sand stack storage site, 291
sand washing site, 291–292
water and electricity supply, 291

Manganese mines, 307–308

Manufactured sand (M-Sand)
aggregates, technical specifications, 325
code states, 324
economics, 326
overburden materials, coal mines, 327
promoting, 327
quarrying activity, 325–327
crushing, 326
drilling and blasting, 326
loading and transportation, 326
sand import, coastal cities, 327
terminology, 324–325

Marshall Mix design, 478

Marxian-Hegelian dialectics, 26

Maturity model, flyrock risk assessment, 237t
blast design and charging, 237–238
mines, comparative maturity status, 238t
predrilling inspection, 237–238
redesign organization, 235
reengineer process, 235
socio-economic impact, 235
stages, 235

Maximum Considered Earthquake (MCE), 54
Allahabad district, 67f
anticipated economic outlay, 67f
homeless, 67f
uninhabitable dwelling units, 68f

Mechanistic-Empirical pavement design, 123–124

Mechanistic-Empirical Pavement Design Guide (MEPDG), 124–125, 129–130

Millennium Development Goals (MDGs), 15

Mine-mill fragmentation system and optimization, 210

Mineral resources, 449

Ministry of Health and Family Welfare (MOHFW), 45–46

Ministry of Local Government, Rural Development and Cooperatives (MLGRDC), 45–46

Min-max normalization technique, 189

Mixed sand, 324

Model building types (MBTs), 56, 60t, 66f

Modernity, 6–7

Modified capacity spectrum method (MADRS), 62

Modified Universal Soil Loss Equation (MUSLE), 112

Mohr–Coulomb constitutive model, 337

Monte Carlo Simulation (MCS) approach, 124, 340

M-Sand, 318

Multi-layer elastic analysis
model, 127
software, 125–126

Multilayer feedforward Perceptron (MLP) network, 169

Multilayer perceptron (MLP) neural network, 187

Multiple linear regression, rutting, 480–483, 481f, 483t

Multivariate fatigue-life control charts
computation, 370b
Hotelling statistics, 369
hypothesis test, 369
process monitoring, 370, 370b

Multivariate log-fatigue-life distribution, 365, 365b

Mymensingh Medical College Hospital (MMCH), Healthcare waste management, 42f
current practice, 47f
data collection
Likert scale, 43
primary data, 43
qualitative data, 43
quantitative methods, 43
research design, 42–43
triad network model analysis, 43–44
feasible measures, 49–50, 50f
healthcare waste transportation, 44–45, 46f

Mymensingh Medical College Hospital (MMCH), Healthcare waste management (*Continued*)
triad network model analysis, 45—49
waste collection, 44, 45f
waste generation, 44

N

Nanoclay
Cloisite 20A, 461—462
modified asphalt binder, 461—462
organo-clay montmorillonite (OMMT), 461—462
Nanoclay modified binder A (NCMB A), 462
Nanoclay modified binder B (NCMB B), 462
Nanoclays, 450
Nano kaolin clay (NKC)
asphalt binder modification. *See* Asphalt binder modification
chemical analysis
energy dispersive X-ray (EDX) investigation, 453, 454f
field emission scanning electron microscopy (FESEM), 452—453
X-ray diffraction (XRD), 451—452
X-ray fluorescence (XRF) test, 451
penetration value, 463f
preparation, 450—451, 450f
size determination
particle size distribution, 455f—459f
TEM image, 453—455, 454f—458f
softening point value, 464f
Nanomaterials, 449—450
Nanosilica (NS) modified porous asphalt
binder mixer, 478, 478f
descriptive statistics, rutting data, 479, 482t
design binder content (DBC), 478
mix design, 478
normality test, 483, 484f
Rut* model
paired T-test, 485, 485t
RMSE, MAE, and MAPE, 485, 485t
scatter plot, 485, 485f
rutting data and parameters, 479—483
rutting parameters
correlation analysis, 479, 480f, 482t

multiple linear regression, 480—483, 481f, 483t
rutting resistance, 478—479
Nanotechnology, 477—478
National Disaster Management Authority report, 54
Natural sand, 324
Neoliberalism, 7
Nonlinear gray Bernoulli models (NGBM), 80
Nonmetallic mineral resources, 449
Nonmodified nanoclay (NMN), 461
Normality test, 483
Normalization, 189
Normalized Difference Vegetation Index (NDVI), 116
Notified or controlled pricing model, 318
np control charts, fatigue-life distribution
average run length (ARL), 367—368
central line (CL), 366
exceedance probability, 366
lower control limit (LCL), 366
nonconforming fraction, 366
parametrization, 366
probability, 367
shift constant, 366—367
upper control limit (UCL), 366

O

Optimally designed pavement structure
deterministic model, 505—506
failure probability, 505—506
probabilistic model, 506
subset simulation method, 506
uncertainty model, 505—506
Organo-clay montmorillonite (OMMT), 461—462
Ozone layer depletion, 445

P

Partial autocorrelation function (PACF), 81
Particle Swarm Optimization (PSO) algorithm, 80—81
Passive Vulnerability Indices (PVIs), 133—134, 145f
Patna, transport system, 487—488
contemporary issues, 487
quadrant analysis approach. *See* Sustainable transportation

Pavement structure
asphalt layer, 499—500
base layer and subbase layer, 499—500
cost items, 502t
cost objective function, 501
dimensions, 500
evolution algorithms, 500
fatigue, 499—500
genetic algorithm, 500
geometry and parameters, 501f
geosynthetic reinforced soil, 499—500
geotechnical structures, 500
integer constraints, 500
laboratory tests, 499—500
lifetime, 502t
modern road network, 500
optimization model
design constraints, 501—503
geotechnical analysis and dimensioning constraints, 502
input data, 502—503
PAVEOPT model, 501
PAVEOPT model, case study, 503—505, 506t
semiempirical AASHTO method, 501—502
quality pavements, 500
road network, 499
subgrade layer, 499—500
upper layers, 499—500
Penetration index (PI), 462
Penetration-viscosity number (PVN), 462
Phase angle, 468—469, 468f—469f
Philosophy-driven doctrines, 11
Pimodel, 117
Planar geosynthetics, 333—334
Policy options, sustainable urban transportation. *See* Quadrant analysis approach, sustainable transportation
Polymer modified nanoclay (PMN), 461
Porous asphalt (PA)
advantages, 477
flexible pavement, 477
gradation, 477
high porosity, 477
nanosilica (NS) modified
binder mixer, 478, 478f
design binder content (DBC), 478

Index

mix design, 478
Rut* model, 485
rutting data and parameters, 479–483
rutting resistance, 478–479
voids content, 477
Portland slag cement (PSC)
oxide composition, 74t
properties, 74t
Positivist, 9
Precautionary principle, 22–23, 381t–382t
Principal component analysis (PCA)
Punalur station, precipitation variability
Eigen values, 33, 33t
methodology, 32
principal components, 32–33, 34f, 34t, 35f
seasonal dataset, 33t
variance percentage, 33t
spatial rainfall data, 31
study area and dataset, 31–32
study area details, 32f
Probabilistic risk factor-based approach
cantilever earth retaining wall
Analysis of Variance F-test, 201
factor of safety, 200t
failure modes, 199, 201t
Monte Carlo simulation, 199
random variables, 202t
safe and economic design recommendation, 202
statistical measures, 200t
gravity retaining wall
deterministic FS, 203
generalized design recommendations, 206
geotechnical variables, 205f
probability of failure, 203–204
random variables, 205t
structure modification, 204, 205t
risk priority number (RPN), 199
Probability-based decision-making tools, 19
Probability density function (PDF), 363
Process capability index (PCI), 361–362, 368–369
Pure kaolin clay powder, 451f
Pythagorean fuzzy environment, 425

Q

Quadrant analysis approach, sustainable transportation
consolidated ranks, 495f
methodology, 489f
policy measures, 491t, 493t
desirability of implementation, 492, 494t
probability of implementation, 492
right mix, 492–496
sustainable transport criteria, 495f
template, 489f
2 x 2 quadrant analysis, 488
Quarry dust, 304

R

Rainfall–runoff erosivity factor (R), 114–115
Random forest, 431–432
Receiver Operating Characteristics (ROCs), 142
Reductive science, 12
Reinforced soil technique, 333–334
Relative Importance Indices (RIIs), 512, 515, 517t–520t
Reliability analysis, 19
Reliability assessments, 18
Reliability-based approaches, 197
cantilever retaining wall, 198
differential sensitivity analysis, 198
failure probabilities, 198
first-order reliability method, 198
Monte Carlo Simulation, 198
partial safety factors, 199
reliability index, 197–198
Renewable energy sources, 443
Renewable technologies, 250
Resilience and geo-infrastructure
cimate change, 403
civil infrastructure threats, 403
infrastructure health, 403
redundancy, 403–404
resourcefulness and rapidity, 403–404
robustness, 403–404
transportation network closure, 403
unforeseen disasters, 403–404
Reversibility Principle, sustainability, 381t–382t
Revised Universal Soil Loss Equation (RUSLE), 112, 114
Rippability rock mass classification system, 260–263, 261t–262t

Risk aggregation, aquifer contamination
Active Vulnerability Indices (AVIs), 133–134, 145f
aquifer risks, 134
Ardabil plain
aggregated risk maps, 146f
ALPRIFT framework, 138–139
dataset preparation, 141–142
DRASTIC framework, 138
geological context, 135–136
hydrogeology, 136–137
inclusive multiple models, 140
InSAR data, 138–139
land use, 137
OSPRC framework, 137–138
performance metrics, 142
physical system, 135
risk aggregation problem, 140–141
risk cells, 137–138
risk mapping, 139–140, 142–144, 143f
supervised learning techniques, 140
vulnerability indices, 141
DRASTIC and ALPRIFT
vulnerability indices, 134
modularized modeling strategy, 135
Passive Vulnerability Indices (PVIs), 133–134, 145f
risk-based decision-making, 133–134
Risk-based decisions, 19–21, 21f
Risk characterization, Human Health Risk Assessment, 168
Risk management model, 419–420
Risk Management Software (TRM1.0), 418
Risk priority number (RPN), 199
River sand mining practices
alternatives, 318
areas/blocks determination, 317–318
bottom ash, 304
business model for allocation, 318–319
Canada (Alberta), 297, 298f
China, 298–300
clearances and endorsements, 319
Colombia, 297–298, 299f
areas of exploitation, 300f
legal framework, 299f

532   Index

River sand mining practices
(*Continued*)
plant type structure of river, 299f
construction and demolition waste,
304
Forest and Climate Change
notification, 320
foundry sand, 304
geological report, 318
guidelines, 320
India, 290–291
Malaysia. *See* Malaysia, river sand
mining practices
vs. manufactured sand, 305–306,
305t–306t
mechanical mining, 319
MoEFCC recommendations, 319
M sand, 304
New Zealand, 300–301
offiine sale provisions, 320
online sale, 319–320
operations, 319
partial sand replacement,
303–304
quarry dust, 304
realistic alternatives, 302
recycling, 302–303
reservoirs, sand deposits, 303
sales and transportations, 319
specific alternatives, 302
specified sand-bearing areas,
317–318
sustainability, 301
United Kingdom, 301
waste rocks, 306–309
Rock mass classification (RMC)
blastability assessment. *See*
Tropically weathered igneous
rocks
excavation system, 258t–259t
blastability, 259–260
geological strength index (GSI),
259–260
natural materials, 259–260
parameters, 259–260
rippability, 261t–262t
parameters, 259–260
slope stability assessment
blastability criteria, 264t–265t
exposure rock mass (ERM),
266–267
rock failure, 266–267
rock stability, 266–267

Slope Mass Rating (SMR), 263,
266–267
slope rock mass (SRM), 266–267
slope stability probability classi?
cation (SSPC) system, 266–267
strata variation, 256
Root Mean Square Errors (RMSE),
171
Runoff erosivity factor, 116–117
Rutting resistance
asphalt binders, 464–466, 465f
nanosilica (NS) mixing process,
478–479

S

Safety risks, underground operations
accidents types and distribution, 417f
deep excavation, 417
excavation process, 416–417
metro construction projects, 415–416
regulatory guidelines, 415–416
risk assessment
algorithms, 420–421
artificial neural network (ANN),
425–426
Bayesian network (BN), 426–428
Building Information Modeling
(BIM), 422–423
Fisher discriminant analysis,
429–431
fuzzy set theory, 424–425
qualitative techniques, 421
quantitative techniques, 421
Random forest, 431–432
structural health monitoring
(SHM) system, 423–424
support vector machine (SVM),
428
threshold-based technique,
421–422
risk identification
deterministic approach, 418–419
Digitalized Tunnel Face Mapping
System (DiTFAMS), 418
IT-based system of managing the
tunneling risk (TURISK), 418
Risk Management Software
(TRM1.0), 418
surrounding environment,
418–419
tunnel construction management,
418
underground cavern construction,
419

vulnerability index, 418
road transport, 415–416
safety management system, 419–420
transportation division, 415–416
transport operations, 415–416
tunnel transportation, 415–416
underground construction accidents,
415–416
urbanization growth, 415–416
Sand availability sources, 315
Sand mining practices. *See also* River
sand mining practices
construction activity, 285
environment accountability, 287
environmental impacts
Anguillan beach (Caribbean
Islands), 289
Ghana coast, 289
Goa (India), 288–289
Karnataka (India), 287
Kerala (India), 287
Kwale coast (Kenya), 289
Maharashtra (India), 288
Middle Eastern countries, 290
singapore, 289–290
Tamilnadu (India), 288
Waikato beaches (New Zealand),
289
fine-grained sand, 286
global sand scenario, 287
marine sand, 286
sand scarcity, 287
yellow sand, 286
Sand Mining Corporation, 320–321
Sand mining process, Malaysia.
See also Malaysia, river sand
mining practices
floodplain mining, 329
geomorphology, 330
guidelines, 328
hydrology, 330
impacts of, 328
in-stream mining, 328–329
river modeling, HEC-RAS, 330
river morphology, 329–330
sand replenishment, 330
Sand mining process, India, 316f
cement usage, 316
illegal curbing
decasting, 323
desilting, 323
monitoring process, 322, 323f
stringent mechanism, 322–324

manufactured sand (M-Sand)
  aggregates, technical specifications, 325
  code states, 324
  overburden materials, coal mines, 327
  promoting, 327
  quarrying activity, 325–327
  sand import, coastal cities, 327
  terminology, 324–325
natural sand substitutions, 316
need approximation methodologies, 316
need-supply evaluation, 315–316
  district survey report, 316–320
  river sand, 317–320
regulatory mechanism
  information technology, 321
  Sand Mining Corporation, 320–321
  sand monitoring committee, 321
replenishment study, 322
Sand washing process, 295f
Schwarz information criteria (SIC), 81–82
Science concepts, 3–6
  bottom-up learning, 25
  evolution, 6
  paradigm, 26
  philosophy role, 4–5
  pivotal scientific concepts, 4
  scientific framing activities, 4
  scientific methodologies, 3–4
  theory, 25
Seismic risk assessment, Central Indo-Gangetic Plain (CIGP). *See* Central Indo-Gangetic Plain (CIGP), seismic risk assessment
Seismic risk, definition, 54
Shallow foundation system
  design, 333
  deterministic numerical analysis, 334
  performance, 333
  planar geosynthetics, 333–334
  reinforced soil technique, 333–334
  risk and vulnerability, 333
  vertical settlement, 333–334
Signal Detection Theory, 171
Slope Mass Rating (SMR), 263, 266–267
Slope rock mass (SRM), 266–267

Slope stability probability classification (SSPC) system, 266–267
Snowball effect, 4
Soapstone mines, 308–309, 308f
Social democracy, 9
Social-learning process, 9
Social networks, 48–49
Soil erodibility factor (K), 115–116
Soil erosion, climate change
  annual soil erosion
    annual soil loss, 114
    crop management factor (C), 116
    land and slope factor (LS), 116
    rainfall–runoff erosivity factor (R), 114–115
    Revised Universal Soil Loss Equation (RUSLE), 114
    soil erodibility factor (K), 115–116
    support practice factor (P), 116, 116t
  anthropogenic activity, 111
  bias correction, 117
  change Intensity (CI), 112, 120t
  general circulation models (GCMs), 111–112
  methodology, 114–117
  projected rainfall, 116–117
  soil loss estimation model, 112
  study area and data description
    digital elevation model (DEM), 112–113
    general circulation models (GCMs), 112–113
    Kosi river basin, Bihar, 112
  for years 2019 and 2070, 117
    change intensity, 117–118, 119f
    priority erosion level maps, 117, 118f
    spatial-temporal changes, 117–118
Soil shear strengths, uncertainties, 334
Solar energy applications
  green innovation, 444–445
  renewable energy sources, 443
  solar farms, 444
  solar thermal water heaters, 444
  sustainable clean energy sources, 444
  thermal turbine system, 444
Steel reinforcement concrete
  casting process, 76f
  compressive strength test, 76t
  preparation, 75

Stress discontinuity conditions, 335–337
Strip footing, reinforced soil slope
  Fast Lagrangian Analysis of Continua (FLAC), 339
    random field and Monte Carlo simulation coupling, 341–342
  lower bound finite element limit analysis
    conic optimization problem, 338–339
    element equilibrium conditions, 335
    modi?ed stress discontinuity conditions, 337
    objective function, 338
    random field and Monte Carlo simulation, 340–341
    reinforcement effect, 335
    shear strength parameters, 335
    stress boundary conditions, 337
    stress discontinuity conditions, 335–337
    three noded triangular element, 336f
    yield condition, 337–338
  Monte Carlo simulation, 340
  probabilistic bearing capacity
    algorithm used, 343f
    boundary conditions and problem domain, 343, 344f
    cohesionless soil, 345–346, 347f
    finite element (FE), 343, 344f
    maximum reinforcing ef?ciency, 342
    Mohr–Coulomb constitutive model, 342
    objective, 342
    soil–reinforcement interface, 342
    spatial distribution, soil strength parameters, 345
  probabilistic load carrying capacity, geocell reinforced soil slope, 349f
    angle of inclination, 349
    boundary conditions, 349–351
    failure mechanism, 354, 355f
    finite difference mesh, 349–351, 350f
    Mohr–Coulomb constitutive model, 349–351
    settlement level of footing, 351–354, 353f

534 Index

Strip footing, reinforced soil slope (*Continued*)
  slope height, 349
  spatial distribution, soil strength parameters, 345, 352f
  three-dimensional modeling, 349–351
  two-dimensional modeling, 349–351
  probabilistic stability analysis
   boundary conditions, 354–356
   finite difference, 354–356
   geotextile, 354
   problem domain, 356f
  random field, 339–340
  soil strength properties, 334
Structural health monitoring (SHM) system, 423–424
Subset simulation method, 506
Subsidence Vulnerability Indices (SVIs), 138
Superplasticizer, 75
Support practice factor (P), 116
Support vector machine (SVM), 83, 428–429, 430f
  vs. Fisher discriminant analysis, 429–431
  proposed approach architecture, 431f
  residual plots, 88f
  scatter plots, 86, 87f
  structural model, 430f
  Taylor diagram, 85–86, 86f
Sustainability
  civil and geotechnical engineering, 382–383
  civil infrastructure, 382–383
  definition, 487
  demands, 379–380
  environment, economy, equity and engineering, 380f
  ethical and technical concepts, 380, 381t–382t
  geo-structures, 383–384
  geotechnical construction, 382–383
  human civilization, 437
  humanity, 379
  infrastructure engineering project, 383–384
  land use policies, 445
  Living Planet Report, 379
  ozone layer depletion, 445
  principles, 380
  science approach, 382

  spatial and temporal variabilities, 383–384
  sustainable agriculture, 438–439
  sustainable development, 379
  and sustainable development, 437–438
  technical concepts, 380
  technological developments, 379–380
Sustainability impact assessment (SIA), trickling filter
  direct and indirect impacts, 99, 100t
  economical aspect, 95–96
  environmental aspect, 95–96
  framework set up, 95–98
  impact quantification and assessment
   decision-making, 104–105, 105t
   impact matrices, 101t–103t
   Leopold matrix method, 96
   qualitative assessment, 101
  qualitative impact analysis, 96
   direct and indirect impacts, 99, 100t
   positive and negative impacts, 98–99, 98t
   reversible and irreversible impacts, 100–101, 100t
   short-term and long-term impacts, 99, 99t
  socio-cultural aspect, 95–96
Sustainability science, 6
Sustainable agriculture, 438–439
Sustainable development, 379–380, 437–439
  air quality, 250
  Brundtland Report, 14
  definition, 249, 487
  energy efficiency, 250–251
  engineers role, 253
  Green Movement, 14
  Millennium Development Goals (MDGs), 15
  recyclable materials, 249
  reductive science, 14
  strategy, 14–15
  water quality, 251–252
Sustainable development goals, environmental systems
  climate change, 441
  environmental citizenship, 439–440
  greenhouse gases, 441
  natural capital, 439–440

  pollution control, 440–441
  species number and diversity, 440–441
  water distress, 440–441
Sustainable energy, 251
Sustainable transportation
  data analysis, 492
  Delphi method
   data collection, 491–492
   feedback-based questionnaire tool, 488
   minimum sample size, 488, 489t
   policy measures, 491t
   questionnaire adequacy, 491
   questionnaire formulation, 490–491
  mid-sized cities, 487–488
  policy measures, 491t, 493t
   desirability of implementation, 492, 494t
   probability of implementation, 492
   right mix, 492–496
   sustainable transport criteria, 495f
  quadrant analysis
   consolidated ranks, 495f
   methodology, 489f
   template, 489f
  2 x 2 quadrant analysis, 488
Systems engineering approach, 382

T

Technological developments, 379–380
Thenmala reservoir, 31–32
Thermal turbine system, 444
Threshold-based technique, 421–422
Time resolutions, 26–27
TOPSIS technique, 424
Triad network model analysis, 41–44
  data collection, 43–44
  economic networks, 47–48
  policy networks, 45–46
  social networks, 48–49
Trickling filter
  construction, 97
  designing, 94–95
  dimensions, 97f
  dismantling phase, 97–98
  energy consumption, 93–94
  filter media, 107–108
  impact matrix, 102–103, 102t
  integrated sustainability impact assessment (SIA)

Index 535

direct and indirect impacts, 99, 100t
economical aspect, 95–96
environmental aspect, 95–96
framework set up, 95–96
impact quantification and assessment, 96, 101–105, 101t
mitigations measures, 105–106
positive and negative impacts, 98–99, 98t
preconditioning, 95
qualitative impact analysis, 96, 98–101
reversible and irreversible impacts, 100–101, 100t
short-term and long-term impacts, 99, 99t
socio-cultural aspect, 95–96
life cycle, 95f
merits, 93
NRC equation, 94–95
operation phase, 97
positive and negative impact, 93–94
pump, 108
rotary distributor, 108
Tropically weathered igneous rocks
Andesite quarry in Cambodia, 274, 274t
Andesite quarry, Java, Indonesia
body geometry, 271
geology, 268–271
quality, 272–273, 273t
rock mass rating (RMR), 271–272, 273t
structural geology, 271–272
classification scheme, 267, 268t–269t
comparison, 276–279, 278t
engineering properties, 267–268
granite and andesite
body geometry, 271
dry unit weight, 270t
porosities, 271t
properties, 270t
thickness, 269
uniaxial compressive strength, 272t
Granite quarry, Thailand
compressive strength, 273–274
exploration data, 273–274
fresh granite, 274t
geology, 273
Malaysia
aggregate production, 275t
Alkali Silica Reaction (ASR), 277t
imported aggregates, Singapore, 276
quarries, 275, 275t
rock mass assessment, 277–279
Two-dimensional numerical model, 351

U

United States Geological Survey (USGS)-based earth explorer portal, 112–113
Univariate fatigue-life distribution, 362–363
cumulative distribution function (CDF), 362–363
maximum likelihood estimators, 363
probability density function (PDF), 363
random variable, 363
R software, 363
Universal Soil Loss Equation (USLE), 112

Utilitarianism, 8–9

V

Value management, construction projects, 512
Void filled with Asphalt (VFA), 478
Void in mineral aggregate (VMA), 478

W

Waste rocks, sand mining, 308–309
Wastewater treatment, 93
Water distress, 440–441
Water quality
contamination sources, 251
dams and reservoirs, 251–252
dual systems, 251
irrigation schemes, 251
land space, 251–252
Weathering of rocks
aggressive reaction, 255–256
definition, 255–256
Weighted Average Score (WS), 43
Wild flyrock, 234
Wilson–Hilferty approximation, 374–375

X

X-ray diffraction (XRD)
asphalt binders, 472–473
kaolin clay powder, 451–452, 452f, 459
X-ray fluorescence (XRF) test, kaolin clay powder, 451, 459

Y

Yellow sand, 286

Printed in the United States
by Baker & Taylor Publisher Services